MW01248203

The Routledge Handbook of Classics and Cognitive Theory

"This is the first book to demonstrate how cognitive theory can be productively applied across a wide range of Classical studies, including linguistics, literary theory, history, art history, religion, theater, and archaeology. The contributors draw on both longstanding cognitivist approaches and the "second wave" of embodied, enactive and distributed cognition. Enriched by a wealth of interdisciplinary bibliography, the volume shows how study of the interaction between the mind-brain and its environment can shed new light on the cultures of Western antiquity."

Jennifer Larson, Kent State University, USA

The Routledge Handbook of Classics and Cognitive Theory is an interdisciplinary volume that examines the application of cognitive theory to the study of the classical world, across several interrelated areas including linguistics, literary theory, social practices, performance, artificial intelligence and archaeology. With contributions from a diverse group of international scholars working in this exciting new area, the volume explores the processes of the mind drawing from research in psychology, philosophy, neuroscience and anthropology, and interrogates the implications of these new approaches for the study of the ancient world.

Topics covered in this wide-ranging collection include: cognitive linguistics applied to Homeric and early Greek texts, Roman cultural semantics, linguistic embodiment in Latin literature, group identities in Greek lyric, cognitive dissonance in historiography, kinesthetic empathy in Sappho, artificial intelligence in Hesiod and Greek drama, the enactivism of Roman statues and memory and art in the Roman Empire.

This ground-breaking work is the first to organize the field, allowing both scholars and students access to the methodologies, bibliographies and techniques of the cognitive sciences and how they have been applied to classics.

Peter Meineck holds the endowed chair of Professor of Classics in the Modern World at New York University, USA. He is also an Honorary Professor of Classics at the University of Nottingham, UK and the Founding Director of Aquila Theatre. His most recent publications include *Theatrocracy: Greek Drama, Cognition and the Imperative for Theatre* (Routledge, 2017), *Combat Trauma and the Ancient Greeks* (ed. with David Konstan, 2014) and a new translation of Aristophanes' *Frogs* (forthcoming). He is also Rescue Captain of the Bedford Fire Department in New York.

William Michael Short joined the Department of Classics and Ancient History at the University of Exeter, UK in 2017, after holding positions at the University of Texas at San Antonio and Loyola University, Maryland, both in the USA. He has edited or co-edited several volumes of papers including *Con i Romani: Studi antropologici del mondo antico* (2014, in Italian), *Embodiment in Latin Semantics* (2016) and *Toward a Cognitive Classical Linguistics* (forthcoming).

Jennifer Devereaux is an advanced PhD candidate in Classics at the University of Southern California, USA. She is a member of the Computational Social Sciences Laboratory at USC's Brain and Creativity Institute and is currently a visiting postgraduate researcher at the University of Edinburgh, UK, in the Department of History, Classics and Archaeology. She has written a number of chapters on the topics of embodied cognition, historiography and rhetoric.

The Routledge Handbook of Classics and Cognitive Theory

Edited by Peter Meineck, William Michael Short and Jennifer Devereaux

LONDON AND NEW YORK

First published 2019
by Routledge
2 Park Square, Milton Park, Abingdon, Oxon OX14 4RN

and by Routledge
711 Third Avenue, New York, NY 10017

Routledge is an imprint of the Taylor & Francis Group, an informa business

British Library Cataloguing-in-Publication Data
A catalogue record for this book is available from the British Library

Library of Congress Cataloging-in-Publication Data
A catalog record has been requested for this book

ISBN: 978-1-138-91352-3 (hbk)
ISBN: 978-1-315-69139-8 (ebk)

Typeset in Bembo
by Swales & Willis Ltd, Exeter, Devon, UK

Contents

Contributors

Rutger J. Allan is Lecturer in Ancient Greek at the Free University Amsterdam, the Netherlands. He has published on a variety of topics in ancient Greek linguistics relating to verbal semantics and discourse pragmatics. He has a special interest in cognitive linguistic and narratological approaches to Greek narrative texts. He is the author of *The Middle Voice in Ancient Greek: A Study in Polysemy* (2002) and co-editor of the volumes *The Language of Literature* (2007) and *The Greek Future and Its History* (2017).

Anna Bonifazi is Professor at the Department of Linguistics of the University of Cologne, Germany (sub-section "Historical and Comparative Linguistics"). Her main research interest is in pragmatic and cognitive approaches to ancient Greek literature. So far she has published three monographs on deixis in Pindar, third-person pronouns in Homeric poetry, and particles in Herodotus and Thucydides.

Laura Candiotto, PhD in Philosophy (Venice-Paris, 2011), is Research Fellow at the Eidyn Centre of the University of Edinburgh, UK. Her area of specialization is philosophy of emotions, in relationship with social epistemology, the history of philosophy, ethics and education. She is editing the volume *Emotions in Plato* (forthcoming) with Olivier Renaut, and *The Value of Emotions for Knowledge* (forthcoming). Among her publications are: "Boosting Cooperation: The Beneficial Function of Positive Emotions in Dialogical Inquiry", *Humana Mente*, 1/2018; "Purification through Emotions: The Role of Shame in Plato's *Sophist* 230b4–e5", *Educational Philosophy and Theory*, 50 (6–7), 2018; "Extended Affectivity as the Cognition of the Primary Intersubjectivity", *Phenomenology and Mind*, 11/2016; "Plato's Cosmological Medicine in the Discourse of Eryximachus in the *Symposium*: Responsibility of a Harmonic *Techne*", *Plato Journal*, 5/2015; "Aporetic State and Extended Emotions: The Shameful Recognition of Contradictions in the Socratic Elenchus", *Ethics & Politics*, XVII, 2015(2). This work was supported by the EU under the Marie Skłodowska-Curie Individual Fellowship for her Project EMOTIONS FIRST (655143), www.emotionsfirst.org.

Joel P. Christensen is Associate Professor in the Department of Classical Studies at Brandeis University, USA. He taught previously at the University of Texas at San Antonio, USA (2007–2016). He received his B.A. and M.A. from Brandeis (2001) in Classics and English and his PhD in Classics from New York University (2007). In addition to articles on language, myth and literature in the Homeric epics, he has published a *Beginner's Guide to Homer* (2013) and *Homer's Thebes* (2019) with Elton T. E. Barker as well as *A Commentary on the Homeric Battle of Frogs and Mice* (2018) with Erik Robinson. He is currently completing a book on the *Odyssey* and modern psychology.

Jennifer Devereaux is an advanced PhD candidate in Classics at the University of Southern California, USA. She is a member of the Computational Social Sciences Laboratory at USC's Brain and Creativity Institute and is currently a visiting postgraduate researcher at the University of Edinburgh, UK, in the Department of History, Classics and Archaeology. She has written a number of volume chapters on the topics of embodied cognition, historiography and rhetoric. Her graduate thesis demonstrates the pre-linguistic structure of Latin texts and develops a cognitive-sociological approach to Latin literature. https://cognitiveclassics.blogs.sas.ac.uk/people.

Paul C. Dilley (PhD Yale, 2008) is Assistant Professor in the Departments of Religious Studies and Classics at the University of Iowa, USA, with a research specialization in the religions of late antiquity, and in particular early Christianity. His monograph, *Monasteries and the Care of Souls in Late Antique Christianity: Cognition and Discipline* (2017), explores the "cognitive disciplines" and "heart work" practiced by ancient Christian monks. He is also a digital humanist, with a focus on text analysis, enhanced manuscript imaging, as well as digital philology more generally, and is co-editor of *Ancient Worlds in Digital Culture* (2016).

Alexander S. W. Forte is Visiting Assistant Professor (2017–2019) in Colgate University's Department of the Classics, USA, after receiving his PhD in Classical Philology (and a secondary field in Historical Linguistics) from Harvard University, USA. His current book project focuses on developing and applying cognitive linguistics to metaphor in early Greek poetry. He is interested in philosophical commonalities between American pragmatism, phenomenology and recent 4EA approaches to cognition. He has published and presented on metaphorical idiom in ancient sources, spatiotemporal metaphor in the *Iliad*, the soundscape of the Eleusinian mysteries, boxing matches in Hellenistic and Virgilian poetry, chariot metaphors in early Greek philosophy, ring-composition in Indo-Iranian, Greek and Roman poetry, and the historical linguistics of θεωρία and νοῦς.

Maria Gerolemou currently works as a Leventis Postdoctoral Research Associate at the Classics and Ancient History Department of the University of Exeter. Her research focuses on ancient drama, primarily through parameters such as those of gender and madness, on paradoxography and, most recently, on ancient mechanics. She is the author of the book *Bad Women, Mad Women: Gender und Wahnsinn in der Griechischen Tragödie* (2011). She is also the editor of the collective volume *Recognizing Miracles in Antiquity and Beyond* (2018).

Luca Grillo is Associate Professor of Classics at the University of Notre Dame in South Bend, USA. He is the author of *The Art of Caesar's Bellum Civile* (2012) and a commentary on Cicero's *De Provinciis Consularibus Oratio* (2015). He co-edited *The Cambridge Companion to the Writings of Julius Caesar* (2018) and has published various articles, especially on Caesar, Cicero and Virgil. He is currently working on a monograph on *Irony in Latin Literature*.

Ahuvia Kahane is Professor of Greek and Co-Director of the Centre for the Reception of Greece and Rome at Royal Holloway, University of London, UK. Continuing his work on *The Chicago Homer* database (online) and early studies of the meaning of metrical form in Homeric epic, he is currently working on the relationship between form and meaning in Homeric verse, incorporating aspects of usage-based linguistics, complexity theory and the politics of aesthetics. His most recent books are *Homer: A Guide for the Perplexed* (2012) and *The Gods in Greek Hexameter Poetry* (2016).

David Konstan is Professor of Classics at New York University, USA. He earned his B.A. in Mathematics and his PhD in Classics at Columbia University, USA. He taught at Wesleyan University, USA from 1967 to 1987 and at Brown University, USA from 1987 to 2010, when he joined the faculty at NYU. His research focuses on ancient Greek and Latin literature, especially comedy and the novel, and classical philosophy. In recent years, he has investigated emotions and value concepts in classical Greece and Rome, and has written books on friendship, pity, the emotions, forgiveness and beauty. He has also investigated ancient physics, atomic theory and literary theory, and has translated Seneca's two tragedies about Hercules into verse. His most recent book on ancient vs. modern conceptions of loyalty, gratitude, love and grief will be published in 2018. He has held visiting appointments in New Zealand, Scotland, Brazil, Argentina, South Africa, Australia and Egypt, among other places. He has been President of the American Philological Association, and is a fellow of the American Academy of Arts and Sciences and Honorary Fellow of the Australian Academy of the Humanities.

Amy Lather has been an Assistant Professor of Classical Languages at Wake Forest University, USA since 2016. She received her PhD in 2016 from the University of Texas at Austin, USA, with a dissertation entitled "Sense and Sensibility: The Experience of *Poikilia* in Archaic and Classical Greek Thought." She has published articles on the aesthetics of music and the construction of sensory experience in ancient Greek literature, and is currently at work on her first monograph, "Matters of the Mind: Materiality and Aesthetics in Ancient Greek Thought."

Jacob L. Mackey is Assistant Professor of Classics at Occidental College, USA. He is completing a book on Roman religion, *Belief and Cult: From Intuitions to Institutions in Republican Rome*, and planning a book with the working title *Ad Incunabula: Perceptions of Children's Cognitive Development in the Roman World.*

Thomas R. Martin is the Jeremiah O'Connor Professor in Classics at the College of the Holy Cross, USA. His teaching and publications concentrate on ancient Greek and Latin historical authors and Greek and Roman history. Prof. Martin is one of the founders of the Perseus Digital Library (www.perseus.tufts.edu/hopper) and the author of its overview of ancient Greek history. His most recent book is *Pericles: A Biography in Context* (2016).

Peter Meineck holds the endowed chair of Professor of Classics in the Modern World at New York University, USA. He is also an Honorary Professor of Classics at the University of Nottingham and the Founding Director of Aquila Theatre. His most recent publications include *Theatrocracy: Greek Drama, Cognition and the Imperative for Theatre* (Routledge, 2017), *Combat Trauma and the Ancient Greeks* (ed. with David Konstan, 2014) and a new translation of Aristophanes' *Frogs* (forthcoming). He is also Rescue Captain of the Bedford Fire Department in New York.

Elizabeth Minchin is Emeritus Professor of Classics at the Australian National University. She has published extensively on the Homeric epics and related topics. One special interest has been the application of research in cognitive studies to the epics in order to illuminate aspects of Homeric composition (*Homer and the Resources of Memory: Some Applications of Cognitive Theory to the* Iliad *and the* Odyssey (OUP, 2001); *Homeric Voices: Discourse, Memory, Gender* (OUP, 2007)). Another research interest is in landscape, myth and memory, especially in the region of the Troad and the Hellespont.

Diana Y. Ng is Associate Professor of Art History at the University of Michigan-Dearborn, USA. She has authored essays on the commemorative purpose of public architecture and spectacle, and is the co-editor of *Reuse and Renovation in Roman Material Culture: Functions, Aesthetics, Interpretations* (forthcoming).

Anne-Sophie Noel is Assistant Professor of Greek at the École Normale Supérieure of Lyon. She is a philologist and cultural historian of ancient Greece, specializing in the performance and reception of Greek drama. She has published widely on Greek tragedy; her current research combines cognitive and affective theories to reconsider the relationships between humans and objects in the ancient world.

Sarah Olsen is Assistant Professor of Classics at Williams College, USA. She has written articles on a wide range of topics, including Homer, Greek vase painting, sympotic entertainers in ancient Greece and the ancient novel. Her current research focuses on dance in archaic and classical Greek literature and culture.

Maggie L. Popkin is the Robson Junior Professor and Associate Professor of Art History at Case Western Reserve University, USA. She is the author of *The Architecture of the Roman Triumph: Monuments, Memory, and Identity* (2016). Her research on Greek and Roman art and architecture has appeared in edited volumes and journals including the *American Journal of Archaeology*, *Hesperia*, the *Journal of the Society of Architectural Historians* and the *Journal of Late Antiquity*.

Jessica Romney is Visiting Assistant Professor in the Department of Classical Studies at Dickinson College, USA. Her work examines the ways in which the ancient Greeks constructed social identities through literary production and performance, whether in reference to the social institution of the *symposion* or, in a new project, in regard to Greek ideas of the world and their place within it. Her work has been supported by the Social Sciences and Humanities Research Council of Canada, and she has published on ancient identity in archaic lyric and Herodotus.

Günther Schörner studied Classical Archaeology, Ancient History, Prehistory, Early Christian Archaeology and Art History at Erlangen, Germany. From 1993 to 2010 he worked at the University of Jena, Germany, as the Chair of Classical Archaeology. From 2010 to 2011 he was Professor of Classical Archaeology at the University of Erlangen-Nürnberg, Germany, and since 2011 he has been Professor of Classical Archaeology (with a focus on Roman Archaeology) at the University of Vienna, Austria. His research interests include religious studies, especially rituals and their visualization in the Roman Empire, material culture studies and culture contact studies. He is the author and editor of several monographs and numerous articles. He has carried out fieldwork in Austria, Germany, Italy, Jordan and Turkey.

William Michael Short joined the Department of Classics and Ancient History at the University of Exeter, UK, in 2017, after holding positions at the University of Texas at San Antonio and Loyola University, Maryland, both in the USA. He earned his PhD in Classics from the University of California, Berkeley in December 2007. Prof. Short's areas of specialization are Latin language and literature, Roman cultural history, cognitive semantics, and cultural anthropology. His research interests rest at the intersection of language, culture and cognition, where he has been pioneering an approach to ancient culture inspired by Lakovian conceptual metaphor theory. He has edited or co-edited several volumes of papers including *Con i Romani: Studi* *antropologici del mondo antico* (2014, in Italian), *Embodiment in Latin Semantics* (2016) and *Toward a Cognitive Classical Linguistics* (forthcoming). His journal articles include studies of Latin's metaphors of time (2016), communication (2013) and mind (2012), as well as an analysis of the Latin preposition *de* from the perspective of image-schema theory.

Antonis Tsakmakis is Associate Professor of Greek in the Department of Classical Studies and Philosophy, University of Cyprus. His research interests are Greek historiography, old comedy, the sophists, Greek stylistics, Greek particles, pragmatic and cognitive approaches to literature, the reception of antiquity in modern times, and Greek in secondary education. He is the author of *Thukydides über die Vergangenheit* (1995), and co-editor of *Brill's Companion to Thucydides, Mnemosyne Supplements* (2006) and *Thucydides between History and Literature* (2013). Recently he has completed a new series of textbooks for teaching Greek in high school. His current research projects include Aristophanes' *Thesmophoriazusae*, Euripides' *Suppliant Women*, the *Hellenica Oxyrhynchia* and Herodotus.

Angeliki Varakis-Martin is Lecturer in Drama and Theater at the University of Kent, UK. She holds an M.A. and PhD from Royal Holloway, University of London, UK. Her main interests and publications are in Greek theater and classical reception. She has published numerous papers on Aristophanic performance (ancient and modern) and the Greek mask. Publications include "Body and Mask in Aristophanic Performance," *BICS* (2010), "Aristophanic Performance as an All-Inclusive Event: Audience Participation and Celebration in the Modern Staging of Aristophanic Comedy" in *Classics in the Modern World: A Democratic Turn?* (2013) and "Encountering the Heroic *Prosopeion* in Fifth Century Theatre Performance" in *The Physicality of the Other* (2017). She is currently preparing a book on theater director Karolos Koun and his work on ancient Greek theater.

Alessandro Vatri (D.Phil. Oxon) is Junior Research Fellow of Wolfson College, Oxford, UK. He is the author of *Orality and Performance in Classical Attic Prose: A Linguistic Approach* (2017) and of several articles on Greek linguistics, ancient rhetoric, Greek oratory and the digital humanities. He is currently working on a computational semantics project on the vocabulary of Greek aesthetics at the Alan Turing Institute.

Foreword

Many years ago, I read an article (I no longer recall where) on the neurology of speech. I learned that uttering a word requires the coordination of several different organs, including the position of the tongue, the shaping of the mouth, the action of the vocal cords, and much more. But this is not the hardest part. The problem is that the nerves that run from the brain to the relevant zones do not all take the same or even the shortest path. Some do run more or less directly to the area to be stimulated, but others dip down well into the chest before turning upward to reach their assigned destination. Now, nerves do not transmit signals instantly; nervous impulses move fast – some travel at 100 meters per second, and even slower ones race along at a quick pace – but they still take time, so much so that as we speak, the brain has to send some signals earlier than others, so that those that run along the shorter circuits arrive at their appointed place at the same time as those that travel more circuitous routes. Sometimes it must simultaneously emit signals for two successive words: the faster one for the first word in the phrase, the slower one for the following word. This is a complex business, and there is, I guessed, a limit to how far in advance the brain can calculate a given neural message. If the limit is a few words, five or six or seven, say, then it may restart the process at given intervals, which suggests that pauses may occur or be needed as the brain refreshes the batches of coordinated signals. This latter conclusion does not necessarily follow; the brain may look ahead continually and keep firing impulses without waiting for the latest to reach their targets in a single bundle with the earliest. But it occurred to me that, if the brain did tarry so that at least some of the multiple signals in a cluster could catch up with each other before launching a new series, it might explain why we tend to speak in phrases, with intervening pauses, however brief. And this in turn might account for the fact that the versification of oral poetry such as the Homeric epics is structured around cola – that is, short phrases, three or more often four to the line – that are not semantic units or pauses to let the reciter catch his breath, but are due to the way the brain packages units of speech. The Homeric line is similar to that of Lucretius, which looks back to early Latin epic versification, for example in Ennius, but is quite different from the enclosed word order pioneered by Catullus and Cicero and then standardized by Virgil and later poets, in which the unit of meaning is the entire verse, with the final word modifying or modified by the first word in the line or one just before or following the caesura. We may see in this latter pattern a consequence of writing as the primary medium of communication rather than oral performance.

I am not sure that the above story is true, nor do I know of any experiments or cross-cultural comparative studies that confirm or disprove my hypothesis. I lay it before the reader simply as an example of how consideration of information derived from physiology or cognitive processes can raise new questions about literature that might not otherwise have suggested themselves.

I was recently struck by the description in one of Seneca's letters (*Moral Epistles* 27) of a wealthy Roman who had a poor memory for poetry, something of an embarrassment for anyone professing to be well educated in ancient Rome. His solution was to purchase well-versed slaves or train his own to memorize authoritative works, and, when the occasion called for reciting excerpts, to have them either prompt him or perform themselves. When someone protested that this was hardly an indication of the man's own competence, he replied that what his slaves know counts as his knowledge. Seneca allows that this may be the case in matters other than philosophy, but does not ridicule the claim itself. I then began to reflect on Aristotle's repeated assertion that a friend is another self, and the comment, attributed to him and repeated by many others, that friends are one soul in two bodies. What if Aristotle meant these assertions not as vivid metaphors of intimacy but rather entirely literally: friends really do share their minds and think with and through one another? I concluded that this was indeed his intention, but I would never have imagined reading him this way had I not been invited to a conference on distributed cognition; that is, the idea that the mind is not the sole locus of thinking but is part of a larger whole that includes the body and, beyond that, the instruments such as paper and pencil or computerized calculators without which we could not solve certain problems, and even further the joint investigations or actions in which friends or coworkers engage. The founder of distributed cognition cited the example of a pilot and a location finder who interact so rapidly that neither can be said to be guiding the plane on their own; they are thinking jointly, neither being the sole cognitive agent.[1] Aristotle may have been a forerunner of this approach, long ago, in part because his culture did not share our notion of the isolated mind as the unique locus of thought.

But it is not only the case that cognitive science offers clues to the interpretation of classical literature and practices. Classics and the humanities generally have lessons to impart in turn to the scientists, including the most deterministic – that is, not just linguists, who obviously stand to benefit from the study of multiple languages, but also neuro-scientists and experimental researchers on the structure of the brain. Efforts to determine the areas of the brain, for example, where emotions such as anger or joy are located, have long been carried on under the assumption that the English vocabulary is a universally reliable expression of the emotions, or in the words of the eminent linguist Anna Wierzbicka, the tendency to "absolutize the English folk-taxonomy of emotions."[2] But what if the experience of the emotions in different cultures is not just a matter of terminology but of substance – that is, the words nearest to English terms constitute distinct kinds of emotion, as different from joy or anger as joy is from happiness in the English vocabulary, or sadness from depression? We might well expect that the apparently comparable feelings in other languages correspond to different regions of the brain, or even that the emotional lexicon has influenced the way the brain itself is organized. In this case, neuro-scientists may well be looking for anger in all the wrong places, and would need to broaden their research to include at least the possibility that not all brains look alike. When Aristotle states in his treatise on rhetoric that it is impossible to experience anger and fear simultaneously in respect to the same person (anger for him is always directed to an individual), we might imagine that the cerebral loci of anger and fear might conflict with each other for an ancient Greek, at least in certain respects, in ways that they do not in the brain of a native speaker of English. Classical writers have left detailed accounts of the passions – accounts that are not subject to the influence of modern English, as even the most remote cultures are today, given the access to multiple forms of dissemination, from the spread of modern psychology to the availability of new media of communication, from literature to film to the internet. For example, foreign terms for "frustration" are virtually universally calques on the English term, whereas no such word, and perhaps no such concept, existed in classical antiquity. In this regard, there is a cultural and linguistic independence to ancient Greek and Latin relative to modern emotional taxonomies that renders them useful complements to even the most anthropologically aware research in modern psychology.

Cognitive science is a mansion with many rooms. The separate compartments include structural linguistics or semantics (more broadly, psycholinguistics), narrative theory and the role of context, the influence of the body on the brain, identification, distributed cognition, cognitive dissonance, the role of imitation in ritual practices, theory of mind, multi-sensory coordination, audience reaction to positive versus negative affect, artificial intelligence, memory retention and false memories, cognitive resource depletion, and more – I have listed here only the approaches adopted in the several chapters of this extraordinary book, which are sufficiently varied in themselves. Cognitive theory is not one field, or at least not yet. What the future will bring is still unknown, but it will surely have taken into account – indeed, is already taking into account – the results of classical scholarship.

David Konstan

Notes

1 Edwin Hutchins, *Cognition in the Wild* (Cambridge, MA: MIT Press, 1995).

2 Anna Wierzbicka, *Emotions across Languages and Cultures: Diversity and Universals* (Cambridge: Cambridge University Press, 1999), p. 171.

Acknowledgments

The first idea for a collected volume on classics and cognitive theory came from Amy Davis-Poynter at Routledge who approached me in the summer of 2014 with the idea for such a book. Following a panel at the Society for Classical Studies in 2015 and a conference on the subject at NYU in 2016, this book has finally come to be. I would like to thank Amy Davis-Poynter and Elizabeth Risch at Routledge for their unfailing support and encouragement throughout this process. Thanks are also extended to the organizing committee of the SCS for agreeing to host a panel on what is a brand new sub-field in classics, to Matthew Santirocco, Maura Pollard, Rebekah Rust, Rebecca Sausville and the NYU Classics department for all their help with the Classics and Cognitive Theory Conference, and to Gabi Starr for her wonderful keynote speech on neuroscience and aesthetics.

David Konstan, my colleague at NYU, has been a source of encouragement and support throughout every stage of this project and I would also like to thank the many scholars, philosophers and scientists who came to speak at my graduate seminar on this subject, including Joe LeDoux, Jessie Prinz, Ed Vessel, Jake Mackey, Kyle Johnson, Bill Short and David Konstan, and the graduate students from NYU, CUNY and Fordham who contributed so much to that new seminar.

Much of the editing of this volume took place at the Bedford Fire Department in New York, where I serve as a volunteer fire captain and EMT. While the endless supply of coffee, a full fridge and sweet treats left by friends and locals are intended to keep us going on calls, they also came in useful as I worked on this book, in between those calls, on the days I was here. If sometimes my editing routine was broken by a fire alarm or an ambulance call, I hope this is not reflected at all in these pages.

My co-editors, Bill Short and Jennifer Devereaux, both put in an enormous amount of work on this volume and their breadth of knowledge of the various fields of the cognitive sciences never ceased to amaze me. The many contributors to this volume were also incredibly open to questions and comments about their own use of cognitive theories, and if we sometimes did not fully agree with their approaches, we are very happy that this volume showcases the diversity of this field and the many ways in which the cognitive sciences are already having an impact.

Final thanks go to my family, Desiree, Sofia and Marina, for their constant love, support and laughter, and for always showing me how many different ways there are to view the world around us.

Peter Meineck
New York

Introduction

Peter Meineck, William Michael Short,
and Jennifer Devereaux

Cognitive theory or cognitive science is the interdisciplinary investigation of the human mind and human behaviour that encompasses several fields of study including psychology, philosophy, linguistics, anthropology, archaeology, neuroscience, artificial intelligence, and robotics. In the past two decades many scholars in the humanities have begun to profitably apply cognitive theories to their work in a diverse range of subject areas such as literature, performance, theatre studies, visual arts, music, history, and cultural studies. Cognitive theories are now being applied to the study of the ancient Greek and Roman worlds and the time is ripe for a collected volume that seeks to offer a snapshot of current work being undertaken in this relatively new sub-field in classical studies. This book, the first of its kind in this area, is not intended to be exhaustive, nor can it include all those currently working with cognitive theory and ancient material. Rather it is intended to offer something of a cross-section of this emerging field as it develops, as well as to act as a basic introduction to the subject, and to encourage further dissemination and development of what we consider an exciting new way to examine antiquity.

Engaging with cognitive theory can sometimes seem like wrestling a particularly slippery methodological octopus: scientific studies are published at a much faster rate than in most areas of the humanities, philosophical debates rage, and across the various fields that study the human mind there is an enormous amount of scholarship in circulation. This can seem even more daunting when this interdisciplinary field is applied to another interdisciplinary field such as classics. But it is the intersecting fluidity of cognitive theory, which is after all a field that strives at nothing less than a better understanding of the human mind, that makes it so useful to classicists, who are also involved in trying to understand human mentalities, albeit ancient ones. Lisa Zunshine characterizes the situation this way: "the boundaries between various subfields of cognitive cultural studies thus remain fluid as a function of the opportunistic nature of cognitive science, that is, of its tendency to grow and change by complementing and enhancing various aspects of other disciplines" (Zunshine 2010). This means that the influence is not just a one-way street and that scholars in the humanities have a great deal to offer cognitive science, especially in areas such as cultural and social cognition, the history of emotions, philosophy, and cognitive cultural and cross-cultural studies. Examples in classics include Chaniotis (2012) and Cairns and Nelis (2017).

The papers collected in this volume are reflective of the kind of fluid cross-disciplinary interaction that tends to characterize the cognitive sciences. Thus, they might best be viewed as active explorations between existing cognitive theories, psychological studies, neuroscience research, phenomenology, and different areas of classical studies. These scholars are seeking new ways to look at ancient material to try to better understand the cultures that produced them. In effect, they are doing what Tom Habinek has described as "articulating a model of thought and action radically different from those taken for granted" which "defamiliarizes the ancient material, opening up new horizons of understanding" (Habinek 2011: 65). Here, Habinek is referring to neuroscience research being applied to Stoic philosophy but his statement applies just as well to cognitive linguistics, cognitive literary theory, social cognition, performance and cognition, artificial intelligence, and cognitive archaeology, the six subject areas covered in this volume.

There have been criticisms of this kind of application of cognitive theory, particularly in the humanities: it has been described as reductionist, unobservable, an oversimplification of human behaviour, and reliant upon inference, and as negating environmental, biological, genetic, and cultural influences and differences. But in the past two decades even the most empirically based areas of cognitive science, such as the brain imaging studies of neuroscience, have started to embrace cultural and social interactions and explanations (Choudhury 2016). Many research findings now demonstrate how environment, genetics, and culture play an essential role in human cognition, with many agreeing on the basic premise that human cognition is extended beyond the confines of the physical brain, connected to the body, and extended out into the environment (Glăveanu 2014). Although there are many different theories of active externalism and the extended mind, they are generally gathered under the term "distributed cognition" (DCog), a framework first developed by Edwin Hutchins (1995, 1995), although surely influenced by the work of philosophers such as Husserl, Heidegger, and Merleau-Ponty.

If there is a unifying element to the diverse studies included here, it is that most of them are influenced by theories of distributed cognition, rather than purely computational models, which posit that the brain is basically a self-contained organism controlling bodily operations, emotions, perception, and thought. Of course, this dichotomy is by no means rigid and many cognitive theories have both computational and externalist commitments. Generally, distributed theories regard the mind as part of a broader cognitive system where the brain is embodied within the entire human corporal structure, embedded in the world, extended out into the surrounding environment, and enacting with what it finds there.[1] Andy Clark called this a cognitive "feedback loop" and explained how external elements can act as "scaffolding" and "turbocharge" our thoughts (A. Clark 1998). Hutchins has maintained that "human cognition is always situated in a complex sociocultural world and cannot be unaffected by it" (Hutchins 1995: xiii). Thus these so-called "second-wave" theoretical models of human cognition that posit distributed models of thought have challenged any divisions between human biology and culture. If, as they contend, all human cognition is distributed, then even the kind of social constructionist models that can be found in literary studies must have a physical and biological basis. Miranda Anderson, who has applied such an approach to Renaissance literature, has stated that "our ability to be constructed by sociocultural forces relates to neurological plasticity . . . at the same time as human extendedness and adaptability (to cultural, physical or linguistic variables) tempers any notion of universals that might be attempted" (Anderson 2015).

If the basic premise of distributed cognition is the notion that the human mind operates both on and in its culture and environment, then its application to what we already know of ancient cultures could be revealing. For example, in the field of cognitive archaeology, Malafouris and Renfrew have proposed that ancient artefacts should be viewed as enactive objects in their own right, as well as the physical remains of an ancient thought process. For example, Malafouris has shown how understanding the mind behind the creation of a Mycenaean Linear B tablet is

> not simply a matter of postulating the putative representational states of events being created inside the head of a Mycenaean when he or she is reading or writing a tablet. It is also a matter of postulating the dynamic interaction between that person and the physical properties of the medium of representation as a material thing – that is, a clay tablet.
> *(Malafouris 2013: 73)*

Thus, the cognitive properties of the exogrammatic writing system are not only a memory device distributed on the tablet, but a product of its original environment, having been crafted by a human for a specific purpose – that is, to enact with the mind of another human in a different time and space. This "distributed" reading of the clay tablet allows us to go beyond the translation of the symbols it contains and to better understand the culture that produced it, as well as its function, value, portability, accessibility, and how it may have changed the culture that produced it. As Malafouris states (2013: 74), "it was not simply the information that was externalized, but also the actual processing of that information."

As the cognitive sciences have developed, many of their sub-disciplines have come to take environment and culture into more careful consideration and in so doing have upended the overly simplistic views of human cognition as universalistic. A good example of this recent research can be found in the area of facial recognition: one fMRI study has shown marked differences in the areas of the brain connected with facial processing between illiterate and literate people, which has suggested that humans may have adapted their existing cognitive mechanisms for face recognition to respond to the new cognitive load demanded by the development of reading and writing (Dehaene-Lambertz et al. 2018). Therefore illiterate or semi-literate people in antiquity would perceive a theatrical mask quite differently than most people today (Meineck 2017). In addition, when facial processing has been studied cross-culturally it has been shown by eye-tracking studies that people from different cultural groups scan faces differently (Caldara 2017). The relatively new fields of cultural and social neuroscience still have a long way to go but there is now general agreement among scientists of the human mind that cognition is culturally dependent.

In the field of classics, David Konstan has sought to demonstrate that differences in ancient Greek culture shaped ancient attitudes to the emotions (Konstan 2006), and recently Douglas Cairns has examined the expression of ancient Greek concepts of emotions in dramatic texts by applying theories of distributed cognition. Cairns has posited that such expressions are products of mind, environmental conditions, and specific cultural forms "embedded in particular religious and philosophical world views" (Cairns 2015). William Short has explored the ways in which Roman society exploited the physical and social environment as a scaffold for extended cognition with a close examination of Latin spatial metaphors (Short 2016), and Esther Eidinow has re-examined concepts of public and private in the context of Greek religion, taking distributed cognition into consideration (Eidinow 2015). These scholars have thus already begun to demonstrate that by applying externalist theories and distributed cognition, classicists can learn more about ancient minds and the cultures within which they were embedded.

Most of the papers included in this volume were originally presented at the conference entitled Classics and Cognitive Theory held at New York University in October 2016 and organized by Peter Meineck. This developed from a desire to start to bring together scholars of the ancient world who were already applying cognitive methodologies to their work. This project began at the 2015 Society for Classical Studies Meeting in New Orleans with a panel including Peter Meineck, William Short, Jennifer Devereaux, Jacob Mackey, and the late Garrett Fagan, with Ineke Sluiter responding. Garrett Fagan was one of the first classical scholars to apply cognitive theory to a substantial work on antiquity with his 2011 monograph *The Lure of the Arena* (Fagan 2011). This ground-breaking book applied social psychology to the behaviour of crowds at the Roman games. Garrett was a driving force in the area of cognitive classics as a community of scholars interested in this new field began to form, and his future work on the emotionality of crowds held a great deal of promise. We had hoped to include his excellent paper, "The Roman Army as a Rabble in Tacitus *Histories 1*", which closed our NYU conference, but Garrett was suddenly struck with pancreatic cancer and passed away far too soon in March 2017. This volume is dedicated to the memory of Garrett Fagan, a superb scholar and a wonderful colleague.

In *The Lure of the Arena*, Fagan laid out several guiding principles to justify the application of scientific studies on modern minds from one culture to ancient minds from another. He started with the premise that the basic physiology of the human brain is the same and that psychological processes are connected to the physical functioning of the brain. Therefore, he posited that we should expect to find similarities in some mental processes on a species-wide basis. He termed this "psychobiological" and stated that this approach did not mean that human actions are genetically determined, but that human psychobiology generates cultural systems and sets limits on just how varied these can be. Fagan also asserted a tenet of distributed cognition: that specific facets of cognition and cultural behaviours exhibited by different groups are "enormously influenced by the environment confronting them" (Fagan 2011: 39). He also maintained that human behaviour is shaped by the interdependence between culturally contextual stimuli and psychological propensity, stating that "how people think and behave is a product of their genetic predispositions acting under the influence of various environmental factors that include but are not limited to their culture" (Fagan 2011: 40). Therefore, culture is not the only shaper of human behaviour but the mind and culture share an implicit bond, each forming and influencing the other. Fagan regarded them as interlocking cogs in a behaviour-generating machine. As an ancient historian, Fagan made a crucial point – if people's psychological functioning varied as widely as their diverse cultural constructs, then the behaviour of people in alien societies ought to remain virtually impenetrable to an outsider. Despite the obvious contention that one culture may project its psychological predispositions upon another, Fagan was correct in that we are able to identify certain common emotions and actions in ancient cultures, such as grief, rage, and joy. Additionally, we can perceive these manifested in ancient material, such as dramas, literary accounts, grave markers, and letters. Finally, Fagan pointed out that the ability to identify with the behaviour of people far removed from us chronologically, geographically, and culturally is, after all, fundamental to the entire business of history.

The breadth of cognitive studies, its interdisciplinary nature, and its application to a field as traditionally diverse as classics means that it would be impossible to include all the ways in which scholars of antiquity are currently applying cognitive theory to their work. Rather, it is our hope that these 24 papers gathered here will encourage scholars and students of the classics to engage with the cognitive sciences and learn more about the ways in which we can profit from them. It is certainly not our assertion that cognitive theory replaces other existing approaches that have been employed by classists, ancient historians, and archaeologists, but rather that the application of cognitive theory can add a new tool to our existing kit of methodologies we might use to analyse ancient material. With this in mind, we have asked our contributors to describe the theoretical models they are using and then demonstrate how they are applied to a piece of text, an artwork, or a facet of ancient cultural practice. Accordingly, the papers in this volume demonstrate a broad range of theoretical applications and a wide diversity of opinions and we hope that all of them go some way to helping to open up new horizons of understanding about the ancient world and its people.

Cognitive linguistics

The first section of this volume contains five papers that apply a variety of theories of cognitive linguistics to Greek and Roman texts. Emerging in the 1970s first as an offshoot of and later as a counterpoint to Chomskyan generative linguistics, cognitive linguistics has become virtually synonymous with "embodied cognition" and has been one of the most powerful engines driving the progress of the so-called "second-wave" cognitive sciences. Rejecting any view of cognition as the computational (algorithmic) manipulation of sets of abstract, amodal symbols, these embodied cognitive sciences take as their point of departure that human thought – and thus also human language – is fundamentally grounded in, and constrained by, the nature of our bodies as well as by our embeddedness in specific environmental (physical, social, cultural) contexts. Instead of treating syntax as an autonomous cognitive "module" governed by its own set of rules and involving its own sort of mental representations (as in Chomsky's approach), cognitive linguistics seeks to investigate the structures of language (including both syntactic and semantic structures) in relation to general principles that can be shown to operate pervasively within the human cognitive faculty and that depend on our sensorimotor capacities. In particular, cognitive linguistics studies language on the basis of what is known from neuroscience, developmental and experimental psychology, psycholinguistics, philosophy, and other related disciplines, about how human beings go about interacting with, categorizing things in, and making sense of the world around them.

For the early stage of its development (through, say, the late 1980s), it is probably something of a misnomer to speak of "cognitive linguistics" as a coherent theoretical paradigm and programme of practical research. Scholars like Ronald Langacker, Ray Jackendoff, Charles Fillmore, George Lakoff, and Mark Johnson worked from sometimes very different perspectives and towards sometimes very different ends: Langacker (1987), for instance, developed an approach to language under the slogan of "grammar is conceptualization," which captured the hypothesis that it was impossible to reduce meanings to simple truth-conditional correspondences ("necessary and sufficient conditions"), grammar being a means of "profiling" specific, humanly determined conceptualizations of the world. Within the Chomskyan tradition, Jackendoff (1985) proposed that conceptual structure characterizes a level of mental representation where linguistic *as well as* sensory and motor information could be integrated. Fillmore (1982) elaborated a theory of meaning known as "frame semantics," which treats words as meaningful only within the larger situation-based or goal-directed groupings or clusters of concepts of which they play a part (what Croft & Cruse 2004 later termed "domains"). Meanwhile, George Lakoff and Mark Johnson (Lakoff & Johnson 1980) articulated a theory of abstract understanding that had metaphor – the cognitive and neural linking or "mapping" of structured conceptual domains – at its heart.

Nevertheless, these figures (and many others along with them) were united by a shared set of theoretical commitments, which Dirk Geeraerts summarizes as follows (Geeraerts 1995):

> Because cognitive linguistics sees language as embedded in the overall cognitive capacities of man, topics of special interest for cognitive linguistics include: the structural characteristics of natural language categorization (such as prototypicality, systematic polysemy, cognitive models, mental imagery and metaphor); the functional principles of linguistic organization (such as iconicity and naturalness); the conceptual interface between syntax and semantics (as explored by cognitive grammar and construction grammar); the experiential and pragmatic background of language-in-use; and the relationship between language and thought, including questions about relativism and conceptual universals.

From these beginnings, cognitive linguistics has come together as a major interdisciplinary research paradigm defined by four main areas of theoretical concern: the image-schematic basis of meaning; the extension of meaning through sensorimotor "construal" operations; the bootstrapping of abstract conceptualization on physico-spatial experience via metaphor; and the treatment of *all* linguistic structures (be they morphemes, words, or larger syntactic units) as conventionalized pairings of form and meaning. A fifth area might be Gilles Fauconnier's (Fauconnier 1994, 1997) "blending theory" or "conceptual integration theory," or the theory that much of conceptualization, including abstract and imaginative conceptualization, can be explained in terms of the highly structured "mental spaces" whose constituent elements are dynamically combined and integrated into emergent organizations of knowledge. Another might be the theory of mental models (see Johnson-Laird 1980; Nuyts 1992), whose influence can be seen especially in Lakoff's notion of "idealized cognitive models" or "cogs," but this may belong more to the field of cultural semantics, which applies insights from cognitive linguistics to anthropological study (see esp. Shore 1998 and Palmer 1996). Except perhaps the latter, which stand somewhat apart in positing extremely general mechanisms of mental functioning, these theories are tightly interconnected in the practice of cognitive linguistics.

Briefly, image schema theory proposes that human concepts (and thus the meanings of words) very often correspond to, or are underpinned by, skeletal cognitive structures that combine, in a gestalt manner, sensorimotor (perceptual and kinaesthetic) information that arises from our bodily experience of the physical world (Johnson 1987; Gibbs 1994). Because of this embodied nature of our conceptualization, semantic structure can therefore be explained in terms of certain "construal" processes that operate on conceptualizations as correlates of perceptual and kinaesthetic experience, such as salience (focus of attention) modifications, vantage point (perspective) shifts, figure/ground alignment, summary or sequential scanning, scalar adjustment, and path- or endpoint focus (Langacker 1990; Talmy 2000). Conceptual metaphor theory then argues that such skeletal experiential structures are projected to abstract conceptualization, enabling the understanding of *abstracta* by grounding them in more immediately and directly comprehensible concrete experiences (Lakoff & Johnson 1980). It is in this sense, then, that cognitive linguistics constitutes a theory of embodied cognition: it proposes that human conceptualization, and thus the inferential processes guiding semantic extension, depends in large part on cognitive structures – i.e., images schemas – and construal operations that arise naturalistically from (indeed are analogues of) perceptual and kinaesthetic experience, and that through mappings of image-schematic structure to domains not directly grounded in experience – i.e., metaphors – literal (physico-spatial) understanding comes to be extended also to abstract reasoning. Finally, the lexicon/syntax continuity hypothesis suggests that these same structures and processes of cognition are relevant to all aspects and levels of linguistic encoding.

The contributions gathered in this volume's section on "cognitive linguistics" fairly represent the central interests of this field, without, however, being derivative or mechanical. Ahuvia Kahane's opening chapter tackles questions of Homeric formulaic language from the usage-based perspective of cognitive construction grammar, arguing that what appear to be "deviations" from the "rules" of epic diction under traditional oral-formulaic theory in fact form part of a "complex adaptive system" that was capable of comprehending strictly rule-governed usages as well as seemingly anomalous usages that emerge through entirely regular analogical processes from the interaction of metrical, lexical, rhythmic, phonetic, semantic, thematic, and narrative elements. Alexander S. W. Forte uses conceptual metaphor theory to investigate Homeric representations of surprise. As Forte demonstrates, the language of Homer reveals several embodied metaphorical conceptualizations of surprise, above all a metaphor in which surprise is conceived as a kind of "holding" or "seizing" or "striking," likely grounded in the phenomenology of the startle reflex. More than this, though, Forte argues that this embodied "striking" metaphor of surprise interacts with the metaphorical representations of other emotion concepts such as anger and of fear (as, for instance, an opponent, or a wild animal), suggesting that the epic tradition in fact included a rich understanding of "multidimensional affective processes." Rutger J. Allan also deals with Homer, but focuses instead on how the epics bring about a feeling of "immersivity" for the audience. As Allan argues, this is achieved through the introduction of specific features of construal and embodiment in the text that tend to heighten the listener's feeling of being immersed in the narration: in particular, elements that relate to specificity and salience, perspective (vantage point), and dynamicity (temporal iconicity and speed). The working of these construal features is illustrated through an analysis of a passage from the Calypso episode in *Odyssey* 5.

Shifting to the Roman world, William Michael Short's chapter presents an approach to the study of Roman culture that derives in equal parts from cognitive linguistics and from Bettinian semiotic anthropology. This approach emphasizes linguistic – and especially metaphorical – structure as a direct "index to culture," and uses patterns of figurative expression in Latin to reconstruct the "folk" mental models that define Roman culture as a distinctive signifying order. After presenting the underpinnings of this approach, the basic methodology is illustrated through an analysis of Latin's metaphors of "meaning" and "ambiguity." As Short demonstrates, although Latin possessed several ways of conceptualizing ambiguity metaphorically, the most conventionalized metaphor for this concept recruited the image of "leading (around) on both sides." A comparative perspective then shows that this metaphor makes Latin something of an outlier. More than just a linguistic curiosity, however, Short goes on to claim that this metaphor in fact contributed to shaping Roman culture's perhaps uniquely positive, even welcoming view of ambiguity. In this sense, the approach he articulates constitutes a method of studying Roman culture "emically."

Alessandro Vatri introduces the field of psycholinguistics to the study of Greek and Latin. Generally speaking, psycholinguistics is concerned with reconstructing the cognitive processes that language speakers rely on during on-line linguistic comprehension (for example, how they disambiguate so-called "garden path" sentences – "The horse raced past the barn fell" – or how they interpret novel figurative expressions). In this chapter, Vatri looks at the method of *metathesis* employed by ancient literary critics, which consisted in rewriting or reordering sentences or whole tranches of texts, usually as corroboration of some positive or negative stylistic judgement. He argues that such paraphrases help shed light on how Greek and Latin speakers actually decoded the sometimes complex syntactic organization of utterances in these languages, and thus on how they comprehended sentences. In the absence of native speakers able to participate in experimental studies, such passages provide empirical evidence of what specific linguistic features might have gone towards native speakers' judgements of what constituted clear (or clearer), more readily comprehensible speech.

Cognitive literary theory

Theory of Mind (ToM) refers to the cognitive ability to attribute mental states to the self and to others. Other names for this capacity include "common-sense psychology," "naïve psychology," "folk psychology," "mindreading," and "mentalizing." All of these names point to the fact that socially engaged individuals have many thoughts and beliefs about their own and others' mental states, not all of which are verbalized. In cognitive science, ToM is used to investigate how it is that people achieve the cognitive capacity to form beliefs and judgements about the mental states of others, states that are not directly communicated. For competing approaches to third-person mentalizing (theory-theory, modularity theory, rationality theory, and simulation theory), see Goldman (2012). For the application of such theories to literature, see Zunshine (2006), who sets forth the argument that evolutionary history is at least as important as personal reading histories in the critical treatment of texts, so the ability for research from cognitive psychology to explain behaviour in terms of ToM can furnish fresh insights into our interactions with literary texts. A more recent and robust overview of the field can be found in Zunshine (2015). In the present volume, Elizabeth Minchin develops these ideas and anchors her chapter on falsehoods in Homer to the work of Jeffrey Walczyk et al. (2014), who have argued that the creativity of lying can be expedient for achieving social goals. Minchin examines in particular the lies told by Odysseus and suggests that the listeners become more engaged, perhaps empathetic, but far more entertained with a narrative when required to expend more cognitive resources on processing the incongruities and inconsistencies between truth and lies.

Anna Bonifazi applies Conceptual Blending Theory (CBT) to Homer and the "compression" of relations such as time and identity, which occurs when humans "blend" multiple mental spaces. CBT was developed by Mark Turner and Gilles Fauconnier (Fauconnier & Turner 2002) and claims that our ability to activate two conflicting mental structures – such as truth and fiction – and blend them creatively into a new "mental space" reflects a basic cognitive operation with constitutive and governing principles. CBT suggests that thought is fundamentally combinatorial and infinitely complex, with ever-evolving and unique thought structures under constant assembly. The complexity of such combinatorial thought is explored by Turner and Fauconnier through what they term a "double scope" story network, wherein input stories with different (and often clashing) organizing frames are blended into a third story organized by a frame that includes parts of each of the input frames. When more than one scenario is integrated, more than one set of actions is compressed into a single conceptual unit. If one scenario is true and another is not, this compression does not change the facts of the first scenario, but it does change their status within the emergent structure of the story-world. That is to say, meaning emerges through the subordination of features to one another within a given "mental space." It is through this emergent structure that readers (or in the case of material culture, viewers) engage with concepts that are otherwise too abstract to find expression through language, which is grounded in embodied experience. Bonifazi asserts that in Homeric poetry compression helps to reduce to human scale concepts that would otherwise be beyond understanding.

Christensen is also influenced by Turner and Fauconnier and applies theories of extended cognition to Homer and how the narrative elements of the *Odyssey* exploit the human cognitive capacity for active prediction. Such predictive brain hypotheses have emerged from computational and cognitive neuroscience. A prevalent account is that of "action-oriented predictive processing" or "active inference" (A. Clark 2013). First developed by Karl Friston et al. (2006), this predictive processing theory hypothesizes that top-down prediction processes rather than bottom-up sensory information govern sensory experience, movement, and energy expenditure. Perception and action are seen to operate within an integrated framework of embodied experience wherein predictive processing allows for "fast and frugal" solutions in the carrying out of tasks across contexts (A. Clark 2015a). Thoughts and actions thus emerge from unconsciously made predictions that are based upon expectations spanning multiple scales of time and space. Sensory signals provide corrective feedback to predictions that emerge from past experience and the greater the predictive-error, the greater the expenditure of energy required to correct for it. Readers interested in a more expansive discussion can look to A. Clark (2015b) and for its application to classics see Meineck (2017).

A departure from the predictive and distributed models of cognition is Antonis Tsakmakis' application of Werth's Text World Theory (Werth 1994, 1999), later refined by Gavins (2007).

This is a cognitive model of discourse processing, which proposes that humans understand discourse by forming internalized mental representations or "story-worlds." According to Gavins, these constructed "text-worlds" are context sensitive and provide for "incrementation," which is described as the "private to public transfer of knowledge." This facet of Text World Theory is put to use by Tsakmakis in examining Aristophanes' *Thesmophoriazusae*, a play where the story-world of the chorus of women at their festival is invaded by a male interloper, who is forced to actively create a series of new story-worlds based on knowledge of the plays of Euripides, in order to be rescued. By using this methodology, his chapter explores the role of knowledge in the appreciation of dramatic performance and literary works, as well as the status of genre conventions and the slippage between the discourse of drama (in this case comedy) and real life. His analysis also suggests an additional insight into the sociocultural conditions of the Athenian classical drama by proposing a new way to consider the vexing question of whether women attended the theatre at this time.

Returning to theories of distributed cognition, Jennifer Devereaux builds on Cognitive (or Conceptual) Metaphor Theory advanced by Lakoff and Johnson (1980), which at its most basic level proposes that language often conceptualizes notions by the understanding of one idea (the conceptual domain) in terms of another. We see this, for example, in phrases such as "love is a journey" or "words are weapons." Devereaux develops this into an innovative proposition she terms *enactive analogies*, suggesting that in Roman literature the body is frequently used to ground abstract social concepts and formulate judgements about the circumstances of others. Devereaux argues that such concepts and judgements are expressed through diffuse linguistic structures that are derived from socially situated bodily knowledge. Taking a particular interest in what she terms "body-as-metaphor," which she identifies in Roman literature through the examination of pre-linguistic information that is situated in the body but is expressed by the integration of internal and external worlds, she identifies paradigms of intercorporeality in a number of ancient texts. She argues that as a consequence of this integration and the intercorporeality it facilitates, readers are placed at the threshold of "undecidability between objective and subjective experience," which fits with Varela et al.'s (2017) theory of enactive cognition, which disputes the notion that the mind creates internalized symbolic representations distinct from the external environment.

Social cognition

Jessica Romney applies theories of discourse dialogue to issues of social identity found in archaic Greek poetry. Van Dijk's (1997) socio-cognitive approach to discourse is concerned with how the properties of social situations of interaction or communication are systematically related to grammar and other discourse properties. In other words, it is concerned with the relations that exist between talk or text and context, with context defined as a socially based but subjective construct based upon a mental model of the properties relevant to the participants. Context thus defined is observed to constrain the production and comprehension of discourse, while at the same time discourse is observed to influence the context models of its participants. Because of this bidirectional influence, a socio-cognitive approach to discourse demands an account of how participants construe social settings, participants, and/or group members, and the social relationships and institutions that act as relevant parameters to the discourse situation. This information is seen as critical to understanding the categorical structuring of the ideologies that characterize the social attitudes that are reproduced by discourse (see Van Dijk 2016).

Laura Candiotto applies distributed theories of social cognition to the works of Plato and argues that the Socratic Method (elenctic inquiry) is a dialogically distributed motivational state that leads to a purification of reasoning (*catharsis*) through cross-examination. Candiotto applies the influential theories of Edwin Hutchins (1995), who, influenced by computational neuroscience, which treats cognition as computation, argues that socially structured activities are a part of a larger cognitive system, which has properties that are distinct from the cognitive properties of the individuals who participate in it. Moving the unit of cognitive analysis "beyond the skin of the individual," Hutchins argues that before discussing the properties of individuals one must describe the culturally constituted worlds in which those properties are manifested. According to Hutchins, when individuals make changes to their agreements about the rules and roles that constitute organizational routines, they simultaneously reorganize the system of which they are a part. Hutchins views this process in terms of evolution, because although individual participants learn solutions after they have come into being, the solution itself must actually be discovered by the cognitive system before it can be discovered by any of its participants. Learning is thus a doubly cultural process, because culturally prescribed behaviours are learned by individual participants by operating within a system structured by learned expectations about learning itself. See also Hutchins (2010).

Thomas R. Martin applies cognitive dissonance theory to decisions made by the Athenian polis after its defeat in the Lamian War of 322 BCE and argues that contemporary religious concepts operating within a cognitive state of dissonance affected the decisions made after the battle. According to cognitive dissonance theory, there is a tendency for individuals to seek consistency amongst their beliefs, opinions, and behaviours. When there is an inconsistency between attitudes or behaviours, dissonance occurs, requiring something to change for the dissonance to be reduced or eliminated. According to Leon Festinger who introduced the theory (Festinger 1957; see also Festinger & Carlsmith 1959), when there is a discrepancy between attitudes and behaviour, it is most likely that the attitude will change to accommodate the behaviour. Proposing that actions can influence subsequent beliefs and attitudes when actors are motivated by dissonance reduction, the theory explains certain human behaviours in terms of actions that are not the result of our beliefs/attitudes, but are rather the cause of them.

Martin also relies on theoretical models of the Cognitive Science of Religion (CSR) which is interested in how we acquire and transmit religious belief through representations, the role of emotion in religious thought and experience, the relationship between ritual action and memory, conceptions and perceptions of the divine, and the body-brain organism as a cultural actor. That is to say, CSR brings theories from the cognitive sciences to bear on why religious thought and action is so common in humans and why religious phenomena take on the features that they do. See Andresen (2001) and Van Slyke (2016), which spotlight religion as an emergent property of human representational systems. Van Slyke focuses on emergent properties because to do so emphasizes the role of feedback and pattern formation in the fluid relationship that exists between internal cognitive constraints and the context in which representations are formed.

Luca Grillo uses Cicero's *Philippics* to test three theories of irony and argues that each has a great deal to contribute to our appreciation of classical texts. In *Logic and Conversation* (Grice 1975) Paul Grice introduced "conversational implicatures," which he uses to refer to phrases that are dependent upon features of the situation or context for their meaning, rather than on the conventional meanings of the words used. Echoic reminder theory asserts that listeners recognize irony when they perceive that a speaker is alluding to some antecedent utterance or state of affairs (Sperber & Wilson 1981). The pretence theory of irony suggests that a speaker pretending to be an injudicious person speaking to an uninitiated audience intends the addressees of the irony to discover the pretence and thereby see his or her attitude towards the speaker, the audience, and the utterance (H. H. Clark & Gerrig 1984). For readers interested in extending the boundaries of the discussion of irony in the ancient world, see Athanasiadou and Colston

(2017). This volume explores irony as more than a verbal phenomenon, delving into questions pertaining to (e.g.) the role of the body in deriving irony, the embodied experience of irony and humour, cultural grounding and perspective taking, fictive elaboration and emotional/evaluative attitudes, and cognitive modelling. Grillo suggests that the application of these theories allows us to account for various passages from the Philippics, where irony can be found but falls outside of the definitions in ancient rhetoric manuals.

Jacob L. Mackey examines the orthopraxy of Roman ritual by applying cognitive theories of overimitation. This is a term used to refer to the tendency of young children, when integrating social and technical intelligence, to copy all the actions of an adult model, even components that are clearly immaterial to the task at hand. Causal accounts of overimitation indicate the pronounced tendency of children observing an adult carrying out intentional actions on a novel object to encode those actions as causally meaningful – even when there is visible evidence to the contrary. Such accounts argue that causal encoding of this nature, despite producing "vivid errors," aids the comprehension of cultural artefacts in the environment and plays an important role in the proliferation of artefact-knowledge (Lyons & Keil 2013).

Normativity accounts claim that overimitation arises when children view causally irrelevant elements as an essential part of an overarching conventional activity. The motivating factor of normative conformity is the avoidance of negative social consequences (see Haun & Tomasello 2011; Haun et al. 2013). Affiliation accounts claim that overimitation derives from an attempt to affiliate with or be like an adult model (Over & Carpenter 2012). According to such accounts, children are aware that irrelevant action is causally immaterial, but they perform it nonetheless to please or otherwise relate with the model. See Lyons (2008), which situates overimitation on a spectrum of rational imitative behaviours.

Paul C. Dilley focuses on early Christian sources and applies the broad cognitive concept of Theory of Mind (see page 7) to ask how a sense of personal cognitive accessibility was developed in early monastic culture with the notion that one's thoughts were accessible to others, human and divine. Dilley builds his thesis from (Luhrmann 2012) theory that there are "local" mentalities and that different practices of attending to mental events have identifiable outcomes. In particular Dilley applies a distinction between the secular and supernaturalist categorization of Theory of Mind with the idea that they can co-exist within a single culture such as the ancient Mediterranean world. Dilley examines the concept of divine omniscience between Olympian belief systems and early Christian thought, how it developed, and the different mentalities that came into contact with the concept.

Performance and cognition

There is already a rich field of inquiry based on the application of cognitive theoretical approaches to theatre and performance studies. The second-wave approaches of cognitive theory (see page 7) with a focus on embodied, enacted, embedded, and extended cognition were particularly well suited to be applied to theatre and performance studies, as is research into the affective sciences that deal with the expression and perception of human emotion. Rhonda Blair and Amy Cook have pointed out that "a cognitive analysis of how any dancer or actor works is likely to find significant similarities in neural, cognitive and kinaesthetic operations, though these operations are understood and applied differently across cultures and 'ecologies'" (Blair & Cook 2016). Blair has used neuroscience research and theories of distributed cognition to further describe the disciplinary methods employed by theatre actors (Blair 2007); Bruce McConachie produced a ground-breaking work on theatre spectating from the perspectives of neuroscience and evolutionary theory (McConachie 2008), and has since produced several works on cognition, evolutionary theory, and the theatre (McConachie 2015). Evelyn Tribble has applied distributed cognition to Shakespeare's theatre and shown how its physical conditions and the ways in which the playbooks were organized aided memory and Elizabethan performance techniques (Tribble 2011). In the world of dance, neuroscientists have worked with dancers and choreographers to explore the human mirror-neuron system, kinaesthetic empathy, proprioception, and inter-group dynamics (Calvo-Merino et al. 2004).

In classics, Felix Budelmann and Pat Easterling (2010) were among the first to explore cognitive approaches. Since then Budelmann has gone on to co-author with Ineke Sluiter an important new collected volume, *Minds on Stage* (forthcoming). Peter Meineck has also applied theories of distributed cognition and neuroscience studies to the experiential aspects of ancient Greek performance anchored by contemporary theories of predictive processing; in his book, Meineck (2017) explores narrative, space, masks, movement, music, and speech to demonstrate the emotional power of ancient drama to generate empathy and suggest alternate perspectives.

Here we have included three papers on ancient performance and cognition and each one applies a different facet of cognitive theory to examine drama from a fresh perspective. All share a common methodological grounding in theories of distributed cognition, the notion that the mind is not confined by the skull or "brainbound" but embodied in the entire human corporeal form and indeed extended out into the environment and back again in a multisensory, predictive, and action-orientated perceptive feedback loop. Sarah Olsen applies Carrie Noland's work on kinaesthesia and agency (Noland 2010) to the performance of archaic Greek poetry with the intention of highlighting the embodied and enacted interactions between performer and audience in antiquity. Referring to an fMRI study on the mirror-neuron system with dancers, and combining this with Noland's work on the enculturation of gestural systems and movements, Olsen surmises that there is a significant difference between viewing an action that one has seen before and one that the viewer has previously performed. This observation, supported by clinical research, has a profound impact on how we might think about chorality in the ancient Greek world, a culture where choral performance was widespread and highly participatory; therefore we can assume the audience members of archaic performance were themselves highly skilled dancers and performers. Olsen takes this premise further and examines the individual agency of the spectator using Deidre Sklar's theories of kinaesthetic empathy (Sklar 2008), which propose that experiencing with the body enables us to "unbraid movement practices from ideological ends" and that this turning inward back to the body "puts cultural conditioning to the test." In applying these theoretical models, Olsen guides us through passages of Alcman and Sappho to indicate some of the ways in which culture, embodiment, and agency were imagined and experienced in ancient Greece.

Anne-Sophie Noel analyses theatrical objects in Aeschylus' *Oresteia* and locates places where visual effects and verbal clues seem to conflict in order to demonstrate the multisensory experience of the ancient live theatrical event. Noel employs recent neuroscience work on the spatial organization of multisensory brain regions and points out that although this research is still in its infancy, all the indications so far point to the flexibility of these systems and complex processes of multisensory integration. This suggests that different sensory modalities (touch, smell, vision, etc.) are not processed in distinct cortical areas and neural pathways but are part of a far more interconnected and fluid cognitive process. Noel maps this notion onto Greek drama with the suggestion that we should "read" them as multimodal artworks rather than favouring the visual or the aural, for example. Noel sets this multisensory notion alongside the perceptual theories of Clark and Meineck and then demonstrates her own approach with specific examples from the *Oresteia*. Her proposal that we view these works within the paradigm of multisensory experience offers a new way to comprehend how the *gesamtkunstwerk* of the Greek theatre would have worked in tandem with the imaginations of the audience, defying the idea that the ancient stage was in any way austere, minimalist, or devoid of special effects.

Angeliki Varakis-Martin examines Aristophanic comedy through the prism of Alice Isen's theories on positive affectivity (Isen 1987), the ways in which people experience positive emotions and situations, and how this affects cognition and social interactions through the facilitation of creative thinking. She links this with Barbara Fredrickson's "broaden and build theory" (Fredrickson 2001) which proposes that mirth and joy increase attention and broaden one's scope and thought actions. In part, this complements Meineck's view of the social and cultural influence of the tragic theatre and the role of the neural transmitter dopamine in helping to open the minds of the audience members and make them more receptive to alternate viewpoints (Meineck 2017). Varakis-Martin offers a sophisticated view of the ways in which an Aristophanic performance may have been received and the role of mirth in promoting cognitive receptiveness to alternate ideas. Like Noel, Varakis-Martin also tackles sensory multiplicity and maintains that the presentation of the violation of social norms and often absurd forms of what she describes as "comic openness" work to promote the creative imaginations of the audience and increase spectator engagement. Thus the broad nature of the frenetic comings and goings of the highly physical Aristophanic stage would have enhanced the broadening of the spectator's mind which then generated original interpretive solutions.

Artificial intelligence

Artificial intelligence (AI), a term for machine learning that became widespread after the 1956 Dartmouth Summer Research Project on Artificial Intelligence (Nilsson 2009), is the theoretical and developmental study of computers to perform the functions that require human intelligence. One area in which AI or "machine learning" is already operating in the field of classics is the development and application of computer software to search classical texts. One such program is the Classical Language Tool Kit, developed by Kyle Johnson, an open source software project that brings natural language processing (NLP) to classical languages. This project leverages machine learning to perform tasks such as sentence splitting, part-of-speech tagging, lemmatization, and word embeddings (http://cltk.org).

In antiquity, we find a number of sources that describe some form of artificial intelligence or machine learning. Pamela McCorduck in her study of the history of artificial intelligence describes the accounts of the automata of Hephaestus in the *Iliad* as one of "the earliest examples of an attempt to make an artificial person" (McCorduck 2004: 4). These ancient Greek examples also famously include Hesiod's account of the creation of Pandora in *Theogony* and *Works and Days*, and the mythical accounts of artificial humanoids such as Talos and Galatea (Paipetis 2008). Amy Lather revisits the accounts of mythical automata in Homer and Hesiod and examines them in relation to the work of A. Clark and Chalmers (1998) on cognitive embodiment. She frames their "parity principal," the concept that cognition held in working memory is equal to cognition that is distributed out into the environment, by showing how Hephaestus' machines are creations of his extended mind and act as higher-order thinking tools to help fulfil the god's aims. By then extending this thesis to Hesiod's descriptions of Pandora, Lather shows us something of how the Greeks thought about thought and how this Theory of Mind conceptualization was projected through literature.

Maria Gerolemou's chapter focuses on representations of artificial life and human thought processes in Greek drama. Meineck has suggested that a culture's dramatic products provide a "mimetic mind" on stage with the theatrical arts recreating and representing human cognition and intelligence within the artificial form of a performed narrative. Gerolemou's chapter makes a similar point and examines in detail examples of artificial reproductions of the human body – its form and functions – as new self-operating agents. These include statues with animate physical attributes and reproductions of body parts and their relationships to external tools, such as the bow of Heracles. She concludes her chapter with a brief examination of the issue of how the artificial reproductions of mental functions on stage are depicted as having a physical effect upon the mimetic body. Gerolemou suggests that Euripides and Aristophanes are perhaps deliberately highlighting the tension that exists between the real physical mind and body and its artificial representation, even going so far as to use their dramas to ridicule and condemn the embracement of artificial intelligence.

Cognitive archaeology

The field of cognitive archaeology is concerned with how the material record of ancient cultures is indicative of their cognitive processes, and investigates to what extent elements of human distributed cognition such as local environments, economies, societies, tool usage, and the adaptation of writing systems affected human minds. Malafouris and Renfrew (2010) have called this "the cognitive life of things" after Appadurai's *The Social Life of Things* (Appadurai 1988), which examined the biological aspects of artefacts and how material objects possess their own agency in their interactions with humans. (For an introduction to the field see Abramiuk 2012). Malafouris has proposed a cross-disciplinary analytical framework with which to examine the ways in which objects are cognitive extensions of the human mind and body, called Material Engagement Theory. This is built on three core concepts; the extended mind: that things are co-substantial, continuous, and co-extensive parts of mind in action; the enactive sign: that things bring forth rather than only represent; and material agency: that things have a causal efficacy in human thought and action.

Diana Y. Ng takes an enactivist approach to Roman public statues and argues that these physical monuments not only functioned as external storage for information about elite people, but were also intended to create active interactions and meanings within the environment and in conjunction with their viewership. Ng applies Kirsh's theories of distributed cognition, particularly how external representations can act as cognitive exograms and reduce cognitive load, thereby allowing for more complex thinking, even facilitating cognition that may be impossible internally without the scaffolding provided by the external means (Kirsh 2000). Ng also links this with Clark's theories of predictive processing (A. Clark 2015b) and Friston's free energy principle. By applying these theories of distributed cognition, Ng proposes that we can find new methods to comprehend how ancient monuments operated interdependently as dynamic enactive agents in their respective social and cultural environments. One way this may have been facilitated was the tendency of Roman statues to undergo physical change during festivals, or by the affixation of announcements, or in their relationship to certain ritual practices. These kinds of visual incongruities provoked the error-correction mechanisms of the human predictive system, eliciting attention and the processing of new forms of sensory information. Ng reveals that ancient monuments were far more dynamic and enactive in their original cultural setting than how they tend to be regarded today, as aesthetic objects or historical artefacts, and in this way her work shares much with Malafouris and Renfrew and their theories of material engagement and the cognitive life of ancient things (Malafouris & Renfrew 2010).

In his chapter, Günther Schörner applies a theoretical model from the field of the cognitive science of religion developed by Schjoedt, Sørensen, and Bulbulia (Schjoedt et al. 2013) called cognitive resource depletion. They contend that as executive functions and attention use the same frontal regions of the brain, they compete for resources in critical situations.

This has been linked to predictive processing and the idea that cognitive resource depletion may disrupt human perceptual abilities. Therefore, external sensory inputs that cause high-arousal states, such as certain religious rituals, may overload the predictive mechanisms and create resource depletion. Schjoedt, Sørensen, and Bulbulia propose that at these times, vivid memories ("flashbulb memories") are encoded and intense emotional responses are provoked as the predictive system urgently searches to assign meaning. In effect these moments can create "cognitive gaps" forced by disruptions to memory and perception and these are sometimes later filled by "culturally shared schemas during a post-ritual consolidation phase" as participants seek to find meaning in what they have experienced.

Schörner applies this approach to two friezes depicting animal sacrifice in Roman Asia Minor and one literary example from the Ephesian tale of Anthia and Habrocomes by Xenophon of Ephesus. He assigns the lack of a singular fixed interpretation of these sacrificial representations to the workings of cognitive resource depletion, whereby no explicit meaning was assigned to the performance of the ritual. Its interpretation was open, and this was partially filled by the iconography of the reliefs or by textual descriptions. Instead of viewing ancient sacrificial depictions only in terms of iconographical and performative congruities, Schörner suggests that we can learn far more about them by acknowledging their important role in provided the kind of shared cultural schemas that form the basis for the human understanding of ancient religious ritual.

Memory processing is also at the heart of Maggie L. Popkin's chapter on how Roman material culture helped perpetuate enhanced and even false memories of events and places for both political and social purposes. Building on the work of Elizabeth Loftus and her work on false memory (Loftus 2005) and Endel Tulving's research on semantic and episodic memory (Tulving 1972), Popkin examines a beaker from Colchester that shows a chariot race in the Roman Circus Maximus that she argues was viewed by people in Britain who had never been there, glassware from Campania depicting a fusion of two nearby cities, Puteoli and Baiae, and the triumphal arch of Septimius Severus in the Roman Forum that depicts a vast triumph that may never have taken place. Popkin contends that these kinds of material objects and structures projected selective images and may have had a great deal of impact on how places and historical events were remembered in antiquity and beyond.

Note

1 See the University of Edinburgh's History of Distributed Cognition project website for excellent descriptions of the main theories of distributed cognition offered by several of the leading theorists, at www.hdc.ed.ac.uk (last visited 20 March 2018).

References

Abramiuk, M. A. (2012). *The Foundations of Cognitive Archaeology*. MIT Press.

Anderson, M. (2015). *The Renaissance Extended Mind*. Springer.

Andresen, J. (2001). *Religion in Mind: Cognitive Perspectives on Religious Belief, Ritual, and Experience*. Cambridge University Press.

Appadurai, A. (1988). *The Social Life of Things: Commodities in Cultural Perspective*. Cambridge University Press.

Athanasiadou, A., & Colston, H. L. (2017). *Irony in Language Use and Communication*. John Benjamins Publishing Company.

Blair, R. (2007). *The Actor, Image, and Action: Acting and Cognitive Neuroscience*. Routledge.

Blair, R., & Cook, A. (2016). *Theatre, Performance and Cognition: Languages, Bodies and Ecologies*. Bloomsbury Publishing.

Budelmann, F., & Easterling, P. (2010). 'Reading minds in Greek tragedy', *Greece & Rome*, 57/2: 289–303.

Cairns, D. (2015). 'Mind, metaphor, and emotion in Euripides (*Hippolytus*) and Seneca (*Phaedra*)'. Presented at the VII Coloquio Internacional del Centro de Estudios Helénicos (La Plata, 2015), June.
Cairns, D., & Nelis, D. P. (2017). *Emotions in the Classical World: Methods, Approaches, and Directions*. Franz Steiner Verlag.
Caldara, R. (2017). 'Culture reveals a flexible system for face processing', *Current Directions in Psychological Science*, 26/3: 249–55. DOI: 10.1177/0963721417710036.
Calvo-Merino, B., Glaser, D. E., Grèzes, J., Passingham, R. E., & Haggard, P. (2004). 'Action observation and acquired motor skills: An FMRI study with expert dancers', *Cerebral Cortex*, 15/8: 1243–9.
Chaniotis, A. (2012). *Unveiling Emotions: Sources and Methods for the Study of Emotions in the Greek World*. Franz Steiner Verlag.
Choudhury, S. (2016). *Critical Neuroscience: A Handbook of the Social and Cultural Contexts of Neuroscience*. John Wiley & Sons.
Clark, A. (1998). *Being There: Putting Brain, Body, and World Together Again*. MIT Press.
—. (2013). 'Whatever next? Predictive brains, situated agents, and the future of cognitive science', *Behavioral and Brain Sciences*, 36/3: 181–204.
—. (2015a). *Embodied Prediction*. Open MIND. Frankfurt am Main: MIND Group.
—. (2015b). *Surfing Uncertainty: Prediction, Action, and the Embodied Mind*. Oxford University Press.
Clark, A., & Chalmers, D. (1998). 'The extended mind', *Analysis*, 58/1: 7–19.
Clark, H. H., & Gerrig, R. J. (1984). 'On the pretense theory of irony', *Journal of Experimental Psychology*, 113/1: 121–6.
Croft, W., & Cruse, D. A. (2004). *Cognitive Linguistics*. Cambridge University Press.
Dehaene-Lambertz, G., Monzalvo, K., & Dehaene, S. (2018). 'The emergence of the visual word form: Longitudinal evolution of category-specific ventral visual areas during reading acquisition', *PLOS Biology*, 16/3: e2004103. DOI: 10.1371/journal.pbio.2004103.
Eidinow, E. (2015). 'Some Ancient Greek theories of (divine and mortal) mind'. Ando C. & Rüpke J. (eds) *Public and Private in Ancient Mediterranean Law and Religion*, Vol. 65, pp. 53–74. De Gruyter.
Fagan, G. G. (2011). *The Lure of the Arena: Social Psychology and the Crowd at the Roman Games*. Cambridge University Press.
Fauconnier, G. (1994). *Mental Spaces: Aspects of Meaning Construction in Natural Language*. Cambridge University Press.
—. (1997). *Mappings in Thought and Language*. Cambridge University Press.
Fauconnier, G., & Turner, M. (2002). *The Way We Think: Conceptual Blending and The Mind's Hidden Complexities*. Basic Books.
Festinger, L. (1957). 'Cognitive dissonance theory', *1989 Primary Prevention of HIV/AIDS: Psychological Approaches*. Newbury Park, CA, Sage Publications.
Festinger, L., & Carlsmith, J. M. (1959). 'Cognitive consequences of forced compliance', *The Journal of Abnormal and Social Psychology*, 58/2: 203–10.
Fillmore, C. (1982). 'Frame semantics'. Kiparsky, P. (ed.) *Linguistics in the Morning Calm*, pp. 111–37. Hanshin.
Fredrickson, B. L. (2001). 'The role of positive emotions in positive psychology: The broaden-and-build theory of positive emotions', *American Psychologist*, 56/3: 218.
Friston, K., Kilner, J., & Harrison, L. (2006). 'A free energy principle for the brain', *Journal of Physiology-Paris*, 100/1–3: 70–87.
Gavins, J. (2007). *Text World Theory: An Introduction*. Edinburgh University Press.
Geeraerts, D. (1995). 'Cognitive linguistics'. Verschueren J., Ostman J.-O., & Blommaert J. (eds) *Handbook of Pragmatics: Manual*, pp. 111–16. John Benjamins Publishing.
Gibbs, R. W. (1994). *The Poetics of Mind: Figurative Thought, Language, and Understanding*. Cambridge University Press.
Glăveanu, V. P. (2014). 'From cognitive to cultural theories of "distribution": A creativity framework'. *Distributed Creativity*, pp. 15–32. Springer.
Goldman, A. I. (2012). 'Theory of Mind'. Margolis, E., Samuels, R., & Stich, S. P. (eds) *The Oxford Handbook of Philosophy of Cognitive Science*, Vol. 1. Oxford Handbooks Online, www.oxfordhandbooks.com/view/10.1093/oxfordhb/9780195309799.001.
Grice, H. P. (1975). 'Logic and conversation'. Cole, P. & Morgan, J. (eds) *Syntax and Semantics, volume 3: Speech Acts*, pp. 41–58. Academic Press.
Habinek, T. (2011). 'Tentacular mind: Stoicism, neuroscience, and the configuration of physical reality'. Stafford, B. M. (ed.) *A Field Guide to a New Meta-Field: Bridging the Humanities–Neuroscience Divide*, pp. 64–83. University of Chicago Press.

Haun, D. B. M., van Leeuwen, E. J. C., & Edelson, M. G. (2013). 'Majority influence in children and other animals', *Developmental Cognitive Neuroscience*, 3: 61–71. DOI: 10.1016/j.dcn.2012.09.003.
Haun, D., & Tomasello, M. (2011). 'Conformity to peer pressure in preschool children', *Child Development*, 82/6: 1759–67.
Hutchins, E. (1995). *Cognition in the Wild*. MIT Press.
Hutchins, Edwin. (2010). 'Cognitive ecology', *Topics in Cognitive Science*, 2/4: 705–15. DOI: 10.1111/j.1756-8765.2010.01089.x.
Isen, A. M. (1987). 'Positive affect, cognitive processes, and social behavior'. *Advances in Experimental Social Psychology*, Vol. 20, pp. 203–253. Elsevier.
Jackendoff, R. (1985). 'Information is in the mind of the beholder', *Linguistics and Philosophy*, 8/1: 23–33. DOI: 10.1007/BF00653372.
Johnson, M. (1987). *The Body in the Mind: The Bodily Basis of Meaning, Imagination, and Reason*. University of Chicago Press.
Johnson-Laird, P. N. (1980). 'Mental models in cognitive science', *Cognitive Science*, 4/1: 71–115. DOI: 10.1016/S0364-0213(81)80005-5.
Kirsh, D. (2000). 'A few thoughts on cognitive overload', *Intellectica*, 1/30: 19–51.
Konstan, D. (2006). *The Emotions of the Ancient Greeks: Studies in Aristotle and Classical Literature*. University of Toronto Press.
Lakoff, G., & Johnson, M. (1980). *Metaphors We Live By*. University of Chicago Press.
Langacker, R. W. (1987). *Foundations of Cognitive Grammar: Theoretical Prerequisites*. Stanford University Press.
—. (1990). 'Subjectification', Cognitive Linguistics (includes Cognitive Linguistic Bibliography), 1/1: 5–38.
Loftus, E. F. (2005). 'Planting misinformation in the human mind: A 30-year investigation of the malleability of memory', *Learning & Memory*, 12/4: 361–6. DOI: 10.1101/lm.94705.
Luhrmann, T. M. (2012). *When God Talks Back: Understanding the American Evangelical Relationship with God*. Knopf Doubleday Publishing Group.
Lyons, D. E. (2008). 'The Rational Continuum of Human Imitation'. *Mirror Neuron Systems*, Contemporary Neuroscience, pp. 77–103. Humana Press. DOI: 10.1007/978-1-59745-479-7_4.
Lyons, D. E., & Keil, F. C. (2013). 'Overimitation and the development of causal understanding'. Banaji, M. R. & Gelman, S. A. (eds) *Navigating the Social World: What Infants, Children, and Other Species Can Teach Us*, pp. 145–9. Oxford University Press.
Malafouris, L. (2013). *How Things Shape the Mind*. MIT Press.
Malafouris, L., & Renfrew, C. (2010). *The Cognitive Life of Things: Recasting the Boundaries of the Mind*. McDonald Institute for Archaeological Research.
McConachie, B. (2008). *Engaging Audiences: A Cognitive Approach to Spectating in the Theatre*. Springer.
—. (2015). *Evolution, Cognition, and Performance*. Cambridge University Press.
McCorduck, P. (2004). *Machines Who Think: A Personal Inquiry into the History and Prospects of Artificial Intelligence*. A.K. Peters.
Meineck, P. (2017). *Theatrocracy: Greek Drama, Cognition, and the Imperative for Theatre*. Taylor & Francis.
Nilsson, N. J. (2009). *The Quest for Artificial Intelligence*. Cambridge University Press.
Noland, C. (2010). *Agency and Embodiment*. Harvard University Press.
Nuyts, J. (1992). 'Subjective vs. objective modality: What is the difference'. Fortescue, M., Harder, P., & Kristoffersen, L. (eds) *Layered Structure and Reference in a Functional Perspective*, pp. 73–97. John Benjamins Publishing.
Over, H., & Carpenter, M. (2012). 'Putting the social into social learning: Explaining both selectivity and fidelity in children's copying behavior', *Journal of Comparative Psychology*, 126/2: 182.
Paipetis, S. A. (2008). *Science and Technology in Homeric Epics*. Springer Science & Business Media.
Palmer, G. B. (1996). *Toward a Theory of Cultural Linguistics*. University of Texas Press.
Schjoedt, U., Sørensen, J., Nielbo, K. L., Xygalatas, D., Mitkidis, P., & Bulbulia, J. (2013). 'Cognitive resource depletion in religious interactions', *Religion, Brain & Behavior*, 3/1: 39–55. DOI: 10.1080/2153599X.2012.736714.
Shore, B. (1998). *Culture in Mind: Cognition, Culture, and the Problem of Meaning*. Oxford University Press.
Short, W. M. (2016). 'Spatial metaphors of time in Roman culture', *Classical World*, 109/3: 381–412. DOI: 10.1353/clw.2016.0038.
Sklar, D. (2008). 'Remembering kinesthesia: An inquiry into embodied cultural knowledge'. Noland, C. & Ness, S. A. (eds) *Migrations of Gesture*, pp. 85–111. University of Minnesota Press.

Sperber, D., & Wilson, D. (1981). 'Irony and the use-mention distinction'. Cole, P. (ed.) *Radical Pragmatics*, Vol. 3, pp. 295–318. Academic Press.
Talmy, L. (2000). *Toward a Cognitive Semantics*. MIT Press.
Tribble, E. (2011). *Cognition in the Globe: Attention and Memory in Shakespeare's Theatre*. Springer.
Tulving, E. (1972). 'Episodic and semantic memory', *Organization of Memory*, 1: 381–403.
Van Dijk, T. A. (1997). *Discourse as Social Interaction*. SAGE.
—. (2016). 'Critical discourse studies: A sociocognitive approach'. Wodak, R. & Meyer, M. (eds) *Methods of Critical Discourse Studies*, pp. 62–85. Sage.
Van Slyke, J. A. (2016). *The Cognitive Science of Religion*. Routledge.
Varela, F. J., Thompson, E., & Rosch, E. (2017). *The Embodied Mind: Cognitive Science and Human Experience*. MIT Press.
Walczyk, J. J., Harris, L. L., Duck, T. K., & Mulay, D. (2014). 'A social-cognitive framework for understanding serious lies: Activation-decision-construction-action theory', *New Ideas in Psychology*, 34: 22–36.
Werth, P. (1994). 'Extended metaphor: A text-world account', *Language and Literature*, 3/2: 79–103.
—. (1999). *Text Worlds: Representing Conceptual Space in Discourse*. Prentice Hall.
Zunshine, L. (2006). *Why We Read Fiction: Theory of Mind and the Novel*. Ohio State University Press.
—. (2010). *Introduction to Cognitive Cultural Studies*. JHU Press.
—. (2015). *The Oxford Handbook of Cognitive Literary Studies*. Oxford University Press.

Part I
Cognitive linguistics

1

Cognitive-functional grammar and the complexity of early Greek epic diction

Ahuvia Kahane

Avant propos

Formal patterns and repetition are prominent features of the diction of Homer's poetry and, in varying degrees, of early Greek epic and other *Homerica*.[1] In the last hundred years, students of oral-formulaic theory have argued that such repetition, structured by formal rules of composition, comprises the essential 'grammar' of Homeric verse and of oral poetry.[2] Whether we accept this view or not,[3] it shares basic methodological premises with many general approaches to language such as the standard, paradigmatic frameworks of Greek and Latin grammar, de Saussure's views of *langue* and *parole*, Chomskyan generative grammar and other transformational grammars. These are all 'principles and parameters' approaches to language which place heavy emphasis on linguistic form and on "a unified set of algebraic rules" that are largely "meaningless themselves and insensitive to the meaning of the elements they algorithmically combine".[4] Most approaches of this type view systems of rules as the core of language and many consider lexicon, idioms, conceptual frameworks, irregular constructions, pragmatic aspects of language and usage in general as peripheral.[5] Oral-formulaic theory approaches epic diction along similar lines. It stresses the importance of formal rules – its key term is the 'formula' – while playing down the idea of contingent lexical, semantic and context-sensitive usage. Singers use an "extensive" but "economical" system of pre-fabricated, traditional formulaic elements which are combined by means of formal 'algorithmic' rules in order to compose well-formed hexameters in performance.[6] To take the familiar example, speech-introductory verses beginning with Τὸν/Τὴν δ' ἀπαμειβόμενος προσέφη [*//→hepth.*] are regularly used in conjunction with several name+epithet formulae such as πολύμητις Ὀδυσσεύς, πόδας ὠκὺς Ἀχιλλεύς, κρείων Ἀγαμέμνων, κορυθαίολος Ἕκτωρ, or νεφεληγερέτα Ζεύς, that fill the slot from the hephthemimeral caesura to the verse-end *[hepth.→//]* to express a general "essential idea" such as "X spoke to him in a certain tone or with a certain gesture" or simply "X [Odysseus; Achilleus; etc.] spoke to him".[7] As with other systematic approaches to grammar, oral-formulaic arguments grasp diction that does not follow the rule as exceptional and anomalous. This view played up to a polarized conception of form vs. meaning and of traditional vs. individual diction.[8]

Of course, oral-formulaic arguments drew strong opposition from a wide range of scholars who were interested in nuanced, context-specific sense, theme, allusion, intertextuality and other aspects of content and meaning.[9] Many scholars also emphasized non-formulaic elements in epic verse as well as inconsistencies and gaps in the formal system of diction in epic.[10] Nevertheless, such opposing views also tended to uphold the methodological polarities of rule and anomaly and of tradition and originality.[11]

Cognitive-functional linguistics

Over the years, there have been various attempts to reconcile form and meaning in epic diction.[12] Some of the more recent of these have incorporated cognitive-functional approaches and what is sometimes known as usage-based grammar. These approaches hold particular promise, above all since they challenge the obdurate dichotomy of form and content at root. Cognitive linguistics argue for a *continuum* of grammar and lexicon driven by communicative practice.[13] As developmental psychologist and linguist Michael Tomasello, for example, says,[14]

> Usage-based theories hold that the essence of language is its symbolic dimension, with grammar being *derivative*.

Verbal communication, argue usage-based linguists, does not depend on an innate, unique, pre-existing linguistic apparatus but on general cognitive skills, which include the ability to share attention with others and to redirect it, the ability to learn the "intentional actions" of others, to recognize, imitate and adapt patterns and to form categories of similar objects and events on the basis of such recognition (hence such patterns are "derivative" rather than pre-existing or primary). As Tomasello points out:[15]

> When human beings use symbols to communicate with one another, stringing them together into sequences, patterns of use emerge and become consolidated into grammatical construction.

We tend to form sensory schemas, patterns of sound, words and phrases, or "constructions", which, as Adelle Goldberg, for example, defines them, are "learned pairings of form with semantic or discourse functions".[16] Such constructions are present at all levels of grammatical analysis including "morphemes or words, idioms, partially lexically filled and fully general phrasal patterns".[17] Through comparison and analogy, usage evolves to create new expressions based on similarities within two or more complex wholes (e.g., in the case of Homer, between the formula πολύμητις Ὀδυσσεύς and πόδας ὠκὺς Ἀχιλλεύς – these are different, but analogous expressions).[18] Here, then, is the crucial innovation within usage-based grammars:[19]

> As opposed to conceiving linguistic rules as algebraic procedures for combining words and morphemes that do not themselves contribute to meaning, this approach conceives linguistic constructions as themselves meaningful linguistic symbols – since they are nothing other than the patterns in which meaningful linguistic symbols are used in communication.

From a usage-based perspective, grammar is an epiphenomenon.[20] Thus, every formal pattern and element of structure can as a matter of principle embody "semantic or discursive" values. This idea, if it can be shown to work in epic diction, offers the prospect of cutting through old critical Gordian knots. It can pave the way to new, less problematic and more meaningful reading of the densely patterned diction of Homeric verse and other early Greek epic.

Homeric scholarship and usage-based analysis

Direct applications of cognitive-functional and usage-based approaches to the grammar and semantics of epic diction are relatively recent, but recognition of the meaning of form and of the importance of usage is evident even in early work within the oral-formulaic scholarly tradition. Albert Lord, for instance, pointed out that:[21]

> When we speak a language, our native language, we do not repeat words and phrases that we have memorized consciously, but *the words and sentences emerge from habitual usage*. This is true of the singer of tales working in his specialized grammar.

Other Homer scholars, for example T. G. Rosenmeyer, had recognized that "the bard regards his poetic phrase as indistinguishable from poetic substance".[22] Views of this type were also often embedded in studies of individual expressions, formulae and formulaic patterns.[23] Consider, for example, Anne Amory Parry's well-known 1973 book *Blameless Aegisthus* which set out to explain the meaning of the epithet ἀμύμων, "blameless" and of other traditional epithets. Ἀμύμων was often thought to be a semantically 'empty' metrical element since in the *Odyssey* (1.29), for example, it describes Aegisthus, who is not "blameless" but famously blamed for his wicked treachery.[24] Looking closely at all the contexts of this epithet's use and at similar epithets such as κρατερός, ἀγλαός and ἄλκιμός, Amory Parry concluded that in Homer Ἀμύμων had no moral connotations. Rather, she argued, it described a heroic quality: its primary connotation was simply "handsome".[25] As might have been expected, among more orthodox exponents of oral-formulaic views, this argument was resisted. In a review of *Blameless Aegisthus*, J. B. Hainsworth, for example, suggested that, "Whatever its emotive colour, what is metrically determined does not impart contextually useful information". Usage in context, he argued, does not indicate the meaning of an epithet but rather "the criteria for its application".[26] Ironically, usage-based linguists who study "learned pairings of form with semantics or discourse functions" could hardly have asked for a clearer endorsement.

It was, nevertheless, mostly in later decades that scholars began to apply usage-based principles more explicitly to systematic studies of Homeric diction. In 1991, Egbert Bakker and Florence Fabricotti had argued for the semantic relation between metrically fixed core elements and semantically resonant peripheral diction in Homeric epic. This was an important attempt to integrate form and meaning on the basis of linguistics, although its underlying dichotomic framework as then phrased was only partially consonant with usage-based grammar.[27] Revising this essay in 2005, Bakker re-focused the argument. Citing Givón, Hopper and other early exponents of the cognitive approach, he pointed out that "grammar is not a constraint but a set of *emergent* rules that make purposeful expression and communication possible". In this revised version, Bakker also suggested that, assuming grammar as communication, the distinction between traditional and original "loses its meaning".[28]

In 1994, in *The Interpretation of Order*, Kahane, following Lord, Rosenmeyer and other Homer scholars who had acknowledged the overlap of formal structure and sense in epic diction, stressed that "the study of patterns is, to a point, an investigation of *usage*". He argued that "certain repetitions of formal features of grammar and metre . . . have semantic and consequently literary significance".[29] Looking closely at grammaticalized elements in Homer, for example at accusative, verse-initial epic 'theme-words' (*andra, mênin*) as well as at large sets of proper-name constructions, for example in the vocative and accusative, Kahane suggested that grammatical case, syntactic function and metrical placement were semantically significant elements within the structure of Homeric hexameter. He outlined a formal metrico-grammatical system that corresponded to the narrative characterization of heroes and gods. In the case of proper-name vocatives, for example, this system was based on three broad categories of usage: verse-initial position as an unmarked "default mode", verse-terminal position marking epic-protagonists, often used in antithesis to initial positioning, and internal positioning as a "semantic/stylistic compromise which reflects the narrator's sympathetic attitude towards particular characters".[30]

More recently, the fuller potential of usage-based approaches on our understanding of the structured diction of Homeric and other epic discourse has begun to be explored. Bakker, for instance, in a brief epilogue to his 2013 book on the *Odyssey*, considered the formation of epic patterns driven by communicative interaction and routinized use. He has argued for the "interformular" potential of traditional diction as meaningful repetition based on the degree to which formulaic usage is attested in specific contexts, on the judgement of performers and audiences and on the degree of similarity between two contexts in which formulaic repetition occurs. The better defined the context in which a formula is re-iterated, the greater the potential for signalling meaningful repetition. His position, nevertheless, is that this does not mark the "likelihood of allusion or quotation" and specific verse-to-verse interaction, but "the specificity of the similarity of scenes to each other" for epic poets and audiences (further below we shall see that within usage-based systems allusion and specific interaction are, in fact, possible).[31] Bakker gives the example of common noun-epithet formulae extending from the hephthemimeral caesura to the end of the verse and speech-introductory formulae. Following observations by Norman Austin, he notes that in the *Odyssey* most instances of the formula πολύμητις Ὀδυσσεύς (62 or 63 if we count inclusively out of 66) are preceded by speech-introductory expressions, often the familiar Τὸν/Τὴν δ' ἀπαμειβόμενος προσέφη . . . Use of the noun-epithet formula, he argues, is metrically conditioned by the first part of the verse but leads to the "epiphany" of the hero as he is about to speak.[32] "The resulting highly formulaic lines are not so much ready-made ways of saying 'and Odysseus answered him' as the performance of a recognizable verbal ritual that is the tradition's coding of a frequently recurring poetic need". "Epic grammar", Bakker concludes, "codes best what epic poets do most".[33] Speakers may remember this kind of usage and re-use it in similar or slightly different situations.

A longer and more technical attempt to redefine repetition in Homeric diction in terms of usage-based grammar was recently put forward by Chiara Bozzone, who, following Goldberg's notion (above) of "learned pairings of form and function", has proposed that we should indeed think of formulae as "constructions".[34] Among Bozzone's examples is the familiar name-epithet formula πόδας ὠκὺς Ἀχιλλεύς, which, she argues, operates at several levels of repetition and abstraction. In the first instance, πόδας ὠκὺς Ἀχιλλεύς is a subject noun-phrase used after the *longum* of the fourth foot (the hephthemimeral caesura in position 7). The same expression is also part of a more abstract construction for a subject noun-phrase in the same metrical position, involving a modifier such as πόδας ὠκὺς, ἑκάεργος, etc., (⏑⏑ __ ⏑) and a noun Ἀχιλλεύς, Ἀπόλλων, etc. (⏑ __ x). These expressions are part of a yet more abstract construction which does not define the internal metrics/structure of the phrase and thus also includes, for example, the construction νεφεληγερέτα Ζεύς. Finally, we could link such expressions to even more general noun-phrase constructions.[35] Each of these patterns, Bozzone suggests, can have semantic, syntactic and discourse functions. In the case of πόδας ὠκὺς Ἀχιλλεύς, the semantic function of the phrase is to designate Achilles and his "thematic connotations" (such as his swiftness of foot); the syntactic function of the construction is to act as a subject noun-phrase in a variety of verse types, for example Τὸν δ' ἀπαμειβόμενος προσέφη πόδας ὠκὺς Ἀχιλλεύς ([—]

object pronoun δ', [⏑⏑—⏑⏑—] subject participle, [⏑⏑—] predicate προσέφη, subject noun-phrase [⏑⏑—⏑⏑—x]); this larger context also has a discourse function: "resuming the old discourse topic 'Achilleus' and possibly marking a scene boundary".[36]

Such semantic characterizations often correspond to our immediate sense of the verse as audiences and readers, but provide a formal, systematic framework for its understanding. In contrast to more rigid oral-formulaic 'grammars', this framework stresses that essential communicative functions should be seen as paired to grammatical usage. As we shall shortly see, arguments of this type can, in fact, reach much further.

The idea of construction grammar as a framework for understanding formulaic diction has recently gained support in further publications, for example by linguists Cristóbal Cánovas and Mihailo Antović in a collection of essays entitled *Oral Poetics and Cognitive Science*.[37] Cánovas and Antović point out that construction grammar and oral-formulaic theory are "congenial approaches". They discuss some examples from Serbo-Croat guslar performances, but, unfortunately offer no detailed examples from Homeric Greek, nor do they deal with the specific difficulties associated with the systematic aspects of Parryan theory such as the assumption of 'economy' and 'extension' or with the emphasis in oral-formulaic theory on composition rather than semantics. In the same volume, Hans C. Boas also sets down the argument for the similarity between construction grammar and epic formulae, although he too does not offer detailed examples from Homer or a discussion of the specific technical restrictions imposed by oral-formulaic theory.[38] Elizabeth Minchin's thoughtful and informed contribution to the same volume also picks up the discussion of cognitive linguistics, applying arguments from Deborah Tannen's "poetics of talk" to an analysis of repetition in Homer. However, the focus of Minchin's discussion is the function of repetition in the production and comprehension of discourse – rather than the continuum of form and content or of signification as it is conceived in construction grammars.[39]

Finally, starting with semantic and poetic arguments and focusing on a discussion of allusion in Homeric epic and its relation to other early Greek epic verse, Bruno Currie has pointed, if only very briefly, to cognitive linguistics, stressing the relation between allusive potential and traditional (epic) art. Assimilating at least some of the principles of construction grammar, Currie suggests that "the linguistic system is built up from . . . lexically specific instances, only gradually abstracting more general representations".[40]

Apart from arguing for the semantics of form, discussions of this type necessarily also entail the assumption of essential grammatical flux and the potential for organic 'evolutionary' mutations of form. Where grammar is not a pre-existing set of universal rules but the epiphenomenal consequence of expressions of meaning in use, there are far more limited constraints on the evolution of form. Form is repeatedly adapted as the communicative acts of users evolve and as such users apply derivative structural patterns and adapt them to new contexts.[41] Furthermore, as we shall see, such adaptations do have the capacity to interact in specific contexts.

Grammaticalized usage in Homer[42]

Let us take a closer look at pairings of form and function in Homeric epic. Consider again the common construction Τὸν/Τὴν δ' ἀπαμειβόμενος προσέφη . . . and its familiar concluding noun-epithet formulae πόδας ὠκὺς Ἀχιλλεύς //, πολύμητις Ὀδυσσεύς //, κρείων Ἀγαμέμνων //, νεφεληγερέτα Ζεύς //, κρατερὸς Διομήδης //, κορυθαίολος Ἕκτωρ //. As we have seen, analogous form here binds several levels of patterns and pairings of form which, at their most abstract, express a general essential idea ('X spoke to him/her'). And yet, as even the hardest Parryan formalist must allow, in epic narrative each hero or god introduced in this way is a unique individualized character with well-developed traits, behaviour, relationships and narratives. These attributes inevitably define the meaning and application, not only of each noun-epithet formulaic variant, which may seem evident (Achilles is "swift-footed", Odysseus is "many-minded"), but also less obvious elements of usage of the larger whole. Thus, for example (as noted from a slightly different angle earlier), the characteristic nominative formula πολύμητις Ὀδυσσεύς, repeated eighty-six times in Homer,[43] appears in the specific construction Τὸν/Τὴν δ' ἀπαμειβόμενος προσέφη . . . in the majority of these (50x), only five of which are in the *Iliad* and the remaining instances in the *Odyssey*. Odysseus' noun+epithet formula also appears in a few other speech-introductory constructions. Bakker is right to suggest that verses of this type stage the "epiphany" of Odysseus as "the man full of *mêtis* precisely and almost exclusively when he takes to the floor in order to speak".[44] Not surprisingly, that epiphany is most distinct in Odysseus' eponymous poem. But, as we have already stressed, Homeric characters are not all the same. Other heroes whose noun-epithet formulae share the same structural pattern [nom. proper name, hepth.→//] have nothing of Odysseus' cunning polytropy. Agamemnon is hypocritical and crass. His speech has no *mêtis*. Achilles speaks with direct, explosive emotion, not with cunning calculation which he famously despises.[45] We may be dealing with a pattern of verbal revelation. Yet each specific pairing of form and content enacts different *kinds* of epiphany at different moments in the narrative. Τὸν δ' ἀπαμειβόμενος προσέφη πόδας ὠκὺς Ἀχιλλεύς (12x, 1.84, etc.) marks the appearance of an impetuous, straight-talking respondent; Τὸν δ' ἀπαμειβόμενος προσέφη κρείων Ἀγαμέμνων (7x, 1.130, etc.) marks the traditional epiphany of an arrogant and overbearing king. Not only will the semantic and narrative uniqueness of each of these speech introductions stand out against the background of their shared formal structures at particular moments, but, more importantly, every time the pattern is re-used, in each individual context, interactive thematic interrelationships will inevitably emerge. Within the dramatic series of exchanges of *Iliad* Book 1, for example, experienced audiences and readers will note contrastive tensions between Achilles' protective response to Kalchas in 1.84 and Agamemnon's selfish and aggressive reply to Achilles in 1.130. We find Achilles' anguished submissive response to Athena in 1.215; Agamemnon's inflexible and damning response to Nestor in 1.285; and so on. We might add that resonating alongside each of these individual verses and parts of verses is an almost open-ended system of partially overlapping imbricated sets of other formally analogous patterns of speech-introductory verses and their individualized constructions;[46] for example, Τὸν δ' ἄρ' ὑπόδρα ἰδὼν προσέφη πόδας ὠκὺς Ἀχιλλεύς (*Il.* 1.148), emphasizing Achilleus' impetuous authority in the quarrel with Agamemnon; or the highly contrastive verse-internal form with Achilleus' single-consonant variant and no epithet, Τὸν δ' Ἀχιλεὺς μύθοισιν ἀμειβόμενος προσέειπεν· (*Il.* 23.794), which emphasizes his harmonious exchange with Antilochos.[47]

When discussing the grammar and semantics of specific constructions and analogous expressions, it is important to take account of the limits of actual usage rather than to assume a complete abstract paradigm of potential grammatical forms.[48] Consider, for example, the usage of the frequent nominative name-epithet formula γλαυκῶπις Ἀθήνη (*Il.* 28x; *Od.* 50x), which is always verse-terminal, and the usage of similarly shaped formulae of other female characters such as λευκώλενος Ἥρη (22x) and πόδας ὠκέα Ἶρις (9x), both also always verse-terminal.[49] Metrically and syntactically, such forms are identical to πόδας ὠκὺς Ἀχιλλεύς, πολύμητις Ὀδυσσεύς, etc. Athena, Hera, Iris and other female goddesses and mortal women certainly converse with other characters. From a formal-paradigmatic perspective, these noun-epithet formulae seem like natural complements to the . . . ἀπαμειβόμενος προσέφη speech-response pattern, requiring only a minor, metrically insignificant change to the grammatical gender of the participle. There are, of course, abundant examples of feminine noun-epithet formulae in speech-response verses such as Τὸν/Τὴν δ' ἠμείβετ' ἔπειτα . . . and related verse types.[50] Yet the specific feminine-gendered speech-introductory variant *Τὸν/Τὴν δ' ἀπαμειβόμενη προσέφη . . . is never attested in Homer or anywhere else in extant epic.[51] Nor, indeed, is the participle *ἀπαμειβομένη ever used in any other construction in epic. In other words, whether any metrical or other formal principle plays any role in the construction, the particular response pattern . . . ἀπαμειβόμενος προσέφη . . . is exclusive to male speakers. Its actual usage is meaningfully 'defective'. To the degree that grammar is built up of semantically specific instances, we are here dealing with one fine detail of the 'gender of grammar' and with specific performative, contextualized usage that defines the pairing of meaning and form.[52]

Grammar, traditional patterns and anomaly

Alongside marked variants of epic phraseology we also often find elements of diction that do not seem to follow directly recognizable traditional patterns and yet other elements which seem to openly exceed or transgress recognized patterns or 'rules', as, for example, sometimes in hiatus and other metrical irregularities.[53] Consider the case of *Iliad* 2.571:

Ὀρνειάς τ' ἐνέμοντο Ἀραιθυρέην τ' ἐρατεινὴν

Discussing the arrangement of this verse's components within the context of the Catalogue, Parry had suggested that we often first find mention of those peoples "who inhabited such and such a town" in the first part of the verse up to the trochaic caesura, followed in the latter half of the line by either "the name of another town governed by the same verb and accompanied by an epithet" (as here), or "two names of towns governed by the same verb".[54] The second half of the verse provides a supplement, "and such and such towns". Preference for this traditional pattern, Parry nevertheless argues, requires that the second element begin with a single consonant. In 2.571, the (acc.) place-name Ἀραιθυρέην, because of its metrical shape [⏑—⏑⏑—], must come directly after the trochaic caesura. Such placement also allows for use of the generic epithet ἐρατεινὴν at the end of the verse, bringing the line conveniently to a close. Yet, Parry says, in order to express this 'essential idea', the poet had to choose between renouncing the [- - - ἐνέμοντο] type of formula and hiatus. According to Parry, the poet chose hiatus.[55] Thus, even from within the perspective of conventional oral-formulaic theory, semantic choices seem to influence grammatical form.

Following Parry, several scholars explored the problem of this verse. In his 1998 study of the Iliadic Catalogue and its patterns, Edzard Visser, for example, considered the composition of *Il.* 2.571, and pointed to several formulaic alternatives that preserved the overall semantic content:[56]

⏓ - ⏓ | Ὀρνειάς τε

*οἵ τ' ἄρ' Ἀραιθυρέην |

⏓ | Ὀρνειάς | ⏑⏑ – x

Incorporating conventional epithets, these options can produce verses that avoid hiatus, such as

*οἵ τ' ἄρ' Ἀραιθυρέην ἐρατεινὴν Ὀρνειάς τε

This solution is certainly possible, although, as Visser notes, both alternatives rely on an unusual localization of the Molossus (- - -) Ὀρνειάς.[57] Visser thus prefers other constructions, using elements attested in other patterns in the catalogue, for example:[58]

ἐρατεινήν.

*Ὀρνειάς τ᾽ εἶχον καί Ἀραιθυρέην |

ἐνέμοντο

We should stress, however, that as an experienced 'user' of Homeric language, Visser is here responding to a local semantic contextual need in precisely the manner expected within the framework of usage-based linguistics. He assumes, implicitly, that there is a specific imperative to mention both Orneiai and Araithyree. This need affects a change of formulaic/grammatical form.

Il. 2.571 was also considered in a later study by Margalit Finkelberg discussing tradition and the individual poet. She too considered the problem of hiatus and noted the alternative construction incorporating . . . ἐρατεινήν.[59] This option is, of course, not actually attested in the tradition. Finkelberg thus concludes that *Iliad* 2.571 in its extant form offers evidence, not of the force of tradition, but of individual poetic choice. The question, says Finkelberg, is why was a verse that resolves the problem of hiatus not used? "The answer that most naturally suggests itself", she says, is that the poet "simply did not think of the appropriate expression while assembling his verses".[60] Whether we prefer Parry's analysis, Visser's or Finkelberg's, the fact of the matter remains: consideration of sense seems to guide the modification of form.

The complexity of epic diction

The examples above and especially Finkelberg's argument nevertheless seem to underscore a methodological and practical divide between tradition and originality, between formal rule-bound repetitive phraseology and anomalous individual usage. Can we, then, bridge these divides and account for all phraseology as part of a single, continuous system? In the concluding part of this chapter I want to emphasize not only that we can, but that it was already Milman Parry himself who provided seminal observations in support of this argument. As we shall see, reframed by a usage-based approach, by discussions of complexity as a basic methodological principle and by the idea of language as a "complex adaptive system", we can establish an organic, systematic link between rule-bound and anomalous usage based on basic cognitive processes which, not surprisingly, is the very essence of epic diction.[61]

Parry's explanation of hiatus in *Iliad* 2.517 points to a relationship between theme and form and suggests that anomaly is the result of conflicting demands within the system. In fact, this idea is far more complicated than it at first appears. Describing the development of formulaic composition and its underlying mechanisms, Parry says:[62]

> the bards, always trying to find for the expression of each idea in their poetry a formula at once noble and easy to handle, created new expressions – insofar as the result was compatible with their sense of heroic style – in *the simplest way possible: they modified expressions already in existence. To this process are due all the series of formulae which we have so far examined* [throughout *TE*]. In each of these series it would be pointless to look for the original or the oldest formula. But in every case there must have been an original expression from which *a series was produced by the system of imitation we call analogy.*[63]

Without this process, epic diction could never have developed. He stresses:

> Analogy is perhaps the single most important factor for us to grasp if we are to arrive at a real understanding of Homeric diction. To understand the role of analogy in the formation of epic language is to understand the interdependence of word, ideas and metre in heroic poetry. It is to see to what extent the hexameter and the genius of the bards influenced epic style.

The importance of analogy in the formation and development of language was, of course, already recognized by linguists in antiquity.[64] However, Parry's thought about this mechanism is more likely to have been influenced by the work of his teacher, Antoine Meillet, who in a seminal article on grammaticalization (1912) wrote:[65]

> Les proceeds par lesquels se constituent les forms grammaticales sont au number de deux; tous les deux sont connus, même des personnes qui n'ont jamais étudié la linguistique . . . L'un de ces procédes est l'analogie; il consiste à faire une form sur le modele d'un autre . . . L'autre proceed consiste dans le passage d'un mot autonome au role délement grammatical.[66]

Later students of oral-formulaic theory recognized this too.[67] Albert Lord, for example, regarded this mechanism as fundamental to the singer's training, to the process of recomposition-in-performance and to the creation and expansion of diction.[68] Certainly among linguists today it is widely acknowledged that "the talent for analogical reasoning constitutes the core of Human cognition" and that analogy is "the core component of linguistic competence".[69]

Parry's understanding of the role of analogy nevertheless involved a crucial further step. Analogy assumes the application of generalized patterns. Yet in a remarkable insight – arguably well ahead of its time – Parry realized that analogy could not be contained within its own rules. It had the power to affect unexpected outcomes, to exceed and sometimes even to contradict the rules of the system. This point requires careful explanation.

Let us very briefly recapitulate: oral-formulaic theory assumes the principles of "extension", i.e., the availability of several metrical variants for expressing the same "essential idea", and of "economy" (or "thrift"), which dictates the avoidance of metrically equivalent variants.[70] Extension provided the bard with ready-made phrases to suit a variety of metrical contexts and was thus essential for rapid composition of well-formed verses in performance. Equivalent expressions offered no further advantage to such composition and were thus regarded as superfluous exceptions to the formal principle of economy. With these principles, Parry sought to characterize the essence of traditional oral diction. Paradoxically, he also realized that precisely such exceptions were an organic part of the system itself. As he says,[71]

> This operation of analogy, the power of which is attested by each artifice of epic diction, *is too powerful to stop once it has created a metrically unique formula.* In the bard's mind, there will always be an association between the words of one unique expression and another, and thus, by analogy, he will draw from two unique formulae one which will repeat the metre of an already existing formula.

One of the clearest illustrations of this idea is found in Parry's discussion of what he regarded as the most elaborate formulaic system of all, the set of 'ship' formulae.[72] By analogy to other noun-epithet formulae and in order to extend their system, says Parry, bards will have created, for instance, the accusative expression εὐεργέα νῆα, beginning with a vowel at the hepthemimeral caesura, as in: *Od.* 9.279 ἀλλά μοι εἴφ', ὅπῃ ἔσχες ἰὼν εὐεργέα νῆα. In the same way, he adds, they created the genitive νεὸς . . . ποντοπόροιο and the dative νήεσσι . . . ποντοπόροισι, extending the clause or sentence from the bucolic diaeresis to the verse-end.[73] The need to create a range of noun-epithet formulae for 'ship' in different metrical forms thus introduces two epithets, ποντοπόροιο and εὐεργέα, into the system. In the oblique cases, each of the expressions formed by these epithets is unique. Yet, as Parry points out, "the nominative of either can serve equally well with νηῦς to make a subject noun-epithet formula":

Il. 24.396 τοῦ γὰρ ἐγὼ θεράπων, μία δ' ἤγαγε νηῦς εὐεργής·

Od. 12.69 οἴη δὴ κείνη γε παρέπλω ποντοπόρος νηῦς

He explains:[74]

> Analogy could equally well lead the bards to choose ποντοπόρος νηῦς on the model of ποντοπόροισι νέεσσι, νήεσσι . . . ποντοπόροισι as νηῦς εὐεργής on the model of εὐεργέα νῆα. And even after one of these expressions had been chosen, the other epithet, closely bound up with νηῦς, remained, ready to spring to mind at any time.

The point is simple: no matter how well ordered epic diction is as a system, it has an inherent capacity to exceed its own boundaries and rules. As Adam Parry says in the Introduction to his father's collected essays:[75]

> apparent deviations from the economy of the system are themselves best explained by the sense of analogy which controls the system as a whole and indeed created it in the first place.

Deviations, we now realize, are not external to the system. They emerge within the system from of the prolific nature of analogy, the mind's most basic cognitive/linguistic process which, as Parry argues, is the most important factor in the formation of epic diction. Analogy generates complex diction that exceeds strict metrical or any other mechanical rule and leaves poets, audiences and readers free to contemplate and refine other, non-mechanical 'symbolic' uses of language.

Scientific method, cognitive linguistics and complexity

Neither Milman nor Adam Parry nor many other students of formulaic discourse treated deviations from economy and similar exceptions as anything but anomalies, nor could they fully acknowledge the implications of the argument about analogy's surplus effect without undermining the methodological framework that assumes essential rules and pre-existing compositional algorithms.[76] However, modern cognitive-functional approaches allow us to make better sense of the evidence, especially when understood as part of the inherent complexity of linguistic usage. If we accept the argument for the primacy of symbolic and communicative functions and for the epiphenomenal character of grammar, it becomes clear that creating diction by the process of analogy can produce prolific, closely related but potentially innovative meaning and mutations of form. Furthermore, where two metrically identical expressions such as νηῦς εὐεργής and ποντοπόρος νηῦς coexist, we may well have an incentive to reanalyse their meaning and interpret these two different constructions individually. We may, for example, view νηῦς εὐεργής as a more general term for ships in contrast to ποντοπόρος νηῦς which emphasizes the ship as a means of crossing *boundaries between worlds* and which in Homeric epic seems to characterize mythological references to the Argo or to the Phaeacian ship at crucial points in Homer's narrative.[77]

The cognitive argument for linguistic complexity represents a fundamental shift in general method and philosophical grasp of the world. Its principles have been drawn from modern science – quantum physics, thermodynamics, meteorology, crystallography, non-linear geometry, topology, etc. – and the shift away from classical ('Newtonian', 'Laplacian') determinism and towards a better understanding of complex, stochastic phenomena within so-called "complex adaptive systems".[78] As in many other scientific disciplines, so in linguistics, complexity[79]

> does not merely mean complicated. Although the agents or components in a complex system are usually numerous, diverse, and dynamic, *a defining characteristic of a complex system is that its behaviour emerges from the interactions of its components.*

Because complex linguistic systems are open, i.e., not fully containable within a closed set of mechanical/algorithmic rules – precisely, let us stress, as in the case of Homer's diction – "what arises may be *in nonlinear relation to its cause.* In other words – just as Parry had suggested when accounting for the effects of analogy – *an unexpected occurrence may take place at any time*".[80] Complex linguistic systems are not static, but dynamic. As linguist Claire Kramsch explains:[81]

> When you learn one additional piece of knowledge, this new knowledge doesn't just add itself to the other things you acquired previously. *The equilibrium you thought you had reached in your prior state of knowledge gets disrupted as one new piece of knowledge reconfigures the whole picture.*

The loss of equilibrium, the reconfiguration of knowledge and its effect on the picture as a whole, allows us to view both rule-based and anomalous diction in early Greek epic as a single, semantically resonant system rather than an amalgam of opposing traditional and individual elements. Metrical and syntactic patterns and *a fortiori* other less formal patterns of meaning (which we recognize by repetition and analogy) can interact with each other and evolve systematically yet unpredictably and at any point, *simply because users use the language in a variety of contexts.* These contexts may be traditional and similar, but they are never fully identical, if only because of the progress of narrative. As we add one piece of knowledge to another, both the relation of that piece of knowledge to the whole and the whole itself change. More rigid repetition does, of course, sometimes occur in epic diction, for example when long formulaic passages are transposed verbatim, when the contents of a speech are cited later in the narrative and so on. But even in such contexts, something will have changed. Making a speech and citing a speech are two very different ways of saying the same thing. They entail, for example, a change of speaker, point of view, temporal point of reference in relation to a narrated event, etc., and, indeed, often changes to grammar and form (for example, from first to third person or from present to past forms).

Language and linguistic usage depend on the systematic processes of analogy and repetition. But these are always complex, dynamic and adaptive processes. The lesson of usage-based linguistics, then, is that no matter how traditional epic diction is, it is always innovative and 'complex'. One cannot step into the same river of discourse twice. Every line of epic hexameter incorporates multiple, often imbricated interacting patterns, metrical, lexical, rhythmic, phonetic, semantic, thematic and narrative. Yet something in the context of such patterns inevitably changes, through nothing more than the process of repeated application, simply through the fact that language is being used.

Notes

1 Russo 1997, esp. 259–60; 2010. Mark-up of repetition (*Iliad, Odyssey*, Hesiod, etc.) online in Kahane, Mueller, Berry *et al.*; further comments in Mueller Online (http://panini.northwestern.edu/mmueller/MyPapers/Homeric_Repetitions.htm).

2 Already Meillet 1923: 61 was speaking of the "grammar" of the traditional hexameter; for Parry, see Foley 1988: 9; Lord 1960: 65, "The poetic grammar of oral epic is and must be based on the formula"; useful discussion in Foley 1988; Schein 2016 (1998).

3 See further below and note 9.

4 Tomasello 2003: 5. Notwithstanding many other fundamental differences between these grammatical approaches.

5 See n. 4, above. Further comments with reference to oral epic in Cánovas and Antović 2016a: 87–8, whose characterization of pre-Saussurean linguistics may nevertheless not sufficiently stress the role of paradigmatic form, e.g., in classical grammar.

6 A. Parry in Parry 1971: 6: "the dialectal and artificial elements of the language of Homer constitute a system characterized at once by great extension and by great simplicity". See also n. 68, below.

7 Parry 1971: 14–16 (a "formula type").

8 See, e.g., in A. Parry's introduction to Parry 1971, *passim*.

9 See Tsagalis 2011: 209–10 on the "great divide" between oralists and Neoanalysis; Montanari 2012; Kullmann 2012, 1984 (an influential early argument); Finkelberg 2012, 2011, 2004; Fowler 2004; Thomas 1992; Schein 2016; etc.

10 Russo 1997: 259–60: "The word formula proved to be a poor thing, hopelessly inadequate to cover the different *kinds* of formulaic realities in Homeric diction. And it is reasonable to assume that the talented traditional poet would always have been capable of some non-formulaic, original language". In a manner of speaking, already Parry himself recognized this in 1932; see 1971: 313: "There are *more general types of formulas*, and one could make no greater mistake than to limit the formulaic element to what is underlined [as formulaic, pp. 301–2]". Shive 1987 is a notable if at times problematic attempt to point out key difficulties within formulaic systems.

11 Fowler 2004: 230, n. 42: "It is difficult to reconcile Neoanalysis with an oral perspective". This long-standing divide led to now well-documented scholarly fatigue; see, e.g., Burgess 2015: 97: "Since the Homeric Question will never be fully resolved, there is surely a point where prolonged speculation gives poor returns for the effort . . . It would be a shame if the Homerist – or the author on introductory books on Homer – did not move on to other subjects".

12 See Bakker 2005 (and earlier work), Bakker 2013 and 'interformularity'; Bierl 2012; Bozzone 2014 and 'constructions'; Burgess 2001, 2006, 2012 and 'meta-Cyclic' poetry; Čolacović 2006 and 'post-traditional' poetry (see also Danek 2005); Currie 2016 on allusion; Danek 1996, 1998, 2002; Finkelberg 1990, 2002, 2003, 2004, etc. and 'meta-epic'; also Finkelberg 2015; Friedrich 2007 (interpreting exceptions to economy) proposing an oral/literate '*tertium*'; Lord's later work (1995); Nagler 1974 and *Gestalt*, 'family resemblance' and '*sphota*'; Montanari, Rengakos and Tsagalis 2012 on orality and Neoanalysis (with overviews by Montanari and Kullmann and with bibliography; Kullmann 2015); Adam Parry 1966 on 'post-orality'; Nagy 2015; Pucci 1987 and Peradotto 1990, pioneering early intertextual readings (discussed in Pedrick 1994); Schein 2016 (1998), 2016; Tsagalis 2008, 2011; 2014 and 'intratraditionality'; Foley's influential notion of 'traditional referentiality' 1991, 1999, etc. (cf. Lord 1960: 148; Danek 2002; Bozzone 2014: 46–9; Currie 2016: 4–9); work on epic and hypertext: Bakker 2001; Balling and Madsen 2003; *Trends in Classics* 2.2 [2010] (but crucial observations on intertextuality and hypertext in Riffaterre 1994 are not accounted for in most studies of Homeric hypertextuality). Earlier studies, general and of individual formulae, are usefully considered in Edwards 1988: 11–42.

13 See Hoffmann and Trousdale 2013: 1; Tomasello 2003: 99: "Usage-based linguists such as Langacker (1987), Bybee (1995), Fillmore (1989), Goldberg (1995), and Croft (2001) . . . recognize a continuum of meaningful linguistic constructions from morphemes to words to phrases to syntactic assemblies"; cf. Goldberg 2006: 220 (11.c: "Lexicon and grammar are not distinct components, but form a continuum of constructions"). For the same in the context of Homer, see Cánovas and Antović 2016b.

14 Tomasello 2003: 5. Emphasis here and below is mine, unless otherwise stated.
15 Tomasello 2003: 5.
16 Goldberg 2006: 5.
17 Goldberg 2006: 5.
18 For analogy and its importance in Homer, see the sixth section, below.
19 Tomasello 2003: 5.
20 I.e., a secondary phenomenon that is dependent on or derivative from other, more basic phenomena. See, e.g., Hopper 1989.
21 Lord 1960: 36; further references in Kahane 1994: 16. For Lord, cf. Bozzone 2014: 13; Cánovas and Antović 2016a: 84.
22 Rosenmeyer 1965: 297.
23 See survey in Edwards 1988: 11–42.
24 Discussions of this formula: Edwards 1988: 30–1; Heubeck 1987; and *s.v.* ἀμύμων in LfgrE.
25 Amory Parry 1973: 157.
26 Hainsworth 1976: 167.
27 Bakker and Fabricotti 1991. For the dichotomy, largely avoided in construction grammar, see Christiansen and Chater 2016: 227–40. In the context of Homeric diction, see Boas 2016: 110: "unlike many other theories of grammar, CxG [Construction Grammar] does not make any theoretical distinctions between different areas of grammar such as core and periphery".
28 Bakker 2005: 21, cf. 1997: 186 n. 3, 208 n. 2. There is, of course, always some phenomenological distinction between groups and individuals, between given verbal *types* and distinct, individual verbal *tokens*. See, e.g., Ellis and Larsen-Freeman 2006: 14–15: "Language exists both in individuals (as idiolect) and in the community of users (as communal language). Language is emergent at these two distinctive but interdependent levels: An idiolect is emergent from an individual's language use through social interactions with other individuals in the communal language, whereas a communal language is emergent as the result of the interaction of the idiolects."
29 Kahane 1994: 15.
30 Kahane 1994: 113. For an intricate recent discussion of the semantic and poetic significance of metrical usage in epic hexameters, see Katz 2013.
31 Bakker's argument, like Burgess 2012, is more cautious on allusion and context-specific intertextuality.
32 Here, nevertheless, Bakker's earlier conceptual division between core and periphery and formal vs. semantic elements (see above) seems to be assumed.
33 Bakker 2013: 158–9 and 162 citing Austin 1975: 28–9; see also Tsagalis 2014: 394–8. Cf. Givón 1979: "today's syntax is yesterday's diction", cited in Tomasello 2003: 14.
34 Bozzone 2014: 4, citing Goldberg 2006; cf. Bozzone 2010.
35 Bozzone 2014: 36.
36 Bozzone 67. I have slightly simplified Bozzone's formulation to avoid technical notation.
37 Cánovas and Antović 2016a; see also Cánovas and Antović 2016c and Cánovas and Antović 2016b. I note also Mocciaro and Short 2018 which, however, was not yet out at the time of writing this chapter and which I have not been able to consult.
38 Boas 2016.
39 Minchin 2016.
40 Currie 2016: 227. For more recent work, see also Kahane Forthcoming-a; Kahane Forthcoming-b; Kahane Forthcoming-c.
41 We should distinguish between linguistic evolution and linguistic change. Cognitive linguistics assumes no fixed formal core, and thus the fundamental evolutionary nature of language. Within mainstream linguistic traditions (e.g., generative grammars), "language change over historical time is assumed to be confined to the non-core aspects of language, such as the grammatical periphery and the lexicon". See Christiansen and Chater 2016: 39.
42 Some conceptions of the process of grammaticalization assume reduced semantic functions ('semantic bleaching'). Usage-based approaches begin with symbolic usage and thus assume the potential for semantic development within different forms (already in the work of Lakoff, Fillmore and others in the 70s and 80s; see Bybee 1985; Givón 1995; Croft 2001; Tomasello 2003; Goldberg 2006) and the process of 'semantic reanalysis'. As Eckhardt (2006: 4–5) explains, evidence from the last decades suggests semantic gain in reanalysis where earlier studies (e.g. Meillet 1912) tended to associate grammaticalization with a weakening of semantic functions. See also Tomasello 2003: 15–17; Beckner and Bybee 2009; Hoffmann and Trousdale 2013: 305–7 (sect. 23.2).

43 Of which, not surprisingly, only eighteen are in the *Iliad* and the rest in the *Odyssey*. Examples, statistics and references in LfgrE, Kahane, Mueller, Berry *et al.* and in the paper concordances.
44 Bakker 2013: 163.
45 *Il* 9.312-3 ἐχθρὸς γάρ μοι κεῖνος ὁμῶς Ἀΐδαο πύλῃσιν // ὅς χ᾽ ἕτερον μὲν κεύθῃ ἐνὶ φρεσίν, ἄλλο δὲ εἴπῃ.
46 For full mark-up of imbricated patterns in Homer and other early epic, see Kahane, Mueller, Berry *et al.*; discussion in Kahane 2005: 78.
47 Following the metrico-semantic framework proposed by Kahane 1994. Cf. Kelly 2007: 419 on usage in this verse, arguing for the authority of final positioning (n. 7) and the semantics of internal positioning.
48 Paradigmatic grammar grudgingly acknowledges such limitations, e.g., in the 'defective' conjugations of certain verbs. See in general Janda 2007: 643–5. The term 'defective' itself points to the methodological premise of paradigmatic analyses.
49 Usage for Iris is, however, more diverse and the patterning is more complex.
50 Τὸν/Τὴν δ᾽ ἠμείβετ᾽ ἔπειτα completed by θεὰ γλαυκῶπις Ἀθήνη 7x; βοῶπις πότνια Ἥρη 5x. *Il.*; θεὰ λευκώλενος Ἥρη 1x, breaking economy, both forms in *Il.* only; θεὰ Θέτις ἀργυρόπεζα 3x and Θέτις κατὰ δάκρυ χέουσα 1x, both forms in *Il.* only; φιλομμειδὴς Ἀφροδίτη 1x; Διώνη, δῖα θεάων 1x; ποδήνεμος ὠκέα Ἶρις 1x; περίφρων Πηνελόπεια 4x, *Od.* only; φίλη τροφὸς Εὐρύκλεια 1x). Cf. also, e.g., *Il.* 8.484 Ὣς φάτο, τὸν δ᾽ οὔ τι προσέφη λευκώλενος Ἥρη, etc. Full mark-up of repetition in Kahane, Mueller, Berry *et al.*
51 Cf. Beck 2005: 33–4 and Beck's Appendix II, 284–5; also Beck 2012.
52 Semantico-grammatical usage-boundaries of this type are likely to be present in all or most constructions in epic, but can easily be missed. As Parry, for example, says (1971: 20): "We are faced with the analysis of a technique which, because the bard knew it without being aware that he knew it, because it was dependent on his memory of *an infinite number of details*, was able to attain a degree of development which we shall never be in a position to perfectly understand".
53 For hiatus, see Ruijgh 2000; Wyatt 1992a. For an extended semantic approach to hiatus, see Fortassier 1989 (and review: Wyatt 1992b).
54 Parry 1971: 207.
55 Parry 1971: 207.
56 Visser 1997: 176–85.
57 Visser 1997: 177 and data in Visser 1987, which slightly revises O'Neill 1942: Pos. 9: 5.5%; pos. 11: 2.4%.
58 Cf. *Il.* 2.574 Πελλήνην τ᾽ εἶχον ἠδ᾽ Αἴγιον ἀμφενέμοντο; 607 καὶ Τεγέην εἶχον καὶ Μαντινέην ἐρατεινὴν; 608 τύμφηλόν τ᾽ εἶχον καὶ Παρρασίην ἐνέμοντο.
59 Finkelberg 2012: 78 . . . ἐρατεινὴν [OCT], also noting *Il.* 2.574, 607 and 608, apparently independently of Visser's work.
60 Finkelberg 2012: 789.
61 'Complexity' is one of the key terms separating the fundamental principles of classical (Newtonian, etc.) and modern science. See n. 78, below.
62 Parry 1971: 68.
63 All further emphasis in citations below is mine, unless stated otherwise.
64 E.g. by Varro (Duso 2006; Castello 2008), Quintilian (Von Fritz 1949) and Herodian (Sluiter 2011); overview in Law 2003.
65 Meillet 1912: 130.
66 Neither process is quite so simple, of course: for analogy, see Blevins and Blevins 2009; for grammaticalization, Hopper and Traugott 2013 (22 on Meillet) and above, n. 42.
67 A. Parry in Parry 1971: xxxii claims it was "overlooked". As we argue below, the paradoxical force of Parry's observations was not fully acknowledged but several scholars had by this time considered the phenomenon. See Lord 1960: 37, 43 (and below); Young 1967; Edwards 1971: 90–3. For a survey, see Edwards 1986: 202–7 (citing work by Notopoulos, Russo, Minton); Nagy 1992: 154; etc.
68 Lord 1960: 37: "I believe that the really significant element in the process is . . . the setting up of various patterns that make adjustment of phrase and creation of phrases by analogy possible".
69 Blevins and Blevins 2009: 2. Cf. Kuryłowicz 1947; Penn, Holyoak and Povinelli 2008. Significantly, contemporary paradigmatic linguists tend to ignore analogy. See, e.g., Chomsky 1986: 32: "analogy is simply an inappropriate concept in the first place".
70 See Parry 1971: 276 and *passim*.
71 Parry 1971: 176.

72 The formulaic system for "ship" is (Parry 1971: 109) "without doubt the most complex of all formulary systems created for common noun; and the *Iliad* and *Odyssey* seem to give us examples of most of the formulae of which the system is made" (see list in 109–13). 'Ship' is the most frequent noun in the Homeric epics; see data in Kahane, Mueller, Berry *et al.* and discussion in Mueller 2009; and Mueller online.

73 E.g.: *Il.* 15.704 Ἕκτωρ δὲ πρυμνῆς νεὸς ἥψατο ποντοπόροιο; *Il.* 3.444 ἔπλεον ἁρπάξας ἐν ποντοπόροισι νέεσσι; *Il.* 2.771 ἀλλ᾽ ὃ μὲν ἐν νήεσσι κορωνίσι ποντοπόροισι.

74 Parry 1971: 176–7.

75 Parry 1971: xxiii.

76 A useful discussion of Parry's relation to scientific method (but without reference to linguistics or complexity studies) is Schein 2016 (1998): 119–20.

77 See *Od.* 12.69–70 οἴη δὴ κείνη γε παρέπλω ποντοπόρος νηῦς, // Ἀργὼ πᾶσι μέλουσα . . .; *Od.* 13.95 τῆμος δὴ νήσῳ προσεπίλνατο ποντοπόρος νηῦς; *Od.* 13.161 ἔνθ' ἔμεν': ἡ δὲ μάλα σχεδὸν ἤλυθε ποντοπόρος νηῦς. For reanalysis, see n. 42, above.

78 Classical science assumed that the whole universe could, in principle, be explained on the basis of algorithmic laws. See Laplace 1951 (1825): 4: "An intellect ['Laplace's Demon'] which at a certain moment would know all forces that set nature in motion, and all positions of all items of which nature is composed . . . would *embrace in a single formula* [orig.: *embrasserait dans la même formule*] the movements of the greatest bodies of the universe and those of the tiniest atom". In contrast, as chemist and Nobel Laureate Ilya Prigogine and physicist Gregoire Nicolis say: "At the end of this [the 20th] century, more and more scientists have come to think, as we do, that many fundamental processes shaping nature are irreversible and stochastic; that the deterministic and reversible laws [of, e.g., Newtonian physics] describing the elementary interactions may not be telling the whole story. This leads to a new vision of matter, one no longer passive, as described in the mechanical world view, but associated with spontaneous activity" (Nicolis and Prigogine 1989: 3); cf. Byrne and Callaghan 2013; in linguistics: Kramsch 2012 and below.

79 Larsen-Freeman and Cameron 2008: 1. For an overview of complexity in science, see Nicolis and Prigogine 1989.

80 Kramsch 2012: 11–12.

81 Kramsch 2012: 11.

References

Amory Parry, A. 1973. *Blameless Aegisthus: A Study of AMYMŌN and Other Homeric Epithets*, Leiden.

Austin, N. 1975. *Archery at the Dark of the Moon: Poetic Problems in Homer's Odyssey*, Berkeley, CA.

Bakker, E. J. 1997. *Poetry in Speech: Orality and Homeric Discourse*, Ithaca, NY.

Bakker, E. J. 2001. 'Homer, Hypertext, and the Web of Myth', in U. Schaefer and E. Spielmann, eds., *Varieties and Consequences of Literacy and Orality / Formen und Folgen von Schriflischkeit un Mündlichkeit: Franz Bäuml zum 75 Geburtstag*, Tübingen: 149–60.

Bakker, E. J. 2005. *Pointing at the Past: From Formula to Performance in Homeric Poetics*, Washington, DC.

Bakker, E. J. 2013. *The Meaning of Meat and the Structure of the Odyssey*, Cambridge.

Bakker, E. J. and F. Fabricotti. 1991. 'Peripheral and Nuclear Semantics in Homeric Diction: The Case of Dative Expressions for Spear', *Mnemosyne* 44: 63–84.

Balling, H. and A. K. Madsen. 2003. *From Homer to Hypertext: Studies in Narrative, Literature and Media*, Odense.

Beck, D. 2005. *Homeric Conversation*, Washington, DC.

Beck, D. 2012. *Speech Presentation in Homeric Epic*, Austin, TX.

Beckner, C. and J. Bybee. 2009. 'A Usage-Based Account of Constituency and Reanalysis', in N. C. Ellis and D. Larsen-Freeman, eds., *Language as a Complex Adaptive System*, Malden, MA: 27–46.

Bierl, A. 2012. 'Orality, Fluid Textualization and Interweaving Themes: Some Remarks on the *Doloneia*: Magical Horses from Night to Light and Death to Life', in F. Montanari, A. Rengakos and C. Tsagalis, eds., *Homeric Contexts: Neoanalysis and the Interpretation of Oral Poetry*, Berlin: 133–74.

Blevins, J. P. and J. Blevins, eds. 2009. *Analogy in Grammar: Form and Acquisition*, Oxford.

Boas, H. C. 2016. 'Frames and Constructions for the Study of Oral Poetics', in M. Antović and C. P. Cánovas, eds., *Oral Poetics and Cognitive Theory*, Berlin: 99–124.

Bozzone, C. 2010. 'New Perspectives on Formularity', in S. W. Jamison, H. C. Melchert and B. Vine, eds., *Proceedings of the 21st Annual UCLA Indo-European Conference*, 27–44.

Bozzone, C. 2014. 'Constructions: A New Approach to Formularity, Discourse, and Syntax in Homer', PhD, University of California Los Angeles.
Burgess, J. S. 2001. *The Tradition of the Trojan War in Homer and the Epic Cycle*, Baltimore, MD.
Burgess, J. S. 2006. 'Neoanalysis, Orality, and Intertextuality: An Examination of Homeric Motif Transference', *Oral Tradition* 21: 148–81.
Burgess, J. S. 2012. 'Intertextuality Without Text in Early Greek Epic', in Ø. Andersen and D. Haug, eds., *Relative Chronology in Early Greek Epic Poetry*, Cambridge: 168–83.
Burgess, J. S. 2015. *Homer*, London.
Bybee, J. 1985. *Morphology: A Study of the Relation Between Meaning and Form*, Amsterdam.
Bybee, J. 1995. 'Regular Morphology and the Lexicon', *Language and Cognitive Processes* 10: 425–55.
Byrne, D. and G. Callaghan. 2013. *Complexity Theory and the Social Sciences*, London.
Cánovas, C. P. and M. Antović. 2016a. 'Construction Grammar and Oral Formulaic Theory', in C. P. Cánovas and M. Antović, eds., *Oral Poetics and Cognitive Science*, Berlin: 79–98.
Cánovas, C. P. and M. Antović. 2016b. 'Formulaic Creativity: Oral Poetics and Cognitive Theory', *Language and Communication* 47: 66–74.
Cánovas, C. P. and M. Antović, eds. 2016c. *Oral Poetics and Cognitive Science*, Berlin.
Castello, L. Á. 2008. 'Analogía y anomalía en Varrón: *De lingua latina* VIII–X', *Docenda Homenaje a Gerardo H Pagés*, Buenos Aires: 185–218.
Chomsky, N. 1986. *Knowledge of Language: Its Nature, Origin and Use*, New York, NY.
Christiansen, M. H. and N. Chater. 2016. *Creating Language: Integrating Evolution, Acquisition, and Processing*, Cambridge, MA.
Čolacović, Z. 2006. 'The Singer Above Tales: Homer, Mededović and Traditional Epics', *Seminari Romani di Cultura Greca* 9: 161–87.
Croft, W. 2001. *Radical Construction Grammar: Syntactic Theory in Typological Perspective*, Oxford.
Currie, B. 2016. *Homer's Allusive Art*, Oxford.
Danek, G. 1996. 'Intertextualität in der Ilias. Intertextualität in der Odyssee', *Wiener humanistische Blätter* 38: 22–36.
Danek, G. 1998. *Epos und Zitat: Studien zu den Quellen der Odyssee*, Vienna.
Danek, G. 2002. 'Traditional Referentiality and Homeric Intertextuality', in F. Montanari, ed., *Omero tremila anni dopo*: 3–19.
Danek, G. 2005. 'Review of Čolacović and Rojc-Čolacović 2004', *Würzburger Studien zur Altertumswissenschaft* 118: 5–20.
Duso, A. 2006. 'L'analogia in Varrone', in R. Oniga and L. Zennaro, eds., *Atti della giornata di linguistica latina*, Venice: 9–20.
Eckhardt, R. 2006. *Meaning Change in Grammaticalization: An Enquiry into Semantic Reanalysis*, Oxford.
Edwards, M. W. 1988. 'Homer and Oral Tradition: The Formula, Part II', *Oral Tradition* 3: 11–60.
Ellis, N. C. and D. Larsen-Freeman, eds. 2006. *Language as a Complex Adaptive System*, Malden, MA.
Fillmore, C. J. 1977. 'The Case for Case Reopened', in P. Cole, ed., *Syntax and Semantics 8: Grammatical Relations*, New York, NY: 59–81.
Fillmore, C. J. 1989. 'Grammatical Construction Theory and the Familiar Dichotomies', in R. Dietrich and C. F. Graumann, eds., *Language Processing in Social Context*, Amsterdam: 17–38.
Finkelberg, M. 1990. 'A Creative Oral Poet and the Muse', *American Journal of Philology* 111: 293–303.
Finkelberg, M. 2002. 'The Sources of the *Iliad*', in H. M. Roisman and J. Roisman, eds., *Essays on Homer (=Colby Quarterly 38)*: 151–61.
Finkelberg, M. 2003. 'Homer as a Foundation Text', in M. Finkelberg and G. G. Stroumsa, eds., *Homer, the Bible, and Beyond: Literary and Religious Canons in the Ancient World*, Leiden: 75–96.
Finkelberg, M. 2004. 'Oral Theory and the Limits of Formulaic Diction', *Oral Tradition* 19: 236–52.
Finkelberg, M. 2011. 'Homer and His Peers: Neoanalysis, Oral Theory and the Status of Homer', *Trends in Classics* 3: 197–208.
Finkelberg, M. 2012. 'Oral Formulaic Theory and the Individual Poet', in F. Montanari, A. Rengakos and C. Tsagalis, eds., *Homeric Contexts: Neoanalysis and the Interpretation of Oral Poetry*, Berlin: 73–82.
Finkelberg, M. 2015. 'Meta-Cyclic Epic and Homeric Poetry', in M. Fantuzzi and C. Tsagalis, eds., *The Greek Epic Cycle and its Ancient Reception: A Companion*, Cambridge: 126–38.
Foley, J. M. 1988. *The Theory of Oral Composition: History and Methodology*, Bloomington, IN.
Foley, J. M. 1991. *Immanent Art: From Structure to Meaning in Traditional Oral Epic*, Bloomington, IN.
Foley, J. M. 1999. *Homer's Traditional Art*, University Park, PA.

Fortassier, P. 1989. *L'hiatus expressif dans l'Iliade et dans l'Odyssée*, Leuven.
Fowler, R. 2004. 'The Homeric Question', in R. Fowler, ed., *The Cambridge Companion to Homer*, Cambridge: 220–32.
Friedrich, R. 2007. *Formular Economy in Homer: The Poetics of the Breaches*, Stuttgart.
Givón, T. 1979. *On Understanding Grammar*, New York.
Givón, T. 1995. *Functionalism and Grammar*, Amsterdam.
Goldberg, A. E. 1995. *Constructions: A Construction Grammar Approach to Argument Structure*, Chicago, IL.
Goldberg, A. E. 2006. *Constructions at Work: The Nature of Generalization in Language*, Oxford.
Hainsworth, J. B. 1976. 'Review of A. Amory Parry, *Blameless Aegisthus* (Leiden, 1973)', *Classical Review* 26: 167–8.
Heubeck, A. 1987. 'Ἀμύμων', *Glotta* 65: 37–44.
Hoffmann, T. and G. Trousdale, eds. 2013. *The Oxford Handbook of Construction Grammar*, Oxford.
Hopper, P. J. 1989. 'Emergent Grammar', *Proceedings of the Thirteenth Annual Meeting of the Berkeley Linguistics Society* 15: 139–57.
Hopper, P. J. and E. C. Traugott. 2013. *Grammaticalization*, Cambridge.
Janda, L. 2007. 'Inflectional Morphology', in D. Geeraerts and H. Cuyckens, eds., *The Oxford Handbook of Cognitive Linguistics*, Oxford: 632–49.
Kahane, A. 1994. *The Interpretation of Order: A Study in the Poetics of Homeric Repetition*, Oxford.
Kahane, A. 2005. *Diachronic Dialogues: Authority and Continuity in Homer and the Homeric Tradition*, Lanham, MD.
Kahane, A. Forthcoming-a. 'The Complexity of Epic Diction', *Yearbook of Ancient Greek Epic* 2.
Kahane, A. Forthcoming-b. 'Oral Theory and Intertextuality: The Case of the Homeric Hymns', in S. Bär and A. Maravela, eds., *Narratology and Intertextuality: New Perspectives on Greek Epic from Homer to Nonnus*, Leiden.
Kahane, A. Forthcoming-c. 'The Politics of the Formula', in P. Vasunia, ed., *The Politics of Literary Form*, Oxford.
Kahane, A., M. Mueller, C. Berry and B. Parod. n.d. *The Chicago Homer*, http://homer.library.northwestern.edu.
Katz, J. 2013. 'The Hymnic Long Alpha: Μούσας ἀείδω and Related Incipits in Archaic Greek Poetry', in S. W. Jamison, H. C. Melchert and B. Vine, eds., *Proceedings of the 24th Annual UCLA Indo-European Conference, Los Angeles, October 26th and 27th, pp 87–101 (Bremen: Hempen, 2013)*, Bremen: 87–101.
Kelly, A. 2007. *A Referential Commentary and Lexicon to Iliad VIII*, Oxford.
Kramsch, C. 2012. 'Why Is Everyone So Excited About Complexity Theory in Applied Linguistics', *Mélanges CRAPEL* 33: 9–24.
Kullmann, W. 1984. 'Oral Poetry Theory and Neoanalysis in Homeric Research', *Greek Roman and Byzantine Studies* 25: 307–23.
Kullmann, W. 2012. 'Neoanalysis between Orality and Literacy: Some Remarks Concerning the Development of Greek Myths Including the Legend of the Capture of Troy', in F. Montanari, A. Rengakos and C. Tsagalis, eds., *Homeric Contexts: Neoanalysis and the Interpretation of Oral Poetry*, Berlin. 12: 13–26.
Kullmann, W. 2015. 'Motif and Source Research: Neoanalysis, Homer and Cyclic Epic', in M. Fantuzzi and C. Tsagalis, eds., *The Greek Epic Cycle and Its Ancient Reception*, Cambridge: 108–26.
Kuryłowicz, J. 1947. 'La nature des process dits "analogiques"', *Acta Linguistica* 5: 15–37 (='The Nature of the So Called Analogical Processes', *Diachronica* 12 [1995]: 113–45).
Lakoff, G. 1987. *Women, Fire, and Dangerous Things: What Categories Reveal About the Mind*, Chicago, IL.
Langacker, R. 1987. *Foundations of Cognitive Grammar, Vol. 1*, Stanford, CA.
Laplace, P.-S. 1951 (1825). *A Philosophical Essay on Probabilities*, New York.
Law, V. 2003. *The History of Linguistics in Europe: From Plato to 1600*, Cambridge.
Lord, A. 1960. *The Singer of Tales*, Cambridge, MA.
Lord, A. 1995. *The Singer Resumes the Tale*, Ithaca, NY.
Meillet, A. 1912. 'L'évolution des formes grammaticales', *Scientia* 12: 130–58.
Meillet, A. 1923. *Les origines indo-européenes des metres grecs*, Paris.
Minchin, E. 2016. 'Repetition in Homeric Epic: Cognitive and Linguistic Perspectives', in C. P. Cánovas and M. Antović, eds., *Oral Poetics and Cognitive Linguistics*, Berlin: 12–29.
Mocciaro, E. and W. Short, eds. 2018. *Toward a Cognitive Classical Linguistics: The Embodied Basis of Constructions in Greek and Latin*, Berlin.
Montanari, F. 2012. 'Introduction: The Homeric Question Today', in F. Montanari, A. Rengakos and C. Tsagalis, eds., *Homeric Contexts: Neoanalysis and the Interpretation of Oral Poetry*, Berlin: 1–12.

Montanari, F., A. Rengakos and C. Tsagalis, eds. 2012. *Homeric Contexts: Neoanalysis and the Interpretation of Oral Poetry*, Trends in Classics, Berlin.
Mueller, M. 2009. *The Iliad*, London.
Mueller, M. n.d. 'About Homeric Repetitions: From Short and Common to Long and Rare', from http://panini.northwestern.edu/mmueller/MyPapers/Homeric_Repetitions.htm.
Nagler, M. N. 1974. *Spontaneity and Tradition: A Study in the Oral Art of Homer*, Berkeley, CA.
Nagy, G. 2015. 'Oral Traditions, Written Texts, and Questions of Authorship', in M. Fantuzzi and C. Tsagalis, eds., *The Greek Epic Cycle and Its Ancient Reception*, Cambridge: 59–77.
Nicolis, G. and I. Prigogine. 1989. *Exploring Complexity: An Introduction*, New York.
O'Neill, E. G. 1942. 'Word-Types in the Greek Hexameter', *Yale Classical Studies* 8: 103–78.
Parry, A. 1966. 'Have We Homer's *Iliad*?', *Yale Classical Studies* 66: 177–216.
Parry, M. 1971. *The Making of Homeric Verse: The Collected Papers of Milman Parry*, Oxford.
Pedrick, V. 1994. 'Reading in the Middle Voice: The Homeric Intertextuality of Pietro Pucci and John Peradotto', *Helios* 21: 75–96.
Penn, D. C., K. J. Holyoak and D. J. Povinelli. 2008. 'Darwin's Mistake: Explaining the Discontinuity Between Human and Nonhuman Minds', *Behavioral and Brain Sciences* 31: 109–78.
Peradotto, J. 1990. *Man in the Middle Voice: Name and Narration in the Odyssey*, Princeton, NJ.
Pucci, P. 1987. *Odysseus Polytropos: Intertexual Readings in the Odyssey and the Illiad*, Ithaca, NY.
Riffaterre, M. 1994. 'Intertextuality vs. Hypertextuality', *New Literary History* 24: 779–88.
Rosenmeyer, T. G. 1965. 'The Formula in Early Greek Poetry', *Arion* 4: 295–311.
Ruijgh, C. J. 2000. 'La genèse du dialecte homérique', *Ziva Antika* 50(1): 213–29.
Russo, J. A. 1997. 'The Formula', in I. Morris and B. Powell, eds., *A New Companion to Homer*, Leiden: 238–60.
Russo, J. A. 2010. 'Formula', in M. Finkelberg, ed., *The Homer Encyclopedia*, Malden, MA. I: 296–8.
Schein, S. L. 2016. 'Ioannis Kakridis and Neoanalysis', *Homeric Epic and Its Reception*, Oxford: 127–37.
Schein, S. L. 2016 (1998). 'Milman Parry and the Literary Interpretation of Homeric Poetry', *Homeric Epic and Its Reception*, Oxford: 117–26.
Shive, D. 1987. *Naming Achilles*, Oxford.
Sluiter, I. 2011. 'A Champion of Analogy: Herodian's *On Lexical Singularity*', in S. Mathaios, F. Montanari and A. Rengakos, eds., *Ancient Scholarship and Grammar: Archetypes, Concepts and Contexts*, Berlin: 291–312.
Thomas, R. 1992. *Literacy and Orality in Ancient Greece*, Cambridge.
Tomasello, M. 2003. *Constructing a Language: A Usage-Based Theory of Language Acquisition*, Cambridge, MA.
Tsagalis, C. 2008. *The Oral Palimpsest: Exploring Intertextuality in the Homeric Epics*, Cambridge, MA.
Tsagalis, C. 2011. 'Towards an Oral, Intertextual Neoanalysis', *Trends in Classics* 3: 209–44.
Tsagalis, C. 2014. 'Γυναίκων εἵνεκα δώρων: Interformularity and Intertraditionality in Theban and Homeric Epic', *Trends in Classics* 6: 357–98.
Visser, E. 1987. *Homerische Versifikationstechnik: Versuch einer Rekonstruktion*, Frankfurt.
Visser, E. 1997. *Homers Katalog der Schiffe*, Stuttgart.
Von Fritz, K. 1949. 'Ancient Instruction in "Grammar" According to Quintilian', *American Journal of Philology* 70: 337–66.
Wyatt, W. F. 1992a. 'Homeric Hiatus', *Glotta* 1992 70: 20–30.
Wyatt, W. F. 1992b. 'Review of Fortassier L'hiatus expressif dans l'Iliade et dans l'Odyssée', *Phoenix* 46: 85–6.

2

The cognitive linguistics of Homeric surprise*

Alexander S. W. Forte

Over the past two decades, research on Classical emotions has experienced a steady bloom, characterized by an increasingly nuanced understanding of both the culturally situated emotional constructions of the modern analyst and certain cross-cultural continuities in emotional experience.[1] Applying cognitive linguistics to emotions in the *Iliad* and *Odyssey*, this chapter argues that various emotions described as "grasping" and "seizing" the body of the patient can be understood as involving surprise. According to this account, the emotion of surprise, metonymically conceptualized through the startle reflex, interacts dynamically with both positively and negatively valenced affective processes in the Homeric poems.[2] This chapter concludes with an account of how cognitive linguistics can provide new perspectives on cross-linguistic terminology of surprise.

Before advocating cognitive linguistics' relevance for the study of the Homeric poems, I will describe my working definition of surprise. To various degrees explicit in phenomenological and pragmatist philosophies,[3] the quotidian, dynamic interplay of surprise and expectation in our daily lives has in recent years become the focus of neuroscientifically informed philosophical theories.[4] In this chapter, I adopt the account of the "neurophenomenological" school, which conceptualizes surprise as a complexly embodied affective process in the emergence of emotion.[5] Specifically, surprise is theorized as a short-lived, neutrally valent, and dynamically interactive process involving the startle reflex,[6] the cardiac defense,[7] and brain activations, all of which are variously enculturated.[8] That pre-existing cardiovascular traits modulate the intensity of the startle reflex (Richter et al. 2009; Melzig et al. 2008) indicates that surprise is a dynamic, interactive process rather than simply a reflex arising from the brain-stem: the heart is involved from the start. This embodied theorization of surprise as an intense feeling of rupture,[9] shock, or lack of control is compatible both with psychological evidence suggesting surprise's ability to amplify co-occurring emotions,[10] and with analyses of emotions in cognitive linguistics,[11] which treats the ways in which speakers express and conceptualize emotions through conceptual metaphors that act upon the body of the patient. These emotions are understood in terms of the following source domains: opponent, captive animal, social superior, fire, or fluid in a container.

In cognitive linguistics, the emotion of anger has received the most sustained attention in both Indo-European and non-Indo-European languages.[12] Comparative studies in English, Chinese (Yu 1995), Japanese (Matsuki 1995), Hungarian (Bokor 1997), Polish (Micholajczuk 1998),

Zulu (Taylor and Mbense 1998), Homeric Greek (Cairns 2003a), and Latin (Riggsby 2015), among others,[13] have demonstrated that the emotion concept of anger is expressed and conceptualized through culturally specific reflections of embodied somatic processes with widespread (but not total) continuity between speakers. Examples of the above source domains applied to anger are: "I fight my temper, but sometimes it wins" (opponent), "The meeting was contentious and they got heated" (fire), "She's going to flip her lid" (container), "He keeps his anger bottled up" (fluid in a container). Raymond Gibbs (1992, 1994) has demonstrated that these are not simply linguistic descriptions, but reflect fully articulated cognitive models that are grounded in the embodied experience of speakers.

The most prevalent conceptual metaphor for an angry person is that of a pressurized container, which ultimately emerges from embodied physiological changes of increased skin temperature and blood pressure. In English, this is elaborated to the metaphor: anger is a hot fluid in a pressurized container,[14] found in phrases such as "she was seething," "he was about to blow his top," and "simmer down." While there are cross-culturally common physiological symptoms of various affective processes (increase in skin temperature, heartrate, a feeling of internal pressure), each of these symptoms is enculturated, or conditioned by and expressed with culturally specific linguistic and conceptual emphases.[15] Zulu, for example, while sharing with Japanese, English, Hungarian, and Chinese a pressurized container metaphor for anger, features a more elaborate fire metaphor, through which one can refer in everyday speech to someone as "extinguishing another's anger by pouring water on them." In contrast, the Chinese pressurized container metaphor for anger (*nu*) does not involve heat at all. Therefore, there are culturally specific developments of basic physiological metonymies: body heat stands for anger, internal pressure stands for anger, and redness in the face and neck area stands for anger, which can develop into elaborate metaphorical systems within a given language. As in the case of anger, there is widespread physiological, conceptual, and linguistic continuity between culturally specific representations of surprise.

The most detailed cognitive linguistic analysis of surprise is Zoltán Kövecses' (2015a) treatment of the English-language folk model. His lexical investigation of surprise renders the conceptual structure (or schematic frame) as follows:[16]

1. cause (of emotion) – 2. causes – 3. emotion
|
4. effect of emotion

This characterizes, according to Kövecses (2015a, 277–8),

> An event or thing (1.) that has the quality of causing surprise (given the appropriate situation) causes the emotion of surprise (2.), and the emotion (3.) causes the emotional self (who is now in the state of surprise) to produce certain effects or responses (4.).

This is an account of the metonymies that proceed from the emotion lexeme "(state or feeling of) surprise," namely a state for the quality of an event or thing that causes that state, "a surprise party," and a state for the result / effect of that state, "a look of total surprise" (or a "surprised expression").

The conceptual metaphors of surprise are:

1 surprising someone is unexpectedly impacting someone[17]

"I was stunned/shocked by . . ."

2 surprising (someone) is an unexpected seizure / attack[18]

"I hope your questions don't take me by surprise."[19]

This second metaphor proceeds from the more generic metaphor that control is holding/possessing, e.g. "I have a handle on the situation" or "This has gotten out of hand."

Both metaphors are actually consistent with the etymology of Modern English "surprise" (< Old French *sorprendre* [and other Romance lexemes] < Medieval Latin *superprendere* < Latin *-praehendere** "grasp"), which indicates that the word "surprise" itself has a lexical history that emerges from these conceptual metaphors. Verbal aspect is crucial in these metaphors: in English at least, it is inapposite to use an imperfective or durative verb, like "The weather was taking me by surprise," unless this is referring to a multiplicity of events. In referring to individual instances of surprise, the verbs all must be aoristic/perfective in aspect.[20] We might then re-conceptualize the metaphors of surprise as:

1 surprising someone is suddenly impacting someone
2 surprising (someone) is a sudden seizure / attack

Kövecses concludes by emphasizing the similarities and differences between surprise and other emotions. The conceptualization of entering an emotional state as a loss of control is a commonality, or in Kövecses' terms, "the cognitive model of surprise overlaps with that of other emotions."[21] However, surprise does not have the common emotional source domains of opponent, captive animal, social superior, fire, and fluid in a container, among others.[22] The conceptual metaphors of surprise simply indicate a lack of control.

In assessing potential overlap between the English-language model of surprise and those of other cultures,[23] cognitive linguistics provides good evidence for cross-linguistic continuity in the embodied feeling of surprise as that which "seizes" or "strikes" one's body. In a variety of Indo-European languages, a subset of nouns and verbs that describe what in English would variously be called "shock," "surprise," or "awe" are, at various stages in their etymological history, related to verbal roots of physical force.[24]

Eng. *surprise* < Old French *sorprendre* (and other Romance lexemes) < Medieval Latin *superprendere* < Latin *-praehendere** "grasp."[25]
Gk. πατάσσειν "strike";[26] παίειν "ibid."; πλήσσειν "ibid."[27] > Gk. ἔκπληξις "surprise."[28]
L. *stupor* "numbness, amazement" < *(s)teup- "strike" [> Gk. τύπτειν "ibid."].[29]
Skt. *abhigrāhyati* "to catch, surprise."[30]

The phenomenological experience of surprise, which holistically includes the startle reflex, the cardiac defense, brain activation is, via a *pars pro toto* metonymy, conceptualized in terms of the startle reflex as a "strike" or a "seizing." This metonymy holds equally well for the phraseological data marshaled by Kövecses in his description of the English-language folk model of surprise.[31] However, the situation becomes more complicated when we turn to the Homeric poems. Whereas a modern investigator using a neurophenomenological approach can take into account both "objective" physiological data and "subjective" interview responses to distinguish between the startle reflex and surprise, we are unfortunately incapable of assessing Hector's heartrate or querying his expectations, and are equally bereft of face-to-face conversation with Homeric poet(s) about what emotions a given character could have felt in the *Iliad* or *Odyssey*.

For teasing out distinctions between the startle reflex and the emotion of surprise in ancient sources, a useful distinction involves an epistemic stance of the subject: the person who is surprised must believe the paradoxical frustration of an expectation ("I believe that I can't believe this."), whereas someone who is startled is not necessarily expecting anything and is shocked ("I can't believe it.").[32] Surprise then is a rehabilitation or reconstitution of an interrupted personal narrative, whereas shock is an embodied feeling of rupture.[33] Nevertheless, knowing where a startle reflex ends and surprise begins in oneself, not to mention in someone else, is difficult in general, and especially for a reader of ancient texts.

Despite these limitations, I will argue below that by taking into account the cross-linguistic evidence of surprise "seizing" us via the startle reflex, we can better understand emotions in the Homeric poems. Although the vast majority of cognitive linguistic research on emotions is based on English-language models, the language of ancient emotion has received welcome attention in recent years.[34] In a recent chapter examining garment metaphors in Greek emotions, Douglas Cairns has collected and analyzed various embodied metaphors that characterize Homeric emotions.[35] Throughout the *Iliad* and *Odyssey*, many different emotions, as well as sleep and death, "grab," "hold," or "seize" the body of the patient.[36] These verbs are rightly taken to reflect the intensely embodied nature of the emotions in question, and I will argue below that a part of this embodiment is the startle reflex. While it is possible that this startle is a component of the emotional process of surprise, concurring with and modulating the subject's affective response, the Homeric poems do not provide the necessary somatic data to confirm anything more than the startle itself. At stake methodologically is a more phenomenologically rich treatment of Homeric emotions: when reading Homer, we are not confronted with emotion-nouns being described one at a time and working in isolation, but with multidimensional affective processes involving the gradual or sudden unfolding of experience.

The most familiar descriptions of "surprise," "awe," or "amazement" in the *Iliad* and *Odyssey* are the nouns θάμβος and θαῦμα, and their denominative verbal formations θαμβεῖν and θαυμάζειν/θαυμαίνειν.[37] A duration of amazement can be expressed either by θάμβος "holding" the body of the patient using present-stem forms of ἔχειν,[38] or by an imperfect of θαμβεῖν.[39] A sudden "surprise" is only once expressed by θάμβος suddenly grabbing the body of the patient using the thematic aorist ἑλεῖν;[40] more commonly, the sigmatic aorist of θαμβεῖν appears.[41] As regards θαῦμα, there is only one example in Homeric poetry where it holds the body of the patient;[42] elsewhere it is the object of verbs of sight, functioning essentially as an internal accusative.[43]

In the Homeric poems there are several instances in which χόλος "anger" seizes or holds the body of the patient.[44] The key semantic difference in the narrated examples is whether the past-tense verb is imperfective or perfective in aspect,[45] in the former case denoting a durative process of anger (variously modulating through time) in response to a provoking event, and in the latter case a sudden onset of anger itself. A case of the former appears in a repeated passage from the *Iliad* that is particularly rich in affect, in which Athena and Hera are variously angry at Zeus' verbal provocation:

> ὣς ἔφαθ', αἳ δ' ἐπέμυξαν Ἀθηναίη τε καὶ Ἥρη·
> πλησίαι αἵ γ' ἥσθην, κακὰ δὲ Τρώεσσι μεδέσθην.
> ἤτοι Ἀθηναίη ἀκέων ἦν οὐδέ τι εἶπε
> σκυζομένη Διὶ πατρί, χόλος δέ μιν ἄγριος ᾕρει·
> Ἥρῃ δ' οὐκ ἔχαδε στῆθος χόλον, ἀλλὰ προσηύδα·
> αἰνότατε Κρονίδη ποῖον τὸν μῦθον ἔειπες.
> (Il. *4.20–5 = 8.457–62)*

So he (Zeus) spoke, and Athena and Hera muttered,
as they sat close to each other and planned evils for the Trojans.
Then Athena was silent and did not say anything,
sulking at father Zeus, but savage anger was grasping her.
Yet Hera's breast did not contain her anger, and she spoke forth,
"Most terrible son of Kronos, what sort of command did you speak!"

In the beginning of book four and in the latter part of book eight, Zeus aggressively broaches the topic of divine intervention in the Trojan war, twice goading Athena and Hera to anger, and the result is the same: Athena is able to resist, responding not with a verbal outburst, but instead sulking in silence, while Hera offers a verbal response. The phrase used to describe Athena's durative and abiding anger involves an imperfective verb of grasping (ᾕρει) and an opponent metaphor, wherein the emotion is conceptualized as an exterior, inimical force which acts upon the body of the patient. In this case, the use of χόλος . . . ἄγριος "savage anger" potentially indicates that this opponent metaphor involves a source domain of a wild animal. The imperfective, durative verbal action does not describe the narrative onset of the startle: Athena's anger is sustained and she is able to control it. On the other hand, Hera's irruptive anger, which leads to her ensuing verbal response, involves an aorist verb (ἔχαδε) and a somewhat underdetermined example of the container metaphor, wherein anger is conceptualized as a substance (or an animate being) that, when controlled, is held within Hera's chest, and when excessively strong, is able to escape from its containment and cause certain behaviors in the patient. Although there is surely some notion of irruptive emotional action here, the moment of startle is likewise not narrated. The perfective verbal aspect is not describing anger's onset, but the point at which it is implicated in further action. The cause of Hera's reply to Zeus is metaphorically represented as anger's breach of her chest, wherein the emotion was contained (and so controlled) until the end of Zeus' speech. As Hera's response indicates, what is in fact "startling" is Zeus' speech (. . . ποῖον τὸν μῦθον ἔειπες), the duration of which provokes a corresponding durative process of anger in Athena and Hera.

Taken together with the contextual evidence of the passages, wherein the goddesses' plans against the Trojans are challenged by Zeus, the nature of Hera's verbal response perhaps indicates something more than a startled anger. As one might expect, verbal expressions of surprise in English feature exclamations and expletives,[46] so Hera's verbal response (. . . ποῖον τὸν μῦθον ἔειπες) ostensibly indicates a dynamic interplay of emotions in this scene. Hera and Athena expected to carry out their devised evils against the Trojans, and are startled, angry, and surprised by Zeus' polemical command.

The imperfective verb of grasping (ᾕρει) does not describe the narrative onset of a startle reflex, instead indicating a durative process of affect understandable in this case as sustained anger in response to an initially startling event. However, in cases wherein the verb of grasping is an aorist, a startle seems contextually appropriate:[47]

αὐτίκ' ἐγὼ πρῶτος κελόμην θεὸν ἱλάσκεσθαι·
Ἀτρεΐωνα δ' ἔπειτα χόλος λάβεν, αἶψα δ' ἀναστὰς
ἠπείλησεν μῦθον ὃ δὴ τετελεσμένος ἐστί.
(Il. *1.386–8)*

I first of all urged then the god's appeasement;
and then anger took hold of Atreus' son, and in speed standing up
he uttered his threat against me and now it is a thing accomplished.[48]

Here Achilles describes in retrospect the quarrel between himself and Agamemnon, and characterizes Agamemnon's anger as a punctual, irruptive reaction to his own proposal to appease Apollo by freeing Chryses' daughter. In this case, Agamemnon presumably expected to go unchallenged, and Achilles' violation of this presupposition caused the leader of the expedition to react with surprise and anger. The form of the aorist in this example represents the emotion's surprising onset and force. If, as it seems, phenomenologically irruptive force is characteristic of startle and its associated physiological processes, then narrated instances of emotions co-occurring dynamically with the startle reflex will act upon the bodies of patients in the Homeric poems via perfective verbs of grasping or seizing in an opponent metaphor.

Below are three further examples from the *Iliad* in which aoristic verbs meaning "to seize/hold" describe a startle, in each case with the grammatical subject of τρόμος "trembling," itself a metonymic representation of an embodied physical symptom. This metonymy describes the emotion(s) and concomitant physical reactions with which that symptom is associated such as fear, terror, and tonic immobility. In these cases, trembling and the startle reflex concur, but the latter is described by the verbs of seizing or holding:

> τὸν τόθ' ὑπ' ὀφρύος οὖτα κατ' ὀφθαλμοῖο θέμεθλα,
> ἐκ δ' ὦσε γλήνην· δόρυ δ' ὀφθαλμοῖο διαπρὸ
> καὶ διὰ ἰνίου ἦλθεν, ὃ δ' ἕζετο χεῖρε πετάσσας
> ἄμφω· Πηνέλεως δὲ ἐρυσσάμενος ξίφος ὀξὺ
> αὐχένα μέσσον ἔλασσεν, ἀπήραξεν δὲ χαμᾶζε
> αὐτῇ σὺν πήληκι κάρη· ἔτι δ' ὄβριμον ἔγχος
> ἦεν ἐν ὀφθαλμῷ· ὃ δὲ φὴ κώδειαν ἀνασχὼν
> πέφραδέ τε Τρώεσσι καὶ εὐχόμενος ἔπος ηὔδα·
> . . .
> ὣς φάτο, τοὺς δ' ἄρα πάντας ὑπὸ **τρόμος ἔλλαβε** γυῖα,
> πάπτηνεν δὲ ἕκαστος ὅπῃ φύγοι αἰπὺν ὄλεθρον.
> (*Il. 14.493–500, 506–7)*
>
> This man (Ilioneus), Peneleus caught underneath the brow, at the base
> of his eye, and pushed the eyeball out, and the spear went clean through
> the eye-socket and tendon of the neck, so that he went down
> backward, reaching out both hands, but Peneleus drawing
> his sharp sword hewed at the neck in the middle, and so dashed downward
> his head, with helm upon it, while still on the point of the great spear
> the eyeball stuck. He, holding it up like the head of a poppy,
> displayed it to the Trojans and spoke vaunting over it . . .
> . . .
> So he spoke, and **trembling seized** all of them by their limbs,
> and each man looked around for a way to escape sheer death.

In this gruesome scene of spectatorship, Peneleus has just killed Ilioneus and has skewered his foe's eyeball on the tip of a spear. Not content with the violence of his kill, he then raises his spear to the sightline of the Trojans and boasts. The collective reaction of the Trojans to this shocking scene is deeply embodied, with trembling (τρόμος) seizing their limbs (ἔλλαβε γυῖα) in an opponent metaphor, the sudden onset of which is expressed by the aorist. The exceptionally grisly nature of the scene causes the internal audience to lose somatic control. As to whether this scene could represent the emotional experience of surprise, presumably soldiers come to battle with certain self-directed narratives that are violated, or surprised, by a given situation. In this case, however, the brutality of Peneleus' mutilation of Ilioneus seems beyond anything that could have been expected. This scenario inclines me to believe that this is a description of shock.

The otherworldly sight of Achilles' divinely forged armor provokes sudden fear in any mortal who looks directly at it, yet his reaction to his own dazzling equipment proceeds from the enraging memory of his two-fold loss, of the former armor and the friend who wore it:[49]

ὡς ἄρα φωνήσασα θεὰ κατὰ τεύχε' ἔθηκε
πρόσθεν Ἀχιλλῆος· τὰ δ' ἀνέβραχε δαίδαλα πάντα.
Μυρμιδόνας δ' ἄρα πάντας **ἕλε τρόμος**, οὐδέ τις ἔτλη
ἄντην εἰσιδέειν, ἀλλ' ἔτρεσαν. αὐτὰρ Ἀχιλλεὺς
ὡς εἶδ', **ὥς μιν μᾶλλον**[50] **ἔδυ χόλος**, ἐν δέ οἱ ὄσσε
δεινὸν ὑπὸ βλεφάρων ὡς εἰ σέλας ἐξεφάανθεν.
τέρπετο δ' ἐν χείρεσσιν ἔχων θεοῦ ἀγλαὰ δῶρα.
(Il. *19.12–17)*

So having spoken the goddess placed the armor
before Achilles, and all elaborate it clattered.
Trembling grabbed all of the Myrmidons, nor did anyone bear
to look straight at it, but they shook. But when Achilles looked at it,
then did anger especially plunge into him, and his eyes shone
terribly out from under his eyelids, as if a blaze.
He rejoiced while holding in his hands the shining gifts of the god.

The sudden onset of the Myrmidons' terror is described by a verb of grasping in the aorist (ἕλε) with the physiological fear response of trembling as its grammatical subject, again conceptualized as a metaphorical opponent. The shocking and terrifying sight of the divine armor is so powerful that the soldiers are unable to gaze directly at it. Achilles, on the other hand, is infuriated by the sight of the armor because it reminds him of that which it replaces, and ultimately of whom it cannot replace, Patroclus. His growing anger is described using a container metaphor, in which anger is conceptualized as an external force that plunges into his body (ὥς μιν μᾶλλον ἔδυ χόλος).[51] In this case, his anger is represented using the source domain of fire, which after entering his body causes his eyes to shine as if with a terrible light. This rage, however, gives way to joy as Achilles handles his new equipment, presumably thinking of the vengeance that these divine gifts will facilitate. Whereas the Myrmidons are gripped by fear, Achilles' intense anger develops from an examination of the new armor and concomitant reminder of loss. Whereas the scenes involving Hera, Athena, and Zeus plausibly involved surprise, the episodes involving τρόμος seem to be more closely aligned with the startle reflex.

One final case of "trembling" in the *Iliad* will speak to its concurrence with a startle, but also potentially with surprise. Achilles' fearsomeness is so notorious that Nestor suggests Patroclus' impersonation of his friend to turn the tide of battle (*Il.* 11.794–803). After Patroclus' death, Achilles' gradual return to the battlefield provokes tremendous fear in the Trojans at every stage, and especially in his duel with Hector:

ὣς ὥρμαινε μένων, ὃ δέ οἱ σχεδὸν ἦλθεν Ἀχιλλεὺς
ἶσος Ἐνυαλίῳ κορυθάϊκι πτολεμιστῇ
σείων Πηλιάδα μελίην κατὰ δεξιὸν ὦμον
δεινήν· ἀμφὶ δὲ χαλκὸς ἐλάμπετο εἴκελος αὐγῇ

ἢ πυρὸς αἰθομένου ἢ ἠελίου ἀνιόντος.
Ἕκτορα δ', ὡς ἐνόησεν, **ἕλε τρόμος**· οὐδ' ἄρ' ἔτ' ἔτλη
αὖθι μένειν, ὀπίσω δὲ πύλας λίπε, βῆ δὲ φοβηθείς·
Πηλεΐδης δ' ἐπόρουσε ποσὶ κραιπνοῖσι πεποιθώς.
(*Il. 22.131–8)*

So he pondered, waiting, but Achilles was closing upon him
in the likeness of the lord of battles, the helm-shining warrior,
and shaking from above his shoulder the dangerous Pelian
ash spear, while the bronze that closed about him was shining
like the flare of blazing fire or the sun in its rising.
And **trembling grabbed** Hector when he saw him, and he could no longer
stand his ground there, but left the gates behind, and fled, frightened,
and Peleus' son went after him in the confidence of his quick feet.

The moment that Hector notices Achilles in his terrifying (and unexpected) new armor, trembling seizes him (ἕλε τρόμος), and he runs with Achilles in pursuit. As seen above, the completely novel and otherworldly sight of Achilles' new armor provokes fear in the Myrmidons and Hector, in the latter case augmenting Achilles' already terrifying presence. Hector awaits and expects Achilles, imagining his foe in the lead-up to this fateful duel, yet even his expectation of a terrifying adversary is outdone by Achilles' actual form, shining and terrible like a war-god clad in divine armor. In this case, one should characterize Hector's affective experience as a dynamic interplay of terror and surprise.

This is not to say that negatively valenced emotions such as fear or anger are alone capable of "seizing" the body of the patient. In one of the famous recognition scenes in the *Odyssey*, Odysseus' nurse Eurycleia realizes her disguised master's identity by seeing and touching the scar on his thigh. Her reaction upon recognizing her long-absent master coincides with a loss of bodily control:

τὴν γρηῢς χείρεσσι καταπρηνέσσι λαβοῦσα
γνῶ ῥ' ἐπιμασσαμένη, πόδα δὲ προέηκε φέρεσθαι·
ἐν δὲ λέβητι πέσε κνήμη, κανάχησε δὲ χαλκός,
ἂψ δ' ἑτέρωσ' ἐκλίθη· τὸ δ' ἐπὶ χθονὸς ἐξέχυθ' ὕδωρ.
τὴν δ' ἅμα χάρμα καὶ ἄλγος ἕλε φρένα,[52] τὼ δέ οἱ ὄσσε
δακρυόφιν πλῆσθεν, θαλερὴ δέ οἱ ἔσχετο φωνή.
(*Od.* 19.467–72)

After taking the scar in her downturned hands, the old woman
recognized it after her touch, and released his foot from her hold.
His shin fell in the basin, and the bronze rang,
and tilted back to the other side, and water spilled on the ground.
Joy and pain together seized her φρήν, her two eyes
were filled with tears, and her quivering voice was held in check.

The shock of Eurycleia's recognition causes her to drop Odysseus' foot into the basin, and her complex emotional response to her master's presence represents the multivalent phenomenal experience of a startling emotion. She simultaneously feels joy and pain (χάρμα καὶ ἄλγος).

Moreover, the suddenness and strength of these emotions cause her to feel as though her body is seized by an external force (ἕλε φρένα), after which she weeps and is unable to speak. Here Eurycleia's unexpected recognition of Odysseus provokes in her a sudden sadness at his long absence and a simultaneous joy at his return, the effect of both expressed by a verb of grasping in the aorist. In this case, the conceptual metaphor perhaps features a more schematic personification than the more negatively valenced opponent metaphor.[53] When she takes the stranger's foot, Eurycleia has no expectation of the stranger's identity, and the sign of Odysseus' true identity is epiphanic in its sudden clarity. Instead of indicating a process of surprise, this episode seems to align more closely with a startle reflex, or a "shock," itself implicated with emotional joy and pain.

In all of these Homeric examples, whether describing anger, fear, or joy, the startle reflex is present, found in verbs of "grasping" or "seizing." In some cases, I have argued that this startle reflex was part of a more overtly epistemic affective process of surprise that interacted with the other emotional processes. As regards the phraseology, the cross-linguistic evidence of the "grasping" lexemes of surprise is enough to forestall any inclination to identify this Homeric language as uniquely "poetic," but it would be correspondingly unsatisfying to attribute the presence of such an elaborate system of affective metaphor in the Homeric poems as a trivial outcome of an anthropologically broad cognitive linguistic tendency. There may, in fact, be cognitive-aesthetic motivation for the selection of such physical metaphors in the depiction of emotion.

Recent neuroscientific and psychological research has indicated that English-language verbal media employing physical metaphors cause corresponding activation in the audience's pre-motor cortices,[54] and those employing emotional metaphors produce activation in emotional centers (e.g. metaphors of disgust produce activation in the anterior insula and pallidum comparable to that caused by disgusting pictures),[55] suggesting that the audience of the Homeric poems may have felt the shock of emotion along with the heroes of the Homeric poems during the performance and reading of these episodes.[56] In addition to enriching our understanding of the Homeric poems' representation of the unfolding, embodied experience of emotion, and the effects of this representation on the audience, the concept of a cognitive linguistic "strike" of the startle reflex also provides a solution to an old and complex problem in Greek lexicography of emotions.

There is no accepted etymology for the neuter s-stem noun Gk. θάμβος, and the set of ostensibly related forms is morpho-phonologically controversial.[57] The earliest forms associated with θάμβος "surprise, wonder" include a denominative stative verbal formation θαμβέω "be surprised, wonderstruck," attested in Homer in the imperfect and sigmatic aorist.[58] This relatively clear situation becomes much more complicated when one considers forms that are semantically and morphologically similar to θάμβος, such as the perfect verbal form τέθηπα (*Il.* 4x, *Od.* 3x), the pluperfect ἐτεθήπεα (*Od.* 6.166), the aorist participle ταφών (*Il.* 6x, *Od.* 16.12), and the noun τάφος (*Od.* 3x). There have been several attempts in the past few decades to arrange these forms into a derivational hierarchy.[59]

The underlying issue is that there are three relevant verbal roots, all reconstructable for the proto-language: *$d^{(h)}eb^{h}$- "to be small, diminished,"[60] *$d^{h}eH_{2}b^{h}$- "to strike,"[61] and *$d^{h}emb^{h}$- "to be stunned." The last of these roots is not widely reconstructed because its nasal is often mistakenly categorized as a synchronic creation via nasal-infixation. The *Lexikon der indogermanischen Verben* (LIV^{2}, 133–4) sets up two separate roots (with reproduced indications of uncertainty): (?) *$d^{h}emb^{h}$- "erstaunen (intr.), in Erstaunen geraten," including aorist ἔταφον "astonished" and perfect τέθηπα "amazed"; and (?) *$d^{h}emb^{h}$- "(zer)schlagen" (?), including two Indo-Iranian iterative-causative formations, Vedic *dambháyati* "smashes" and Khoresmian *δnby-* "to strike."[62] These two roots are in fact one and the same, and the combined evidence indicates that the nasal is old, and not a recent, synchronically introduced infix.[63] The semantics of the root must be something like "to be stunned," with the causatives meaning "cause someone to be stunned → strike." In terms of the phonology, the beta of θάμβος is expected according to the Greek sound-law wherein stops are deaspirated after a nasal and an accented syllable (θάμβος ~ ταφών, στρόμβος ~ στρέφω, θρόμβος ~ τρέφω).[64] In Indo-Iranian, regular dissimilation of aspirates (Grassmann's law) occurred in the verbal forms, as it did in the case of Gk. ταφών. The vocalism is unexpected. The full-grade is expected of any Indo-European s-stem, yet Attic and Ionic demonstrate widespread replacement of the full-grade with a zero-grade,[65] and θάμβος is in effect an ersatz zero-grade with a retained nasal (the "real" zero-grade [< $*d^{h}\underset{\circ}{m}b^{h}$-] is found in the s-stem τάφος at *Od.* 21.122).[66] The aorist participle ταφών "stunned," from the zero-grade $*d^{h}\underset{\circ}{m}b^{h}$-,[67] is likely a stative-intransitive primary formation from the onomatopoeic root. Despite repeated attempts to integrate τέθηπα into this verbal paradigm,[68] the absence of any nasal renders the phonologies incompatible,[69] and the perfect formations should instead be understood as remodeled from $*d^{h}eH_{2}b^{h}$- "to strike."[70] The situation in the Homeric poems is difficult, but potentially explainable by the combination of two separate but phonologically and semantically similar roots into a single, morphologically suppletive paradigm.[71]

One can associate both roots, $*d^{h}emb^{h}$- and $*d^{h}eH_{2}b^{h}$-, with other verbal roots that are onomatopoeic of "thumping" or "tapping." Examples include Eng. *thud* < OEng. *þyddan* "to strike"; Skt. *tudáti*, L. *tundo* < *(s)teud- "to strike";[72] Proto-Germanic *stamp- > Eng. *stamp*,[73] and other forms with an internal nasal: Eng. *thump*, EFris. *dump* "thump," Sw. dial. *dumpa* "to make a noise." Other examples lack the final stop, indicating a resonant sound consistent with the percussive instruments indicated: Skt. *ḍamarin* "drum," *ḍāmara* "terrifying"[74] (cf. English *drum*, Germ. *Trommel*, Danish *tromme*, Swedish *trumma*).

The lexical history of Gk. θάμβος "surprise, wonder" is rooted in an onomatopoetic representation of a beating sound. From a cognitive linguistic perspective, one would posit a metonymic semantic progression of $*d^{h}emb^{h}$- from a pure onomatopoeia (cf. Eng. "thump") to that which causes the sound (i.e. a "strike" or "blow"). This would then fit into the cross-linguistic data marshaled above, in which surprise is represented as an external force that "strikes" or "seizes" unexpectedly upon the body.[75] Under this analysis, the phenomenological experience of surprise, which holistically includes the startle reflex, brain activations, and changes in the autonomic nervous system (including heartrate and respiration), is, via a *pars pro toto* metonymy, conceptualized in terms of the startle reflex, conceptualized as a "strike," θάμβος, upon the body.[76] The historical similarities between the etymologies of lexemes of surprise, including Eng. *surprise* and Gk. θάμβος, speak to common, diachronically reoccurring experiences of emotional embodiment.

In short, cognitive linguistics enables the analysis of cross-cultural continuities in the embodiment of emotions, allowing one to appreciate how the *Iliad* and *Odyssey* represent the dynamic processes of affective experience, in the stability of fulfilled expectation and the unexpected shock of emotion. Evidence for these dynamic affective processes exists at the levels both of noun, in the case of the θάμβος, and of verb, wherein emotions are represented as grasping and seizing the body of the patient. The metaphors of emotion in the Homeric poems link audience, poet, and character via phenomenologically rich and fundamentally embodied representations of emotional experience that, despite nearly three millennia, remain wondrously familiar.

Notes

* Many thanks to Peter Meineck, Bill Short, and Jennifer Devereaux for their helpful suggestions and keen editing. David Konstan provided characteristically enlightening feedback on surprise as emotion, and Michael Weiss and Jeremy Rau were greatly helpful with Indo-European *arcana*. All remaining blunders are my own.

1 The following volumes demonstrate the state of the art: Cairns and Fulkerson (eds.) 2015 (with an especially good introduction on the difficulties of translating emotions); Cairns and Nelis (eds.) 2017; and Caston and Kaster (eds.) 2016. For a comparison of the Greek and Roman terms and concepts which do not quite correspond to "forgiveness," see Konstan 2010; for a similar project focused on "remorse" and "regret," see Fulkerson 2013. An earlier work that gives a good impression of its own contemporaneous state of the field is Marincola 2003.

2 Affectivity in a more general sense can include bodily affects, moods, dispositions/temperaments, and character traits, all of which are dynamical and multidimensional in terms of both the biological systems involved and their contextual expressions. See Colombetti 2014, 53–82 on emotions as dynamical, diachronic patterns (or processes) rather than synchronic states. See also Thelen and Smith 1994; Tschacher and Dauwalder (eds.) 2003; Werner 2011.

3 On surprise in the works of Adam Smith, Husserl, Peirce, Heidegger, Merleau-Ponty, Ricœur and others, see Dastur 2000 and Depraz 2014. See also the forthcoming volume on surprise edited by Anthony Steinbock and Natalie Depraz.

4 A potential future task for a neuroscientifically informed philosophy of the mind is to integrate the phenomenological and pragmatic accounts of surprise (and associated English-language terms such as "awe") with the predictive theories of human cognition (variously termed "Predictive Processing," "Prediction Error Minimizing," and "Predictive Processing Accounts of Cognition") including recently Clark 2016a, 2016b; Fabry 2017; Gallagher 2017; Hutto and Kirchhoff 2016 (and the following commentary in *Constructivist Foundations* 11[2]); Kirchhoff 2017.

5 Desmidt et al. 2014.

6 The startle reflex is an involuntary (top-down) defensive motor reflex which commonly involves muscular contraction of the face (hence a blink), neck, torso, and limbs (hence the startle). Already in sections 70–7 of *On the Passions* (1649/1985), Descartes theorizes wonder (*admiration*) as the Ur-passion emerging from the brain and spirit, secondarily affecting the muscles, and not involving the heart and blood. In the work of Paul Ekman, the father of "Basic Emotions Theory," there are already some doubts about the evidence for cross-culturally identifiable emotions of surprise and contempt (Ekman 1992, 176); cf. Baron-Cohen et al. 1993 and Reisenzein et al. 2006. Some treatments of emotions from the perspective of evolutionary biology operate with a model of "basic emotions" that are narrowly identified with anatomical structures (e.g. Plutchik 2003). This approach, even if one largely accepts it, requires modification, on which see LeDoux 2015. For an extensive reworking of "Basic Emotion Theory" that is more compatible with constructionist accounts, see Scarantino 2015.

7 The generally neglected cardiac defense involves a complex pattern of accelerations and decelerations of heartrate in response to an unexpected stimulus, for further on which see Vila et al. 2007.

8 I am currently of the opinion that these are probable, and not necessary, components of surprise, but this is based upon my own recollection of "surprises" without a startle reflex. In addition to being biologically constrained, emotions are conceptualized, performed, and expressed in culturally and individually specific ways. For a history of conceptualizing emotions, see Plamper 2015. The psychological constructionist approach sees each emotional term in a given language as referring to a category of highly variable instances, and not to a stable, natural kind. See Barrett 2006; Barrett 2016; Barrett and Russell (eds.) 2015; Gendron, Roberson, and Barrett 2015; Touroutoglou, Lindquist, Dickerson, and Barrett 2015; Barrett and Gendron 2016. For componential criticisms of constructivist approaches, see Scherer 2015. One must not assume a dichotomy between biological and cultural approaches to emotions. Even proponents of embodied affectivity concede that nuance in the conception of emotions is methodologically necessary, e.g. recently Eickers, Loaiza, and Prinz 2017.

9 In Noordewier, Topolinksi, and Van Dijk 2016, surprise is again conceptualized as an emotional process, but in this case it is an "initial interruption," featuring activation in the anterior cingulate cortex (ACC) and a family of P300 responses or *novelty* P3. They allege that this interruption is negatively valenced, concluding that surprise is a "metacognitive feeling." They theorize that this initial negative valence, followed by an organism–environment interaction that is positively valenced, can actually intensify and amplify the positive.

10 Kahneman and Miller 1986; Mellers et al. 1997; Oliver 1997; Elster 1998; Valenzuela et al. 2010. One can easily speculate about the evolutionary and survival advantages of such a process, which results in increased memorability of the surprising moment in question, on which see Clark 2016b, 79. See also Wessel 2017 for surprise's facilitation of inhibitory motor control.

11 Cognitive linguistics is generally characterized by an emphasis on pragmatic (or context-oriented) semantics, the encyclopedic nature of linguistic knowledge, and the notion of language as a structuring tool, on which see Geeraerts 2006. The branch of cognitive linguistics most famous for its theory of

embodiment as the foundation of linguistic meaning is that which proceeds from Lakoff and Johnson [1980] 2003 and Lakoff 1987. A series of publications applying cognitive linguistics to emotions are Kövecses 1986, 1990, 2000, 2008, 2015a, 2015b, 24–6, 42–8, 80–96, 157–9.

12 See Geeraerts and Grondelaers 1995; Gevaert 2001, 2005; Lakoff 1987, 380–415; Lakoff and Kövecses 1987; Kövecses 1986, 11–37; Kövecses 2010; Riggsby 2015.

13 Some of these studies are summarized with further examples in Kövecses 2000.

14 Kövecses 2010, 165.

15 Whether the use in one's native language of certain emotion vocabulary, including the word "emotion," biases the investigator is a complex issue. See Cairns 2008 for an evaluative, Classics-oriented treatment of the works of Anna Wierzbicka, who has attempted to establish a Leibnizian meta-language to avoid such bias, and of Paul Griffiths, who has sought to link culturally specific emotion-concepts to objective psychological states. Recent works such as Lakoff 2016 and Barrett 2016 both point to fundamental disagreements about the definition of "emotion." On defining emotion see also the October 2012 (4.4) edition of *Emotion Review*, and Wassmann 2016. Even within a language, somewhat synonymous terms may have unintended experimental effects: see Noordewier and Breugelmans 2013 on potential word-bias involving "surprise" (positive) and "unexpected" (negative).

16 Kövecses 2015a, 277. In an earlier work, summarizing the work of Kendrick-Murdock 1994, Kövecses claimed that most of our understanding of surprise derives from three conceptual metaphors: surprise is a physical force, a surprised person is a burst container, and surprise is a natural force. He concluded (2000, 33), "Second, not surprisingly, surprise is the least metaphorically comprehended concept on our list. The reason possibly is that surprise is not a socially very complex phenomenon, and, consequently, there is not a great amount of conceptual content to be associated with it."

17 "In the source domain there is a physical or psychological force that impacts a person suddenly and unexpectedly and it results in certain physical or psychological responses in that person" (Kövecses 2015a, 281).

18 Ibid., 282.

19 Cf. Estonian *kedagi ootamatult tabama* "take someone by surprise," with *tabama* "take" potentially being a borrowing from an old Germanic word related to English *dab*.

20 See Ponsonnet 2013, 191–2 on Dalabon, "Successful tests clarified some simple cases, for instance confirming the punctuality of kangu-barrh(mu) ['belly+crack'] 'be surprised, undergo emotional shock,' which indeed cannot combine with the durative adverb munguyh (54)."

21 Kövecses 2015a, 284.

22 Ibid., 285.

23 Soriano, Fontaine, and Scherer 2015 analyzes a wide-variety of "surprise" words (and their cognitive prototypes) according to the GRID project (www.affective-sciences.org/grid), in which emotion terms are cross-linguistically rated according to valence, power, arousal, and novelty. All indigenous profiles of "surprise," in addition to being internally homogenous, were statistically quite close to the mean cross-cultural profile (with the exception of Burmese, an important reminder that universals are tricky things). The novelty measure, specifically, distinguishes surprise from other emotions by three standard deviations above the mean. Surprise also features a medium-sized arousal score (Soriano, Fontaine, and Scherer 2015, 438).

24 Cf. MGer. *Überraschung* "surprise" ~ English "over-rushing"; Sp. *de repente* "suddenly" < L. *repente* < *repo* "creep." Gk. θαῦμα "wonder, awe" is probably derived from a root *dheH$_2$u-, meaning "to see," perhaps distantly cognate with Skt. *dhī-* < *dheiH$_2$- "to see, imagine." Words of emotion roughly corresponding to English "surprise" or "awe" are sometimes metonymic from verbal roots whose prototypical meaning seems to be "to look at" or "to open one's mouth," cf. L. *admiratio* < *miror* "to gaze at, wonder at" < *smei-, underlying Eng. "smile."

25 For additional cross-linguistic examples, see Buck 1949, 1093–6.

26 For metaphors of emotions "striking" (πατάσσειν) a part of a person, see Soph. *Ant.* 1097; Ar. *Ra.* 54.

27 Usually the aorist passive form is used to indicate that someone is "dumbstruck." For example, the Trojan charioteers are "dumbstruck" when they see the flame above Achilles' head: ἡνίοχοι δ' ἔκπληγεν, ἐπεὶ ἴδον ἀκάματον πῦρ, *Il.* 18.225.

28 Frequently associated with negatively valenced emotions such as fear, e.g. Aesch. *Pers.* 606: τοία κακῶν ἔκπληξις ἐκφοβεῖ φρένας. See also *DGE* s.v. ἔκπᾰγλος, ἐκπλᾰγής, ἔκπληκτος, ἐκπλήσσω, and *LfgrE* s.v. ἔκπαγλος (M.A. Harder) and πλήσσω (J.N. O'Sullivan), esp. s.l. BII2 (ἐκ-). Associated verbs are δείδω, ἐκπατάσσω, ὀρίνω, ταράσσω, ταρβέω, and τρομέω. On the metaphor of "being struck by" something as an indication

of surprise in ancient Greek, English, German, and Italian, see Devereaux 2016, 244–7. In German, *Zorn* "anger" is often the grammatical subject of *packen* "grab" acting upon the body of the patient.

29 At Hdt. 3.64, τύπτειν (in its sigmatic aorist) describes surprise; in fact, nearly all of these verbs of striking appear in descriptions of surprise during the Cambyses episode, on which see Devereaux 2016, 246.

30 See *KEWA* I 343–4 and *LIV*^{2} 201, s.v. *g^{h}rebH$_2$- "ergreifen."

31 If one can be surprised without having a physical reflex of the startle, the cognitive linguistic metonymy can still be explained by reference to memorability and ensuing categorical prototypicality (see Lakoff 1987) of "shocking" surprises. Surprise cross-culturally tends to be expressed metonymically through language of startle, shock, and onrush because these are the most embodied, somatically intense (hence memorable) instances of surprise.

32 This is Anthony Steinbock's (2017, 16–19) formulation. I note that if one is currently surprised at one's earlier surprise, this might be represented as "I believe that I can't believe that I believed that I couldn't believe that it happened."

33 David Konstan (p.c.) points out to me that this narrative approach is consistent with the approach involving "emotional scripts" found in Kaster 2005.

34 Recent works are Cairns 2015, 2016a, 2016b, 2016c; Díaz Vera 2011; and Riggsby 2015. On metaphor in Homeric poetry, see Nieto Hernández 2011.

35 See Cairns 2016a, particularly 26n.4; 27n.6, on emotions that "take," "hold," or "seize" citing the following Homeric passages: *Il.* 4.421, 5.812, 817, 7.479, 8.77, 11.402, 13.224, 470, 581, 14.387, 475, 15.657–8, 16.599, 17.67; *Od.* 2.81, 3.123, 4.75, 142, 596, 6.140, 161, 8.384, 11.43, 279, 633, 12.243, 14.144, 21.299, 22.42, 24.438, 450, 533, to which I add *Il.* 21.221, *Od.* 3.227, 16.243.

36 See Clarke 1999, 231–3 for a catalogue of textual examples pertaining to personified Death and Sleep. See also Cairns 2016a, 27n.8 on "grasping."

37 The noun ἔκπληξις "surprise" is unattested in Homeric poetry and the verb ἐκπλήσσω is found in the Homeric scholia as a gloss on θάμβος; see b(BCE^{3}E^4)T scholia ad *Il.* 3.342.

38 *Il.* 3.342, 4.79, 23.815, 24.482.

39 *Od.* 4.638, 10.63, 17.367; *HH. Ap.* (3.)135.

40 *Od.* 3.372.

41 *Il.* 1.199, 3.398, 8.77, 23.728–881 (θηεῦντο τε θάμβησάν τε), 24.483–4; *Od.* 1.323, 2.155, 16.178, 24.101. It appears in participial form more rarely: *Od.* 1.360, 21.354; *HH. Dem* (2.)15. For a phenomenological treatment of wonder in the Homeric poems, see Prier 1989. On the importance of wonder to Archaic and Classical aesthetics, see Neer 2010 (esp. 21–69 on textual sources). On the sublime in antiquity, see Porter 2016.

42 *Od.* 10.326: θαῦμά μ' ἔχει. Additionally, there appear to be craft metaphors, *HH. Herm.* (4.)196: περὶ θαῦμα τέτυκται ~ *Il.* 18.549: περὶ θαῦμα τέτυκτο; *HH. Dem.* (2.)240: μέγα θαῦμ' ἐτέτυκτο, 403: ἄνει μέγα θαῦμα.

43 *Il.* 13.99, 15.286, 20.344, 21.54; *Od.* 19.36, *HH. Herm.* (4.)219: . . . ἦ μέγα θαῦμα τόδ' ὀφθαλμοῖσιν ὁρῶμαι. θαῦμα ἰδέσθαι: *Il.* 5.725, 10.439, 18.83, 18.377; *Od.* 3.306, 7.45, 8.366, 13.108; Hes. *Th.* 575, 581, [*Asp.*] 140, 224; *HH. Dem.* (2.)428, *HH. Aph.* (5.)90. θαῦμα ἰδεῖν: Hes. [*Asp.*] 318; *HH. Aph.* (5.)205.

44 *Il.* 4.23, 8.460; *Od.* 8.304: χόλος δέ μιν ἄγριος ᾕρει. *Il.* 18.322: μάλα γὰρ δριμὺς χόλος αἱρεῖ. *Il.* 9.675: χόλος δ' ἔτ' ἔχει μεγαλήτορα θυμόν. See Latacz, Nünlist, Stoevesandt 2000 ad *Il.* 1.387 (. . . χόλος λάβεν . . .) "Emotionale Regungen werden häufig so dargestellt, als ob sie von außen die Figur ergriffen," citing *Il.* 4.23, 8.460, 18.322, the *LfgrE* s.v. λάζομαι (G.C. Wakker), esp. s.l. B1b, and Snell [1939] 1975. On χόλος, see Walsh 2005, 205–25 and Cairns 2003a, 2003b (esp. 71–5) reviewing Clarke 1999.

45 An instance such as . . . ἐπεὶ χόλος ἔμπεσε θυμῷ (*Il.* 9.436, 14.207 = 14.306) is more complex because it is only spoken by characters describing the anger of others (Patroclus of Achilles, Hera of Oceanus, and Tethys). In each case that anger is represented as overwhelming, thereby disrupting expected behavior, yet surprise is absent.

46 Krawczak and Glynn 2015.

47 The optative formation at *Il.* 16.30 (μὴ ἐμέ γ' οὖν οὗτός γε λάβοι χόλος, ὃν σὺ φυλάσσεις . . .) has Patroclus' imagining Achilles' anger in these embodied metaphorical terms.

48 Translations are from Lattimore 1951, slightly adapted.

49 The commentary of Coray 2009 is very useful on this passage, with extensive references to earlier scholarship.

50 I adopt here the paroxytone of West's text from Coray's edition (2009).

51 See also *Il.* 9.553, 22.94, cf. *Il.* 9.239, 17.210, 19.367; *Od.* 18.348, 20.286 and especially *DGE* s.v. 1 δύω and *LfgrE* s.v. δύνω.

52 It is essentially impossible to translate φρήν into English, because the current English-language folk model of mind posits a brain–body dualism as well as an emotion–reason dualism, which is decidedly

not the situation in Homeric poetry, on which see Holmes 2010, 58–64 with references to earlier scholarship. On the body in Homer, the Hippocratic texts, and Plato, see also Holmes 2017.

53 If surprise in Homeric Greek is expressed by verbs of grasping, then we might expect the opposite of surprise, expectation, to be expressed in a completely different way. As Douglas Cairns has pointed out recently, the vocabulary in early Greek poetry for hope and expectation (ἐλπίς, ἔλπειν/ἔλπεσθαι) does not participate in the metaphorical conceptualization of emotions as violent, sudden forces or agents that act upon the body of the patient (Cairns 2016c). This confirms what seems intuitive: that the felt, phenomenological experience of expectation cannot be surprising, for us as much as for the ancient Greeks. On suspense as an emotion in narrative, see Konstan 2008.

54 Lakoff 2016.

55 Aziz-Zadeh and Gamez-Djokic 2016.

56 The metaphors of emotion could then implicate character, performer, and audience in a cognitively extended affective process. See Colombetti 2015 for an overview of an enactive, affective analysis of cognition. See Colombetti and Roberts 2014; Krueger 2014; Slaby 2014; Roberts 2015; Krueger and Szanto 2016; and Carter, Gordon, and Palermos 2016 for accounts of extended affectivity.

57 See Frisk 1960–72 s.v. and Chantraine 1999 s.v. for the phonological difficulties. Both entries cite *IEW* 233 *d^{h}ab^h- "to strike." Recourse to "Pre-Greek" (Beekes 2010 s.v. θάμβος) to explain the unexpected vocalism and inconsistencies in nasals/labials is an explanation from *aporia*, but is not impossible. Szemerényi 1954 and Barton 1993 collect all of the associated forms.

58 If one does not want to posit a primary adjective that is unattested, this pattern is probably analyzable as being synchronically extended by Greek speakers who thought of κράτος "power" as making a denominative verb κρατέω "be powerful."

59 Tucker 1990, 42–3; Barton 1993; Hackstein 2002, 237–8.

60 PIE *d$^{(h)}$eb^h- "be diminished" > Hitt. *tepu*- "small," Indo-Iranian *dabh- "deceive" (KEWA II 17–18 s.v. *dabhnóti*, EWA I: 694 ff.).

61 PIE *d^{h}eH$_2$b^h- > Lith *dóbti* "to beat," Latv. *dâbt* "to strike," Norw. *dabba* "to stomp," MEng. *dabben* "to strike" > Eng. *dab*. Old Fr. *taper* "strike" is probably a borrowing from an early Continental West Germanic form of this root, and eventually gives Eng. *tap*. The Germanic forms are probably iteratives built to an already transitive root, given the Baltic evidence as well as rarely adduced forms from Tocharian: TB *tsāp^ā*- "to crush" (Class VI pres.), and TB *tsop*- "to strike" (Class I pres.), see Malzahn 2010, 976. The same root in the zero-grade (*d^{h}H$_2$b^h-) is probably found in L. *faber* (<*d^{h}H$_2$b^h-ro-) "blacksmith, craftsman" and Arm. *darbin* (<*d^{h}H$_2$b^h-r-īno-) "blacksmith." See also Kroonen 2013, 88 and *IEW* 233. It should be noted that *LIV*^{2} explains PIE *d^{h}eb^h- (132) as primary and the Balto-Slavic forms as secondarily lengthened, although with hesitation.

62 The Khoresmian form speaks against the approach in Insler 1969, analyzing the nasal as secondary. This would mean that the nasal was independently added in Indic, Iranian, and Greek, which is certainly not impossible, but seems unlikely. A more believable analysis of the Indic situation is in Narten 1988–90.

63 See also the primary adjective PGmc. *dumba- > Eng. *dumb* "speechless." Szemerényi's (1954) connection of the etymology of Germanic *dumba- with θάμβος as deriving from a zero-grade of a root *d^{h}m̥b^h- meaning "to strike" is plausible, although the root would have to mean "to be stunned." His subsequent attempts to subsume as many roots as possible under the proto-root *d^{h}en- (plus root-extensions ad inf.) strain even my capacious credulity. See Heidermanns 1993, 166 for the full entry of Germanic cognates. The attempt of Kroonen (2013, 108) to relate *dumba- to a verb *dimban- ~ *dimpan- meaning "to smoke" is semantically implausible.

64 But note γόμφος, not γόμβος*, on which see most recently Kümmel 2012, which argues that the voiced stop in anlaut blocked deaspiration via contrastive and dissimilatory effect.

65 In general, zero-grade vocalism (which often appears as a-vocalism), is widespread in s-stem nouns in Attic-Ionic, but a smattering of forms in other dialects show the older full-grade of the root (κρέτος "power" (Alc. 141.3), θέρσος (Alc. 206.2) "courage" → Att.-Ion. κράτος, θάρσος). Cognate neuter s-stems in other Indo-European languages with full-grades allow the reconstruction of this as the earlier state of affairs. The zero-grade in these Attic-Ionic forms is an innovation, which occurred under the influence of corresponding u-stem adjectives and denominative-stative verbs in the zero-grade (κρατύς "strong" → κρατέω "be strong," θρασύς "bold" → θαρσέω "be bold") from which the vocalism presumably spread to the adjective abstracts. This derivational system primarily involves verbal roots that have property-concept meaning, hence forming certain types of primary adjectives, and is commonly designated the "Caland System" (on which see Rau 2009, 65–186). This system expanded variously in the Indo-European languages, and within Greek there is a set of Caland-adjacent emotion terms, such as τάρβος and θάμβος, that were probably integrated into the derivational system at an early stage.

66 See Hackstein 2002, 227–8, 233, 237–8. The integration of θάμβος into this system may have been spurred by a core group of Caland roots that yielded adjective abstracts associated with emotion, such as animate amphikinetic s-stems such as Ved. *bhiyás-* "fear" (: *bhīrú-* "afraid, fearful") and Gk. ἡ αἰδώς, -όος "shame" (: αἰδοῖος "having a claim to reverence, shamefaced, ashamed"), on which see Rau 2009, 133, 154. While τάρβος has a co-occurring adjective ταρβαλέος (*HH. Herm.* [4.]165), θάμβος has no early associated adjective (θαμβαλέος is only attested in the imperial period). This lack of a primary adjective is troubling insofar as it potentially indicates that θάμβος is not an archaic member of the Caland system; however, we do find the s-stem compound adjective ἀθαμβής attested quite early (Iby. fr. 5.11; Bacch. *Dith.* 1.58; Phryn. Trag. fr. 2).

67 For stative-intransitive, zero-grade root aorists from Caland roots, cf. ἔτραφον against Ved. *átr̥pam AV* (Rau 2009, 152n.81). The archaic paradigm might have included a full-grade present *θέμβω to fill out a paradigm *θέμβω, ἔταφον (perf. *τέθεμβα) parallel to λήθω, ἔλαθον (perf. λέληθα, cf. λελάθων Alc. fr. 73.8), with (belatedly attested) s-stem noun λᾶθος (Theoc. *Id.* 23.34) and ἥδομαι, εὔαδον/ἕαδον (e.g. *Od.* 16.28 and A scholia ad *Il.* 14.340[c1]), with s-stem ἦδος. It seems as though forms of θαμβέω have replaced any potential primary formations (e.g. τεθάμβηκ' [Soph. *Ant.* 1246]).

68 The root *θηπ-/*θαπ- must be secondary because voiced aspirates and voiceless stops are never found in the same Indo-European root. Olav Hackstein (2002, 237–8), in addition to collecting earlier treatments, has explained the missing nasal and lengthened vowel as a secondary lengthening process of *θαπ- (underlying θάμβ-) > θηπ-. In addition to the somewhat sporadic nature of this remodeling, what makes this account uncertain is the Indo-European evidence for a totally separate root *d^{h}eH$_2$b^h-, on which see n.61 *supra.*

69 On the other hand, the root *d^{h}eH$_2$b^h- contains no nasal. If it were inherited into Greek, it would realize as *θηφ-/θαφ- (Grassmann's law would yield *τηφ-/ταφ-.). The remodeling of *θηφ-/θαφ- to *θηπ-/*θαπ- potentially proceeds from the *nomina agentis* root-noun θώψ (<*d^{h}óH$_2$b^h-s), θωπός, ὁ, "flatterer (← he who strikes)." On the "deceitful" semantics of this root and their potential relation to verbs of physical force, note that Eng. *deceive* < MEng. *deceyven* < OFr. *decevoir* < L. *dēcipere* "to catch, deceive." Evidently from θώψ, θωπός, ὁ, speakers extracted the root θωπ- as indicated transparently by θώπτω "to flatter" (Aesch. *Prom. Vinc.* 937). For the root shape, cf. κλώψ "thief" ~ κλέπτω "to steal." Less believably, if an early (pre-Grassmann's) *-ye/o-* present were made to the zero-grade of *d^{h}eH$_2$b^h- (full-grade found in Lith *dóbti* "to beat," Latv. *dâbt* "to strike"), it would yield *θάπτω, from which speakers could easily extract a secondary root *θηπ-/*θαπ-. Whether this hypothetical formant is related to θάπτω "to bury" is too complicated to address here, but the connection between "amazement" and "burial" was folk etymologized in antiquity (see e.g. the *scholia vetera* ad *Od.* 16.12).

70 Related forms such as those found in the Hesychian glosses Θ510 (θήπει· ψεύδεται), 513 (θήπω· ἐπιθυμῶ. θαυμάζω), and 514 (θήπων· ἐξαπατῶν, κολακεύων. θαυμάζων.), which attest thematic present forms of a verb θήπω, potentially indicate that Greek speakers extracted the full verbal ablaut θωπ-/θηπ-/θαπ- appropriate to a root containing the second laryngeal (for the zero-grade, see Hesych. Θ101 θάπαν· φόβον; Θ102 θάπτρα· μνῆμα. Κρῆτες). West's edition (*IEG*) of Hippon. 12.1 emends the form to an epsilon-contract verb on the authority of Tzetzes *Posthomerica* 687: "θηπῶν" θήπεον ἐθαύμαζον. τὸ θέμα θήπω. καὶ Ἱππῶναξ. I doubt the antiquity of Tzetzes' verb, remaining agnostic on its authenticity; an epsilon-contract verb that has a full-grade of the root and is transitive must be deadjectival/denominative (μετρέω ~ μέτρον). While the Hesychian gloss Θ512 does give evidence for a full-grade primary thematic adjective θηπόν· καταθύμιον. θαυμαστόν, the semantics of a derived verb θηπέω do not easily yield "to deceive/flatter."

71 Two inherited roots of similar shape, one meaning "be stunned" (*d^{h}emb^h-) and another meaning "to strike" (*d^{h}eH$_2$b^h-), could easily merge within Greek and form a suppletive paradigm. The question remains how a root *d^{h}eH$_2$b^h- "to strike" that curiously shows both transitive ("to flatter") and intransitive ("to be amazed") semantics in Greek is found only as a stative-intransitive in Homer (e.g. τέθηπα "I am amazed"). A potential explanation is that if *d^{h}eH$_2$b^h- were to make any zero-grade verbal formation, for example a thematic aorist *d^{h}H$_2$b^h-om, it would yield *τάφον "I struck," precisely the same phonological outcome of a zero-grade form of the stative-intransitive *d^{h}emb^h-, as in the aorist τάφον (<*d^{h}m̥b^h-om) "I was struck." This point of overlap left Greek speakers with a choice: to generalize either the transitive or intransitive semantics throughout forms that synchronically appeared to "share" an aorist. At this point, it was a coin-toss, and speakers generalized the intransitive semantics, effectively causing the perfects from *d^{h}eH$_2$b^h- "to strike" to act in morphological suppletion to the present and aorist forms of *d^{h}emb^h- "to be stunned." For suppletion in Greek, see Kölligan 2007.

72 See *LIV*^{2} 601. See also *LIV*^{2} 148 s.v. *d^{h}eug^h- "to strike."

73 On the basis of the transitivity of the formations, perhaps not to be connected with PIE *stemb^{h}H$_x$- (*LIV*^{2} 595).

74 On these and associated Sanskrit terms see *KEWA* I 459–61, cf. Bhavabhūti, *Mālatīmādhava* 5.13.

75 See Kövecses 2000, 61–86.

76 Another way of conceptualizing this onomatopoeia would be to do away with the metonymic shift from the pure onomatopoeia to the "strike" or "blow" which causes the sound. Instead, θάμβος would simply represent a sonic representation of the heartbeat and, instead of metaphorically representing a "strike" upon the body, would metonymically represent surprise via the beating (or thumping) of one's heart during the cardiac defense. Under this analysis, θάμβος and ἔκπληξις would metonymically represent surprise via two different symptoms, the former through the embodied thump of the cardiac defense, and the latter through the "strike" of the startle reflex.

References

Aziz-Zadeh, L. and V. Gamez-Djokic. 2016. "Comment: The Interaction Between Metaphor and Emotion Processing in the Brain." *Emotion Review* 8(3), 275–6.

Baron-Cohen, S., A. Spitz, and P. Cross. 1993. "Do Children with Autism Recognise Surprise?" *Cognition and Emotion* 7(6), 507–16.

Barrett, L.F. 2006. "Are Emotions Natural Kinds?" *Perspectives on Psychological Science* 1, 28–58. doi:10.1111/j.1745-6916.2006.00003.x.

Barrett, L.F. 2016. "Navigating the Science of Emotion." In *Emotion Measurement*, H.L. Meisselman (ed.). Boston, MA: Elsevier, 31–63.

Barrett, L.F. and M. Gendron. 2016. "The Importance of Context: Three Corrections to Cordaro, Keltner, Tshering, Wangchuk, and Flynn (2016)." *Emotion* 16(6), 803–6.

Barrett, L.F. and J. Russell (eds.). 2015. *The Psychological Construction of Emotion*. New York, NY: Guilford.

Barton, C.R. 1993. "Greek τέθηπα etc." *Glotta* 71, 1–9.

Beekes, R.S.P. 2010. *Etymological Dictionary of Greek*. 2 vols. Leiden: Brill.

Bokor, Z. 1997. *Body-Based Constructionism in the Conceptualization of Anger* (C.L.E.A.R. Series, No. 17). Budapest: Department of English, Hamburg University and the Department of American Studies, ELTE.

Buck, C.D. 1949. *A Dictionary of Selected Synonyms in the Principal Indo-European Languages*. Chicago, IL: University of Chicago Press.

Cairns, D.L. 2003a. "Ethics, Ethology, Terminology: Iliadic Anger and the Cross-Cultural Study of Emotion." In *Ancient Anger: Perspectives from Homer to Galen* (Yale Classical Studies 32), S.M. Braund and G.W. Most (eds.). Cambridge: Cambridge University Press, 11–49.

Cairns, D.L. 2003b. "Review: Myths and Metaphors of Mind and Mortality." *Hermathena* 175, 41–75.

Cairns, D.L. 2008. "Look Both Ways: Studying Emotion in Ancient Greek." *Critical Quarterly* 50(4), 43–62.

Cairns, D.L. 2015. "The Horror and the Pity: *Phrike* as a Tragic Emotion." *Psychoanalytic Inquiry* 34, 75–94.

Cairns, D.L. 2016a. "Clothed in Shamelessness, Shrouded in Grief: The Role of 'Garment' Metaphors in Ancient Greek Concepts of Emotion." In *Spinning Fates and the Song of the Loom*, G. Fanfani, M. Harlow, and M.-L. Nosch (eds.). Oxford; Philadelphia, PA: Oxbow Books, 25–41.

Cairns, D.L. 2016b. "Mind, Body, and Metaphor in Ancient Greek Concepts of Emotion." *L'Atelier du Centre de recherches historiques* [En ligne], 16, 2016, mis en ligne le 26 mai 2016, consulté le 08 juin 2016. URL: http://acrh.revues.org/7416. DOI: 10.4000/acrh.7416.

Cairns, D.L. 2016c. "Metaphors for Hope in Archaic and Classical Greek Poetry." In *Hope, Joy, and Affection in the Classical World*, R. Caston and R. Kaster (eds.). Oxford: Oxford University Press, 13–44.

Cairns, D.L. and D.P. Nelis (eds.). 2017. *Emotions in the Classical World: Methods, Approaches, and Directions*. Stuttgart: Franz Steiner Verlag.

Cairns, D.L. and L. Fulkerson (eds.). 2015. *Emotions between Greece and Rome*. London: ICS.

Carter, J.A., E.C. Gordon, and S.O. Palermos. 2016. "Extended Emotion." *Philosophical Psychology* 29(2), 197–218.

Caston, R. and R. Kaster (eds.). 2016. *Hope, Joy, and Affection in the Classical World: Emotions of the Past*. Oxford; New York: Oxford University Press.

Chantraine, P. 1999. *Dictionnaire Étymologique de la Langue Grecque*. New edition with a supplement by A. Blanc, C. de Lamberterie, and J.-L. Perpillou. Paris: Klinksieck.
Clark, A. 2008. *Supersizing the Mind: Embodiment, Action, and Cognitive Extension*. Oxford; New York: Oxford University Press.
Clark, A. 2016a. "Busting Out: Predictive Brains, Embodied Minds, and the Puzzle of the Evidentiary Veil." *Nous* [early version], 1–27. doi: 10.1111/nous.12140.
Clark, A. 2016b. *Surfing Uncertainty*. Oxford; New York: Oxford University Press.
Clarke, M. 1999. *Flesh and Spirit in the Songs of Homer: A Study of Words and Myths*. Oxford: Clarendon Press.
Colombetti, G. 2014. *The Feeling Body: Affective Science Meets the Enactive Mind*. Cambridge, MA; London: MIT Press.
Colombetti, G. 2015. "Enactive Affectivity, Extended." *Topoi*, 1–11.
Colombetti, G. and T. Roberts. 2014. "Extending the Extended Mind: The Case for Affectivity." *Philosophical Studies*, 1–21. doi: 10.1007/s11098-014–0347-3.
Coray, M. 2009. *Homers Ilias: Gesamtkommentar*, vi.ii. Berlin: De Gruyter.
Dastur, F. 2000. "Phenomenology of the Event: Waiting and Surprise." *Hypatia* 15(4), 178–89.
Depraz, N. 2014. "The Surprise of Non-Sense." In *Enactive Cognition at the Edge of Sense-Making: Making Sense of Non-Sense*, M. Cappuccio and T. Froese (eds.). London: Palgrave Macmillan, 125–52.
Descartes, R. 1649/1985. "Passions of the Soul." In *The Philosophical Writings of Descartes: Volume I*, J. Cottingham, R. Stoothoff, and D. Murdoch (trans.). Cambridge: Cambridge University Press, 325–404.
Desmidt T., M. Lemoine, C. Belzung, and N. Depraz. 2014. "The Temporal Dynamic of Emotional Emergence." *Phenomenology and the Cognitive Sciences* 13, 557–78.
Devereaux, J.J. 2016. "Embodied Historiography: Models for Reasoning in Tacitus's *Annales*." In *Embodiment in Latin Semantics*, W.M. Short (ed.). Amsterdam; New York: John Benjamins, 237–68.
DGE = *Diccionario Griego-Español*, accessible at http://dge.cchs.csic.es/xdge.
Díaz Vera, J.E. 2011. "Reconstructing the Old English Cultural Model for Fear." *ATLANTIS. Journal of the Spanish Association of Anglo-American Studies* 33(1), 85–103.
Eickers, G., J.R. Loaiza, and J. Prinz. 2017. "Embodiment, Context-Sensitivity, and Discrete Emotions: A Response to Moors." *Psychological Inquiry* 28(1), 31–8.
Ekman, P. 1992. "An Argument for Basic Emotions." *Cognition and Emotion* 6, 169–200.
Elster, J. 1998. "Emotions and Economic Theory," *Journal of Economic Literature* 36(2), 47–74.
EWA = Mayrhofer, M. 1992–2001. *Etymologisches Wörterbuch des Altindoarischen*. 3 vols. Heidelberg: Winter Verlag.
Fabry, R.E. 2017. "Transcending the Evidentiary Boundary: Prediction Error Minimization, Embodied Interaction, and Explanatory Pluralism." *Philosophical Psychology*. doi: 10.1080/09515089.2016.1272674.
Frisk, H. 1960–72. *Griechisches etymologisches Wörterbuch*. 3 vols. Heidelberg: Winter Verlag.
Fulkerson, L. 2013. *No Regrets: Remorse in Classical Antiquity*. Oxford; New York: Oxford University Press.
Gallagher, S. 2017. *Enactivist Interventions: Rethinking the Mind*. Oxford; New York: Oxford University Press.
Geeraerts, D. 2006. "Introduction: A Rough Guide to Cognitive Linguistics." In *Cognitive Linguistics: Basic Readings*, D. Geeraerts (ed.). Berlin; New York: Mouton de Gruyter, 1–28.
Geeraerts, D. and S. Grondelaers. 1995. "Looking Back at Anger: Cultural Traditions and Metaphorical Patterns." In *Language and the Cognitive Construal of the World*, J. Taylor and R. MacLaury (eds.). Berlin: de Gruyter, 153–79.
Gendron, M., D. Roberson, and L.F. Barrett 2015. "Cultural Variation in Emotion Perception is Real: A Response to Sauter, Eisner, Ekman, and Scott (2015)." *Psychological Science* 26, 357–9. http://dx.doi.org/10.1177/0956797614566659.
Gevaert, C. 2001. "Anger in Old and Middle English: A 'Hot' Topic?" *Belgian Essays on Language and Literature*, 89–101.
Gevaert, C. 2005. "The ANGER IS HEAT Question: Detecting Cultural Influence on the Conceptualization of Anger through Diachronic Corpus Analysis." In *Perspectives on Variation: Sociolinguistic, Historical, Comparative*, N. Delbecque, J. van der Auwera, and D. Geeraerts (eds.). Berlin; New York: Mouton de Gruyter, 195–208.
Gibbs, R.W. 1992. "What Do Idioms Really Mean?" *Journal of Memory and Language* 31(4), 485–506.
Gibbs, R.W. 1994. *The Poetics of Mind: Figurative Thought, Language, and Understanding*. Cambridge; New York: Cambridge University Press.
Hackstein, O. 2002. *Die Sprachform der homerischen Epen*. Wiesbaden: Dr. Ludwig Reichert Verlag.

Heidermanns, F. 1993. *Etymologisches Wörterbuch der germanischen Primäradjektive*. Berlin; New York. Walter de Gruyter.

Holmes, B. 2010. *The Symptom and the Subject: The Emergence of the Physical Body in Ancient Greece*. Princeton, NJ: Princeton University Press.

Holmes, B. 2017. "The Body of Western Embodiment: Classical Antiquity and the Early History of a Problem." In *Embodiment: A History*, J.E.H. Smith (ed.). New York: Oxford University Press, 17–49.

Hutto, D.D. and M.D. Kirchhoff. 2016. "Never Mind the Gap: Neurophenomenology, Radical Enactivism, and the Hard Problem of Consciousness." *Constructivist Foundations* 11(2), 346–53.

IEG = West, M.L. 1989–92. *Iambi et Elegi Graeci*. Oxford; New York: Oxford University Press.

IEW = Pokorny, J. 1959–69. *Indogermanisches etymologisches Wörterbuch*. Berne; Munich: Francke Verlag.

Insler, S. 1969. "Vedic *dambháyati*." *Indogermanische Forschungen* 74, 11–31.

Kahneman, D. and D.T. Miller. 1986. "Norm Theory: Comparing Reality with Its Alternatives." *Psychological Review* 93(2), 136–53.

Kaster, R. 2005. *Emotion, Restraint, and Community in Ancient Rome*. Oxford; New York: Oxford University Press.

KEWA = Mayrhofer, M. 1956–80. *Kurzgefaßtes etymologisches Wörterbuch des Altindischen*. 4 vols. Heidelberg: Winter Verlag.

Kirchhoff, M.D. 2017. "Predictive Processing, Perceiving and Imagining: Is to Perceive to Imagine, or Something Close to it?" *Philosophical Studies*. doi: 10.1007/s11098-017-0891-8.

Kölligan, D. 2007. *Suppletion und Defektivität im griechischen Verbum*. Bremen: Hempen Verlag.

Konstan, D. 2008. "In Defense of Croesus, or Suspense as an Aesthetic Emotion." *Aisthe* 3, 1–15.

Konstan, D. 2010. *Before Forgiveness: The Origins of a Moral Idea*. Cambridge; New York: Cambridge University Press.

Kövecses, Z. 1986. *Metaphors of Anger, Pride, and Love: A Lexical Approach to the Study of Concepts*. Amsterdam: Benjamins.

Kövecses, Z. 1990. *Emotion Concepts*. Berlin; New York: Springer-Verlag.

Kövecses, Z. 2000. *Metaphor and Emotion*. Cambridge: Cambridge University Press.

Kövecses, Z. 2008. "Metaphor and Emotion." In *The Cambridge Handbook of Metaphor and Thought*, R.W. Gibbs (ed.). New York: Cambridge University Press, 380–96.

Kövecses, Z. 2015a. "Surprise as an Emotional Category." In *Expressing and Describing Surprise*, A. Celle and L. Lansari (eds.). [*Review of Cognitive Linguistics* 13:2], 270–90.

Kövecses, Z. 2015b. *Where Metaphors Come From: Reconsidering Context in Metaphor*. Oxford; New York: Oxford University Press.

Krawczak, K. and D. Glynn. 2015. "Operationalizing Mirativity: A Usage-Based Quantitative Study of Constructional Construal in English." In *Expressing and Describing Surprise*, A. Celle and L. Lansari (eds.). [*Review of Cognitive Linguistics* 13:2], 353–82.

Kroonen, G. 2013. *Etymological Dictionary of Proto-Germanic*. Leiden; Boston, MA: Brill.

Krueger, J. 2014. "Varieties of Extended Emotions." *Phenomenology and the Cognitive Sciences* 13, 533–55.

Krueger, J. and T. Szanto. 2016. "Extended Emotions." *Philosophical Compass* 11(12), 863–78. doi: 10.1111/phc3.12390.

Kümmel, M.J. 2012. "The Distribution of IE Roots Ending in IE *ND." In *The Sound of Indo-European 2. Papers on Indo-European Phonetics, Phonemics and Morphophonemics*, R. Sukač and O. Šefčík (eds.). Munich: LINCOM Europa, 159–76.

Laborde S. and E. Mosley. 2016. "Commentary: Heart Rate Variability and Self-Control: A Meta-Analysis." *Frontiers in Psychology* 7.653. doi: 10.3389/fpsyg.2016.00653.

Lakoff, G. 1987. *Women, Fire, and Dangerous Things: What Categories Reveal about the Mind*. Chicago, IL: University of Chicago Press.

Lakoff, G. 2016. "Language and Emotion." *Emotion Review* 8(3), 269–73.

Lakoff, G. and M. Johnson. [1980] 2003. *Metaphors We Live By*. Updated edition with afterword. Chicago, IL: University of Chicago Press.

Lakoff, G. and Z. Kövecses. 1987. "The Cognitive Model of Anger Inherent in American English." In *Cultural Models in Language and Thought*, D. Holland and N. Quinn (eds.). Cambridge: Cambridge University Press, 195–221.

Latacz, J.R. Nünlist and M. Stoevesandt. 2000. *Homers Ilias: Gesamtkommentar*, i. ii. Munich: Saur.

LeDoux, J.E. 2015. "Emotional Construction in the Brain." In *The Psychological Construction of Emotion*, L.F. Barrett and J.A. Russell (eds.). New York: Guilford, 459–63.

LfgrE = Snell, B., H.J. Mette, M. Meier-Brügger, et al. 1979–2010. *Lexikon des frühgriechischen Epos*. 25 fascicles. Gottingen: Vandenhoeck & Ruprecht.

Malzahn, M. 2007. *The Tocharian Verbal System*. Leiden: Brill.

Marincola, J. 2003. "Beyond Pity and Fear: The Emotions of History." *Ancient Society* 33, 285–315.

Matsuki, K. 1995. "Metaphors of Anger in Japanese." In *Language and the Cognitive Construal of the World*, J.R. Taylor and R. MacLaury (eds.). Berlin: Mouton, 137–51.

Mellers, B.A., A. Schwartz, K. Ho, and I. Ritov. 1997. "Decision Affect Theory: Emotional Reactions to the Outcomes of Risky Options." *Psychological Science* 8(6), 423–9.

Melzig, C.A., A.I. Weike, A.O. Hamm, and J.F. Thayer. 2008. "Individual Differences in Fear-Potentiated Startle as a Function of Resting Heart Rate Variability: Implications for Panic Disorder." *International Journal of Psychophysiology*. doi:10.1016/j.ijpsycho.2008.07.013.

Micholajczuk, A. 1998. "The Metonymic and Metaphoric Conceptualization of Anger in Polish." In *Speaking of Emotions: Conceptualization and Expression*, A. Athanasiadou and E. Tabakowska (eds.). Berlin: Mouton, 153–91.

Narten, J. 1988–90. "Die vedischen Verbalwurzeln *dambh* und *dabh*." *Die Sprache* 34, 142–57 (= *Kl. Schr.* 380–95).

Neer, R. 2010. *The Emergence of the Classical Style in Greek Sculpture*. Chicago, IL: University of Chicago Press.

Nieto Hernández, P. 2011. "Metaphor." In *Homer Encyclopedia (3 Vol.)*, M. Finkelberg (ed.). Chichester; Malden, MA: Wiley-Blackwell.

Noordewier M.K., S. Topolinksi, and E. Van Dijk. 2016. "The Temporal Dynamics of Surprise." *Social and Personality Psychology Compass* 10.3, 136–49, 10.1111/spc3.12242.

Noordewier, M.K. and S.M. Breugelmans. 2013. "On the Valence of Surprise." *Cognition and Emotion* 27, 1326–34.

Oliver, R.L., R.T. Rust, and S. Varki. 1997. "Customer Delight: Foundations, Findings, and Managerial Insight." *Journal of Retailing* 73(3), 311–36.

Pessoa, L. 2015. "Précis on *The Cognitive-Emotional Brain*." *Behavioral and Brain Sciences* 38. doi: 10.1017/S0140525X14000120.

Plamper, J. 2015. *The History of Emotions: An Introduction*. K. Tribe, trans. Oxford; New York: Oxford University Press.

Plutchik, R. 2003. *Emotions and Life: Perspectives from Psychology, Biology, and Evolution*. Washington, DC: American Psychological Association.

Ponsonnet, M. 2013. *The Language of Emotions in Dalabon (Northern Australia)*. A thesis submitted for the degree of Doctor of Philosophy of the Australian National University.

Porter, J.I. 2016. *The Sublime in Antiquity*. Cambridge; New York: Cambridge University Press.

Prier, R.A. 1989. *Thauma Idesthai: The Phenomenology of Sight and Appearance in Archaic Greek*. Tallahassee, FL: FSU Press.

Rau, J. 2009. *Indo-European Nominal Morphology: The Decads and the Caland System*. Innsbruck: IBS.

Reisenzein, R., S. Bördgen, T. Holtbernd, and D. Matz. 2006. "Evidence for Strong Dissociation between Emotion and Facial Displays: The Case of Surprise." *Journal of Personality and Social Psychology* 91(2), 295–315.

Richter, S., A. Schulz, J. Port, T.D. Blumenthal, and H. Schächinger. 2009. "Cardiopulmonary Baroreceptors Affect Reflexive Startle Eye Blink." *Physiology & Behavior* 98(5), 587–93.

Riggsby, A.M. 2015. "Tyrants, Fire, and Dangerous Things." In *Roman Reflections*, G. Williams and K. Volk (eds.). Oxford; New York: Oxford University Press, 111–28.

Rix, H. 2001. *Lexikon der indogermanischen Verben*. 2nd edition. Wiesbaden: Reichert.

Roberts, T. 2015. "Extending Emotional Consciousness." *Journal of Consciousness Studies* 22(3–4), 108–28.

Scarantino, A. 2015. "Basic Emotions, Psychological Construction, and the Problem of Variability." In *The Psychological Construction of Emotion*, L.F. Barrett and J.A. Russell (eds.). New York, NY: Guilford, 334–98.

Scherer, K. 2015. "The Component Process Model of Emotion, and the Power of Coincidences." Interview by Andrea Scarantino. *Emotion Researcher, ISRE's Sourcebook for Research on Emotion and Affect*, Andrea Scarantino (ed.), http://emotionresearcher.com/the-component-process-model-of-emotion-and-the-power-of-coincidences. Web. 20 Oct. 2016.

Scheve, C. von and M. Salmela (eds.). 2014. *Collective Emotions*. Oxford; New York: Oxford University Press.

Slaby, J. 2014. "Emotions and the Extended Mind." In *Collective Emotions*, C. von Scheve and M. Salmela (eds.). Oxford; New York: Oxford University Press, 32–46.

Snell, B. 1975. "Die Auffassung des Menschen bei Homer." In *Die Entdeckung des Geistes. Studien zur Entstehung des europäischen Denkens bei den Griechen*. Gottingen: Vandenhoeck & Ruprecht, 13–29. (Originally: Snell, B. 1939. "Die Sprache Homers als Ausdruck seiner Gedankenwelt." *NJAB* 2, 393–410.)

Soriano, C., J.R.J. Fontaine, and K.R. Scherer. 2015. "Surprise in the GRID." In *Expressing and Describing Surprise*, A. Celle and L. Lansari (eds.). *Review of Cognitive Linguistics* 13(2), 436–60.

Steinbock, A.J. 2017. "La sorpresa como emoción: Entre el sobresalto y la humildad." *Acta Mexicana de Fenomenología* 2, 13–30.

Steinbock, A.J. and N. Depraz (eds.). Forthcoming. *Surprise: An Emotion*. Dordrecht: Springer.

Szemerényi, O. 1954. "Greek τάφων - θάμβος - θεάομαι." *Glotta* 33, 238–66.

Taylor, J. and T. Mbense. 1998. "Red Dogs and Rotten Mealies: How Zulus Talk about Anger." In *Speaking of Emotions: Conceptualization and Expression*, A. Athanasiadou and E. Tabakowska (eds.). Berlin: Mouton, 191–226.

Thelen, E. and L.B. Smith. 1994. *A Dynamic Systems Approach to the Development of Cognition and Action*. Cambridge, MA: MIT Press.

Touroutoglou, A., K.A. Lindquist, B.C. Dickerson, and L.F. Barrett. 2015. "Intrinsic Connectivity in the Human Brain Does Not Reveal Networks for 'Basic' Emotions." *Scan* 10, 1257–65. doi: 10.1093/scan/nsv013.

Tschacher, W. and J.-P. Dauwalder (eds.). 2003. *Dynamical Systems Approaches to Embodied Cognition*. Singapore: World Scientific.

Tucker, E. 1990. *The Creation of Morphological Regularity: Early Greek Verbs in -éō, -àō, -óō, -úō and -íō*. Goettingen: Vandenhoeck & Ruprecht.

Valenzuela, A., J. Strebel, and B. Mellers. 2010. "Pleasurable Surprises: A Cross-Cultural Study of Consumer Responses to Unexpected Incentives." *Journal of Consumer Research* 36, 792–805.

Vila, J., P. Guerra, M.A. Muñoz, C. Vico, M. I. Viedma-del Jesús, L.C. Delgado, and S. Rodríguez. 2007. "Cardiac Defense: From Attention to Action." *International Journal of Psychophysiology* 66(3), 169–82.

Walsh, T.R. 2005. *Fighting Words and Feuding Words: Anger and the Homeric Poems*. Lanham, MD: Lexington Books.

Wassmann, C. 2016. "Forgotten Origins, Occluded Meanings: Translation of Emotion Terms." *Emotion Review* 1, 1–9. doi: 10.1177/1754073916632879.

Werner, G. 2011. "Viewing the Extended Mind Hypothesis (Clark & Chalmers) in Terms of Complex System Dynamics." In *Decision Making: A Psychophysics Application of Network Science*, P. Grigolini and B.J. West (eds.). Singapore: World Scientific, 21–38.

Wessel, J.R. 2017. "Perceptual Surprise Aids Inhibitory Motor Control." *Journal of Experimental Psychology: Human Perception and Performance* 43(9), 1585–93.

Yu, N. 1995. "Metaphorical Expression of Anger and Happiness in English and Chinese." *Metaphor and Symbolic Activity* 10, 223–45.

3

Construal and immersion

A cognitive linguistic approach to Homeric immersivity

Rutger J. Allan

1. Introduction: Homer and immersion[1]

> All great poetry at its best transports us to the realms of gold. If for the moment we can put reason in abeyance, we are 'enthralled'. The spell of poetry can make the hearer forget both himself and the poet and the real world about him. It can banish all awareness that an image of life is being presented, because of its magic power to make the image seem the only reality. The spell is the poetic illusion.
>
> *(Bassett 1938: 26)*

> The story [in Homer] seems almost to tell itself. The words which transport us to the world of the heroes come from a source so submerged from view that the heroic life seems to move of its own vitality.
>
> *(Bassett 1938: 27)*

With these words Samuel Bassett described the well-known effect of Homer's style to make the listener feel as if mentally present in the world of the story. This famous quality of the Homeric language has been praised by ancient as well as modern literary critics, and it is often associated with the term *enárgeia*, 'the power of bringing the things that are said before the senses of the audience', as Dionysius of Halicarnassus defines it.[2] In this chapter, I will approach this Homeric quality through the modern concept of *immersion*, a concept that has emerged in the study of virtual reality but has also found its way into cognitive approaches to art, drama, and film studies, literary studies, stylistics, and linguistics.[3] Immersion is the feeling of being transported to a virtual world to the extent that one experiences it - up to a point - as if it were the actual world.

The concept of immersion was introduced into the field of literary studies by the cognitive narratologist Marie-Laure Ryan, who defines immersion as 'the experience through which a fictional world acquires the presence of an autonomous, language-independent reality populated by live human beings' (Ryan 2015: 9). 'Immersion is a corporeal experience . . . it takes the projection of a virtual body . . .to feel integrated in an art world' (Ryan 2015: 13). In her book of 2001, of which an updated edition appeared in 2015, Ryan identifies four aspects of the 'Poetics of Immersion': *spatial* immersion, a narrative's ability to immerse readers in a sense of place; *temporal* immersion, the story's creation of interest (suspense, curiosity, surprise) through the dynamics of the temporal unfolding of the told events; a combined *spatio-temporal* immersion, the reader's imaginative transportation ('recentering') into the story world; and *emotional* immersion, which relates to the emotional involvement (such as empathy) with the fate of the characters.

Ryan also discusses a number of narrative strategies facilitating the experience of being immersed into the storyworld. In order to give a narrative an immersive quality, a narrator will tend to opt for the following narrative devices rather than their counterparts: (1) *scene* narration, rather than summary narration, (2) *internal and variable focalization* (representing characters as subjects), rather than external focalization (looking at characters as objects), (3) *dialogue* (direct discourse) and *free indirect discourse*, rather than indirect discourse, (4) *prospective first-person* narration, rather than third-person retrospective narration, (5) a totally *effaced narrator*, rather than a visible narrator, and (6) *mimesis* ('showing'), rather than diegesis ('telling'). Ryan also discusses a number of specific linguistic devices, such as adverbial deictic shift, speech and thought report, and (present) tense-marking, which serve to shift the deictic center toward the storyworld and assign it to the perspective of a character.[4]

Immersion should be thought of as a gradable phenomenon: the experience of being immersed can be more or less intense, depending on the presence of particular immersive features in the text or text segment: the intensity of the immersive experience may also vary through time while reading or hearing a text. The degree of immersion will be dependent on the number and diversity of textual features conducive to immersion (and the absence of features detrimental to immersion). Ryan distinguishes four degrees of immersion in the act of reading. From a low to high degree of readerly absorption these are: concentration, imaginative involvement, entrancement, and addiction (Ryan 2015: 68–69).

Often metaphors come into play to describe the immersive experience (the term 'immersion' itself being an obvious example). A well-known metaphor is that of a reader being 'transported' or 'traveling' to the world of the story, a metaphor also used by Bassett in the citation above. This metaphor has been the starting-point for the psychologists Victor Nell (1988) and Richard Gerrig (1993) in their pioneering work on the psychology of the immersed reader. Other common metaphors that appear in connection with narrative immersion involve a reader being 'entranced' (cf. also Bassett's use of the words 'enthralled', 'magic power', 'spell'), 'absorbed', 'caught up' in a story, or being 'lost in a book'.

The immersive effect of literary texts is, of course, not a novel phenomenon: it has already been described by the ancient literary critics, in connection with such notions as *enárgeia*, *enagṓnios*, and *ékstasis*. *Enárgeia* is 'the power of bringing the things that are said before the senses of the audience', thereby creating the illusion of actually perceiving the objects and events described by the text (cf. Ryan's spatial immersion). It is achieved, according to the ancient rhetoricians and commentators, by detailed sensory descriptions, by presenting character speech, and by turning the listener into a virtual eyewitness. As Ruth Webb (among others) has pointed out, *enárgeia* also has an emotional component (cf. Ryan's emotional immersion): 'Inseparable from this representational and informative function of *enárgeia* is its ability to move the audience and to make them feel the emotions appropriate to the events described' (Webb 2009: 90).[5] A narrative is *enagṓnios* if it is 'vivid', 'actively involving', 'engaging', and 'full of suspense' (Ooms and De Jonge 2013). According to Longinus (25–27), it can be achieved by using devices such as the historical present tense or by a switch to direct speech, and it gives the listener the feeling of being himself in the middle of danger. In other words, the listener is mentally transported to the narrated scene. *Ékstasis* ('movement outward', 'displacement') is the mental state, produced by sublime literary texts, of being out of one's senses and no longer oneself ('out of oneself').[6]

The Homeric narrator also seems already well aware of the immersive power of narrative, using the metaphor of enchantment – which since then has become very prolific – to describe the effect of story-telling involving a loss of self-control and self-consciousness. The effect of Odysseus' tale is that his audience is 'spellbound' (κηληθμῷ δ' ἔσχοντο, *Od.* 11.334, 13.2) and Odysseus, who is compared to an epic singer (17.518–521), is able to enchant the listener with his tale (θέλγοιτό κέ τοι φίλον ἦτορ, 17.514; ὣς ἐμὲ κεῖνος ἔθελγε, 17.521).

It is clear that the Greeks were no stranger to the phenomenon that literary texts are capable of making the audience feel as if present at the scene, and they describe this phenomenon in terms that are very similar to the modern concept of immersion. It is therefore worthwhile to see whether the modern concept of immersion can contribute to a better understanding of how texts are able to bring about this particular effect. Using the theory of immersion as a framework for the analysis of Ancient Greek (and, more specifically, Homeric) narrative may offer a number of benefits. First, since the notion of immersion is firmly grounded both in cognitive linguistics and in cognitive narratology, it opens up a large and diverse arsenal of well-established linguistic and narratological notions that can be brought to the analytical table. Linguistic categories that are relevant to the immersive qualities of a text are: tense-aspect, modality, deixis, and cognitive schemas/frames. Narratological concepts important to immersion include speed, order, focalization, narratorial visibility, suspense, and genre conventions. A second attractive feature of immersion is that it is also the subject of experimental psychological research, which offers insights into how immersion is rooted in our general cognitive, emotive, and sensorimotor capacities.[7]

Elsewhere I have proposed an inventory of linguistic and narratological features of texts that are conducive to the reader's experience of immersion and I have discussed a number of issues relevant to immersion, such as the persuasive power of immersion, the relationship between immersion and experientiality, and the difference in immersivity between Herodotus' and Thucydides' narrative styles.[8] In this chapter, the focus will be on the underlying cognitive phenomena on which the experience of narrative immersion hinges. Central to my approach to immersion will be *construal* and *embodiment*, which I will discuss in sections (2) and (3), respectively. In section (4), I will present an analysis of *Odyssey* 5.59–73 (Calypso's Cave) and examine, from a cognitive linguistic point of view, which specific linguistic and narratological features of the text contribute to the listener's feeling of being immersed in the described scene.

2. Construal

A central idea in cognitive linguistics is *construal*, 'our ability to conceive and portray the same situation in alternate ways' (Langacker 2015: 120). The idea of construal can be illustrated by comparing it to vision. Since vision and conceptualization show a number of important parallels, vision can be seen as a subtype of conceptualization.[9] When we look at a scene or object in the world around us, what we actually see is crucially determined by the distance and the angle from which we view it, at what part of it we choose to direct our gaze, and what elements we pay most attention to. In the same way, a linguistic expression construes a described situation from a certain distance and a specific perspective, and it selects and focuses on some elements of the situation while backgrounding others. Construal is not only fundamental to lexical semantics, it also pervades grammatical structure: each lexical as well as grammatical item in a language is associated with a particular way of construing ('viewing') its conceptual content as part of its conventional semantics.

In cognitive linguistics, grammar is not seen as an abstract formal system that is separated from other domains of human cognition and life. Instead, cognitive linguistics sees grammar as meaningful: grammatical categories such as tense and aspect are devices employed by speakers to conceptualize ('view') the world in a certain way; grammatical constructions are used to combine component concepts into more complex and elaborate conceptualizations. In cognitive linguistics, in other words, grammar is 'an essential aspect of the conceptual apparatus through which we apprehend and engage the world' (Langacker 2008: 4).

A number of more specific processes of conceptualization, or *construal operations*, have been distinguished, such as *specificity*, *salience*, *perspective*, *dynamicity*, and *attention focus*. These construal operations are fundamental dimensions of linguistic conceptualization in general and they are therefore also helpful in understanding through which linguistic means texts are able to generate an immersive experience.[10]

(i) *Specificity* and *salience*. The construal operation of *specificity* concerns the level of precision and granularity with which a situation is portrayed. Expressions may vary in their degree of specificity. An example given by Langacker (2008: 56) is the following scale of specificity ranging from more schematic to more specific:

Something happened. →

A person perceived a rodent. →

A girl saw a porcupine. →

An alert little girl wearing glasses caught a brief glimpse of a ferocious porcupine with sharp quills.

The more specific and concrete an expression is, the more immersive it will be. A text that describes a scene in highly fine-grained (perceptual) detail will be more effectively able to tap into the rich experiential (sensorimotor and emotional) resources stored in the listener's or reader's semantic memory (cognitive schemata and scripts) and it will thus evoke a more intense mental simulation of the described scene. Immersive texts will therefore tend to focus on psychologically salient entities: concrete, physical, preferably moving, visible, or otherwise perceivable animate beings or objects. I will return to the topic of mental simulation in section (3).

(ii) *Perspective*. An important aspect of perspective has to do with the relationship between the subject of conceptualization ('viewer') and the object of conceptualization ('viewed'). When a listener or reader is immersed in the world of the story (cf. the arrow in Figure 3.1 below), the relationship between the listener or reader (conceptualizing subject) and the entities and events in the 'onstage' storyworld (object of conceptualization) is maximally asymmetric; that is to say, being completely absorbed by the story, maximal attention goes out to the storyworld and its inhabitants (indicated by the thick circle), while the listener or reader is only minimally aware of him or herself, the narrator (or his/her real-world counterpart, the author), and the real world surrounding him or her – they remain 'offstage' in the periphery of the listener's or reader's awareness (dashed circles).[11] In immersive texts, in other words, the narrator remains, as much as possible, invisible: he or she does not draw attention to him or herself or to the fact that he or she is narrating. The story seems to tell itself.

Perspective is also relevant to immersion in a different respect: immersive texts typically show a shift in vantage point. Two main types of shifts in vantage point that are associated with immersion can be distinguished. The first type involves a deictic shift ('recentering'), in which the ground as the deictic center is moved from an external, distanced, retrospective, spatio-temporal location with respect to the storyworld onto the described scene. The effect of this

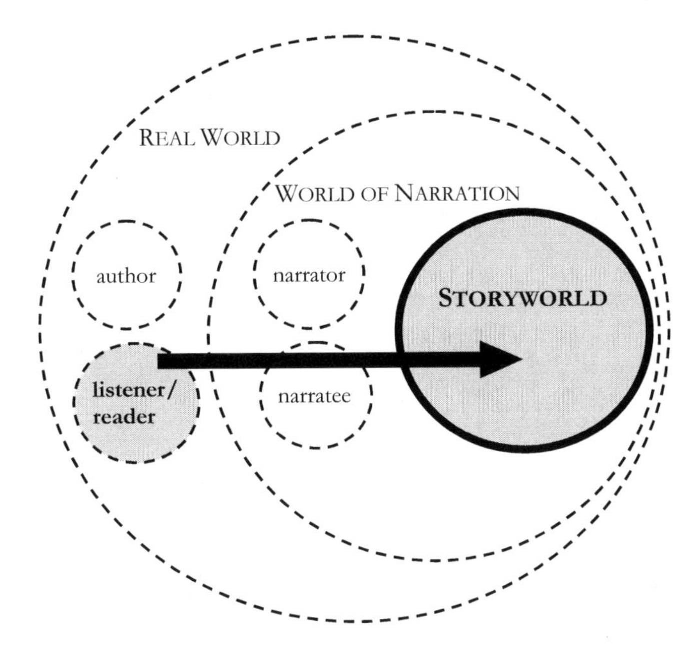

Figure 3.1 Immersion: maximal attention to storyworld

deictic shift is that the narrator and narratee are transported into the scene, as if observing the narrated events as they unfold ('pseudo-eyewitness effect'). In this perspectival configuration, the usual spatio-temporal distance between the world of narration and the storyworld is (construed as being) collapsed. Typical linguistic signals of this deictic shift are the use of the historical present and the use of proximal ('here and now') deictics, instead of the past tense and distal ('there and then') deictics, to refer to the storyworld.[12]

The other type of shift in vantage point, which is much more frequent in Greek narrative, does not feature a complete deictic shift but involves the setting-up of a secondary vantage point within the storyworld. The distance between the world of the narration and the storyworld is not reduced completely, as in the first type, but it is 'bridged'. The listener or reader is invited to view the described scene *via* a vantage point within the scene. This secondary vantage point can either be a 'camera standpoint', a depersonalized standpoint registering the narrated events in a detached way, or it can be a story character or an anonymous spectator through whose eyes we are observing the events and with whom we can identify ourselves and, preferably, also empathize.[13] In narratological terms, the latter type of shift in vantage point is called secondary (or embedded) focalization (De Jong 2014: 50–56).[14]

Both types of vantage point shifts serve to enhance the listener's or reader's feeling of being transported and they stimulate the listener or reader 'to project one's virtual body into the fictional world and onto the scene of the events' (Ryan 2015: 95). Apart from the shifted use of tense and deictic adverbs, another important linguistic phenomenon indicative of a shift in vantage point toward the storyworld is the use of direct or free indirect discourse, both of which present the character's words or thoughts as if they are directly accessible to the narratee, without the interference of the narrator's mediating voice.[15]

(iii) *Dynamicity*. A construal operation which is closely related to perspective is that of dynamicity, which has to do with the way in which the conceptualization unfolds over time.

Conceptualization is an inherently dynamic process: it occurs through time. More precisely, in the process of conceptualization, *two* times are involved: *processing time*, the time that is required for a conceptualizing subject to process an experience of some sort, and *conceived time*, time as the *object* of conception. There are two main types of relationship between processing and conceived time: coincidence and non-coincidence. When they coincide, we are dealing with direct observation (Figure 3.2a): 'When an event is directly observed, its apprehension coincides with its occurrence: its temporal phases are accessed serially, each being fully activated just when it is manifested' (Langacker 2015: 133). In normal language use (and certainly in narration), it is more usual that an event's occurrence and its conception are non-coincident. In this case, processing and conceived time have to be clearly distinguished (Figure 3.2b). Since the occurrence and conception of the event are independent, they will tend to differ in duration. In Figure 3.2b, processing time is shorter than conceived time, which is indicated by the different distances between the dots on line *t* and those on line *T*.

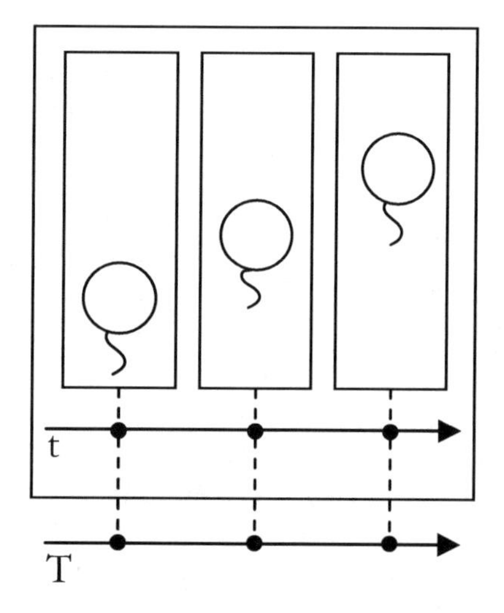

Figure 3.2a Coincidence of processing and conceived time: direct observation

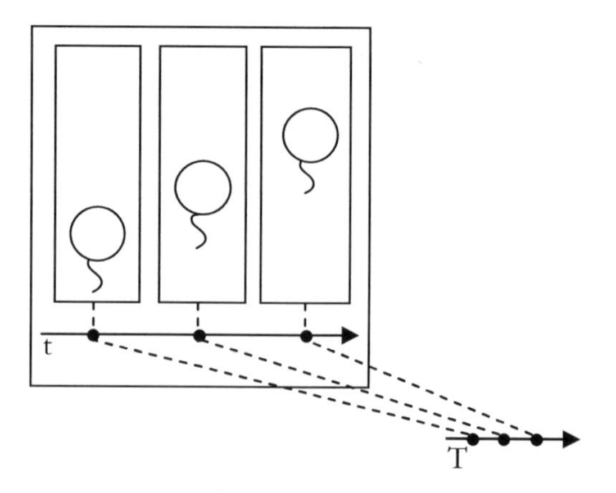

Figure 3.2b Non-coincidence of processing and conceived time: retrospective narration

In everyday language, there is a natural tendency for 'temporal iconicity', in which processing and conceived time proceed in a parallel development; that is, the order in which the events are processed (narrated and conceptualized) corresponds to the order in which the conceived events have occurred (are 'located on the time line'). This natural way of apprehending events is called 'sequential scanning' in cognitive grammar: temporal phases of an event are accessed ('scanned') sequentially so that conceived time correlates with processing time. 'The sequential scanning of the event constitutes the mental simulation of its observation' (Langacker 2015: 133).

Immersive texts tend to adhere to this natural tendency for temporal iconicity. The order in which events are narrated (narrating time) is aligned with the order in which the narrated events have occurred (narrated time). Anachronies, deviations from the chronological order such as flashbacks (analepses) or flash-forwards (prolepses),[16] are avoided since they disturb the natural order, causing an increased processing effort, and, instead, draw the listener's or reader's attention to the activity of the narrator – which is also detrimental to immersion.[17]

Another aspect of the relation between processing and conceived time relevant to narration is *speed*.[18] In narratology, various relations between narrating and narrated time are distinguished: if narrating time (which can be equated with processing, i.e. hearing or reading, time) progresses at a lower speed than narrated (conceived) time, we are dealing with *summary* narration; if narrating time advances at a higher speed than narrated time, it is a *slowing down* (slow motion) or *retardation*; if narrating time proceeds while narrated time does not, it is a *pause*; if narrating time and narrated time (roughly) coincide and advance at an equal pace, we are dealing with *scenic* narration. Scenic narration is the most common narration type in immersive texts since it presents the events at a speed at which we also experience (observe) them in our everyday lives.

The scenic type of narration can be represented by Figure 3.2c:

In Figure 3.2c, the durations of the narrating time (T) and the narrated time (t) are (approximately) equal (note the equal distances between the dots on line *t* and *T*), as is typical of scenic narration; the narrated time is anterior to the narrating time (retrospective narration).

Besides specificity, salience, perspective, and dynamicity, other construal operations are relevant to immersion such as *focus of attention*. These additional construal types will be addressed in the analysis of the Homeric passage (section 4).

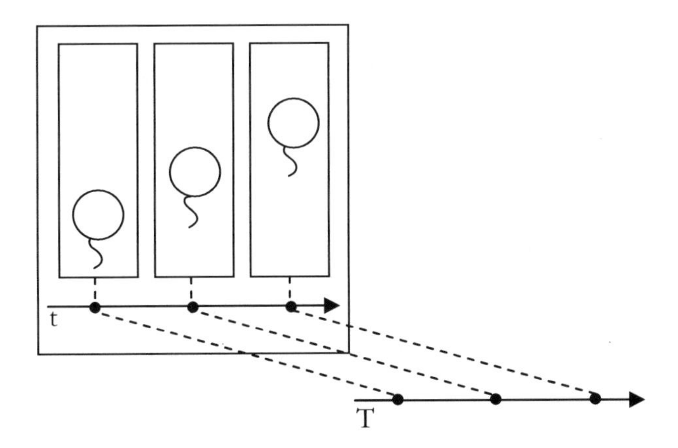

Figure 3.2c Retrospective scenic narration

3. Embodiment

The notion of immersion is also central to the Immersed Experiencer Framework, one of the leading cognitive theories in the research of language comprehension. In this theoretical framework, developed by the cognitive psychologist Rolf Zwaan (2004), language is conceived of as

> a set of cues to the comprehender to construct an experiential (perception plus action) simulation of the described situation. In this conceptualization, the comprehender is an immersed experiencer of the described situation, and comprehension is the vicarious experience of the described situation.[19]

In Zwaan's theory of language comprehension, the psychological phenomenon of immersion is not restricted to the particular experience of feeling present in a virtual world as triggered by a (literary) text – it is a mental process that is fundamental to language comprehension *in general*.

The Immersed Experiencer Framework is an *embodied* theory of language comprehension: it revolves around the idea that hearing or reading words activates the same brain regions that are active during the real-life perception of, and the motoric interaction with, the words' referents. Language comprehension involves the construction of a mental representation of the described situation which is firmly grounded in the sensorimotor and emotional experiences ('functional webs') active in the experiencer's brain. Thus, on an embodied view such as Zwaan's, hearing or reading a narrative will activate experiential resources which are then used to construct a mental simulation of the narrated scene.[20] Hearing or reading the word *cat*, for example, will activate our memory of previous experiences with cats and trigger an internal simulation of what a cat looks like, the sound it makes, how it moves, and how it feels to stroke it or to be scratched by it. Of course, the *strength* with which each of the various experiential aspects are activated is dependent on the particular context in which the word is used.

There is also empirical evidence that grammatical structures influence the strength of the mental simulation. For example, there is experimental evidence that aspectual marking and speech representation are sensitive to embodiment effects. Progressive (imperfective) sentences such as *John is closing the drawer* (portraying the action in progress) drives comprehenders to mentally simulate the process in action more strongly than perfect sentences like *John has closed the drawer* (Bergen and Wheeler 2010). Direct speech is more likely to give rise to perceptual (auditory) simulations than indirect speech (Yao, Belin, and Scheepers 2011). It is no coincidence that both imperfective aspect and direct speech are also important linguistic features of immersive texts.[21]

Zwaan's Immersed Experience Model distinguishes three components of language comprehension: *activation*, *construal*, and *integration*.

> (1) *Activation*: a word activates a functional web (i.e. 'the totality of experiences with a certain entity or event' [Zwaan 2004: 39]) that is also activated when the referent is experienced in real life.
>
> (2) *Construal*: the functional webs activated by subsequent words are immediately and incrementally integrated into a mental simulation of a specific event.[22] The linguistic unit on which construal operates is the intonation unit. Intonation units are basic units of discourse, identifiable by phonological cues such as acceleration-deceleration, changes in overall pitch level, terminal pitch contour, pauses,[23] and/or changes in voice quality (Chafe 1994).[24]

> Intonation units also have conceptual import: they express information that is, at some moment in time, active in a speaker's or listener's consciousness. Zwaan adopts Langacker's conception of the intonation unit as an attentional frame: 'Its conceptual value resides in the very act of making a single attentional gesture - imposing a single window of attention for the simultaneous viewing of conceptual content' (Langacker 2001: 155). According to Langacker, each of these successive attentional frames which make up discourse serves as an instruction to update the current discourse space.[25]

Intonation units show a tendency to coincide with grammatical clauses (and therefore they tend to express one state of affairs), but they may also contain just one word, a noun phrase, or be fragmentary.[26] The observation that intonation units do not clearly correspond with one particular type of semantic or syntactic unit has led Hannay and Kroon (2005) to the idea that intonation (or punctuation) units can be understood better when you see them as realizations of *discourse acts*, which often consist of a single word or a noun phrase. I will return to the role of discourse acts in section (4).

Zwaan's notion of construal has a number of components, which show some similarities with the aspects of Langacker's construal. In Zwaan's model, a construal of a state of affairs involves the following:

- *Time interval*: comprehenders keep events active in working-memory as long as the state of events is ongoing.
- *Spatial region*: the state of affairs occurs in a certain spatial region, a section of space delimited by the human senses (vision, audition, smell, touch) and by their actions.
- *Perspective*: a spatio-temporal vantage point (location, distance, orientation) from which the construed state of affairs is experienced.
- *Entities and features*: grammatical markers such as word order or case markers signal which entities are construed as *foregrounded* entities and those which are *backgrounded*. Foregrounded entities are typically marked as subject (topics) or objects, while backgrounded entities tend to be referred to by prepositional phrases or oblique (genitive, dative, etc.) cases. Features are typically referred to by adjectives.

> (3) *Integration*: when processing discourse, a language comprehender will have to mentally integrate successive construals of the state of affairs. Relevant components from a previous event construal will remain activated in working-memory and influence the current construal. Integration refers to the transition of one construal to another. Zwaan's thesis is that these transitions are grounded in human everyday life experience (Zwaan 2004: 46, 48–49).

In descriptions of static scenes, the regulation of attention (through the flow of intonation units as attentional frames) is perceptual (typically visual) in nature, simulating the sensory (visual) experience of the scene. Examples of such transitions are zooming, panning, scanning, and fixating. This mode of discourse comprehension will also be relevant to our analysis of *Odyssey* 5.59–73 below. In descriptions of dynamic scenes/action sequences, transitions typically involve *cause–effect* or *motivation–action* relations. A language comprehender, as an immersed experiencer, will by default be invited to take a protagonist's perspective (Zwaan 2004: 47–48).

Besides concordance with human experience, according to Zwaan, other factors influence the integration of event construals, such as the amount of conceptual *overlap* between successive construals - overlaps in time, space, causation, perspective, entities, and features, the degree of *predictability* of the transition from one event construal to another, and, finally, *linguistic cues*

(Zwaan 2004: 48–51). Word order and cases signal which entities have to be construed as foregrounded and which as backgrounded. Tense markers locate the event on a time line. Aspect markers construe the event as (un)bounded within a temporal frame.[27] Prepositions indicate the location of entities within the mental simulation. Discourse markers specify the way in which one state of affairs (as expressed by an intonation unit/clause) has to be integrated with another into a more complex event sequence.

Zwaan's model of language comprehension is summarized in Table 3.1, which shows the components of the process, the linguistic and representational units on which they operate, and the denoted referential units:

Let me sum up some of the intermediate conclusions regarding the relation between immersion, construal, and embodiment. The immersivity of a text can be defined as the degree to which a text has the capacity to evoke a mental simulation of the described situation in the mind of the listener or reader, and to transport the listener's or reader's virtual body into the scene. The intensity of the experience of immersion is dependent on the close interplay of a number of construal phenomena. In order to be maximally immersive, the described situation should possess the following features:

(1) *specificity*: it shows a high level of specificity (granularity) of perceptual details;
(2) *space*: it provides a strong sense of the spatial dimensions;
(3) *time*: it shows an iconic temporal organization (no deviations of chronological order or time compressions);
(4) *perspective*: it invites one to experience ('view') the situation from a spatio-temporal and emotional (affective-evaluative) vantage point (embedded focalizer) located in the scene;
(5) *focus of attention*: a strong attentional focus on the described situation: no distractions to 'offstage' elements such as the narrator and the world of narration;
(6) *experientiality*: it shows a general concordance with human everyday experience.

4. Calypso's cave

These immersive features can be seen as conceptual dimensions that cooperate with and reinforce one another in order to engender in the listener or reader a feeling of being present at the narrated scene. It is therefore more insightful to approach these dimensions as a closely connected unity and analyze their synergetic interplay than to consider each of these dimensions in isolation. As a typical example of Homer's immersive quality, I will discuss the description of Calypso's cave in *Odyssey* (5.59–73). Hermes arrives at the cave and finds Calypso.[28]

πῦρ μὲν ἐπ' ἐσχαρόφιν μέγα καίετο, τηλόσε δ' **ὀδμὴ**
κέδρου τ' εὐκεάτοιο θύου τ' ἀνὰ νῆσον ὀδώδει (60)
δαιομένων· **ἡ** δ' ἔνδον ἀοιδιάουσ' ὀπὶ καλῇ
ἱστὸν ἐποιχομένη χρυσείῃ κερκίδ' ὕφαινεν.

Table 3.1 Comprehension process

Process component	*Linguistic unit*	*Representational unit*	*Referential unit*
Activation	Word/morpheme	Functional webs	Objects and actions
Construal	Clause/intonation unit	Integrated webs	Events
Integration	Connected discourse	Sequence of integrated webs	Event sequences

ὕλη δὲ σπέος ἀμφὶ πεφύκει τηλεθόωσα,
κλήθρη τ' αἴγειρός τε καὶ εὐώδης κυπάρισσος.
ἔνθα δέ τ' **ὄρνιθες** τανυσίπτεροι εὐνάζοντο, (65)
σκῶπές τ' ἴρηκές τε τανύγλωσσοί τε κορῶναι
εἰνάλιαι, τῇσίν τε θαλάσσια ἔργα μέμηλεν.
ἡ δ' αὐτοῦ τετάνυστο περὶ σπείους γλαφυροῖο
ἡμερὶς ἡβώωσα, τεθήλει δὲ σταφυλῇσι.
κρῆναι δ' ἑξείης πίσυρες ῥέον ὕδατι λευκῷ, (70)
πλησίαι ἀλλήλων, τετραμμέναι ἄλλυδις ἄλλη.
ἀμφὶ δὲ **λειμῶνες** μαλακοὶ ἴου ἠδὲ σελίνου
θήλεον.

A great fire was burning in the hearth, and far over the island there was spread the scent of the cedar logs and citron-wood burning there. And she was inside, singing in a lovely voice and she worked to and fro at her loom weaving with a golden shuttle. Round the cave there was abundant growth of trees, alder and poplar and fragrant cypress and in the trees roosted wide-winged birds, owls and hawks and long-tongued cormorants, sea-birds whose work is on the water. And right there round the mouth of the hollow cave ran a golden vine in full glory, rich with clusters and grapes. And there were four springs running with bright water, close at their source then turning to flow each in its own direction. On either side soft meadows grew thick with violet and wild celery.

Specificity. The cave and its scenery are described in full detail. The description focuses on perceivable physical entities and it contains an abundance of nouns, adjectives, and verbs providing sensory (visual, olfactory, as well as auditory) information activating experiential information from semantic memory, which enable the listener to conjure up a fine-grained mental simulation of the described scene: the fire is large, there is a smell of cedar and juniper, Calypso is singing with a beautiful voice, her shuttle is golden, alder, poplar, and sweet-smelling cypresses are growing abundantly, the birds are wide-winged, the cormorants are long-tongued,[29] the vine is in its prime, full of clusters of grapes, there are four springs with bright water, the meadows are soft and rich in violets and celery.

Space. The passage belongs to the text type *description* (descriptive mode); that is, the progression of the text is structured in terms of the dimension of space, rather than the dimension of time, as is the case in narrative proper.[30] Descriptions are typically devoted to an object or a global theme, of which successively a number of component parts or subtopics are attended to. Of each of these component parts or subtopics an ongoing activity or property is described.

The spatial dimensions of the scene are specified by means of a considerable number of adverbs and adverbial phrases (59 τηλόσε, 60 ἀνὰ νῆσον, 61 ἔνδον, 63 σπέος ἀμφί, 65 ἔνθα, 68 αὐτοῦ, 68 περὶ σπείους, 70 ἑξείης, 72 ἀμφί). These adverbial expressions serve to orient and guide the listener's attention carefully through the space of the scene. I will return to the issue of space below when I discuss perspective, to which it is intimately connected.

Time. The various activities and states which are described do not reach an endpoint, which is marked by the use of imperfects (59 καίετο, 62 ὕφαινεν, 65 εὐνάζοντο, 70 ῥέον, 73 θήλεον) and pluperfects (60 ὀδώδει, 63 πεφύκει, 68 τετάνυστο, 69 τεθήλει) forms. These ongoing states and activities are to be understood as occurring simultaneously: there are no temporal adverbial expressions that indicate a shift in time. This means that, on the level of the described scene, there is no sequence of events and therefore no explicit temporal progress. However, on a different level, that of the *observation* of the scene, time does progress as the observer (that is, Hermes as focalizer) fixes his gaze on each of the entities that are part of the scene. Since this touches on the issue of perspective, I will address it further below.

Perspective. As is usual in Homeric epic with characters arriving at a place, the description is focalized by the arriving visitor. In this passage, Hermes' focalization is explicitly referred to in 75–76: ἔνθα στὰς θηεῖτο διάκτορος Ἀργεϊφόντης.[31] The verb θηεῖτο cues us to the wonderful spectacle Hermes is witnessing, engages our attention and emotional involvement, and thus primes the audience's mind to an immersive experience. We are invited to identify with Hermes' amazement and to view the scene through his eyes, to smell with his nose, and to hear with his ears.

A second, more indirect, indication that the description is focalized by Hermes rather than the narrator is the use of the past tense. Since the topography of the cave, being the abode of a goddess, is of an everlasting and unchanging nature,[32] it would have been a natural option for the narrator to use the omnitemporal-habitual present tense: the cave and its surrounding scenery are, after all, in the same state at the time of narration as they were when Hermes visited the cave.[33] However, by the use of the past tense the narrator explicitly signals that the scene is not viewed from an omniscient narratorial perspective, but through the eyes of Hermes, at a specific moment in time located in the past.

The description is structured in such a way that we follow Hermes' visual, olfactory, and auditory perception as it moves through the scene. A number of jointly operating linguistic devices are deployed to convey the impression that we are observing the scene through the sensory channels of a viewer present at the scene. These are: segmentation into intonation units, subject marking, and aspectual marking.

As we have seen, intonation units are attentional frames: they comprise the conceptual content to which a conceptualizer's attention is directed at a given moment in time. For Homeric discourse, it has been shown by Egbert Bakker and Simon Slings that verse end and caesurae can be identified as boundary markers between intonation units.[34]

In Table 3.2 I present a segmentation of the passage into intonation units.[35]

As already briefly mentioned earlier, it is useful to analyze intonation units not only as basic cognitive units but also, with Hannay and Kroon (2005), as basic communicative units or *discourse acts*. In Functional Discourse Grammar, discourse acts are defined as 'the smallest identifiable units of communicative behavior' (Kroon 1995: 65; Hengeveld and Mackenzie 2008: 60).

Discourse acts may have various illocutions, such as assertion, question, and order, and they can show a dependence relation; that is, some discourse act are treated as *central* to the communicative purposes of the speaker, while others perform a *subsidiary* (supportive) function with respect to the central discourse act.[36] Relevant for my analysis of the Homeric passage are two types of subsidiary discourse acts: (1) Orientation, which as an instruction as to how to integrate the subsequent central discourse act into the current discourse space, and (2) Elaboration, which provides additional specifying or clarifying information about some element of the preceding discourse act.[37]

Our passage contains eight main clauses. These are the central discourse acts, each of which is accompanied by a number of subsidiary discourse acts. In Functional Discourse Grammar, a constellation of a central discourse act with its dependent subsidiary discourse act is called a *move*.[38] Each of these eight moves is introduced by the particle μέν, which prepares the listener for a subsequent equivalent unit with δέ, or by the particle δέ, which marks a switch to a new focus of consciousness (Bakker 1997: 63). The discourse units start off with an intonation unit with the function of orientation (πῦρ μὲν ἐπ' ἐσχαρόφιν, τηλόσε δ' ὀδμή, ἡ δ' ἔνδον, ὕλη δὲ σπέος ἀμφί, ἔνθα δέ τ' ὄρνιθες, ἡ δ' αὐτοῦ, κρῆναι δ' ἑξείης, ἀμφὶ δὲ λειμῶνες) serving a double function: to direct the listener's (visual) attention to a location in space (spatial adverb) and to focus it on a

Table 3.2 Intonation units and their discourse functions

Intonation units	*Function*
[μέν] A fire on the hearth, [T]	Orientation
it was burning fiercely. [B]	Central Discourse Act
[δέ] Far across a smell, [VE]	Orientation
[τε] namely of split cedar [T]	Elaboration1
[τε] and citron-wood, [H]	Elaboration2
it smelled over the island, [VE]	Central Discourse Act
namely of them (i.e. logs) being burned. [Tri]	Elaboration
[δέ] She inside, [T]	Orientation1
singing with a lovely voice, [VE]	Orientation2
going to and fro at her loom, [P]	Orientation3
she was working with a golden shuttle [VE]	Central Discourse Act
[δέ] Trees around the cave, [T]	Orientation
they were growing abundantly, [VE]	Central Discourse Act
[τε] alder, [1F]	Elaboration1
[τε] and poplar, [T]	Elaboration2
[καί] and fragrant cypresses. [VE]	Elaboration3
[δέ] And there birds, [P]	Orientation
wide-winged ones, [B]	Elaboration
they were roosting, [VE]	Central Discourse Act
[τε] owls [1F]	Elaboration1
[τε] and hawks [T]	Elaboration2
[τε] and long-tongued crows, [VE]	Elaboration3
namely those of the sea, [Tri]	Elaboration3A
whose work is on the water. [VE]	Elaboration3B
[δέ] And that right there, [Tri]	Orientation
it was spread out, [T]	Central Discourse Act
namely around the cave, [VE]	Elaboration
the abundantly growing vine, [T]	Elaboration (Tail)
[δέ] and it was rich with clusters of grapes. [VE]	Central Discourse Act
[δέ] Springs in a row, [P]	Orientation
four of them, [H]	Elaboration1
they flowed with bright water. [VE]	Central Discourse Act
close to each other, [P]	Elaboration2
turned in different directions. [VE]	Elaboration3
[δέ] On either side meadows, [P]	Orientation
soft ones, [H]	Elaboration1
of violet and celery [VE]	Elaboration2
they were growing luxuriantly. [1F]	Central Discourse Act

perceptually salient entity (grammatical subject).[39] In cognitive linguistics, the grammatical subject is seen as the default topic: subject marking signals that the entity is in the center ('spotlight') of attention, serving both as a cognitive point of attachment to the already established discourse space and as a starting-point for the conceptualization of the subsequent information.[40]

The order of the topical entities attended to corresponds to the path of Hermes' visual (and olfactory) attention. After Hermes arrives, his gaze first focuses on the fire in the hearth first, which is central to the scene and is the most visually salient entity. The visual cue of the fire makes him realize that this fire is the source of the scent of split cedar and citron-wood which he had already smelled as he came to the island flying through the sky and traveled the distance from the shore to the cave by foot.

Then, as she is illuminated by the fire, Hermes distinguishes Calypso, who is singing and working at her loom. At this point, his gaze starts to move from the center of the scene to the periphery: first he focuses on the trees around the cave, next he zooms in on the birds in the trees, then he notices the vine, the springs, and, finally, the meadows around the cave.

The successive descriptions of these topical entities also simulate the way in which Hermes' visual observation is structured. A recurrent pattern is that a general entity is first focused on, followed by intonation units providing specifying information (elaborations), iconically representing the observer's cognitive process as he zooms in on further perceptual details in the scene.[41] For example, in lines 63–64, Hermes first directs his attention to the trees around the cave, of which he observes that they were growing abundantly; then he zooms in on them and identifies them as alders, poplars, and fragrant cypresses.[42] The description of the vine (lines 68–69) constitutes a similar case. In 68, the observer's attention is first drawn to the unnamed entity near the trees (ἡ δ' αὐτοῦ 'And that thing there'), which is spread around the cave. Only then is the species determined ('an abundantly growing vine'), by means of an elaboration placed in enjambment.[43] An additional visual detail about the vine is given in the next clause/intonation unit ('it was rich in clusters of grapes').

The path of Hermes' visual attention through the space of the scene also has a temporal dimension. The temporal structure of the passage can be analyzed as a combination of the two types of narration mentioned in section 3. On the one hand, we are dealing with retrospective narration, which is signaled by the use of past tenses. On the other hand, as we have seen, there are many linguistic and narratological cues signaling that the passage should be understood as simulated observation. One of these cues is the pervasive use of imperfects and pluperfects, which are aspectual forms that suggest that the described states and activities are viewed from a temporal vantage point *within* the scene. This combination of narration is depicted in Figure 3.3.

The narrated time (t) is anterior to the narrating time (T): retrospective narration. The speed of the narrating time and the narrated time is (approximately) equal, which is typical

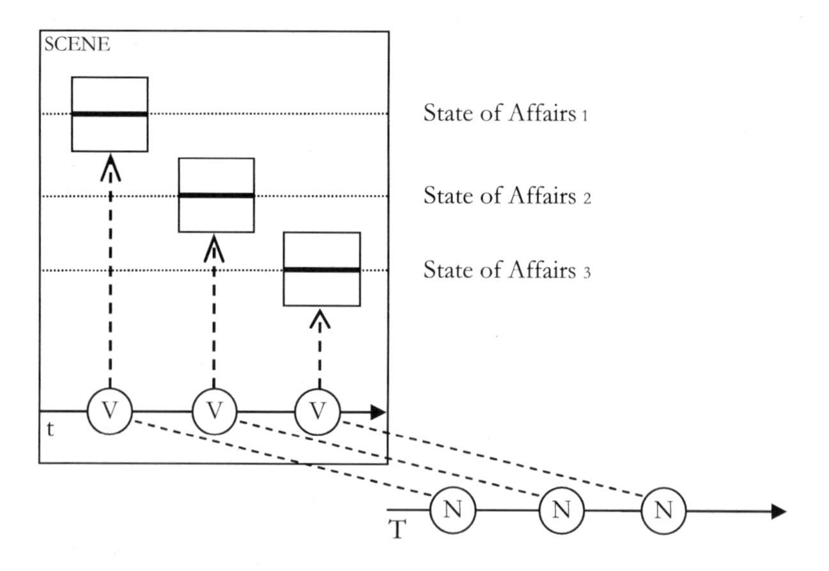

Figure 3.3 Retrospective simulated observation

of scenic narration and, more specifically, of simulated observation. The narrator (N) portrays (dashed lines) subsequent states of affairs as they are observed by the story-internal viewer (V). In the scene, the viewer (V) focuses his/her attention from one state of affairs to another at successive moments in time (upward-pointing arrows). For example, in our passage Hermes first observes the fire burning fiercely (State of Affairs $_1$); he then focuses his attention on the smell of the burning logs (State of Affairs $_2$); then he watches Calypso going to and fro while she is working at her loom (State of Affairs $_3$), etc. Each of the perceptions of the states of affairs has a certain duration. In the terminology of cognitive grammar, this is the *temporal scope*: the window of attention selected for focused viewing of the state of affairs (represented by the square boxes).[44] Within this window of attention, the state of affairs is put in focus (indicated by the thickness of the line). The imperfect and pluperfect aspectual marking expresses that states of affairs are seen as unbounded: they had been going on before the temporal scope of view and continue afterwards (represented by the dotted lines left and right of the temporal scope). This special spatio-temporal organization of the text encourages the listener to position him/herself within the described scene and experience it as if actually present. It goes without saying that this type of simulated observation strongly contributes to the experience of being immersed in the scene.

Conclusion

A well-known quality of Homeric epic, as well as of many other literary works, is its capacity to make the audience experience a feeling of being immersed in the storyworld. To achieve this effect on the listener, Homeric discourse deploys a rich and varied arsenal of linguistic and narrative devices, such as lexical choices, tense-aspect marking, deixis, discourse particles, word order, intonation units, case marking, spatio-temporal organization, and perspective (focalization), in order to bring the storyworld to life and to make us feel part of it. Cognitive approaches to narrative that have emerged in the fields of linguistics, narratology, and psychology help us to better understand how these highly diverse textual phenomena tightly interact and jointly contribute to Homer's immersive power.

Notes

1 I thank the anonymous reviewer for the many helpful comments on an earlier version of this chapter.
2 *On Lysias* 7.1–2. There is a rich literature on the visual aspect or 'vividness' of the Homeric style, of which recent examples are Bakker (1993), Bakker (1997), Bakker (2001), Bakker (2005), Clay (2011), Bonifazi (2012), Tsagalis (2012), and Allan, De Jong, and De Jonge (2017).
3 In art, film, and literary studies, a number of concepts have emerged that are closely related to immersion, such *as aesthetic illusion* (Wolf 1993, 2004; Wolf, Bernhart, and Mahler 2013) and *transportation* (Gerrig 1993; Green and Brock 2000).
4 Ryan (2001: 120–157); Ryan (2015: 85–114).
5 Cf. also Plutarch's characterization of Thucydides' style in *De gloria Atheniensium* 347A, where he notes that Thucydides' *enárgeia* produces in the readers the astonishment and confusion which were experienced by those who witnessed the actual events.
6 Cf. Longinus 1.3. The relationship between *enárgeia*, *enagṓnios*, and *ékstasis* and the modern concept of *immersion* is discussed in Allan, De Jong, and De Jonge (2017).
7 Examples of empirical research relating to immersion and transportation are: Green and Brock 2000 (Transportation Theory) and Zwaan (2004). Zwaan's Immersed Experiencer Framework will be discussed in more detail below. A brief overview of recent cognitive psychological research on immersion is given in Chapter 3 of Ryan (2015).
8 Allan (forthc. a), Allan (forthc. b), Allan (forthc. c).
9 Cf. Langacker (2000: 203–245); Langacker (2008: 261).

10 Helpful discussions of construal operations in cognitive linguistics are: Croft and Cruse (2004: 40–73), Langacker (2008: 55–92), and Langacker (2015).

11 As Egbert Bakker has rightly pointed out, in Homeric poetry in performance the roles of the historical author and the fictional narrator collapse: the text-internal 'I' of the narrator refers to the text-external 'I' of the performing poet (Bakker 2009). This does not affect the point I make here: immersion requires that neither the performing poet nor the text-internal narrator draw the audience's attention to themselves (rather than directing it to the narrated world) – whether they can be clearly conceptually separated or not. This entails, incidentally, that the epic proems (in which the singer-narrator explicitly refers to himself) are not immersive (see also Allan, forthc. c).

12 For the displacement of the deictic center to the past in Greek narrative, see Bakker (2005), Ch. 9.

13 On the role of empathy in immersion reading, see Keen (2007).

14 Linguistically, this type of vantage point shift is often (but not necessarily) accompanied by the use of the imperfect tense (see Rijksbaron 2012). It is worth noting that there is empirical psychological evidence that imperfective sentences show a significant embodiment effect in that they encourage the language comprehenders more strongly to mentally simulate the process in action than perfect sentences. I will return to this role of the imperfect in mental simulation and immersion below.

15 For the relationship between speech and thought representation (especially free indirect discourse) and immersion, see also Allan (forthc. b).

16 For the terms anachrony, analepsis, and prolepsis, see De Jong (2014: 78–87).

17 There are anachronies in Homeric epic. However, they do occur only infrequently and almost always briefly interrupt the general continuity of the main storyline, which typically proceeds in chronological order and is of a scenic character (De Jong 2007). Moreover, two important aspects of Homeric anachronies ensure that they do not generally affect the immersive quality of the narrative: (1) Homeric anachronies are very often part of character speech (e.g. Odysseus' tale in book 9–12) rather than interventions by the narrator himself (see De Jong 2007), i.e. they do not draw attention away from the storyworld to the primary narrator's activity. (2) If anachronies in narrator speech do occur, they very often serve to produce a pathetic effect or to create suspense, e.g. the 'obituaries' of killed warriors, anticipations of the deaths of the Patroclus, Hector, and Achilles, and the fall of Troy (see De Jong 2007: 21–22, 25–26). In other words, these anachronies importantly contribute to the hearer's *emotional* immersion in the storyworld.

18 For the narratological concept of speed (also called rhythm) and the various types of narration associated with it, I refer to De Jong (2014: 92–98).

19 Zwaan (2004: 36).

20 Psychological empirical evidence for embodied effect in language (and narrative) understanding is strong and ever growing. Some examples are: (motor simulation) Glenberg and Kaschak (2002), Taylor and Zwaan (2008), Bergen and Wheeler (2010); (perceptual simulation) Zwaan, Madden, Yaxley, and Aveyard (2004), Yao, Belin, and Scheepers (2011); (emotional simulation) Havas, Glenberg, and Rinck (2007). Overviews of the research on embodiment and language comprehension are: Barsalou (2010), Sanford and Emmott (2012: 132–160), and Kaschak, Jones, Carranza, and Fox (2014).

21 That both lexical items and grammatical structure can generate embodiment effects and thus are meaningful is, as we have seen, in accordance with a cognitive linguistic view on grammar. The idea that conceptual structure ultimately derives from embodied cognition is also a central tenet of cognitive linguistic theory from its inception (e.g. Lakoff 1987; Langacker 1987, 2008).

22 Note that Zwaan uses the term *construal* in a different but partially related sense to Langacker. An important difference is that Zwaan's construal seems to relate more emphatically to the (automatic) processing of incoming information (hearer/reader-oriented), while Langacker's construal seems to leave room for a (conscious) choice of presenting information in one way rather than another (speaker-oriented).

23 Contrary to what one might think, pauses are not very reliable indicators of intonation units: they often occur within intonation units, and they are often absent at boundaries of intonation units (Chafe 1994: 57).

24 For intonation units in Greek, see Devine and Stephens (1994: 432–433) and especially Scheppers (2011).

25 The current discourse space is the cognitive common ground between the interlocutors 'comprising those elements and relations construed as being shared by the speaker and hearer as a basis for communications at a given moment in the flow of discourse' (Langacker 2001: 144).

26 Based on data from five languages, Croft (2007: 11–12) observes that around 50% of the intonation units are clauses. The rest of the intonation units are single words, noun phrases, and other grammatical units.

27 For Greek aspect, boundedness, and construal, see Allan (2017).

28 **Subjects**, spatial adverbial expressions, imperfects/pluperfects. The translation is by Hammond (2000).

29 Cf. Ameis-Hentze's remark *ad loc.*: 'τανύγλωσσοι zungenstreckend malt die Seekrähen im Moment des Schreiens für das Auge'.

30 For a typology of the main text types in Greek (i.e. the Immediate/Displaced Diegetic, the Descriptive, and the Discursive Mode), see Allan (2009) and Allan (2013).

31 Cf. De Jong (2001); De Jong and Nünlist (2004). Whether or not Hermes could actually see all the items standing near the entrance of the cave is a matter of some controversy. For discussion, see De Jong and Nünlist (2004) (with references). I do not see any compelling reason to assume Hermes could not see the springs or the meadows from his standpoint outside the entrance of the cave.

32 Note that the permanent character of the scene is also signaled by the use of 'epic τε' in 65 (Ruijgh 1971: 270).

33 Only μέμηλεν in 67, which also refers to an omnitemporal fact (cf. the 'epic τε'), is a present (perfect) tense. A possible explanation for this shift to the narrator's focalization may be that the sea-related activities of the cormorants are not visible at the moment Hermes is viewing the scenery.

34 See Bakker (1990), Bakker (1997), and Slings (1992). For verse lines as prototypical intonation units (major phrases), see also Devine and Stephens (1994: 424–425). See more recently also Bonafazi, Drummen, and de Kreij (2016: Ch. II.2.2).

35 Caesurae: 1F = diaeresis at the end of first foot, Tri = Trithemimeral, Tro = Trochaic, P = Penthenimeral, H = Hepthemimeral, B = Bucolic diaeresis, VE = verse end. According to West (1982: 36), these are the caesurae that most frequently coincide with sense pauses. The segmentation of the passage is based on the default assumption that verse end, the middle caesura, and the caesurae in the first and second half of the verse (if present) indicate boundaries between intonation units. Also the position of a particle (Wackernagel's Law) can help to identify boundaries. Occasionally, a caesura does not seem to correspond with a conceptual (semantic or pragmatic) boundary ('sense pause'). In that case, I have not represented it in my table. An example is line 67 (τῇσίν τε // θαλάσσια / ἔργα μέμηλεν), in which the trochaic caesura and the bucolic diaeresis do not seem to indicate any clear semantic or pragmatic boundary.

36 For a more detailed discussion of the discourse act, I refer to Hengeveld and Mackenzie (2008: 60–68).

37 That an analysis of Homeric discourse in terms of discourse acts can help us to understand the communicative dynamics of Homeric discourse has already been shown by Bonifazi and Elmer (2012) and Bonifazi, Drummen, and De Kreij (2016: esp. Chapter II.2).

38 A move is defined by Kroon (1995: 66) as 'the minimal free unit of discourse'; cf. also Hengeveld and Mackenzie (2008: 50–60).

39 In Functional Grammar terminology, these initial extra-clausal constituents perform two more specific orienting functions, that of a referentially orienting Theme (i.e. left-dislocated topic) and that of a spatio-temporally orienting Setting (see Dik 1997: 388; Allan 2014). An Orientation intonation unit may also consist of a subordinate finite or participial clause preceding the main clause (as in lines 61–62). These have the function of Settings serving to ground the subsequent central discourse act by specifying the spatio-temporal location and the other background circumstances.

40 Cf. Chafe (1994), (Langacker 2001), Langacker (2008: 512–517).

41 In Homeric discourse, elaborations typically have the form of appositions or participial clauses following the main clause.

42 Note that each of these three types receive their own intonation unit (κλήθρη τ'/ αἴγειρός τε // καὶ εὐώδης κυπάρισσος). This analysis is supported not only by the location of the caesura, but also by cross-linguistic considerations. First, assigning each item to a separate intonation unit is in accordance with Chafe's (1994) One New Idea Constraint, which states that every intonation unit should not contain more than one item of new information. Second, there is a very strong cross-linguistic tendency to divide parallel, coordinated items (note τε and καί) over separate intonation units, even if they consist of only one word (Croft's *parallelism constraint*; see Croft 2007).

43 In technical terms, this subtype of elaboration is a tail, a right-dislocated extra-clausal constituent (i.e. a separate intonation unit, cf. the enjambment), used to clarify the identity of a referent mentioned in the preceding clause (Dik 1997: 401–403; Bakker 1990: 13; Allan 2014).

44 See e.g. Langacker (1987: 258–262) and Langacker (2000: 224); for a cognitive grammar approach to Greek aspect, see Allan (2017).

References

Allan, R.J., "Towards a Typology of the Narrative Modes in Ancient Greek: Text Types and Narrative Structure in Euripidean Messenger Speeches", in *Discourse Cohesion in Greek*, ed. S.J. Bakker and G.C. Wakker (Leiden and Boston, MA: Brill, 2009), 171–203.

Allan, R.J., "History as Presence: Time, Tense and Narrative Modes in Thucydides", in *Thucydides Between History and Literature*, ed. A. Tsakmakis and M. Tamiolaki (Berlin and New York: De Gruyter, 2013), 371–390.

Allan, R.J., "Changing the Topic: Topic Position in Ancient Greek Word Order", *Mnemosyne* 67 (2014): 181–213.

Allan, R.J., "The Imperfect Unbound: A Cognitive Linguistic Approach to Greek Aspect", in *Language Variation and Change, Tense, Aspect and Modality in Ancient Greek*, ed. K. Bentein, M. Janse, and J. Soltic (Leiden and Boston, MA: Brill, 2017), 100–130.

Allan, R.J., "Persuasion by Immersion", in *The Language of Persuasion*, ed. T. Liao and A. Vatri (Leiden and Boston, MA: Brill, forthc. a).

Allan, R.J., "Herodotus and Thucydides: Distance and Immersion", in *Textual Strategies in Greek and Latin War Narrative*, ed. I.J.F. de Jong, L. van Gils, and C.H.M. Kroon (Leiden and Boston, MA: Brill, forthc. b).

Allan, R.J., "Narrative Immersion: Some Linguistic and Narratological Aspects", in *Narrative and Experience*, ed. J. Grethlein and L. Huitink (Berlin and New York: De Gruyter, forthc. c).

Allan, R.J., I.J.F. de Jong, and C.C. de Jonge, "From Enargeia to Immersion: The Ancient Roots of a Modern Concept", *Style* 51 (2017): 34–51.

Bakker, E.J., "Homeric Discourse and Enjambement: A Cognitive Approach", *Transactions of the American Philological Association* 120 (1990): 1–21.

Bakker, E.J., "Discourse and Performance: Involvement, Visualization, and 'Presence' in Homeric Poetry", *Classical Antiquity* 8 (1993): 5–20.

Bakker, E.J., *Poetry in Speech: Orality and Homeric Discourse* (Ithaca, NY and London: Cornell University Press, 1997).

Bakker, E.J., "Similes, Augment, and the Language of Immediacy", in *Speaking Volumes: Orality and Literacy in the Greek and Roman World*, ed. J. Watson (Leiden: Brill, 2001), 1–23.

Bakker, E.J., *Pointing at the Past: From Formula to Performance in Homeric Poetics* (Washington, DC and Cambridge, MA: Center for Hellenic Studies, 2005).

Bakker, E.J., "Homer, Odysseus, and the Narratology of Performance", in *Narratology and Interpretation: The Content of Narrative Form in Ancient Literature*, ed. J. Grethlein and A. Rengakos (Berlin and New York: de Gruyter, 2009), 117–136.

Barsalou, L.W., "Grounded Cognition: Past, Present, and Future", *Topics in Cognitive* Science 2 (2010): 716–727.

Bassett, S.E., *The Poetry of Homer* (Berkeley and Los Angeles, CA: University of California Press, 1938).

Bergen, B. and K. Wheeler, "Grammatical Aspect and Mental Simulation", *Brain & Language* 112 (2010): 150–158.

Bonifazi, A., *Homer's Versicolored Fabric: The Evocative Power of Ancient Greek Epic Wordmaking* (Washington, DC and Cambridge, MA: Center for Hellenic Studies, 2012).

Bonifazi, A. and D. F. Elmer, "Composing Lines, Performing Acts: Clauses, Discourse Acts, and Melodic Units in a South Slavic Epic Song", in *Orality, Literacy and Performance in the Ancient World*, ed. E. Minchin (Leiden: Brill, 2012), 89–109.

Bonifazi, A., A. Drummen, and M. de Kreij, *Particles in Ancient Greek Discourse: Five Volumes Exploring Particle Use across Genres* (Washington, DC: Center for Hellenic Studies, 2016). http://nrs.harvard.edu/urn-3:hul.ebook:CHS_BonifaziA_DrummenA_deKreijM.Particles_in_Ancient_Greek_Discourse.2016.

Chafe, W., *Discourse, Consciousness, and Time: The Flow and Displacement of Conscious Experience in Speaking and Writing* (Chicago, IL and London: The University of Chicago Press, 1994).

Clay, J.S., *Homer's Trojan Theater: Space, Vision, and Memory in the Iliad* (Cambridge: Cambridge University Press, 2011).

Croft, W., "Intonation Units and Grammatical Structure in Wardaman and in Cross-Linguistic Perspective", *Australian Journal of Linguistics* 27 (2007): 1–39.

Croft, W. and D.A. Cruse, *Cognitive Linguistics* (Cambridge: Cambridge University Press, 2004).

De Jong, I.J.F., *A Narratological Commentary to the Odyssey* (Cambridge: Cambridge University Press, 2001).

De Jong, I.J.F., "Homer", in *Time in Ancient Greek Literature*, ed. I.J.F. de Jong and R. Nünlist (Leiden and Boston, MA: Brill, 2007), 17–37.

De Jong, I.J.F., *Narratology & Classics: A Practical Guide* (Oxford: Oxford University Press, 2014).

De Jong, I.J.F. and R. Nünlist, "From Bird's Eye View to Close-Up: The Standpoint of the Narrator in the Homeric Epics", in *Antike Literatur in neuer Deutung*, ed. A. Bierl, A. Schmidt, and A. Willi (Munich and Leipzig: Saur, 2004), 62–83.

Dik, S.C., *The Theory of Functional Grammar. Part 2: Complex and Derived Constructions*, ed. K. Hengeveld (Berlin and New York: Mouton De Gruyter, 1997).

Gerrig, R. J., *Experiencing Narrative Worlds. On the Psychological Activities of Reading* (New Haven, CT: Yale University Press, 1993).

Glenberg, A.M. and M.P. Kaschak, "Grounding Language in Action", *Psychonomic Bulletin & Review* 9 (2002): 558–565.

Green, M.C. and T.C. Brock, "The Role of Transportation in the Persuasiveness of Public Narratives", *Journal of Personality and Social Psychology* 79 (2000): 701–721.

Hammond, M., *Homer: The Odyssey* (London: Duckworth, 2000).

Hannay, M. and C.H.M. Kroon, "Acts and the Relationship Between Discourse and Grammar", *Functions of Language* 12 (2005): 87–124.

Havas, D.A., A.M. Glenberg, and M. Rinck, "Emotion Simulation During Language Comprehension", *Psychonomic Bulletin & Review* 14 (2007): 436–441.

Kaschak, M.P., J.L. Jones, J. Carranza, and M.R. Fox, "Embodiment and Language Comprehension", in *The Routledge Handbook of Embodied Cognition*, ed. L. Shapiro (London and New York: Routledge, 2014), 118–126.

Keen, S., *Empathy and the Novel* (Oxford: Oxford University Press, 2007).

Kroon, C.H.M., *Discourse Particles in Latin: A Study of* nam, enim, autem, vero *and* at (Amsterdam: Gieben, 1995).

Lakoff, G., *Woman, Fire and Dangerous Things: What Categories Reveal about the Mind* (Chicago, IL: The University Press of Chicago, 1987).

Langacker, R.W., *Foundations of Cognitive Grammar, Vol. I: Theoretical Prerequisites* (Stanford, CA: Stanford University Press, 1987).

Langacker, R.W., *Foundations of Cognitive Grammar, Vol. II: Descriptive Application* (Stanford, CA: Stanford University Press, 1991).

Langacker, R.W., *Grammar and Conceptualization* (Berlin/New York: Mouton De Gruyter, 2000).

Langacker, R.W., "Discourse in Cognitive Grammar", *Cognitive Linguistics* 12 (2001): 143–188.

Langacker, R.W., *Cognitive Grammar: A Basic Introduction* (Oxford: Oxford University Press, 2008).

Langacker, R.W., "Construal", in *Handbook of Cognitive Linguistics*, ed. E. Dąbrowka and D. Divjak (Berlin: De Gruyter,2015), 120–142.

Nell, V., *Lost in a Book: The Psychology of Reading for Pleasure* (New Haven, CT: Yale University Press, 1988).

Ooms, S. and C.C. de Jonge, "The Semantics of ΕΝΑΓΩΝΙΟΣ in Greek Literary Criticism", *Classical Philology* 108 (2008): 95–110.

Rijksbaron, A., "The Imperfect as the Tense of Substitutionary Perception", in *Hyperboreans: Essays in Greek and Latin Poetry, Philosophy, Rhetoric and Linguistic*, ed. P. da Cunha Corrêa, M. Martinho, J.M. Macedo and A. Pinheiro Hasegawa (São Paulo: Humanitas/CAPES, 2012), 331–371.

Ruijgh, C.J., *Autour de τε épique. Etudes sur la syntaxe grecque* (Amsterdam: Hakkert, 1971).

Ryan, M.-L., *Possible Worlds, Artificial Intelligence, and Narrative Theory* (Bloomington and Indianapolis, IN: Indiana University Press, 1991).

Ryan, M.-L., *Narrative as Virtual Reality: Immersion and Interactivity in Literature and Electronic Media* (Baltimore, MD and London: Johns Hopkins University Press, 2001).

Ryan, M.-L., *Narrative as Virtual Reality 2: Immersion and Interactivity in Literature and Electronic Media* (Baltimore, MD and London: Johns Hopkins University Press, 2015).

Sanford, A.J. and C. Emmott, *Mind, Brain and Narrative* (Cambridge: Cambridge University Press, 2012).

Scheppers, F., *The Colon Hypothesis: Word Order, Discourse Segmentation and Discourse Coherence in Ancient Greek* (Brussels: VUB Press, 2011).

Slings, S.R., "Written and Spoken Language: An Exercise in the Pragmatics of the Greek Sentence", *Classical Philology* 87 (1992): 95–109.

Taylor, L.J. and R.A. Zwaan, "Motor Resonance and Linguistic Focus", *Quarterly Journal of Experimental Psychology* 61 (2008): 896–904.

Tsagalis, C., *From Listeners to Viewers: Space in the Iliad* (Washington, DC: Center for Hellenic Studies, 2012).

Webb, R., *Ekphrasis, Imagination and Persuasion in Ancient Rhetorical Theory and Practice* (London/New York: Routledge, 2009).

West, M., *Greek Metre* (Oxford: Clarendon Press, 1982).

Wolf, W., *Ästhetische Illusion und Illusionsdurchbrechung in der Erzählkunst. Theorie und Geschichte mit Schwerpunkt auf englischem illusionsstörenden Erzählen* (Tübingen: Niemeyer, 1993).

Wolf, W., "Aesthetic Illusion as an Effect of Fiction", *Style* 38 (2004): 325–351.

Wolf, W., W. Bernhart, and A. Mahler, eds., *Immersion and Distance: Aesthetic Illusion in Literature and Other Media* (Amsterdam and New York: Rodopi, 2013).

Yao, B., P. Belin, and C. Scheepers, "Silent Reading of Direct Versus Indirect Speech Activates Voice-Selective Areas in the Auditory Cortex", *Journal of Cognitive Neuroscience* 23 (2011): 3146–3152.

Zwaan, R.A., "The Immersed Experiencer: Toward an Embodied Theory of Language Comprehension", in *The Psychology of Learning and Motivation*, Vol. 44, ed. B.H. Ross (New York: Academic Press, 2004), 35–62.

Zwaan, R.A., C.J. Madden, R.H. Yaxley, and M.E. Aveyard, "Language Comprehenders Mentally Represent the Shape of Objects", *Psychological Science* 13 (2004): 168–171.

4

Roman cultural semantics

William Michael Short

In this chapter, I outline an approach to the study of ancient Roman culture that I call "Roman cultural semantics".[1] This approach owes much to the school of Roman anthropology pioneered especially by Maurizio Bettini in Italy, in particular its semiotic view of culture as a global ordering of signs, texts, and meanings; its emphasis on language as probably the most immediate index to culture; and its privileging of the "native's" point of view in ethnographic description.[2] My approach builds on this foundation, however, by adopting certain theories and methods of the so-called "second-generation" cognitive sciences.[3] Drawing especially on the theory of conceptual metaphor developed in cognitive linguistics, and the theory of "folk" models in modern cognitive anthropology, my method is to use patterns of conventionalized figurative expression in Latin to reconstruct the conceptual models that characterize the everyday thinking of Roman society and that organize and shape its symbolic activities (including, but not limited to, language). From conceptual metaphor theory, I adopt the idea that figurative sense relations tend to characterize *not* any individual word's semantic structure, but a language's overall conceptual system: in other words, that metaphors manifest supra-lexical structures of meaning that consist in mappings between conceptual domains at the level of cognition (and indeed neuronally).[4] From cognitive anthropology, I take the view that metaphors often operate as a society's "working theory" of experience, delivering inferences and constituting an automatically relied-upon basis for reasoning and behavior vis-à-vis certain domains.

This kind of approach therefore provides a realistic, plausible account of how Latin speakers' language, thought, and behavior across seemingly unrelated contexts of social practice fit together under a coherent worldview. By showing how culture-specific ways of understanding are scaffolded on human-universal processes of cognition, this approach in fact affords exactly the sort of etic framework needed to produce emic analyses of Roman culture, especially when incorporating a comparative perspective. While sketching out the theoretical underpinnings of my approach, I illustrate its analytic potential through a study of Roman culture's metaphors of "meaning" and "ambiguity".

What kind of semantics is a Roman cultural semantics to begin with? In my view, it is one committed to a theory of "embodied" meaning like that developed in cognitive linguistics and allied disciplines. In cognitive linguistics, thought and language are held to be fundamentally dependent on the nature of the human brain and grounded in the character of human bodily interaction with the social and physical environment.[5] Instead of viewing meanings as defined by lists of necessary and sufficient conditions expressed in amodal propositional format (as in traditional philosophical and linguistic semantics), an embodied semantics emphasizes how human categorization and conceptualization is instead deeply constrained by the way we encounter the world, and thus situationally embedded sensory and motor capacities are responsible for the intuitive core of concepts.[6]

Cognitive linguistics argues that the meanings of words very often in fact correspond to gestalt experiential structures or "image schemas".[7] Image schemas are low-level structures of cognition which capture regularities in human bodily experience and provide the conceptual building blocks used in our understanding and reasoning. Though almost always treated as being basically visual in nature, image schemas are theorized as preconceptual cognitive representations that depict the structural contours of sensorimotor experience and indeed integrate information from multiple different modalities. Thus, along with schemas deriving from visuospatial experience – IN and OUT, CONTAINMENT, or PATH, for instance – that are easily pictured "in the mind's eye" (or represented graphically, as below), there are also more abstract orientational schemas such as UP and DOWN, CENTER and PERIPHERY, as well as force schemas like MOTION or BLOCKAGE. Image schemas can also be combined together to create more complex scenarios such as BALANCE or RESISTANCE.[8] Because the meanings of words correspond to image schematic conceptualizations, their semantic structure can therefore be explained through body-based construal processes such as salience (focus of attention) modifications and figure/ground re-alignment (conceptual foregrounding or backgrounding), vantage point (perspective) shifts, summary or sequential scanning (conjunction or disjunction in temporal phasing), scalar adjustment (extent of spatial dimensionality), and path or endpoint focus (following the trajectory of fictive motion or highlighting its goal) that may operate on image schemas as correlates of sensorimotor experience.[9]

The classic example of construal at the level of lexical organization in English is probably Charles Fillmore's account of the difference between *shore* and *coast*.[10] Although both words denote the boundary separating a body of water from a contiguous land mass, they differ in the visual perspective taken on this referent: *shore* refers to this boundary from the water's point of view, whereas *coast* refers to it from the land's. An example of how construal can organize a word's internal semantic structure, meanwhile, is provided by Claudia Brugman and George Lakoff's analysis of *over*, whose basic meaning they describe as an imagistic scenario portraying a dynamic trajector located above, and following a path across, a static landmark.[11] Different construal operations effecting either the trajector or the landmark, or both, within the imagined scene then account for *over*'s range of meanings: so, for instance, the difference between "The painting is over the mantle" and "The plane is flying over the hill" emerges from a static or dynamic construal of the trajector, while the difference between "Sam walks over the hill" and "Sam lives over the hill" can be accounted for by the so-called "path to end-of-path" transformation, where conceptual focus narrows from the trajector's imagined path of motion to the endpoint of this motion.

In Latin, the difference in meaning between *ora* and *litus* captures a similar variation in construal, as *ora* is typically used in contexts where the water's edge is referred to from the point of view of the inhabitation or occupation of land (e.g., Cic. *Fam.* 12.5.1, *a prima enim ora Graeciae usque ad Aegyptum optimorum civium imperiis muniti erimus et copiis*, "For we will be fortified by the commands, and troops, of the best citizens from the nearest coast of Greece all the way to Egypt"), while *litus* occurs largely in sea-faring contexts and where the perspective of someone sailing is focalized (e.g., Verg. *Aen.* 7.24, *quae ne monstra pii paterentur talia Troës / delati in portus neu litora dira subirent*, "Lest the pious Trojans undergo these monstrous changes, if carried to that port or land on those cursed shores . . ."; Hor. *Ep.* 16.40, *Etrusca praeter et volate litora*, "fly alongside Etruscan shores"). As defined by the *Digest* (50.16.96), "*litus* is (the point) up to which the highest tide from the sea reaches". In point of fact, *litus* denotes the boundary that separates the land from the sea as an abstract topographical reference line, while *ora* denotes the area of land that borders on the sea and in geographical reference to the adjacent land, and so is almost always opposed to inland territory. This is why it is possible to say in Latin *litoris ora* "the coast of the shore", as *ora* describes the real tract of land that extends along the imaginary boundary defined by *litus*: cf. Plin. *Ep.* 5.6.2, *gravis et pestilens ora Tusculorum, quae per litus extenditur*, "the harmful and disease-ridden coast of Etruria, which extends along the shore".

By the same token, my analysis of *de* and Luisa Brucale and Egle Mocciaro's of *per* illustrate the operation of various kinds of construal in determining the range of senses characterizing prepositional semantics in Latin. In the case of *de*, this word's highly variable spatial senses can all be accounted for by positing dynamic shifts in salience (conceptual prominence) over the elements that constitute its core meaning: a region of destination, a region of origin, and a trajectory connecting these. Similarly, *per*'s possible spatial senses – especially the difference between its restrictive ("through") and extensive ("throughout") uses, as in *liberos homines per urbem modico magis par est gradu ire*, "Free men should proceed through the city at a moderate pace" (Plaut. *Poen.* 522) and *per corium, per viscera, perque os elephanti transmineret brachium*, "Your arm would have pierced through the elephant's skin, innards, and bone" (*Mil.* 29-30) – emerge out of alternately uniplex and multiplex construals of an underlying path schema.

If in an embodied cognitive semantics the literal meanings of words correspond directly to sensorimotor schemas, abstract conceptualization comes about through the figurative interpretation of image schematic structure.[12] According to the theory of conceptual metaphor, it is through the regular metaphorical projection or "mapping" of embodied image schemas onto concepts not directly grounded in experience that human abstract thought is in fact possible. In this view, people talk about abstract concepts in metaphorical terms of concrete physical concepts because they conceptualize them metaphorically. Metaphors, that is, much more than simply imaginative or creative ways of speaking about certain abstract concepts, actually constitute people's ways of understanding and reasoning about those concepts. Metaphorical mappings are not arbitrary and unconstrained, however: not anything can serve as a source for conceptualizing anything else. Rather, metaphors are motivated by, or grounded in, embodied experience. Typically, this grounding is provided by systematic correlations in phenomenal experience (that is, when the pairing of domains occurs repeatedly in experience) or by perceived structural correspondences between experiences (where they are seen as having a similar inherent cognitive "topology"). Moreover, metaphorical mappings do not involve the wholesale transfer of conceptual material from the more concrete to the more abstract domain. Instead, mappings are always partial, preserving conceptual structure from the source domain only so long as it is consistent and compatible with the existing structure of the target domain.[13] Abstract concepts are thus "embodied" to the degree that they piggyback on structures of cognition (image schemas) that emerge from our bodily interaction with the world, and make sense to us by virtue of their grounding in sensorimotor experiences that are directly meaningful.[14]

Consider the conceptualization of fear captured by Latin *horror*.[15] The etymology of this word from *horreo* "stand erect; bristle" indicates this conceptualization centrally includes the idea of hair standing on end. As suggested by expressions like *mihi frigidus horror / membra quatit*, "Cold fear shakes my limbs" (Verg. *Aen.* 3.29–30), *mihi gelidus horror ac tremor somnum excutit*, "Cold fear and shaking knocks the sleep out of me" (Sen. *Tro.* 457), and *me luridus occupat horror*, "Pale fear overwhelms me" (Ov. *Met.* 14.198), fear is additionally conceived by Latin speakers as cold, shivering, and paleness. *Horror* is also something that "seizes" (*percapere*), "strikes" (*excutere*), or "penetrates" (*pervadere*) the body. These images are not simply fancy ways of talking about fear, however, nor are they selected at random. Instead, they are (part of) Latin speakers' absolutely regular and ordinary vocabulary of fear. Moreover, the metaphor itself constitutes the understanding that Latin speakers have of fear: for speakers of Latin, fear is – not simply "is like" – cold, shivering, and paleness. Moreover, what makes sense of these metaphors (that is, what makes them easily interpretable as being about fear) is that they are grounded in concrete physical experience: very specifically, the autonomic bodily responses – especially sharp fluctuations in body temperature, along with perspiration, shivering or shuddering, pallor, and piloerection – that characterize, and systematically correlate to, the lived experience of fear in human beings universally.[16]

In an embodied metaphorical conceptualization like *horror* in Latin, it is not one single element of the experience of fear (cold, pallor, piloerection) that delivers abstract understanding: rather, it is the inferential structure of this experience (i.e., that certain physiological reactions occur together and have certain kinds of knock-on effects in the body, like sleeplessness or the sensation of being wholly overwhelmed by emotion) that provides structure to conceptualization and thus to language. This is why fear (esp. *metus*) is said in Latin to "break" (*frangere*) (cf. Quint. *IO*. 12.5.1; Plin. *Pan*. 93.1) someone, carrying through the imagistic logic of cold, brittleness, and shaking. Because the mappings in a conceptual metaphor are systematic and project relational structures as well as conceptual content, inferential patterns in fact regularly carry over to the metaphorically defined domain to constitute its logic. In this sense, metaphors function as the mental models that form the basis for reasoning and behaving vis-à-vis abstract domains.[17]

Because metaphorically defined mental models of this kind emerge naturalistically, however – rather than being constructed explicitly for description and explanation – they can be characterized as "folk" models.[18] In cognitive anthropology, a folk model is any non-technical or naïve understanding that functions as the unconscious and automatic operating theory of some domain of experience. Furthermore, because folk models are intuitively meaningful (because based on figurative patterns highly conventionalized in thought and speech), they tend to be widely distributed and intersubjectively shared, and so mediate the behaviors of those who possess them across different contexts of social practice. Cognitive linguists take as given that complex, culturally situated concepts may in fact be modeled via whole networks of metaphors. Where a concept is characterized by several distinct metaphorical images, these metaphors work together to provide an overall coherent understanding, each metaphor targeting the understanding of some dimension of the metaphorically defined concept.[19] Though the metaphors may yield competing or even contradictory inferences, they provide a basis for reasoning that is adequate to most everyday situations.[20]

Consider the concept of "meaning" in Latin.[21] As Figure 4.1 is meant to illustrate, Latin speakers' conceptualization of meaning emerges from several different metaphorical mappings. Constructions with *sibi vult*, for instance, like Terence's *quid volt sibi, Syre, haec oratio?* (*Heaut*. 615), suggest a personification metaphor in which what a linguistic expression "wills" or "wants (for itself)" is metaphorically understood as its meaning. Plautus's *ut litterarum ego harum sermonem audio* (*Ps*. 98), where Pseudolus likens understanding the message of a letter he has just read aloud to hearing its "speech", suggests that meaning could sometimes also be construed in terms of vocal utterance. Otherwise, meaning could be understood as the "(state of) mind" (*mens*) a word conveys or as a kind of "force" (*vis*, and later *potestas*) residing in words (cf., e.g., Cic. *Fam*. 6.2.3; *Fin*. 2.2.6; Gell. *NA*. 10.29.1). Cicero even once seems to refer to the meaning of a word as its "house" (*Fam*. 16.17.1, *cui verbo domicilium est proprium in officio*). At the same time, *significatio* and *significatus* employ the image of "marking" or "stamping" (*signum*). Finally, the etymology of *sententia* and *sensus* from PIE root **s(e)nt-*, which cognates indicate has to do literally with spatial motion

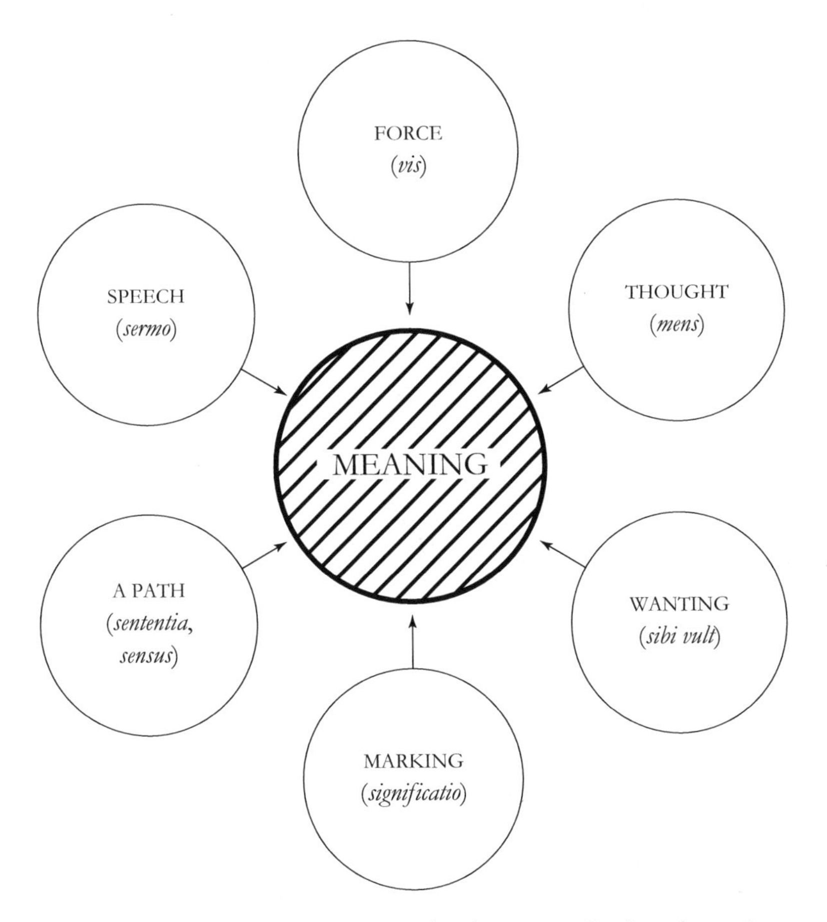

Figure 4.1 Metaphors converging on Latin speakers' conceptualization of meaning

(cf., e.g., OIr. *sét* "road"; PGerm. **sandjan*- "send", **sinþa*- "road; way", **sindō*- "travel"), as when traveling on a journey, suggests that meaning can also be conceptualized as a path.

Each of these metaphors targets the understanding of a somewhat different aspect of meaning. The personification metaphor emphasizes the human intentions that words express: what a word "wants" – its meaning – is an extension of the intentionality of the speaker (in fact, the formulation appears to originate in expressions having the speaker as subject: cf., e.g., Ter. *Eun.* 45, *ut pernoscatis quid sibi Eunuchus velit*). The speech metaphor again stresses the connection with human intention, but this time focusing, via a metonymy of "produced for producer", on the whole speech act as an expression of meaning. The state-of-mind metaphor operates via a similar metonymy, where the meaning of a word is directly equated with the mental representation (*mens*) constituting the proposition the human speaker's utterance is intended to convey. The "force" metaphor, meanwhile, probably has to do with the kinds of physical effects that words were felt to produce: their meaning is the way in which they impinge, directly and bodily, on a listener. (As Bettini 2008 has shown, Roman culture in fact viewed certain utterances as capable of bringing about tangible effects on the social and material environment.) The metaphorical notion of meaning as a word's "house", on the other hand, seems to be a way of understanding that words have their proper contexts of usage (just as a word has its proper "citizenship" by belonging to a particular language: cf., e.g., Quint. *IO.* 8.1.3). And the "stamping" metaphor is an attempt to describe the relation of the linguistic sign to its referent: in this image, a word's meaning is the thing in the world that it "marks" or "points out" against everything else.

These metaphors appear to be somewhat circumscribed, however. For instance, the "stamping" metaphor is largely confined to the jargon of the Latin grammarians. *Sibi vult* is attested above all in dramatic contexts and seems to belong to a mostly conversational register. *Vis*, meanwhile, is mostly found in Cicero and is probably a gloss on Greek *dúnamis*, which is used in the same sense from the fifth century onward. The house metaphor is very likely similarly inspired by Greek: in Aristotle's *Poetics*, a word's domain of reference is often called its *οἶκος* "house" and its literal sense is characterized as *οἰκεῖον* "belonging to the household" (Membrez 2018). The path metaphor, by contrast, represents Latin speakers' most general as well as most conventionalized conceptualization of meaning. By "most general" I mean that, rather than targeting some specific dimension of meaning through detailed imagery (as in Cicero's image of a word's *domicilium*) or focusing on a particular dimension of linguistic experience (as when an utterance's meaning is conceived in terms of the intentions of its human speaker) or on the effects of utterances (as in *vis*-meaning), this metaphorical conceptualization seems to have the broad function of picking out the domain of "meaning" and of providing it with minimal conceptual definition. It is the "most conventionalized" in the sense that it is the one metaphor most regularly used by Latin speakers to convey the concept of meaning and most deeply entrenched in Latin's semantic system. In fact, in addition to formulas like *(accipere) in* or *ad sententiam (partem, intellectum)*, where the sense in which a word is understood is expressed as directed motion along a path (*in/ad* + acc.),[22] the metaphor is embedded etymologically in the two commonest and the most contextually unmarked terms with this meaning, namely *sententia* and *sensus*. That Sanskrit *gati*, lit. "going" and *sañjñā* (cf. **snt-*), lit. "direction" can both be found in the sense of "meaning" suggests the path metaphor may in fact belong to Indo-European.

But how exactly do *sententia* and *sensus* (or *sañjñā* and *gati*, for that matter) capture an understanding of meaning in terms of a "path"? From a cognitive linguistic perspective, the etymological root sense of these words indicates that Latin speakers understood the concept of meaning via something like the SOURCE-PATH-GOAL schema – an image schema that, in Mark Johnson's description, emerges from our primary experience of learning to focus on and track objects moving through our visual field, as well as from repeated bodily activities, beginning in infancy, that involve intentional movement from one location to another (reaching for a toy, crawling, walking, or running toward a caregiver, or indeed traversing space to reach any desired destination). As he writes,

> Our lives are filled with paths that connect up our spatial word. There is the path from your bed to the bathroom, from the stove to the kitchen table, from your house to the grocery store . . . Some of these paths involve an actual physical surface that you traverse . . . Others involve a projected path, such as the path of a bullet shot into the air. And certain paths exist, at present, only in your imagination, such as the path from Earth to the near star outside our solar system.
>
> In all of these cases there is a single, recurring image schematic pattern with a definite internal structure. In every case of paths there are always the same parts: (1) a source, or starting point; (2) a goal, or endpoint, and (3) a sequence of contiguous locations connecting the source with the goal. Paths are thus routes for moving from one point to another . . .
>
> This definite internal structure for our path schema provides the basis for a large number of metaphorical mappings from concrete, spatial domains onto more abstract domains.
>
> *(Johnson 1987: 113–14)*

As a cognitive representation that generalizes over these and myriad similar activities, the path schema organizes and makes sense of experience by affording a certain conceptual structure or "topology" to perceptions, images, and events. This structure can be represented as in Figure 4.2, where the schema consists of a moving entity or "trajector" (marked as TR) that traces a linear path from a source-point to a fixed endpoint.

Yet I believe the interest of this image schematic metaphor goes well beyond just linguistic curiosity – that Latin speakers happen to have understood meaning, as captured by the terms *sententia* and *sensus*, in terms of linear spatial motion, "going". As Latin speakers' most conventionalized conceptualization of meaning, the path metaphor can also be seen to have systematic organizing effects on conceptualization. This can be easily detected not only in how the metaphor structures Latin's concepts of ambiguity – that is, multiplicity of meaning in language – but also, I suggest, in how this metaphor informed the kinds of values that Latin speakers attach to ambiguity as part of cultural experience.

As Claude Moussy has shown, Latin's normal way of conveying the concept of "ambiguity" is given by the terms *ambiguum* and *ambiguitas* (and, more rarely, *ambages*).[23] Ambiguity was thus understood by wholly metaphorical means – since, etymologically, these words are all constructed on the basis of the preposition *ambi-* "round; about" and *agere* "lead", whose combined literal meaning is something like "leading (around) on two sides".[24] Latin speakers' metaphorical conceptualization of ambiguity can therefore be represented as something like Figure 4.3, which suggests that what is "ambiguous" is understood in Latin through the complex image of paths diverging.

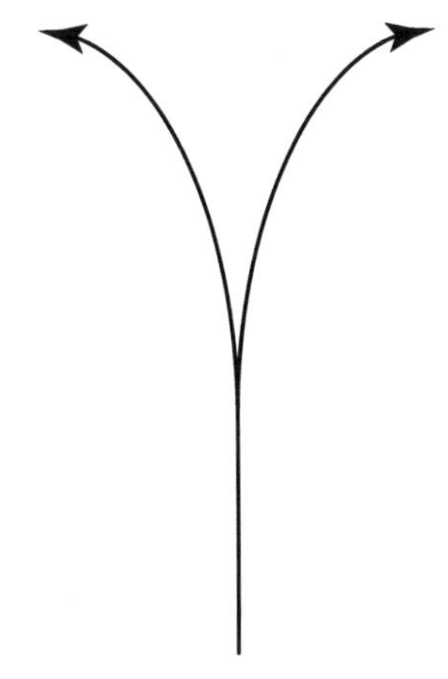

Figure 4.2 Path schema constituting the concept of meaning captured by *sententia* and *sensus*

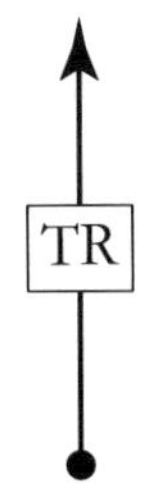

Figure 4.3 Image schematic understanding of ambiguity (*ambages, ambiguum, ambiguitas*) in terms of paths diverging (< *amb-* "around" + *agere* "lead")

It is reasonable to ask, however, what an analysis of the image schematic and metaphorical structuring of Latin's conceptualization of meaning and ambiguity offers in terms of insight into the Latin language and Roman society. From a narrowly linguistic perspective, I would suggest that this kind of analysis provides a model of meaning capable of explaining (not only of describing) the polysemy of words in a systematic fashion: that is, it provides a motivated account for why a word whose etymological sense has to do – historically speaking – with motion along a path has the meaning of "meaning", and also why the image of diverging paths means "ambiguous" (i.e., "having multiple different meanings") for Latin speakers. These two facts of Latin's semantic system are not independent. Rather, they make sense together and they make sense of each other. If spatial motion ("going": *sententia, sensus*) metaphorically delivers the logic of meaning, then Latin speakers have good reason for conceiving multiplicity of meaning – ambiguity – in terms of a multiplicity of paths ("leading around": *amb-ig-*). Latin's conceptualizations of meaning in terms of a path and of ambiguity in terms of paths diverging are, in other words, internally motivated and mutually supporting. Paying close attention to the image schematic structuring of words thus helps shed light on the way in which Latin's system of meanings is interconnected and self-consistent.

Still, this does not go much beyond what any mainline structuralist semantics could probably tell us.[25] What an image schematic account along cognitive linguistic lines allows us to see, in addition, is that these metaphorical patterns are also motivated by features that stand outside of the semantic system proper and are grounded in embodied experience. What I mean is that Latin speakers' metaphorical representation of meaning in terms of spatial motion (and so too any conceptualization entailed by this metaphor) arises out of definite patterns of human embodiment, and thus "makes sense" because it fits with, and within, a meaning structure that characterizes Latin's vocabulary in a systematic way. If for Latin speakers meanings are equated with intentions (as expressions like *sibi vult* and the use of *mens* to mean "meaning" confirm), meanings can be spoken of metaphorically as "goings" because of recurring experiences in which intentions (to do something, say something, or whatever) are actually coupled with spatial motion. For human beings, intentions in fact normally involve doing something *somewhere* and therefore moving our bodies in space in order to achieve those intentions. Our experience of having intentions, that is, is correlated with going to places and this strong experiential correlation motivates the metaphorical representation of meaning (intention) directly as a "going" in both thought and language.[26] In this sense, a cognitive linguistic account allows us to explain both the systematic nature of linguistic forms and the motivation for these forms in the very first place (and so to obviate the problem of *l'arbitraire du signe*, i.e., of language as an ungrounded symbol system).

What's more, this kind of account allows us to extend beyond the confines of language by offering a model of cultural meaning.[27] In treating figurative associations pervasive and highly conventionalized in the linguistic system as reflecting conceptual (not narrowly linguistic) structures, such an approach can help identify the sorts of meanings that link together Roman society's various imaginative activities into a cohesive signifying order. For instance, alongside representations of the concept of ambiguity in spatial terms embedded in the Latin language, we can think of the many times that ambiguity comes to be thematized in explicitly linear spatial terms in literary representation. Catullus, Vergil, Ovid, and Petronius, for instance, all deploy the image of the labyrinth to symbolize a whole series of thematic uncertainties and paradoxes, including or especially relating to poetic composition.[28] In all these cases, the metaphor of ambiguity as paths diverging functions as a symbolic framework that suggests the association of two conceptual domains and affords specific imagery for imagining ambiguities of different kinds.

An image schematic account, especially when combined with a comparative perspective, can in fact help provide the sort of "experience-near" perspective that anthropologists take for granted as the aim of ethnographic study, by highlighting how different languages and cultures may capture the "same" concept through different metaphorical images.[29] Indeed, comparing Latin's linear spatial metaphor of ambiguity with that of Greek suggests that even subtle shifts in the underlying imagery of metaphorical concepts can have significant implications for understanding. Greek's vocabulary of ambiguity, especially terms like *ἀμφιβολία* (*ἀμφιβολητικός*), *ἀμφίγλωσσος*, and *ἐπαμφότερος*, and likewise *διχόγνωμος* where the prepositions *ἀμφί* ("around") and *δίχα* ("on both sides") deliver figurative meaning through spatial images, appears very similar to Latin's. However, the image metaphorically underpinning these concepts turns out to be quite unlike that of the Latin metaphor. Usage suggests that the literal sense of *ἀμφί* actually has to do with notions more of containment and even concealment than of paths diverging, and that *δίχα* refers in particular to static position (rather than to linear motion) "on two sides".[30]

This seemingly minor shift in metaphorical imagery appears to have, however, significant implications for the sorts of inferences that speakers of Greek and Latin make about ambiguity and indeed on the value they attribute to it. Consider the ontology of ambiguity that emerges from the image of the Greek metaphor. In this metaphor, ambiguity is seen as a kind of enclosure that surrounds or encloses or conceals another object. As such, ambiguity renders its "object" temporarily inaccessible, and, being something placed on top of and "(completely) around" and "on both sides" of its object, will also require additional effort to be removed in order to uncover whatever lies beneath. Metaphorically interpreted as "meaning", this image implies a theory of how meanings relate to words. It implies, first, that any expression has both an "outer" meaning, which is in some way contingent, and an "inner" meaning, which represents the true sense of the expression. Second, that ambiguity belongs squarely to the level of "outer" meaning and is, as something obstructing access to "inner" meaning by "covering" it, simply detrimental to truth. No doubt this helps explain why in the Greek world, as Catherine Atherton has written, "Ambiguity was . . . regarded . . . as a difficulty or defect, something to be coped with, not courted, and eliminated if possible". Certainly Aristotle's *Topics* (e.g., 1.13.105a23, 1.15.106a1, 1.18.108a18) is full of suggestions on how to use ambiguity resolution in argumentation (for clarity or deceit or simply as a test of relevancy).

In the image of the Latin metaphor, on the other hand, ambiguity is a kind of divergence from the singular meaning-trajectory of an expression, which presents the availability of alternate sense-paths "around" or "on both sides of" that trajectory. Under this image, ambiguity does not actually seem to preclude the discovery of true meaning at all. In fact, just as when traveling on a journey a detour can eventually bring a traveler back to his or her initially set-upon path, or provide simply different – but no less feasible – ways of reaching the originally intended destination, under this metaphor ambiguity constitutes a different, even if somehow indirect and unforeseen, means of discovering something's true meaning. In fact, like a fork in the road, ambiguity emerges not so much as an obstruction on the way to true meaning that must be avoided at all costs, but rather as a naturally occurring and perhaps even essential part of any meaning-journey. Unlike a garment or net or other covering, ambiguity in this sense cannot be considered (in terms of the metaphor) separate at all from something's "true" meaning. Ambiguity thus turns out to be a mode of truth-finding that is seen as equally acceptable as what may otherwise have seemed the most direct route. There may even be something appealing and worthwhile about ambiguity, just as stepping off the beaten path can often bring the traveler unexpected discoveries.

What makes this metaphor a truly cultural model is that the theory of ambiguity implied by the image of paths diverging – that ambiguity is in no way preclusive of something's "true" meaning – appears to characterize Roman society's default ways of valuing ambiguity across different areas of social life. We know that Roman authors, far from avoiding this kind of interpretive uncertainty, frequently employed ambiguity as a part of their imaginative literary expression: as Karl Galinsky has written, "The Romans' concept of *ambiguitas* . . . is more akin to a polysemy which is deployed quite intentionally, and not just by the poets".[31] Following the Greek (Aristotelian) view, the author of the *Rhetorica ad Herennium* in fact criticizes writers who, in a kind of fad, always "are on the lookout for any double meanings, even where one of the meanings renders nonsense" (2.16, *omnes enim illi amphibolias aucupantur, eas etiam quae ex altera parte sententiam nullam possunt interpretari*). We also recognize that ambiguity was an essential element of Roman religious thought, actually typifying belief about the nature of divinity, which could be represented almost indifferently as singular or plural, or even as male or female. Roman society readily introduced ambiguities of identity between "master" and "slave" and "public" and "private" in festival contexts as well, especially in the role-inversions of the Saturnalia and Compitalia. Such welcoming of ambiguity in so many circumstances suggests that Latin speakers appreciated ambiguity as a feature of their symbolic world generally, viewing it as not only not inconsistent with but also in fact determinative of a kind of truth.[32]

To my claim that Roman culture's largely positive valuation of ambiguity emerges as a direct entailment of the "paths diverging" metaphor (and that Greek's largely negative valuation of the same concept emerges equally from their "enclosure" metaphor), the objection will inevitably arise that I am advocating a Whorfian view of the relationship between language and thought: that I am proposing, in other words, that the Latin language itself determines Latin speakers' possible ways of thinking and reasoning about ambiguity. This is not at all what I am arguing, though. Still, because the Sapir-Whorf Hypothesis (or "linguistic relativity hypothesis") is normally seen as coming in two varieties – a "strong" and a "weak" variety – it may be instructive to consider where my claim sits with respect to these ideas.[33] The "strong" version of linguistic relativity, as we know, states that the grammatical structures of a language – how it encodes time, gender, or number, for instance – sets a limit to how its speakers understand the world. Sapir had claimed that individuals are in fact "at the mercy of" their language, which exerts a "tyrannical hold" over their minds. Likewise, for Whorf, people are parties to an "absolutely obligatory" agreement to view the world in a certain way, which is provided by the "thought-grooves" laid down by their language. In the "weak" version of the hypothesis, the idea of motivation is substituted for that of determination, with languages seen as merely providing the pathways of thought its speakers will normally follow, all other things being equal. Dan Slobin (1996) has recently advocated something very much like this with his notion of "thinking for speaking", which proposes that because most of the time we are not just "thinking" for thinking's sake but in fact planning for verbal communication, we tend to rely on the categories provided by our language as a convenient or default conceptual framework, even if we are not absolutely bound by them.

It should be clear that what I have been claiming about Latin's metaphors of meaning and ambiguity – and what a Roman cultural semantics claims about Latin's metaphors overall – is not at all of a piece with the strongest version of the Sapir-Whorf Hypothesis, and is only partially in line with the weaker form of this hypothesis. Indeed, I have argued that Latin speakers' ability to conceptualize a complex, multi-dimensional concept like MEANING rests precisely in their possessing multiple different metaphors for this one concept: in this sense, the metaphors create, rather than inhibit, understanding and do so by working together as a system. It is the simultaneous availability of the different metaphorical ways of imagining meaning that in fact enables Latin speakers' understanding of this concept, even if it is also true that the "path" metaphor represents their preferred or privileged conceptualization. But, more importantly, my claim has been that the metaphors underpinning Latin speakers' concepts of meaning and ambiguity do not in the first instance belong to language, but to their overall conceptual system. The metaphors are mappings that operate at a neural, cognitive, conceptual, and cultural level all at once. The Latin language reflects these metaphors because this language is a manifestation of the cognitive and conceptual categories – including metaphorically structured concepts – that systematically frame and organize the way its speakers think and talk. The language is metaphorical because the thinking is metaphorical.

What I am saying, then, is that metaphorical linguistic expressions in Latin are capable of revealing to us the also metaphorical structure of the concepts that speakers of this language possess, and of pointing us toward seeing ways in which these metaphorically structured concepts may function as the "everyday" mental models that pervasively and persistently guide Roman society's symbolic expression. Like all human beings, Latin speakers rely on metaphor as a major mechanism of abstract conceptualization and reasoning. Thus, we can expect that the inherently metaphorical structure of many of their concepts will manifest itself not only in their conventionalized linguistic expression, but also in their literature, in their rituals, and in their belief systems. The fact that their thinking rests partly on a certain number of systematic conceptual metaphors means that their language will activate those metaphors, which we can often very easily observe in the meanings of words, idiomatic phrases, or even large-scale syntactic constructions and then trace out as they work their effects on other forms of imaginative activity. Where metaphors are traceable across different levels of linguistic encoding, across different authors, genres, and periods of the language, we should not hesitate to ascribe these meanings to the conceptual system – in a word, to the culture – that Latin speakers share, and which they rely on when constructing and interpreting texts of all kinds.

Notes

1 The term "cultural semantics" comes especially from Kövecses 2006, where the author lays out a similarly metaphor-based approach to culture. This approach differs from the cultural semantics of Jay 2012 in that it stresses a cognitive conception of culture, and eschews Jay's strongly Whorfian view that language use *per se* mediates and shapes our experience. It also differs from the cultural semantics formulated in Clifford and Wierzbicka 2014, which stresses "scripts" as a basic unit of culture analysis. It is therefore closest to the "cultural linguistics" of Sharifian and Palmer 2007, also heavily influenced by cognitive linguistics and cognitive anthropology.

2 See esp. Bettini and Short 2014.

3 If the "first-generation" cognitive sciences viewed cognition largely in information-processing terms, the "second-generation" cognitive sciences emphasize mental processes as embodied, embedded, enacted, and extended: see esp. Rowlands 2010; Boden 2008; and Wilson 2002.

4 Thus, for instance, the Latin word *locus*, literally, "a place", can be used metaphorically in the sense of "an idea" (as in Cic. *Div*. 2.1.2, *perpurgatus est <u>is locus</u> a nobis quinque libris*, "I have treated that idea in five books"). But this figurative construal of thought in unambiguously *spatial* terms can also be detected in dozens of other expressions, such as where different kinds of mental processes are represented in terms of "entering" or "exiting" or "being in" locations.

5 See esp. Lakoff and Johnson 1980 and 1999; Johnson 1987; Gibbs 1994; and Grady 1997.

6 So, under this view, even categories with seemingly fixed criteria for membership and stable boundaries like BIRDS, TOOLS, or FRUIT are seen as organized by principles of fuzzy classification, where membership may be graded and based on similarity to a best example or most salient member ("prototype"), or conditioned by prior experience, ad-hoc contextual considerations, overall visual appearance, or patterns of behavioral interaction: see Rosch 1973 and 1978; Barsalou 1983; Lakoff 1987; Taylor 1989; Atran 1993.

7 On image-schematic conceptualization, see esp. Hampe and Grady 2005; Talmy 2003; Taylor 2002; Gibbs and Colston 1995; Langacker 1987; Lakoff 1987; Johnson 1987; Lakoff and Johnson 1980.

8 Cf. Tyler and Evans 2003, who refer to these as "protoscenes".

9 For construal processes, see Langacker 1990: 315–42.

10 Fillmore 1982: 121.

11 Brugman and Lakoff 1988.
12 Kövecses 2006 and 2005; Lakoff 1993; Lakoff and Johnson 1980; Lakoff 1987; Johnson 1987; cf. also Feldman 2006.
13 The Invariance Principle: see Lakoff 1990; Turner 1990; Brugman 1990.
14 Recent studies suggest that metaphors are in fact instantiated neuronally through the recruitment of domain-specific sensory cortex during figurative language processing: cf., e.g., Desai et al. 2011, 2013; Lacey et al. 2012.
15 For extended discussion, see Wharton 2011.
16 Cf. Panksepp 2010, 1989, 1988; LeDoux 2016; and Gray 1987.
17 For mental models generally, see Johnson-Laird 1983, and now Held et al. 2006.
18 The concept of "folk model" has been developed particularly by Holland and Quinn 1987; D'Andrade 1990; and D'Andrade and Strauss 1992. Of course, this is not to suggest that "expert" technical models are not also normally metaphorical constructed: see Mischler 2013 and Gentner 1983.
19 On the character and function of such metaphor systems, see Danesi and Perron 1999 and Kövecses 2006.
20 Gentner and Stevens 1983 presents the classic study of the effect of competing metaphorical models on inference making.
21 For fuller discussion of this material, but from a somewhat different perspective, see Short, forthcoming.
22 E.g. Liv. *AUC.* 3.67.1, *ibi in hanc sententiam locutum accipio*, "I take it that he (sc. T. Quinctius Capitolinus) spoke on that occasion in this sense"; Plaut. *Eun.* 867, *equidem pol in eam partem accipioque et volo*, "For my own part I am quite willing to accept it in that sense"; Quint. *IO.* 10.1.11, *alia . . . ad eundem intellectum feruntur, ut ferrum et mucro*, "Others are understood in the same sense, like *ferrum* and *mucro*"; cf. also Cic. *Part.* 108.
23 See Moussy 2007; cf. also Christol 2007.
24 The derivation proposed by Pucci 2014: 220 n. 6, **amb- + gerere*, i.e., "bearing on both sides", is linguistically impossible.
25 Cf. O'Halloran 2003.
26 This is also why in English the "be going to" construction encodes not only spatial motion or – via a typical mapping of space to time – futurity, but also intentionality: see Jakobi 2006.
27 See Agar 1996 and 1995.
28 Cat. *Carm.* 64; Verg. *Aen.* 6.24–30; 5.588–95; Ov. *Met.* 8.157–68; Petr. *Sat.* 7374. Cf., e.g., Gaisser 1995; Martindale 1993; Doob 1990.
29 Cf. Detienne 2000; Bettini 2009; and Bettini and Short 2014.
30 E.g., *ἥρως Ἰδομενεύς ῥῆξεν δέ οἱ ἀμφὶ χιτῶνα*, "The warrior Idomeneus clave his coat of bronze round about him (sc. Alcathous)" (Hom. *Il.* 13.439); *ἐν δὲ κρήνη νάει, ἀμφὶ δὲ λειμών*, "In it (sc. the grove of Athena) a spring wells up, and round about is a meadow" (*Od.* 6.292); *κεραίαν μεγάλην δίχα πρίσαντες ἐκοίλαναν ἅπασαν*, "They sawed in two and scooped out a great beam from end to end" (Thuc. *Hist.* 4.100.2); *καὶ δόγμα ἐποιήσαντο, ἐάν τις τοῦ λοιποῦ μνησθῇ δίχα τὸ στράτευμα ποιεῖν, θανάτῳ αὐτὸν ζημιοῦσθαι*, "They passed a resolution that if any man from this time forth should suggest dividing the army, he should be punished with death" (Xen. *Anab.* 6.4.11).
31 Galinsky 1994: 305.
32 This is not to suggest, of course, that ambiguity would *always* have been valued by Latin speakers in every context of discourse: Quint. *IO.* 7.9 is a description of different types of (lexical, syntactic) ambiguity – including the famous *Lachetem audivi percussisse Demeam*, where the subject of the infinitive could be interpreted either as Lachetes or as Demea – and provides remedies through syntactic variation.
33 For more recent consideration of the Sapir-Whorf Hypothesis, see esp. Lucy 2005 and the essays collected in Gumperz and Levinson 1996.

References

Agar, M. 1995. *Language Shock: Understanding the Culture of Conversation.* New York.

Agar, M. 1996. *The Professional Stranger: An Informal Introduction to Ethnography.* New York.

Atherton, C. 1993. *The Stoics on Ambiguity.* Cambridge.

Atran, S. 1993. *Cognitive Foundations of Natural History.* Cambridge.

Barsalou, L. 1983. "Ad hoc categories". *Memory & Cognition* 11: 211-27.

Bettini, M. 2009. "Comparare i Romani. Per un'antropologia del mondo antico". *SIFC* 7 (4): 1–47.

Bettini, M. and W. M. Short. 2014. *Con i Romani. Antropologia della cultura antica.* Bologna.

Boden, M. 2008. *Mind as Machine: A History of Cognitive Science.* Oxford.

Bramble, J. 1970. "Structure and ambiguity in Catullus LXIV". *PCPhS* 16: 22-41.

Brugman, C. 1990. "What is the invariance hypothesis?" *Cognitive Linguistics* 1 (2): 257–66.

Brugman, C. and G. Lakoff. 1988. "Cognitive topology and lexical networks". In G. Cottrell, S. Small, and M. Tanenhaus, eds., *Lexical Ambiguity Resolution*, San Mateo, CA. 477-508.

Christol, A. 2007. "Du latin *ambiguus* à l'ambiguïté des linguistes". In C. Moussy and A. Orlandini, eds., *L'ambiguïté en Grèce et à Rome: approche linguistique*, Paris. 9-22.

D'Andrade, R. 1990. "Some propositions about relations between culture and cognition". In J. Stigler, R. Shweder, and G. Herdt, eds., *Cultural Psychology: Essays on Comparative Human Development*, Cambridge. 65–129.

D'Andrade, R. and C. Strauss, eds. 1992. *Human Motives and Cultural Models*. Cambridge.

Danesi, M. and P. Perron. 1999. *Analyzing Culture*. Bloomington, IN.

Deignan, A. 2003. "Metaphoric expressions and culture". *Metaphor and Symbol* 18 (1): 255–71.

Desai, R., J. Binder, L. Conant, Q. Mano, and M. Seidenberg. 2011. "The neural career of sensory-motor metaphors". *Journal of Cognitive Neuroscience* 23: 2376–86.

Desai, R., L. Conant, J. Binder, H. Park, and M. Seidenberg. 2013. "A piece of the action: Modulation of sensory motor regions by action idioms and metaphors". *NeuroImage* 83: 862–9.

Detienne, M. 2000. *Comparer l'incomparable*. Paris.

Doob. P. 1990. *The Idea of the Labyrinth from Classical Antiquity to the Middle Ages*. Ithaca, NY.

Feldman, J. 2006. *From Molecule to Metaphor: A Neural Theory of Language*. Cambridge, MA.

Fillmore, C. 1982. "Frame semantics". In Linguistic Society of Korea, ed., *Linguistics in the Morning Calm*, Seoul: Hanshin. 111-38.

Fillmore, C. 1985. "Frames and the semantics of understanding". *Quaderni di Semantica* 6 (2): 222-54.

Gaisser, J. 1995. "Threads in the labyrinth: Competing views and voices in Catullus 64". *AJP* 116: 579-616.

Galinsky, K. 1994. "Reading Roman poetry in the 1990s". *Classical Journal* 89: 297-309.

Gallagher, S. 2005. *How the Body Shapes the Mind*. Oxford.

Geertz, C. 1973. *The Interpretation of Cultures*. New York.

Geertz, C. 1983. *Local Knowledge: Further Essays in Interpretive Anthropology*. New York.

Gentner, D. 1983. "Structure-mapping: A theoretical framework for analogy". *Cognitive Science* 7: 155–70.

Gentner, D. and A. Stevens. 1983. *Mental Models*. Hillsdale, NJ.

Gibbs, R. 1994. *The Poetics of Mind: Figurative Thought, Language, and Understanding*. Cambridge.

Gibbs, R. and H. Colston. 1995. "The cognitive psychological reality of image schemas and their transformations". *Cognitive Linguistics* 6: 347-78.

Goddard, C. and A. Wierzbicka. 2014. *Words and Meanings: Lexical Semantics Across Domains, Languages, and Cultures*. Oxford.

Grady, J. 1997. *Foundations of Meaning: Primary Metaphors and Primary Scenes*. PhD dissertation, University of California, Berkeley, CA.

Grady, J. 1999. "A typology of motivation for conceptual metaphor". In R. Gibbs and G. Steen, eds., *Metaphor in Cognitive Linguistics*, Amsterdam. 79-100.

Gray, J. 1987. *The Psychology of Fear and Stress*. Cambridge.

Gumperz, J. and S. Levinson. 1996. *Rethinking Linguistic Relativity*. Cambridge.

Hampe, B. and J. Grady, eds. 2005. *From Perception to Meaning: Image Schemas in Cognitive Linguistics*. Berlin.

Held, C., G. Vosgerau, and M. Knauff. 2006. *Mental Models and the Mind: Current Developments in Cognitive Psychology, Neuroscience and Philosophy of Mind*. Amsterdam.

Holland, D. and N. Quinn, eds. 1987. *Cultural Models in Language and Thought*. Cambridge.

Jakobi, T. 2006. *The English Be Going to Construction and Its Grammaticalization Process*. German National Library.

Jay, M. 2012. *Cultural Semantics*. Amherst, MA.

Johnson, M. 1987. *The Body in Mind*. Chicago, IL.

Johnson, M. 1993. *Moral Imagination*. Chicago, IL.

Johnson-Laird, P. 1983. *Mental Models: Towards a Cognitive Science of Language, Inference and Consciousness*. Cambridge, MA.

Kelley, E. 1992. *The Metaphorical Basis of Language: A Study in Cross-Cultural Linguistics*. New York.

Kövecses, Z. 2005. *Metaphor in Culture: Universality and Variation*. Cambridge.

Kövecses, Z. 2006. *Language, Mind and Culture*. Oxford.

Lacey, S., R. Stilla, and K. Sathian. 2012. "Metaphorically feeling: Comprehending textural metaphors activates somatosensory cortex". *Brain and Language* 120 (3): 416–21.

Lakoff, G. 1987. *Women, Fire and Dangerous Things: What Categories Reveal about the Mind*. Chicago, IL.

Lakoff, G. 1990. "The invariance hypothesis: Is abstract reason based on image-schemas?" *Cognitive Linguistics* 1 (1): 39–74.
Lakoff, G. 1993. "The contemporary theory of metaphor". In A. Ortony, ed., *Metaphor and Thought*, Cambridge. 202–51.
Lakoff, G. and M. Johnson. 1980. *Metaphors We Live By*. Chicago, IL.
Lakoff, G. and M. Johnson. 1999. *Philosophy in the Flesh: The Embodied Mind and Its Challenge to Western Thought*. New York.
Langacker, R. 1987. *Foundations of Cognitive Grammar: Theoretical Prerequisites* (Vol. 1). Stanford, CA.
Langacker, R. 1990. *The Foundations of Cognitive Grammar: Descriptive Application* (Vol. 2). Stanford, CA.
Langacker, R. 2008. *Cognitive Grammar*. Oxford.
LeDoux, J. 2016. *Anxious: Using the Brain to Understand and Treat Fear and Anxiety*. Penguin.
Lucy, J. 1992. *Grammatical Categories and Cognition: A Case Study of the Linguistic Relativity Hypothesis*. Cambridge.
Lucy, J. 2005. "Through the window of language: Assessing the influence of language diversity on thought". *Theoria* 54: 299-309.
Martindale, Charles. 1993. "Descent into hell: Reading ambiguity, or Vergil and the critics." *Proceedings of the Virgil Society* 21: 111-50.
Maturana, H. and F. Varela. 1980. *Autopoiesis and Cognition*. Dordrecht.
Maturana, H. and F. Varela. 1987. *The Tree of Knowledge*. Berkeley, CA.
Membrez, G. 2018. "Metaphor by any other name: The reception of Aristotle in cognitive linguistics." In E. Mocciaro and W. M. Short, eds., *Toward a Cognitive Classical Linguistics*, Berlin. 207–227.
Mischler, J. 2013. *Metaphors Across Time and Conceptual Space*. Amsterdam.
Moussy, C. 2007. "*Ambiguus, ambiguitas, anceps, utroqueversus* dans le vocabulaire de l'ambiguïté". In C. Moussy and A. Orlandini, eds., *L'ambiguïté en Grèce et à Rome: approche linguistique*, Paris. 57-64.
O'Halloran, K. 2003. *Critical Discourse Analysis and Language Cognition*. Edinburgh.
Palmer, G. 1996. *Toward a Theory of Cultural Linguistics*. Austin, TX.
Panksepp. J. 1988. "The neurobiology of emotions". In H. Wagner and T. Manstead, eds., *Handbook of Psychophysiology*, London.
Panksepp. J. 1989. "The psychoneurology of fear". In *Handbook of Anxiety*, vol. 3, Amsterdam.
Panksepp. J. 2010. "Affective neuroscience of the emotional brain". *Dialogues in Clinical Neuroscience* 77: 2905–7.
Pucci, J. 2014. "Order, ambiguity and authority in Venantius Fortunatus, *Carm.* 3.26". In J. Martinez, ed., *Fakes and Forgers of Classical Literature*, Leiden. 219-30.
Rosch, E. 1973. "Natural categories". *Cognitive Psychology* 4: 328-50.
Rosch, E. 1978. "Principles of categorization". In E. Rosch and B. Lloyd, *Cognition and Categorization*, Hillsdale, NJ. 27-48.
Rowlands, M. 2010. *The New Science of the Mind: From Extended Mind to Embodied Phenomenology*. Cambridge, MA.
Sharifian, F. and G. Palmer. 2007. *Applied Cultural Linguistics*. Amsterdam.
Shore, B. 1996. *Culture in Mind*. Oxford.
Short, W. M. Forthcoming. "Spatial metaphors of ambiguity in Latin". In M. Fontaine, C. McNamara, and W. M. Short, eds., *Quasi Labor Intus: Ambiguity in Latin Literature*, Berlin.
Slobin, D. 1996. "From 'thought and language' to 'thinking for speaking'". In J. J. Gumperz and S. C. Levinson, eds., *Rethinking Linguistic Relativity*, Cambridge. 70-96.
Spencer, D. 2011. "Movement and the linguistic turn". In R. Laurence and D. J. Newsome, eds., *Rome, Ostia, Pompeii: Movement and Space*, Oxford. 57-80.
Talmy, L. 2003. *Toward a Cognitive Semantics*. Cambridge, MA: MIT Press.
Taylor, J. 1989. *Linguistic Categorization: Prototypes in Linguistic Theory*. Oxford.
Taylor, J. 2002. *Cognitive Grammar*. Oxford.
Tuggy, D. 1993. "Ambiguity, polysemy and vagueness". *Cognitive Linguistics* 4 (3): 273-90.
Turner, M. 1990. "Aspects of the invariance hypothesis." *Cognitive Linguistics* 1 (2): 247-55.
Tyler, A. and V. Evans. 2003. *The Semantics of English Prepositions*. Cambridge.
Versnel, H. 1992. *Inconsistencies in Greek and Roman Religion*. Leiden.
Wharton, D. 2011. "Linguistic semantics and the representation of word meaning in Latin dictionaries". In R. Oniga, R. Iovino, and G. Giusti, eds., *Formal Linguistics and the Teaching of Latin*, 255–78. Newcastle-upon-Tyne.
Wilson, M. 2002. "Six views of embodied cognition". *Psychonomic Bulletin and Review* 9: 625–36.

5

Psycholinguistics and the classical languages

Reconstructing native comprehension

Alessandro Vatri

Introduction

Psycholinguistics is the cognitive science that studies the production and comprehension of language. This means that, on the one hand, it aims to understand how utterances are produced *beyond* 'grammar' by pinpointing the psychological processes that lie *behind* 'grammar'. On the other hand, it aims to understand how the linguistic input is processed by the human mind—that is, to understand what cognitive mechanisms are activated when words are perceived in hearing or reading, alongside the concomitant paralinguistic information (e.g. prosody and gestures in oral communication, and graphic features in writing).[1] Much experimental research on language comprehension focuses on specific features of the linguistic input that affect the way in which sentences are represented, parsed, and processed in the mind and seeks to identify cognitive mechanisms and principles at the highest level of generalization. In other words, scientific endeavor in this field aims to cast light on comprehension not only as far as individual languages are concerned (e.g. understanding the native comprehension of, say, English, Mandarin, or French), but extending at least to typologically defined classes of languages (e.g. free- vs. fixed-word-order languages) and, ideally, to the universal working of the 'human parser'.[2]

Approaches based on the theoretical insights and analytical results of research in psycholinguistics provide an original and powerful toolbox to classical scholars and philologists. This chapter will show how methods based on experimental research on language comprehension can help us 'crack the code' of ancient stylistic and rhetorical theory and practice—how we can reconstruct the cognitive mechanisms that underlie native linguistic perception as it surfaces in the remarks and precepts of ancient critics and rhetoricians—in order to learn lessons about how native speakers of classical Greek and Latin processed their own language and so that *we* modern critics can attempt to replicate to some extent *their* way of reading an ancient text. Apart from its intrinsic interest, the reconstruction of the native perception of the classical languages opens up new ways for scholars to address tantalizing questions in a methodologically sound way.[3] At the same time, ancient rhetorical materials occasionally reveal native sensitivity for types of linguistic structures whose cognitive effects in modern languages have not yet been fully studied experimentally. Such materials may, and ideally will, form the basis for collaboration between classicists and psycholinguists and contribute to setting the agenda for experimental research.

Reading the critics' mind: ancient rhetoric clarity and language comprehension

Ancient literary critics build on concepts and precepts from the rhetorical tradition in order to identify and describe the psychological and aesthetic effects of language. In general, they ascribe such effects to figures and their combinations, as well as to structural and phonetic features of stretches of language. If we wish to read and use their work as evidence for the native perception of classical Greek or Latin, we might not always be in the position to take their remarks and observations at face value. First and foremost, we need to take into account the potential 'prescriptive bias' of ancient rhetorical and critical literature (de Jonge 2007: 219). That is, we may often suspect that critics framed their observations in such a way as to fit their own theoretical or didactic purposes, which would have determined their prescriptive, rather than purely descriptive, character. Such problems can be controlled for and overcome, if the ancient sources are interpreted appropriately (see Vatri 2017: 132–7). At the same time, we are in a better starting position if we focus on the ancient examples and demonstrations of the rhetorical qualities of a text.

Critics address audiences of native (or at least highly proficient) speakers or readers of classical Greek or Latin, and the points they make are meant to rely on and be convincing to the 'native ear': native speakers should be able to actually 'feel' the effects that critics claim to be produced. In this perspective, examples may be construed as 'experimental material' designed by critics for this purpose, whether they are entirely made up—as is often the case with examples in rhetorical handbooks—or they consist of manipulations of passages from literary texts. This is precisely what the ancient critical method of *metathesis* consists in (de Jonge 2008: 367–90). Ancient critics rewrote sentences or passages to show either how bad (if *metathesis* was meant as an improvement) or how good (if no improvement was shown to be possible) the style of that passage was. In some cases, they just aimed to provide an alternative without implying that it was better or worse than the original. Quite importantly, *metathesis* was supposed to call on the *empirical* judgment of the readers. This point is explicitly made by Dionysius of Halicarnassus (*Dem.* 19):

> εἰ δὲ ὀρθῶς ἐπιλογίζομαι ταῦτ' ἐγὼ καὶ ἔστιν ἐν ταύταις ταῖς ἀρεταῖς ἐνδεέστερος ὁ ἀνήρ, πάρεστι τῷ βουλομένῳ σκοπεῖν ἐπὶ τῆς ἀρτίως παρατεθείσης λέξεως ποιουμένῳ τὴν ἐξέτασιν. εὐθέως γοῦν τὴν πρώτην διάνοιαν ὀλίγοις ὀνόμασιν ἐξενεχθῆναι δυναμένην μακρὰν ποιεῖ κυκλογραφῶν καὶ δὶς ἢ τρὶς τὰ αὐτὰ λέγων. ἐνῆν μὲν οὖν ἐν τῷ πρώτῳ κώλῳ τῷ 'τίς γὰρ ἂν ἄλλοθεν ἐπελθών' τὸ 'καὶ μὴ συνδιεφθαρμένος ἡμῖν, ἀλλ' ἐξαίφνης ἐπιστὰς τοῖς γινομένοις'· δυνάμει γὰρ ἄμφω ταὐτά. καὶ ἐν τῷ 'οἳ φιλοτιμούμεθα μὲν ἐπὶ τοῖς τῶν προγόνων ἔργοις' τὸ 'καὶ τὴν πόλιν ἐκ τῶν τότε πραχθέντων ἐγκωμιάζειν ἀξιοῦμεν'· τὸ γὰρ αὐτὸ φιλοτιμεῖσθαί τε καὶ ἐπαινεῖν. καὶ ἐν τῷ 'οὐδὲν δὲ τῶν αὐτῶν ἐκείνοις πράττομεν' τὸ 'ἀλλὰ πᾶν τοὐναντίον'· ἤρκει γὰρ αὐτῶν εἰρῆσθαι θάτερον. ἐξῆν δέ γε μίαν ἐκ τοῖν δυοῖν ποιῆσαι περίοδον καὶ συντομωτέραν καὶ χαριεστέραν· 'τίς γὰρ ἂν ἄλλοθεν ἐπελθὼν οὐκ ἂν μαίνεσθαι νομίσειεν ἡμᾶς, οἳ φιλοτιμούμεθα μὲν ἐπὶ τοῖς τῶν προγόνων ἔργοις, οὐδὲν δὲ τῶν αὐτῶν ἐκείνοις πράττομεν;'

Any reader can judge for himself whether my argument is sound and Isocrates is inferior in these qualities by examining the passage which I have just quoted. The very first idea

> could have been expressed in a few words, but he spins it out by circumlocution and by saying the same thing two or three times. Thus in the first sentence, beginning 'Now what if a stranger from abroad' we have '. . . and suddenly find himself embroiled in our affairs, before having the time to become corrupted by our depravity': both clauses expressing the same idea. And the clause 'When we glory in the deeds of our ancestors' is followed by 'and think it right to sing the city's praises': 'glory' and 'praise' mean the same thing. The clause '. . . and yet act in no way like them' is followed by 'but do the exact opposite': only one of these was necessary. It would have been possible to make one period out of two, and a more elegant one at that, in the following way: 'What stranger from abroad would not think us insane, when we glory in the deeds of our ancestors, but act in no way like them?'
> *(Transl. S. Usher)*

It is hard to imagine that texts selected for the exercise of *metathesis* would be chosen following a purely 'ideological' criterion. Such a selection can be described as consisting of the following logical steps:

a one assumes that certain stylistic features are 'good' or 'bad',
b one identifies such features in a text, and argues on their basis that that text is 'good' or 'bad',
c one rewrites the text altering the features and turning it into a worse or better version.

An empirical selection process, by which the critic starts from the aesthetic/cognitive effect of the text, is easier to envisage:

a a text is selected because it 'feels' 'good' or 'bad',
b certain features that are supposed to be 'good' or 'bad' are recognized in that text,
c once again, one rewrites the text altering the features and turning it into a worse or better version.

If the assumption that *metathesis* relies on empirical processes is correct, we can conclude that this critical method was only effective if both the original and the rewritten texts had the intended effect on the reader in the first place. Only thus would the critic's argument be persuasive.

Examples and demonstrations of the rhetorical qualities of texts and passages may be analyzed in an appropriate framework for the purpose of reconstructing the native comprehension of classical Greek and Latin. If we are able to read into the cognitive processes involved in the ancient critics' description of the effect of a stretch of language, we may be able to extract well-defined principles that we may then apply to other ancient texts in order to approximate the native perception of the linguistic material of which they consist. In many ways, we are in a similar position to that of a student of ancient epidemiology who might want to understand what disease an ancient author talks about. If we were dealing with ancient descriptions of the symptoms of some medical condition, modern medical science would in many cases allow us to identify the disease beyond the shortcomings of ancient diagnostics. Analogously, when ancient mother-tongue rhetoricians discuss the rhetorical effects of stretches of language they give as examples, they provide us with a cognitive symptomatology of those linguistic materials.

Such a symptomatology may be interpreted in the light of modern research on language comprehension in order to reach a 'diagnosis' of the cognitive processes they triggered in native readers or listeners, and thus extrapolate general analytical principles and approximate native reading of ancient texts.

The rhetorical quality that relates most directly to language comprehension is clarity (*sapheneia, perspicuitas*) and, as a consequence, ancient discussions and demonstrations of this quality are a natural starting point for the examination of ancient rhetorical literature in the light of psycholinguistic research on language comprehension. Clarity was recognized throughout the ancient rhetorical tradition as essential to any well-composed text—it can almost be understood as a requirement for a text to be 'rhetorically grammatical' (cf. Innes 1985: 255; Russell 1981: 135–7). However, the notion of clarity is anything but a simple one. Clarity can be conceived of as an 'end-state' of communication: clarity is a mental state which the sender of a linguistic message should aim to generate in the receiver (cf. Leech 1983: 105). At the same time, clarity may be conceived of as a feature of the message itself. In this respect, it may be found to operate at different levels. The ancient rhetoricians themselves distinguished between clarity concerning words (which I will call 'processability') and clarity concerning the subject-matter (which I will call 'understandability').[4] On the one hand, understandability is intrinsic to the content of a text. On the other, it may be optimized or disrupted by the way in which content is organized and presented in discourse. Less than optimal understandability does not entail the 'unprocessability' of the verbal input, but makes it difficult for the receiver to get a clear grasp of the subject-matter. This, in turn, need not result in 'mental' obscurity: the end-state of communication may still be 'clarity'. Reticence and allegory (cf. Demetr. *Eloc.* 99–100, 254), whose obscurity concerns the subject-matter and is intended to trigger cognitive processes leading to a deeper understanding of the message, are good examples of deliberate lack of clarity (see Kustas 1973 on obscurity as a rhetorical quality; cf. also Sluiter 2016). Obscurity is helplessly disruptive, instead, when it affects 'processability' of a stretch of language: no text can be called well composed unless each of its discourse units is fully comprehensible from a syntactic, phonological, lexical, and semantic point of view (see Vatri 2017: 101–8 for a fuller discussion).

'Processability' is precisely what sentence-processing models in psycholinguistics aim to assess, which means that their toolbox can be used to 'reverse engineer' the ancient critics' observations and examples of clarity and obscurity of form (as opposed to content). This exercise rests on the assumption that the 'human parser' (the way in which humans are 'wired' to process language) is the same for contemporary speakers of any language as well as for ancient speakers of dead languages. This does not mean that the comprehension processes are to be assumed to be exactly the same for all languages: as a matter of fact, theories that account well for the comprehension of English, for instance, are often challenged when tested experimentally on other languages (Konieczny 2000).

An uncontroversial and universal feature of human language comprehension is the fact that it is an incremental process (see e.g. Dąbrowska 2004: 23; Schlesewsky and Bornkessel 2004; van Gompel and Pickering 2007; Levy 2008: 1129; Levy, Fedorenko, and Gibson 2013). As we encounter words, either in reading or listening, we tend to extract as much linguistic information as possible and to maximize the interpretation of the part of the stretch of language we have already perceived, without waiting for the sentence to be complete. This means that when we read or hear each word in a sentence, we retrieve its meaning from the mental lexicon (Aitchison 2012: 132–3; Traxler and Tooley 2007: 63), we connect it to a referent (i.e. to the entity it evokes, cf. Kaiser and Trueswell 2004; Warren and Gibson 2002), and we generate a mental representation of its syntactic relationships with the other words in the sentence (cf. e.g. Warren and Gibson 2005: 753; Gibson 2000: 102–5). As new words are encountered, we try to integrate them into the partially parsed structure as soon as possible.

In certain languages (such as Hindi, German, and Japanese), this cognitive build-up appears to be driven especially by probabilistic expectations: the more expected an element is at the point where it occurs in the sentence, the easier it is to process. Elements can be more or less expected from the point of view of their syntactic category (e.g. a direct object is anticipated by a transitive verb), semantics (e.g. nouns denoting a human being anticipate verbs denoting actions or states in which humans can play a role), or even morphological features (e.g. a plural subject anticipates a verb in the plural). In other languages (such as English and French), a more prominent role is played by working memory (Levy, Fedorenko, and Gibson 2013: 465; Jäger et al. 2015: 116), and processing difficulty is determined by the distance between constituents that 'go together'. The longer the distance, the harder the integration of new words into the mental representation of the partially interpreted linguistic structure. Long distances also make it possible that intervening words sharing morphological and semantic features with those that are supposed to 'go together' interfere with the integration process (see Warren and Gibson 2005: 754–5; Fedorenko, Gibson, and Rohde 2006; Gordon et al. 2006; Martin and McElree 2009). Such effects are observed, for instance, in the native comprehension of an English sentence like *The worker was surprised that the resident who said that the neighbor was dangerous was complaining about the investigation*. In this sentence, the verb *was complaining* triggers the retrieval from memory of a syntactic element to be constructed as its subject. The verb itself provides cues for retrieval: its subject must be a noun in the singular denoting an entity that could plausibly perform the action of complaining. The nearest syntactic element matching these cues is the noun phrase *the neighbor*, and its presence in a medial position between the verb and its real subject (*the resident*) can interfere with the retrieval process and smooth comprehension of the sentence (see Van Dyke 2007).

Typological features of languages predict to a good extent whether expectation or memory play a larger role in sentence processing. Expectation effects show especially in verb-final, high-morphological marking languages, while memory effects show most prominently in fixed-word-order languages. However, such mechanisms are not mutually exclusive, as a study of sentence processing in Russian indicates (Levy, Fedorenko, and Gibson 2013). Crucially for our purposes, Russian is probably the most similar language to classical Greek and Latin from a typological point of view among those on which such theories have been tested, and we may reasonably argue that it is likely that native sentence-processing difficulty in classical Greek and Latin would be predicted by expectation-based models, provided that a memory effect such as interference is taken into account.

These principles can be illustrated by examining a Latin example. In the eighth book of the *Institutio Oratoria*, Quintilian discusses how obscurity can be caused by less than optimal sentence structure, as is the case, for instance, with complex hyperbaton resulting in *mixtura verborum* (*Inst.* 8.2.14). Such a phenomenon is exemplified with a syntactically intricate verse from the *Aeneid* (1.109):

> saxa vocant Itali mediis quae in fluctibus aras
>
> Rocks the Italians call in the midst of the waves altars.
>
> *(Transl. D. A. Russell)*

Let us attempt a dynamic, word-by-word parsing of this verse and build up an incremental interpretation of its linguistic structure. The initial noun *saxa* may be interpreted either as a nominative or an accusative, and raises the expectation for a verb to form a complete sentence (a noun alone could in principle represent a complete utterance e.g. in answers to *wh*-questions). Such a verb would have to be semantically able to take 'rocks' as its subject or direct object: the linguistic experience of a speaker of any language would suggest that it is more probable that a word denoting 'rocks' is an argument of verbs meaning, for instance, 'fall' or 'cast' than of verbs meaning 'calculate' or 'diagnose'. When *vocant* is encountered, third-person plural morphology does not immediately disambiguate the morpho-semantics of *saxa*. This noun can still be interpreted as a nominative, and all the more so because in literary Latin, rocks *can* speak (cf. for instance Cicero, *Arch.* 19: *saxa atque solitudines voci respondent*); as experimental research has shown, such fictional humanizations tend to override semantic expectations based on ordinary linguistic experience (see Nieuwland and Van Berkum 2006; Roland et al. 2012). Only when *Itali* is encountered does it become clear that *saxa* is the direct object of *vocant*. *Mediis* raises the expectation for a noun in the ablative plural. This expectation is reinforced by *in*, but before the preposition is encountered, *quae* intervenes, which could be interpreted as a pronoun introducing a relative clause with a verb to follow—a verb that would eventually govern the phrase introduced by *in*. According to this possible parsing, the initial interpretation of the verse would translate as 'Italians call rocks which in the midst of the waves . . .'. As one reaches the end of the verse, the accusative *aras* occurs but the verb anticipated by *quae* does not, which would prompt a reader or listener to reanalyze the whole verse and reinterpret its syntax. This is a cognitively demanding task, and it is not surprising that Quintilian regarded this verse as obscure.

In his commentary, Servius rewrites the verse by simply reordering its words in such a way as to minimize the distance between words that 'go together' and to avoid false starts. His paraphrase reads *quae saxa in mediis fluctibus Itali aras vocant*, with *quae* as a relative adjective modifying *saxa* and referring back to the same noun in the previous line of the *Aeneid* (*tris Notus abreptas in saxa latentia torquet*, *Aen.* 1.108). The position of *Itali* clarifies the semantics of *(quae) saxa* before the verb is encountered, and the position of *aras* anticipates that the upcoming verb must govern a double accusative (*aras* and *saxa* are expected to be accommodated in the same government pattern). In short, both Quintilian's disapproval and Servius' paraphrase indicate that Virgil's offending original wording disrupted native comprehension for reasons that can be explained in the light of modern research on language comprehension.

Instructive *metatheseis* are especially presented by Greek critics. A good example is the paraphrase of a sentence of Xenophon (*Anab.* 1.2.21) made by 'Demetrius' (*Eloc.* 198):

> φεύγειν δὲ καὶ τὰς πλαγιότητας· καὶ γὰρ τοῦτο ἀσαφές, ὥσπερ ἡ Φιλίστου λέξις, συντομώτερον δέ. παράδειγμα πλαγίας λέξεως καὶ διὰ τοῦτο ἀσαφοῦς τὸ παρὰ Ξενοφῶντι, οἷον
>
> 'καὶ ὅτι τριήρεις ἤκουεν περιπλεούσας ἀπ' Ἰωνίας εἰς Κιλικίαν Τάμον ἔχοντα τὰς Λακεδαιμονίων καὶ αὐτοῦ Κύρου'.
>
> τοῦτο γὰρ ἂν ἐξ εὐθείας μὲν ὧδέ πως λέγοιτο·
>
> 'τριήρεις προσεδοκῶντο εἰς Κιλικίαν πολλαὶ μὲν Λάκαιναι, πολλαὶ δὲ Περσίδες, Κύρῳ ναυπηγηθεῖσαι ἐπ' αὐτῷ τούτῳ. ἔπλεον δ' ἀπ' Ἰωνίας· ναύαρχος δ' αὐταῖς ἐπεστάτει Τάμος Αἰγύπτιος'.
>
> μακρότερον μὲν οὕτως ἐγένετο ἴσως, σαφέστερον δέ.

Avoid also the use of dependent constructions (*plagiotes*), since this too leads to obscurity, as Philistus' style shows. A shorter example of how the use of dependent constructions (*plagia lexis*) causes obscurity is this passage of Xenophon:

> 'and that he had heard that triremes were sailing round from Ionia to Cilicia commanded by Tamus, ships belonging to the Spartans and to Cyrus himself.'
>
> This sentence could be redrafted without dependent constructions (*ex eutheias*) in the following sort of way:
>
> 'Triremes were expected in Cilicia, many of them Spartan, many of them Persian and built by Cyrus for this very purpose. They were sailing from Ionia, and the commander in charge of them was the Egyptian Tamus.'
>
> *(Transl. D. C. Innes)*

'Demetrius' maintains that the sentence is unclear because of *plagiotes* and promises to make its form clearer by rewriting it *ex eutheias*. This expression should indicate that the obscure 'dependent constructions' corresponding to the participles περιπλεούσας and ἔχοντα are turned into clearer 'direct constructions' (clauses with the verb in the indicative).[5] Now, there is no evidence that non-finite verbs are intrinsically harder to understand than finite verbs in any language, even though it is true, as we shall see, that they may be 'hotspots' for difficult syntactic patterns (see Vatri 2017: 162–3); the impression reported by 'Demetrius' must be due to some other feature of the sentence.

In order to identify such a feature, we can attempt to reconstruct the native incremental parsing of this sentence, starting from the initial conjunctions καὶ ὅτι. These immediately anticipate a verb, and such an expectation is strengthened by the upcoming noun τριήρεις, which could be interpreted either as a nominative or as an accusative. The expectation for a verb is satisfied by ἤκουεν, which, by virtue of being in the third-person singular, reveals that τριήρεις cannot be its subject and is in the accusative. Now, ἀκούω can govern an accusative of the thing or event about which one hears, and τριήρεις does not designate an entity that can be heard (such as a speech or a story) but only a potential source of sound, which would be expressed by a genitive. As a consequence, a predicative participle or an infinitive specifying the event in which the triremes are involved and whose report can be heard would not be surprising for a reader to encounter. As a matter of fact, a predicative participle follows immediately (περιπλεούσας ἀπ' Ἰωνίας εἰς Κιλικίαν). From a syntactic and semantic point of view, this partially perceived sentence can be considered complete here ('and that he had heard that triremes were sailing round from Ionia to Cilicia'). However, a new word (Τάμον) follows, which might be unexpected and would remain uninterpreted until ἔχοντα is encountered. This participle depends directly on ἤκουεν and governs τριήρεις περιπλεούσας, which gives the meaning 'he had heard that Tamus commanded triremes that were sailing round'. This interpretation must replace the wrong one that a reader would have built from the partially input sentence (up to Κιλικίαν). Once again, this cognitively demanding task would be the basis of the linguistic difficulty experienced by 'Demetrius'. In his paraphrase, 'Demetrius' breaks up the sentence and replaces the participles περιπλεούσας and ἔχοντα with finite clauses, avoiding all false starts and making syntactic relationships more transparent to begin with. This analysis indicates that difficulty in this sentence did not necessarily stem from 'dependent constructions', even though they were identified as the source of difficulty, and indeed greatly reduced in the paraphrase, by 'Demetrius'. At the same time, the fact that a native reader such as 'Demetrius' found this sentence obscure makes good sense in the light of our current knowledge of language comprehension processes.

The same approach can be applied to *metatheseis* of longer passages. In the treatise *On Thucydides*, for instance, Dionysius of Halicarnassus rewrites a long sentence in order to show how Thucydides could have made it 'clearer and more pleasant' (σαφεστέρα δὲ καὶ ἡδίων), and we can assess how he carried this out by comparing the two versions and focusing on the processing differences entailed by either wording. Thucydides' text and Dionysius' paraphrase are printed below as parallel texts:

Th. *4.34*	D.H. *Th. 25*
τῶν δὲ Λακεδαιμονίων οὐκέτι ὀξέως ἐπεκθεῖν ᾗ προσπίπτοιεν δυναμένων, γνόντες αὐτοὺς οἱ ψιλοὶ βραδυτέρους ἤδη ὄντας τῷ ἀμύνασθαι, καὶ αὐτοὶ τῇ τε ὄψει τοῦ θαρσεῖν τὸ πλεῖστον εἰληφότες πολλαπλάσιοι φαινόμενοι καὶ ξυνειθισμένοι μᾶλλον μηκέτι δεινοὺς αὐτοὺς ὁμοίως σφίσι φαίνεσθαι, ὅτι οὐκ εὐθὺς ἄξια τῆς προσδοκίας ἐπεπόνθεσαν, ὥσπερ ὅτε πρῶτον ἀπέβαινον τῇ γνώμῃ δεδουλωμένοι ὡς ἐπὶ Λακεδαιμονίους, καταφρονήσαντες καὶ ἐμβοήσαντες ἁθρόοι ὥρμησαν ἐπ' αὐτοὺς . . .	τῶν δὲ Λακεδαιμονίων οὐκέτι ἐπεκθεῖν, ᾗ προσπίπτοιεν, δυναμένων, γνόντες αὐτοὺς οἱ ψιλοὶ βραδυτέρους ἤδη, συστραφέντες καὶ ἐμβοήσαντες, ὥρμησαν ἐπ' αὐτοὺς ἀθρόοι· ἔκ τε τῆς ὄψεως τὸ θαρρεῖν προσειληφότες, ὅτι πολλαπλάσιοι ἦσαν, καὶ ἐκ τοῦ μηκέτι δεινοὺς αὐτοὺς ὁμοίως σφίσι φαίνεσθαι καταφρονήσαντες, ἐπειδὴ οὐκ εὐθὺς ἄξια τῆς προσδοκίας ἐπεπόνθεσαν, ἣν ἔσχον ὑπόληψιν, ὅτε πρῶτον ἀπέβαινον τῇ γνώμῃ δεδουλωμένοι ὡς ἐπὶ Λακεδαιμονίους.
Since the Lacedaemonians were no longer able to dash out rapidly where they attacked before, the light troops, finding that they were by then slower as they were defending themselves, and having themselves gained most of their courage for the sight, since they seemed many more themselves, and having become more familiar with the fact that they no longer seemed to them as terrible, since they had not suffered things worth what they feared, as when they first disembarked subjugated by the idea that they were to fight the Lacedaemonians, with disdain and shouting they rushed in crowds against them.	Since the Lacedaemonians were no longer able to dash out rapidly where they attacked before, the light troops, finding that they were by then slower, formed in a mass and shouting, rushed in crowds against them: having gained their courage from the sight, since they were many more, being disdainful since they no longer seemed to them as terrible, because they had not suffered things worth what they feared, which prejudice they had, when they first disembarked subjugated by the idea that they were to fight the Lacedaemonians.

We can start by noting that Dionysius eliminates the digression ranging from καὶ αὐτοὶ τῇ τε ὄψει to ἐπὶ Λακεδαιμονίους and moves its contents to the end (ἔκ τε τῆς ὄψεως . . . ἐπὶ Λακεδαιμονίους), as he himself points out:[6]

> ἥδ' ἡ περιοχὴ ὤφελε μὲν κατεσκευάσθαι μὴ τοῦτον ὑπ' αὐτοῦ τὸν τρόπον, ἀλλὰ κοινότερον μᾶλλον καὶ ὠφελιμώτερον, τοῦ τελευταίου μορίου τῷ πρώτῳ προστεθέντος, τῶν δὲ διὰ μέσου τὴν μετὰ ταῦτα χώραν λαβόντων. ἀγκυλωτέρα μὲν οὖν ἡ φράσις οὕτω σχηματισθεῖσα γέγονε καὶ δεινοτέρα, σαφεστέρα δὲ καὶ ἡδίων ἐκείνως ἂν κατασκευασθεῖσα· . . . ὑπεξαιρουμένης δὲ τῆς περιγραφῆς πάσης, τἆλλα πάντα ὠνόμασταί τε τοῖς προσφυεστάτοις ὀνόμασι καὶ περιείληπται τοῖς ἐπιτηδειοτάτοις σχηματισμοῖς.
>
> This group of events should not have been arranged thus, but in a more normal and helpful way, making the final part follow upon the first, with the intervening parts coming after these. Thucydides's arrangement has produced a more compact and striking sentence, but it would have been clearer and more pleasing if it had been arranged thus: . . . Thus all circuitous structure is entirely removed, and everything that is left is given its most natural name and is expressed in the most suitable figures.
>
> *(Transl. S. Usher)*

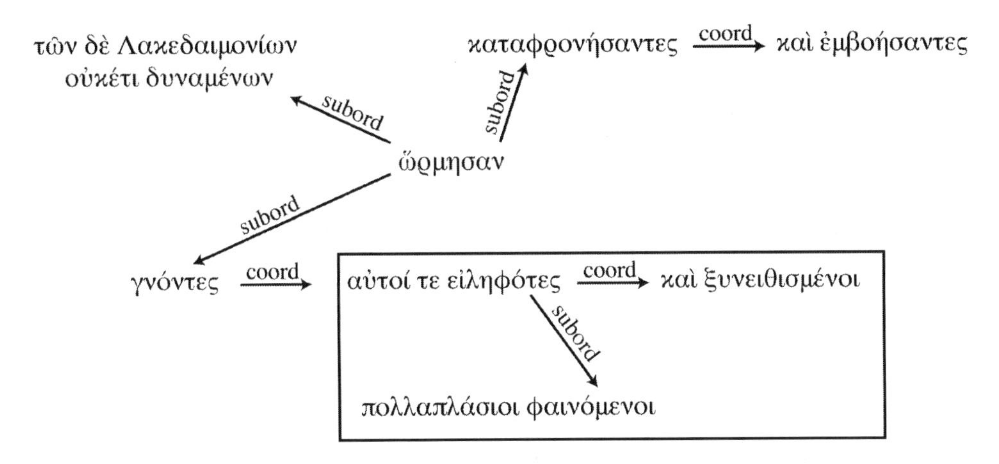

Figure 5.1 Syntactic dependencies between participles in Th. 4.34.1

This type of alteration affects clarity in the communication of the subject-matter (understandability), but Dionysius' version also alters clarity of form (processability). In the original text, a long series of participles in the nominative (γνόντες, εἰληφότες, φαινόμενοι, ξυνειθισμένοι, καταφρονήσαντες, ἐμβοήσαντες) embedded in the main clause follows the initial genitive absolute (τῶν δὲ Λακεδαιμονίων . . . δυναμένων) and precedes the main verb (ὥρμησαν). Two of them, εἰληφότες and ξυνειθισμένοι, are coordinate to one another by means of τε and καί and, as a group, they are coordinate to γνόντες through previous καί. Φαινόμενοι, instead, is subordinate to εἰληφότες. This complex hierarchical structure is visualized in Figure 5.1 below:

Arguably, the incremental parsing of this hierarchically complex structure must have been a rather complex task. The first occurrence of καί may well raise the expectation for a participle coordinated to γνόντες. The connective particle τε, which follows shortly, does not anticipate in an obvious way that it coordinates the expected participle (εἰληφότες), more than any other element of the participial clause, to some other element to follow (τε, in principle, could have scope over, say, the noun phrase τῇ ὄψει, for instance). Since καὶ ξυνειθισμένοι immediately follows φαινόμενοι, the chances are that this participle could be interpreted as coordinate to φαινόμενοι itself, which, in turn, is unequivocally subordinate to εἰληφότες. Dionysius' paraphrase makes all such hierarchical relations explicit by substituting a finite verb causal clause (ὅτι πολλαπλάσιοι ἦσαν) for the participle φαινόμενοι.

Furthermore, Thucydides uses αὐτοί both as an anaphoric pronoun referring to the Lacedaemonians (αὐτοὺς . . . βραδυτέρους ἤδη ὄντας, αὐτοὺς ὁμοίως σφίσι φαίνεσθαι, ἐπ' αὐτούς) and as a demonstrative pronoun referring to the skirmishers (οἱ ψιλοί . . . αὐτοὶ . . . φαινόμενοι). Dionysius instead only uses αὐτοί to refer to the Lacedaemonians, removing all potential ambiguities in the assignment of a referent to this pronoun.

Thucydides also embeds clauses with a finite verb in the third-person plural (ὅτι . . . ἐπεπόνθεσαν, ὥσπερ ὅτε . . . ἀπέβαινον) between the subject of the main verb (οἱ ψιλοί) and the main verb itself (ὥρμησαν). The subject of these verbs is the same as that of the main verb, and it is likely that the intervening verbs might have interfered with the integration of the main verb in the partially processed sentence, as a memory-based account would predict. This potential hindrance to smooth processing is also taken care of by Dionysius, who turns this nested structure into a linear one, as he moves all center-embedded clauses to the end of the sentence: (καταφρονήσαντες,) ἐπειδὴ . . . ἐπεπόνθεσαν, and (ἔσχον ὑπόληψιν,) ὅτε πρῶτον ἀπέβαινον.

Just as we observed about the *metathesis* presented by 'Demetrius', Dionysius need not have been aware of these types of changes. Arguably, this would make them all the more significant, in that they would be even more likely to reflect what he *empirically* considered clearer than what he *theorized* to be clearer. Once again, these alterations can be explained in the light of psycholinguistic research on language comprehension and confirm the applicability of its insights to the study of native comprehension of ancient texts.

The other way around: classics and the experimental research agenda

Certain ancient rhetorical materials show effects of comprehension phenomena that can be described in the conceptual framework of modern psycholinguistics, but that have not yet been the object of experimental research. The comment of Dionysius of Halicarnassus (*Th.* 31) on the following sentence of Thucydides (3.82.6) is a case in point:

> καὶ μὴν καὶ τὸ συγγενὲς τοῦ ἑταιρικοῦ ἀλλοτριώτερον ἐγένετο διὰ τὸ ἑτοιμότερον εἶναι ἀπροφασίστως τολμᾶν.
>
> Moreover, ties of kindred became less binding than partisan ties, because the latter induced a greater readiness for unstinted action.
>
> *(Transl. S. Usher)*

Dionysius comments as follows:

> τό τε 'ἀπροφασίστως τολμᾶν' ἄδηλον, εἴ τε ἐπὶ τῶν φίλων κεῖται νῦν εἴ τε ἐπὶ τῶν συγγενῶν. αἰτίαν γὰρ ἀποδούς, δι' ἣν τοὺς συγγενεῖς ἀλλοτριωτέρους ἔκρινον τῶν φίλων, ἐπιτίθησιν, ὅτι τόλμαν ἀπροφάσιστον παρείχοντο.
>
> It is uncertain whether the phrase 'for unstinted action' is here applied to one's friends or to one's relatives; for to supply the reason why they thought their kinsmen were less closely attached to them than their friends, he adds 'because they showed readiness for unstinted action'.
>
> *(Transl. S. Usher)*

In modern terms, Dionysius notes that the semantic subject of the verb τολμᾶν, which should be inferred from the superordinate infinitive εἶναι, is 'the partisan' (τὸ ἑταιρικόν), but that Thucydides' wording is ambiguous in this respect. Dionysius goes on to paraphrase the whole sentence, adding:

> σαφὴς δ' ἂν ἦν ὁ λόγος, εἰ τοῦτον ἐξήνεγκε τὸν τρόπον κατὰ τὴν ἑαυτοῦ βούλησιν σχηματίζων·
>
> 'καὶ μὴν καὶ τὸ ἑταιρικὸν οἰκειότερον ἐγένετο τοῦ συγγενοῦς διὰ τὸ ἑτοιμότερον εἶναι ἀπροφασίστως τολμᾶν'.
>
> The argument would have been clear if he had expressed it in the following manner, using his chosen figure:[7]
>
> 'Again "comradely" became more intimate than "kindred", owing to a greater readiness for unstinted action'.
>
> *(Transl. S. Usher)*

In the *metathesis*, Dionysius swaps around the object and standard of comparison of the sentence (from τὸ συγγενές—τοῦ ἑταιρικοῦ to τὸ ἑταιρικόν—τοῦ συγγενοῦς) and inverts the semantics of the comparative head (from ἀλλοτριώτερον 'less binding' to οἰκειότερον 'more intimate'). Dionysius' alteration seems to indicate that the inference of the correct semantic subject of τολμᾶν is more sensitive to syntactic parallelism than to the short-distance availability of a candidate subject. In other words, in Thucydides' original formulation, readers are supposed to infer the semantic subject of τολμᾶν from the most recent possible candidate in the linear order of the sentence (τοῦ ἑταιρικοῦ). Dionysius regards this as confusing, and reorganizes the sentence in such a way that the semantic subject of τολμᾶν (and εἶναι) must be inferred from the subject of the main verb (τὸ ἑταιρικὸν ἐγένετο). The facilitating effect of syntactic parallelism is well established in experimental research (see e.g. Knoeferle and Crocker 2009; Sturt, Keller, and Dubey 2010), but whether it would override the effect of distance, as Dionysius' remark suggests, is a hypothesis that calls for experimental testing on typologically adequate modern *comparanda*.

Another interesting phenomenon that requires further research is the attachment of relative clauses whose relative pronoun can refer to more than one potential antecedent in the superordinate clause. Sentences containing structures of this kind were identified as ambiguous by ancient rhetoricians, as an example given by Aelius Theon shows (*Prog.* 82.13–18):

> δῆμον Ἐρεχθῆος μεγαλήτορος, ὅν ποτ' Ἀθήνη
> θρέψε Διὸς θυγάτηρ, τέκε δὲ ζείδωρος ἄρουρα.

> People of great-hearted Erechtheus, who once Athena, daughter of Zeus, nourished, and the corn-bearing earth generated.
>
> (*Il.* 2.547–8)

In this passage, the relative pronoun ὅν may be taken to refer back to either the genitive Ἐρεχθῆος, which is syntactically governed the noun δῆμον, or to δῆμον itself; as Theon puts it, either 'Erechteus' or 'the people' may be construed as the object of the verbs θρέψε and τέκε (*Prog.* 82.13–18 Patillon). If we accepted that, once again, the distance between words in the linear order of the sentence affects comprehension, we could suppose that ὅν would preferentially be interpreted as referring back to the nearest possible antecedent (Ἐρεχθῆος). However, Theon's remark indicates that this is not to be taken for granted. Experimental research has shown that in a number of modern languages (e.g. Spanish), when a relative pronoun can refer either to a nearer antecedent or to a more distant one by which the nearer candidate is governed (just as is the case with Theon's example, with δῆμον being syntactically superordinate to Ἐρεχθῆος), native speakers tend to attach the relative clause to the superordinate antecedent (Hemforth et al. 2015). If this were the case with classical Greek, we would suppose that ὅν was preferentially interpreted by native readers as referring to δῆμον rather than Ἐρεχθῆος. However, comprehension experiments on this phenomenon show a high degree of variability in results both across and within languages. Attachment to syntactically superordinate elements ranges between about 30% and 60% of the time, with higher scores for long relative clauses than for short ones. This phenomenon has been tentatively explained by hypothesizing that long relative clauses attach to distant potential antecedents of the relative pronoun because of a balancing principle: long constituents are preferably attached to constituents of the same prosodic size (see Hwang and Steinhauer 2011 for a summary; cf. Hemforth et al. 2015: 54). In languages in which syntactically superordinate elements tend to precede the elements they govern, such a prosodic feature (the size of the prosodic unit to which the relative clause is attached) corresponds to a syntactic one (the higher hierarchical position of the syntactic element to which the relative clause is attached). In a free-word-order language like classical Greek relative clauses may attach to far constituents that are not higher in the syntactic hierarchy than near ones, as in the following sentence of Lysias (1.49):

> πολὺ γὰρ οὕτω δικαιότερον ἢ ὑπὸ τῶν νόμων τοὺς πολίτας ἐνεδρεύεσθαι, οἳ κελεύουσι μέν, ἐάν τις μοιχὸν λάβῃ, ὅ τι ἂν οὖν βούληται χρῆσθαι, οἱ δ' ἀγῶνες δεινότεροι τοῖς ἀδικουμένοις καθεστήκασιν ἢ τοῖς παρὰ τοὺς νόμους τὰς ἀλλοτρίας καταισχύνουσι γυναῖκας.
>
> This would be more just than to let the citizens be trapped by laws, which on the one hand command that one, if he catches an adulterer, may treat him as he pleases, while on the other hand the trials are more dangerous for the victims than for those who, against the law, dishonor other people's wives.

In this sentence, the masculine accusative plural noun πολίτας competes with the more distant νόμων as possible antecedent of the relative pronoun οἵ. The phrase of which νόμων is part expresses the agent of the verb ἐνεδρεύεσθαι, whereas τοὺς πολίτας is its grammatical subject. From a syntactic point of view, the competing elements are not part of the same noun phrase and are not subordinate to one another. If we believe that prosodic balance is sufficient to determine a preference for far attachment (a matter of linear distance and prosodic size), the preference for syntactically high attachment in head-first languages may be an epiphenomenon. Once again, this question amounts to a new research hypothesis that ought to be addressed experimentally through research on typologically adequate (free-word-order) languages.

Conclusion

Psycholinguistics is a relatively young science driven by experimental research on living languages, whereas classical philology has an incomparably long tradition and focuses on written texts in ancient languages without any surviving native speaker. However, there is far more room for exchange between these two disciplines than one would initially think. Psycholinguistics provides the tools to reconstruct the cognitive processes involved in native language comprehension, and ancient rhetorical literature can profitably be read as a surrogate, if not as a replacement, for experimental material and native judgment on linguistic perception. As we have seen, we might be able to go some way toward reading an ancient critic's mind and approximate the cognitive processes experienced by native speakers of classical Greek and Latin as we read and analyze ancient texts. Moreover, the exchange between psycholinguistics and classics need not only go in one direction. The ancient rhetoricians' concern for clarity and attention to ambiguity provides insights that can form the basis for new research questions in experimental psycholinguistics as well as raise interest for comprehension phenomena in free-word-order languages on which much work still needs to be done.

Notes

1 For a useful general survey of the main research areas within this discipline see most recently Spivey, Joanisse, and McRae (2012).
2 See van Gompel (2013) for summaries of current theories and research trends.
3 E.g. the question of the connection between style and intended mode of reception of ancient literary texts (as a way of assessing the suitability of a text for oral communication), on which see Vatri (2017).
4 See e.g. *Rh.Al.* 30.6 1438a26–7, D.H. *Lys.* 4, Theo *Prog.* 80.9–11 Patillon, Anon.Seg. 80–4 in the Greek tradition; in the Latin tradition cf. *Rh.Her.* 1.15 and Cic. *Inv.* 1.29. Another level is that of communicative intentions: one may express oneself clearly by using language in a comprehensible

and unambiguous way, or because what one intends to communicate is easy to understand. A reticent statement may be perfectly clear from the point of view of the form and content, but less clear—or altogether obscure—from the point of view of what one means (cf. already Quintilian, *Inst.* 2.20–1: *quae verbis aperta occultos sensus habent*).

5 The term *plagiotes* is potentially confusing: for the grammarians this word refers to oblique cases (see e.g. D.H. *Comp.* 6.6, cf. de Jonge 2008: 158), but 'Demetrius' mentions elsewhere that the accusative is acceptable at the beginning of a sentence (*Eloc.* 201, see Chiron 1993: 122–3). Besides, grammarians use *eutheia* to refer to the nominative (Dickey 2007: 239), but the term used by 'Demetrius' is *orthe* (*Eloc.* 201). Moreover, rhetoricians can refer to oblique cases as *enkeklimena* (e.g. [Hermog.] *Prog.* 2.5, see de Jonge 2008: 304–8), whereas *plagiasmos* is used to refer to participial constructions by pseudo-Aristides (1.55, cf. Patillon 2002: 121 n. 131) and Hermogenes (especially to indicate the genitive absolute, e.g. *Id.* 1.11 288.13–25 and 2.1 317.15–20 Rabe). See Vatri (2017: 128–30) for a full discussion.

6 The detrimental effect of digressions on clarity is noted by Dionysius in *Amm* 2.15, cf. also Arist. *Rh.* 1407b21–5, Demetr. *Eloc.* 196, Hermog. *Id.* 1.4 239.14–240.9 Rabe; see Vatri 2017: 121–3.

7 That is, the substitution of neuter substantivized adjectives for abstract nouns (τὸ γὰρ 'συγγενές' καὶ τὸ 'ἑταιρικόν' <ἀντὶ τῆς συγγενείας καὶ τῆς ἑταιρίας> κείμενον μετείληπται).

References

Aitchison, Jean. 2012. *Words in the Mind: An Introduction to the Mental Lexicon.* 4th ed. Chichester/Malden, MA: Wiley-Blackwell.

Chiron, Pierre. 1993. *Démétrios: Du Style.* Paris: Les Belles Lettres.

Dąbrowska, Ewa. 2004. *Language, Mind and Brain: Some Psychological and Neurological Constraints on Theories of Grammar.* Edinburgh: Edinburgh University Press.

de Jonge, Casper Constantijn. 2007. "From Demetrius to Dik: Ancient and Modern Views on Greek and Latin Word Order." In *The Language of Literature. Linguistic Approaches to Classical Texts*, edited by Rutger J. Allan and Michel Buijs, 211–32. Leiden/Boston, MA: Brill.

—. 2008. *Between Grammar and Rhetoric: Dionysius of Halicarnassus on Language, Linguistics and Literature.* Leiden/Boston, MA: Brill.

Dickey, Eleanor. 2007. *Ancient Greek Scholarship.* New York/Oxford: Oxford University Press.

Fedorenko, Evelina, Edward Gibson, and Douglas Rohde. 2006. "The Nature of Working Memory Capacity in Sentence Comprehension: Evidence against Domain-Specific Working Memory Resources." *Journal of Memory and Language* 54 (4): 541–53.

Gibson, Edward. 2000. "The Dependency Locality Theory: A Distance-Based Theory of Linguistic Complexity." In *Image, Language, Brain*, edited by Alec Marantz, Yasushi Miyashita, and Wayne A. O'Neil, 95–126. Cambridge, MA/London: MIT Press.

Gompel, Roger P. G. van, ed. 2013. *Sentence Processing.* Current Issues in the Psychology of Language. London/New York: Psychology Press.

Gompel, Roger P. G. van, and Martin J. Pickering. 2007. "Syntactic Parsing." In *The Oxford Handbook of Psycholinguistics*, edited by M. Gareth Gaskell, 284–307. Oxford: Oxford University Press.

Gordon, Peter C., Randall Hendrick, Marcus Johnson, and Yoonhyoung Lee. 2006. "Similarity-Based Interference during Language Comprehension: Evidence from Eye Tracking during Reading." *Journal of Experimental Psychology: Learning, Memory, and Cognition* 32 (6): 1304–21.

Hemforth, Barbara, Susana Fernandez, Charles Clifton, Lyn Frazier, Lars Konieczny, and Michael Walter. 2015. "Relative Clause Attachment in German, English, Spanish and French: Effects of Position and Length." *Lingua* 166 (October): 43–64.

Hwang, Hyekyung, and Karsten Steinhauer. 2011. "Phrase Length Matters: The Interplay between Implicit Prosody and Syntax in Korean 'Garden Path' Sentences." *Journal of Cognitive Neuroscience* 23 (11): 3555–75.

Innes, Doreen C. 1985. "Theophrastus and the Theory of Style." In *Theophrastus of Eresus: On His Life and Work*, edited by William W. Fortenbaugh, Pamela M. Huby, and A. A. Long, 251–67. Rutgers University Studies in Classical Humanities, v. 2. New Brunswick, NJ: Transaction Books.

Jäger, Lena, Zhong Chen, Qiang Li, Chien-Jer Charles Lin, and Shravan Vasishth. 2015. "The Subject-Relative Advantage in Chinese: Evidence for Expectation-Based Processing." *Journal of Memory and Language* 79–80 (February): 97–120.

Jun, Sun-Ah, and Jason Bishop. 2015. "Prominence in Relative Clause Attachment: Evidence from Prosodic Priming." In *Explicit and Implicit Prosody in Sentence Processing: Studies in Honor of Janet Dean Fodor*, edited by Lyn Frazier and Edward Gibson, 217–40. Studies in Theoretical Psycholinguistics 46. Cham: Springer.

Kaiser, Elsi, and John C. Trueswell. 2004. "The Role of Discourse Context in the Processing of a Flexible Word-Order Language." *Cognition* 94 (2): 113–47.

Knoeferle, Pia, and Matthew W. Crocker. 2009. "Constituent Order and Semantic Parallelism in Online Comprehension: Eye-Tracking Evidence from German." *The Quarterly Journal of Experimental Psychology* 62 (12): 2338–71.

Konieczny, Lars. 2000. "Locality and Parsing Complexity." *Journal of Psycholinguistic Research* 29 (6): 627–45.

Kustas, George L. 1973. *Studies in Byzantine Rhetoric.* Analekta Vlatadōn. Thessaloniki: Πατριαρχικό Ἵδρυμα Πατερικῶν Μελετῶν.

Leech, Geoffrey N. 1983. *Principles of Pragmatics.* London/New York: Longman.

Levy, Roger. 2008. "Expectation-Based Syntactic Comprehension." *Cognition* 106 (3): 1126–77.

Levy, Roger, Evelina Fedorenko, and Edward Gibson. 2013. "The Syntactic Complexity of Russian Relative Clauses." *Journal of Memory and Language* 69 (4): 461–95.

Martin, Andrea E., and Brian McElree. 2009. "Memory Operations that Support Language Comprehension: Evidence from Verb-Phrase Ellipsis." *Journal of Experimental Psychology: Learning, Memory, and Cognition* 35 (5): 1231–9.

Nieuwland, Mante S., and Jos J. A. Van Berkum. 2006. "When Peanuts Fall in Love: N400 Evidence for the Power of Discourse." *Journal of Cognitive Neuroscience* 18 (7): 1098–111.

Patillon, Michel. 2002. *Pseudo-Aelius Aristide: Arts rhétoriques. Tome I, Livre I: Le discours politique.* Paris: Les Belles Lettres.

Roland, Douglas, Hongoak Yun, Jean-Pierre Koenig, and Gail Mauner. 2012. "Semantic Similarity, Predictability, and Models of Sentence Processing." *Cognition* 122 (3): 267–79.

Russell, Donald A. 1981. *Criticism in Antiquity.* Berkeley, CA: University of California Press.

Schlesewsky, Matthias, and Ina Bornkessel. 2004. "On Incremental Interpretation: Degrees of Meaning Accessed during Sentence Comprehension." *Lingua* 114 (9–10): 1213–34.

Sluiter, Ineke. 2016. "Obscurity." In *Canonical Texts and Scholarly Practices: A Global Comparative Approach*, edited by Anthony Grafton and Glenn W. Most, 34–51. Cambridge/New York: Cambridge University Press.

Spivey, Michael, Marc Joanisse, and Ken McRae, eds. 2012. *The Cambridge Handbook of Psycholinguistics.* Cambridge/New York: Cambridge University Press.

Sturt, Patrick, Frank Keller, and Amit Dubey. 2010. "Syntactic Priming in Comprehension: Parallelism Effects With and Without Coordination." *Journal of Memory and Language* 62 (4): 333–51.

Traxler, Matthew J., and Kristen M. Tooley. 2007. "Lexical Mediation and Context Effects in Sentence Processing." *Brain Research* 1146 (May): 59–74.

Van Dyke, Julie A. 2007. "Interference Effects from Grammatically Unavailable Constituents during Sentence Processing." *Journal of Experimental Psychology: Learning, Memory, and Cognition* 33 (2): 407–30.

Vatri, Alessandro. 2017. *Orality and Performance in Classical Attic Prose. A Linguistic Approach.* Oxford Classical Monographs. Oxford: Oxford University Press.

Warren, Tessa, and Edward Gibson. 2002. "The Influence of Referential Processing on Sentence Complexity." *Cognition* 85 (1): 79–112.

—. 2005. "Effects of NP Type in Reading Cleft Sentences in English." *Language and Cognitive Processes* 20 (6): 751–67.

Part II
Cognitive literary theory

6

The cognition of deception

Falsehoods in Homer's *Odyssey* and their audiences

Elizabeth Minchin

Literary theory and cognitive theory

What is it about a story that holds our interest? Why do we find some stories more engaging than others? How to explain the pleasure we find as we listen to or read stories?[1] Certainly, a critical factor in any story's success is its themes: stories about love, death, and personal relationships, for example, have always caught our attention.[2] We know from experience, too, that much depends on the telling. There are some basic requirements: a story should not be longer than it need be—nor shorter; no matter how complex its plot, we expect its language to be clear and unambiguous; and we expect a certain orderliness in its telling.[3] And yet there is more to a story than this. None of these presentational attributes explains the capacity of a story to *engage* its audiences.

Literary theory has for some time, at least from the 1970s, proposed that the reader, or the audience, has a role to play in the creation of meaning in a text. The most significant contributions in this respect have come from reader-response criticism, and the observations of scholars, such as Wolfgang Iser, who regard the audience as an active agent in a participatory relationship with the author.[4] Iser, for example, claimed that a text to some extent controls the audience's responses but that it contains 'gaps' that listeners or readers are required to fill.[5]

From that same period there has been strong interest amongst cognitive psychologists in the mechanisms through which audiences engage with stories.[6] This is an area where important advances have been made. In this chapter I respond to research currently emerging from cognitive studies that throws light on the mental activities performed by audience members as they fill 'gaps' and process complex narrative: I refer here to Theory of Mind (ToM). This cognitive capacity relies on a resource of neural networks that each of us has developed from our early years, which enable us to understand our own mental states and, on the basis of this understanding, to develop intuitions about the intentions and actions of those around us, even though this information has not been formally shared.[7] This ability to 'read' the minds of others enables individuals to explain to themselves why others behave as they do, and thus supplies the link in the chain of causality that connects motivation and action.[8] Just as people rely on this function in everyday life, they rely on it also as they process the stories that they encounter, where neither all the motives of the storyteller nor all the desires and intentions of the characters within the tale are made explicit. This research in cognitive studies, therefore, allows us to think not only about how the listeners or readers who make up a story's audience participate in these 'gap-filling' activities but also how the work they undertake shapes their experience of the tale. By way of testing these ideas, I propose to observe what happens when a storyteller such as Homer throws some serious challenges at his audience and how his listeners might respond.

An ideal locus in which we might explore the experience of the audience in these terms is Homer's *Odyssey*, a story in which disguise is central to the action.[9] Disguise, of course, generates complication; and complication teases and tests audiences both internal (within the narrative) and external (in the real world). My discussion of the complications within the *Odyssey* will focus on a repeated scenario, as the hero, from the moment he arrives on Ithaca, complements his disguise with a string of false identities and a series of false tales.

Telling false tales: a cognitive framework

In 2014 psychologist Jeffrey Walczyk and his colleagues proposed a framework for understanding what they termed 'serious' lies—lies told in high-stakes situations.[10] Through this framework, they argue, the course of any serious lie may be tracked, from the moment an individual is asked for information that he or she does not wish to reveal, through his or her evasive answer, to the response of the addressee. The framework itself is built up from well-established cognitive constructs, principally 'working memory', the 'central executive', and, as described above, 'ToM'. None of these cognitive functions is exceptional; they use precisely the same brain areas as other mental processes.[11] This ability, to deceive others through false information and to maintain that deception, is a skill we acquire in our early years: it first appears at about age three; and, except in extraordinary circumstances, it is a skill we never lose.[12]

So let me talk briefly about the processes of generating a false tale, a serious lie. When individuals are put 'on the spot', so to speak, when they face a request for information, the truth itself is automatically retrieved from their long-term memory and, on transfer to working memory, is immediately available.[13] But, as they weigh up the likely impact of the truth on their addressee, they may judge that an honest response is not an appropriate response. And they will assess what their addressee is likely to believe should they construct a false tale. These two functions are the task of ToM, that learned capacity to infer the state of mind of those around us in the real world.[14] The central executive function of the brain, which integrates metacognition, working memory, and inhibition, supports and keeps track of these ToM inferences, and, at the same time, its inhibitory processes overcome unhelpful interference by the truth as the false tale is constructed.[15]

As they construct their false tales, successful liars try to simplify the information they share. Although they wish to repress the truth, they may draw, strategically, on elements of it; they consult long-term memory to locate personal memories or events 'vicariously experienced' on which they can build; and they turn to episodic memory as a source of vivid, and authentic, detail.[16] This strategy has three advantages: by staying close to the truth speakers find the deception easier to remember for the longer term; through their use of authentic experiences they are able to maximize the plausibility of their tale; and, critically, they are able to minimize the additional cognitive load that high-stakes deception generates. Whereas everyday deception, the white lie, has been shown to make minimal demands on cognitive resources, activities in connection with deception in high-stakes contexts—the decision to lie, the careful construction of the lie, tracking the response of the addressee, and recalling the lie—are more of a burden, particularly on working memory.[17] Nevertheless, practice is helpful: a well-rehearsed lie can reduce that burden significantly.[18]

These functions are all represented by the poet of the *Odyssey* in the person of Odysseus, for whom he (the poet) composes false tales to obscure the 'truth' about the hero's identity and his past.[19] Like any teller of a serious lie, the poet himself must inhibit the 'truth' (or what he has persuaded us is the real story); and yet he must not lose sight of it. He must construct a lie that is plausible and that will be favourably received by its audience; and he must keep track of that lie, so that he can construct answers that are plausible, in terms of the events of the story, to any follow-up questions. Odysseus' lies are generated in what are depicted as high-risk situations in the storyworld;[20] and his addressee is, in most cases, taken in by the fiction.[21] But the context in which the lies will *actually* be judged is that of the story-realm, that zone in which the poet and his audience interact. It is in this realm that the plausibility of the fabrications attributed to Odysseus and the success of the tale as a whole will be measured.

External audience members have the advantage, so to speak, over members of the internal audience, in that, even as they listen to one of the hero's false tales, they *know* the 'real' story of Odysseus' identity and his recent history. But they are not omniscient. They are left to infer the intentions of the poet himself, who can be as reticent about the desires and intentions of the characters within the tale as he is about his intentions for his story as a whole. As David Herman observes, the ability to comprehend narrative '*requires*' us to situate participants in the tale 'within networks of beliefs, desires, and intentions'—which we must supply from what the storyteller has told us and from our own resources, in particular our ToM.[22] And, I add, we must be able to situate the storyteller too within those same networks.[23]

My aim in this present exercise is to draw on these ideas in cognitive studies in order to outline a new way of analysing some of Homer's most challenging scenes. Ultimately, I aim to illuminate aspects of cognitive activity in the members of Homer's external audience as they try to track the hero's lying tales and their effect; and I shall show how, as external audience members are teased and tested by the poet (as they observe internal audience members being teased and tested by the hero), their ToM-led processing activities (and their pleasure in resolving these lower-level exercises) foster their engagement with the tale as a whole.

Before I turn to my selection of the 'serious lies' that Odysseus tells the people he encounters on Ithaca, let us consider a singular moment in Odysseus's account of his wanderings, a moment when the hero, throwing caution to the winds, tells the truth about himself. Odysseus and his men have escaped from Polyphemus; they are on board their vessel and are leaving the land of the Cyclops (9.462–472). Their safety seems assured. At this point Odysseus taunts the giant, not once, but three times. Ignoring the pleas of his companions, he reveals to Polyphemus his identity (502–505). This is indeed unusual behaviour in our hero, who has been associated by tradition with the stealth of the ambush (*Il.* 10.242–247, 338–531), the deceit of the Wooden Horse (*Od.* 4.269–289), and an ingenious escape only a little earlier from that same Cyclops' cave (*Od.* 9.424–467).

What appears to have happened can be explained in cognitive terms: at the very moment when the hero should have been responding to an ambush-situation, where caution is crucial, he exhibited the behaviour of a triumphant Iliadic hero on the battlefield.[24] At this crucial moment, as he defies Polyphemus, Odysseus' central executive, operating on its default 'Iliadic' setting, has ignored its normal inhibitory processes, which would have urged silence on him or, at least, would have led him to suppress the truth.[25] It is this unhappy moment of truth-telling that brings the Cyclops' curse and Poseidon's wrath upon the hero, and this in turn prolongs his journey and leads to the special circumstances of his homecoming.[26]

Odysseus tells a false tale to Athena

The first of the serious lies the hero tells is the false tale he composes shortly after he wakes in his own fatherland (13.187–188). At this point Odysseus does not recognize Ithaca (οὐδέ μιν ἔγνω, 188), because Athena has poured a mist over the island. She does so, the poet suggests, to create the opportunity to advise her protégé of the situation on the island and to change his appearance (190–193). To complement the unfamiliar setting, Athena has disguised herself as well, as gods are wont to do. She has assumed the form of a young shepherd boy.

Odysseus addresses this young stranger. Approaching him as a suppliant (231), the hero asks where he is and the name of the people who live there. As the shepherd announces that the island is Ithaca, Odysseus rejoices. The truth—that this is his homeland—springs automatically to his lips (250–251). He is at the point of betraying himself. But he has learnt his lesson from that earlier lapse of attention with the Cyclops. Once again, much is at stake: the hero does not know what his reception in his homeland or in his palace will be. Furthermore, when he was in the Underworld he had heard Agamemnon's tale of his homecoming (11.405–434) and he had absorbed Agamemnon's advice (that he should return without attracting notice, κρύβδην, and not openly, μηδ' ἀναφαδνά, 11.454–456). This, then, is the motivation for his subsequent actions. As the poet makes clear, Odysseus' inhibitory processes engage: he 'pushes back' what he was about to say (πάλιν δ' ὅ γε λάζετο μῦθον, 254) and, relying on ToM and working memory (we assume), he prepares a false tale.[27] His goal at this point is to protect himself and the gifts he had been given by the Phaeacians.[28]

Let us engage again with the framework for serious lies that I described above. As we follow the false tale that the poet has created for Odysseus, we infer the mental activities the poet has attributed to his hero. First, we understand that Odysseus has read the social context, has concluded from the young shepherd's appearance what his listener might know of the world, and has guessed at what might impress him. He shapes his story around a number of well-established 'facts' that will serve this purpose, constructing a believable timeline and inventing only as much detail as is necessary to achieve his goals. In his new guise, as hero of his own false tale, Odysseus comes from Crete, a location so remote, he hopes, that his account cannot be disproved.[29] He has heard of Ithaca, he says. He is a hero of the Trojan campaign, and a friend of Idomeneus, his fellow countryman; he is the kind of man, he implies, who is awarded spoils after a successful campaign, who will kill to protect what is rightfully his, and who will not submit himself to others (13.256–271). Such an economical combination of fact and fiction underpins any successful lie. At the heart of this false tale we find another figure familiar to us (and, apparently, to the internal audience also) from the Troy-story, and some familiar themes: Idomeneus is a leading hero in the action of the *Iliad*; Odysseus' claimed association with him, even as a subordinate, is a claim to status.[30] Furthermore, the division of spoils and the quarrels that arise over them represent a theme familiar to the internal audience from the Troy-story more generally and to the external audience from the version of the Troy-story that we know from the *Iliad*.

Odysseus goes on to say that he has been brought to Ithaca by Phoenicians, who, having promised to take him to Pylos or Elis, have tricked him—in accord with their reputation (272–286).[31] His lie is carefully tailored to the situation as he sees it: he seeks sympathy and respect—respect for his status and for his readiness to defend his own interests, and sympathy for his current plight.[32]

But this false tale does not persuade his addressee. As a goddess, Athena, being omniscient, knows the 'real' story of Odysseus. What is interesting here is that in this episode, in which Odysseus tells the first of his lies, Athena, in the storyworld, models for us, the external audience, the response the poet expects in his listeners as they hear such a tale, and as they hear the other false tales that Odysseus later tells on Ithaca.[33] As she listens, Athena smiles (287), recognizing the hero's natural propensity for deception and, I suggest, taking some pleasure in her own easy ability to detect the lie.[34] Indeed, she compliments her protégé on his deviousness (291–295). And this, the poet hints, is what we should do too. We should enjoy the experience of detection; we should admire Odysseus for his cunning; and we should congratulate the poet who has created this episode. Our enjoyment and our admiration will not be limited to this tale alone. There are more spectacular manipulations of the truth ahead, as Odysseus defers the moment of recognition until he has tested even his wife.

But what kind of mental activities have we, the 'knowing' external audience, been engaged in? Mark Turner has long been interested in the capacity of the human mind to activate and run different, even incompatible, stories at the same time.[35] Some stories—jokes, for example—do not make sense unless we are able to process them from more than one perspective.[36] Likewise, the cognitive processes associated with lies and deception (for a knowing audience) require activity at two levels: first, with regard to the memory of the actual situation as we understand it and, second, with reference to the false tale that we hear. These dual-processing activities require us, first, to shift from perspective to perspective, overlaying each lie on what we already know of the epic account of Odysseus; second, to compare the two narratives (the 'real' and the false);[37] and, finally, to reach some understanding of the goals of the hero—and of the poet.

Odysseus' second false tale to Eumaeus

I turn now to a tale that Odysseus tells Eumaeus, his swineherd host, on his first evening in his dwelling. Odysseus, disguised now as a beggar, has already complemented his disguise with an elaborate Cretan tale that takes up many of the elements of his false tale to Athene, but which goes further, announcing the imminent return of Odysseus.[38] As the evening chill makes itself felt, the beggar asks his host for a blanket. Indeed, the poet tells us that the hero was testing (πειρητίζων) Eumaeus, to see whether he would give him his own cloak or the cloak of one of his men (14.459–461).[39] In putting his request he tells a story of a venture that, he boldly claims (εὐξάμενός, 463), he undertook in the course of the Trojan War. Naming two of the great heroes of the Trojan campaign, Odysseus and Menelaus, as leaders of an ambush, he includes himself (now the beggar) as a third (470–471).[40] Bad weather came on (475–477); he (the beggar) was caught without a cloak for warmth (478–482). He remarked on his condition to Odysseus (486–489), who, promptly devising a plan to deal with the problem, incited the hero Thoas to leap up in haste and run off to take a message to Agamemnon (495–498). Thoas's cloak now became available for our storyteller, the beggar (499–502).

As the beggar offers this second false tale to Eumaeus the poet persuades us that he (the beggar) has carefully tracked his host's reactions to his earlier concoction and is confident that his host has accepted its principal elements—except for the information about Odysseus' return (363–365). In reusing elements that are 'real' or, at least, consistent with his earlier false tales, he (more accurately, the poet) reduces his cognitive load.[41] The cloak-tale is set in the context of the Trojan expedition. So much is plausible. It includes the big names from Troy: plausible again; and it reflects quite accurately the wily mind of its hero, Odysseus.

The poet, of course, has told the external audience in advance why the beggar will tell this story. And the beggar himself gives a strong although indirect hint to the swineherd that a cloak is what he needs (504–506). Now the swineherd responds promptly. He commends the tale as a good story (508); he compliments the teller for the way in which his request was presented (509–510);[42] and he lends him a cloak for the night (510–512). What has impelled him to do so? In pointing out that there are no extra cloaks for the swineherds Eumaeus makes it clear nevertheless that there has been something about this story that has moved him to make a concession. The impetus, I suggest, is a sense of the presence of his master in the story itself, both in the telling and in the quick thinking that is at its heart. It is in response to this authenticity, and in honour of the absent Odysseus, that Eumaeus accedes to his visitor's request. And in doing so he passes the test that his master has set him.[43]

So much for the internal audience. As we, the external audience, follow this tale we engage in two simultaneous processing activities. First, we hear and process the tale as it is told to Eumaeus by the 'beggar', who, on the one hand, creates a fake-Odysseus, distinct from himself, brilliantly evoking the hero's resourcefulness and his agile mind. He brings him to life. At the same time, however, we know that the beggar *is* Odysseus: so, as a parallel activity, we process the tale as it is told *by Odysseus himself* to the swineherd, in which he (in beggar's guise) tells a story about himself as a quick thinker and 'fixer' who comes to the aid of a completely fictitious character (the 'beggar'/*not*-Odysseus) who had forgotten his cloak on a cold night not unlike the present.

But our processing activities do not stop there. How can we explain the motive for the poet's remarkable inclusion of Odysseus as a player in the story the beggar tells? When Odysseus/the beggar includes the figure of Odysseus as the hero of this tale, we propose to ourselves, as I suggested above, that this choice has been made in order to increase the chance of a favourable response from Eumaeus. But what is happening in the story-realm, where the poet plays to his external audience? What are the implications, at this level, of the poet's decision to represent Odysseus in multiple guises, as storyteller in disguise and as hero of his own story?

Some audience members might propose that the poet has his hero include himself in his tale because Odysseus enjoys playing with his own identity (he is, after all, πολύτροπος [*Od.* 1.1]);[44] others might suggest that he takes the opportunity to sing his own praises and to compliment himself;[45] it is possible too that the hero in disguise is anxious not to lose his identity. All these proposals are valid, and are supported by events elsewhere in the narrative. But I propose in addition a fourth motive, which feeds directly into the relationship of poet and audience: the poet himself, through his representation of his hero's devices, aims to impress us with his ability to make agile shifts between parallel and all but intersecting stories, between 'real' and fictitious identities, and between 'the real story' and truthlike fiction.[46] For us, the task is to keep pace with him as we handle these cognitive challenges. As to why the poet has included Odysseus as a player in this tale, there is, I think, no single 'correct' answer. But I come back to the notion that the poet's desire to win the admiration of his audience represents, for him, a high-stakes situation: hence the complications that he builds into his tale.

Odysseus tells a 'Cretan lie' to his wife

At 19.104–105 Penelope asks the beggar the questions that are put to any stranger at the first encounter: who are you? where do you come from? who are your parents? Now that their meeting, which the beggar had deferred (17.561–573), has at last begun, he is again reluctant, unwilling to respond. He fends off her questions at first (115–120) but eventually yields, in apparent exasperation (165–166),[47] promising to tell her what she wants to know. And yet he stalls, giving a long geographical account of Crete (172–180), which he claims is his home. His father, he says, is king Deucalion, grandson of Minos (178–181); he is the younger brother of Idomeneus, who went to Troy (181–184). His own name is Aethon (183). He had met Odysseus, he says, on his way to Troy (185). The hero, blown off course, had landed in Crete and had asked for Idomeneus, a guest-friend. Aethon claims to have entertained Odysseus for twelve days before sending him on his way to Troy (185–202).

The beggar recycles elements of the earlier false tales: the association with Crete, his relationship to Idomeneus, and news of Odysseus. In now claiming noble birth, he has adjusted his fictional status, as younger brother of Idomeneus, to match that of his addressee, Penelope, and to win her trust.[48] And, for the first time in these lying tales, he gives himself a name, Aethon.[49] The use of proper names normally enhances the authenticity of any narrative: indeed, the success of the tale that he is at present constructing depends more than ever on its plausibility.[50]

Indeed, on passing on information that Odysseus had been driven off course as he passed Cape Malea, he refers to an acknowledged sailing risk in this world. In fact, he draws on his own experience, as the most accomplished liars do, of his return journey from Troy (9.80–81).[51]

As the poet tells us at 203, 'he knew how to tell many lies that were like the truth' ('ἴσκε ψεύδεα πολλὰ λέγων ἐτύμοισιν ὁμοῖα).[52] Penelope, at this point, takes in the story in the terms that it was told, and she weeps as she recalls her husband (204–209). But, ever-cautious, she seeks (215–217) to test the accuracy of that part of the tale that relates to Odysseus; *that* she will be able to verify. She asks what garments Odysseus was wearing when he arrived in Crete (218–219). Here is the advantage of using elements of the truth (and a plausible timeline) in the construction of a lie. Odysseus is, unsurprisingly, able to give a full and accurate account 'as my heart imagines' (ὥς μοι ἰνδάλλεται ἦτορ, 224)—after an appropriate disclaimer, in which he laments the lapse of time (221–223). Penelope confirms that he has spoken the truth (249–250); she declares that the beggar will be her friend in the palace (253–254). But at no point in this scene (257–260, 312–316) or in subsequent scenes (560–581) will she accept that Odysseus is still alive.[53]

At the heart of the hero's well-constructed tale are two important messages. The first is that, just as his *alter ego* Aethon had not forgotten Odysseus, neither should Penelope. After Aethon has won Penelope's trust, at least to the extent that he persuades her that he met the hero twenty years before (the second part of this false tale, 221–248), he announces that the hero is at hand (the third element, 268–307).[54] Aethon's references to Odysseus are not idle. As in the cloak-tale, Odysseus has cast his real self as the hero of his false tale to catch Penelope's attention and to engage her goodwill.[55] But Penelope, in reply, denies that Odysseus will ever return to his home: she is unpersuaded (312–313). And yet, with respect to his internal audience, this is not his only motive for putting Odysseus at the centre of his tale. Recall the Odyssean qualities that I identified above: the hero who enjoys playing with the truth, the hero who is not reluctant to sing his own praises, and the hero who is struggling to maintain his sense of self. These qualities are in play again here and now, even when the hero is, at last, in his own palace in the company of his wife.

The second message lies in Aethon's account of the hospitable treatment he had offered Odysseus years before (194–201). This account, I suggest, is designed to remind Penelope of the hospitality appropriate for a guest-friend or a stranger, such that Odysseus himself used to dispense in the palace.[56] In his generation of this fictitious but plausible 'memory' we recognize another Odyssean characteristic: we see the same indirectness in these hints to his wife that we observed in his cloak-tale to Eumaeus. Indeed, responding to Aethon's suggestion, Penelope will offer the hospitality that the stranger seeks (317–324), apologizing—ironically—for not being able to match Odysseus in what she can offer her guest (313–316).

As we, the external audience, process the narrative, we observe the tantalizing scenario devised by the poet:[57] a distressed wife seeks information about her long-absent husband from that very man; and he is not yet prepared to reveal himself to her.[58] By including the figure of Odysseus as a character in the story Aethon tells Penelope, the poet again requires us, his external audience, to acknowledge his skill in tracking the knowledge states of his characters (in this case, Penelope, Odysseus the hero, and Odysseus as Aethon)—and to recognize his readiness to challenge our own.

Conclusions

We have sampled the remarkable series of false tales told by a master-tale-teller, Odysseus, to a series of addressees on Ithaca.[59] As a strategic measure in a potentially dangerous environment, Odysseus has refrained from revealing his identity. Repressing the 'real story' at each encounter, he has created an appropriate false tale for each of his addressees. Some scholars have discussed the remarkable plausibility of these tales;[60] others have argued that these false tales suggest alternative oral traditions of Odysseus' return to his homeland;[61] yet others have discussed the place of the lie in the value system of the ancient Greeks.[62] What I am interested in at this point is the authenticity of the poet's representation of *both* the process of telling a false tale, from the moment the hero decides to repress the truth, *and* the mental responses he attributes to each of the hero's addressees.

I have tracked these so-called serious lies from a cognitive perspective—following the often simultaneous operations of ToM, working memory, and the central executive. The consistent threads we observe in Odysseus' string of Cretan tales speak to any serious liar's efforts to reduce the consequent cognitive load; and the variation we note in those lies is his ToM-led response to the expectations of his respective audiences. But, in fact, it is not Odysseus who is actually drawing on these mechanisms or labouring under these cognitive burdens. It is the poet, not Odysseus, who, like any teller of false tales in the everyday world, represses the 'real story' in order to construct a string of plausible lies for the hero's internal audience.

The poet's activities are not limited, however, to interactions in the storyworld. He is concerned also about his relationship with his external audience. Unlike the teller of false tales within the *Odyssey*, the poet is not using this string of lies to deceive his listeners; his aim is to achieve their commendation for his inventiveness.[63] Think back to Athena's response to Odysseus' first lying tale: she had smiled in pleasure, complimenting her protégé and delighting in her own cunning intelligence. So, even as the lies Odysseus tells must be plausible and persuasive in the world of the story, the poet's representation of a liar at work and of the lies he tells must be plausible, persuasive, and involving in the eyes of his real-world audiences.

Let me turn now to the ways in which the poet engages his listeners—and where this engagement leads. As members of his audience follow the story of Odysseus' arrival on Ithaca and as they observe his negotiations with Athena, Eumaeus and Penelope, they, like the poet, are aware that the hero is in disguise; they do not forget that the beggar is, in fact, Odysseus; they are aware that he is the son of Laertes, and that Ithaca, not Crete, is his homeland; and they know him as a schemer. So, as they listen to the hero's fabrications, addressed to Athena, to Eumaeus, or to Penelope, this other bundle of information is at call in working memory. The task before them is to follow the story that the beggar presents, comparing it with his 'real story', identifying points of correspondence (the expedition to Troy, for example, or his association with Idomeneus) and points of difference (his relationship to Idomeneus; Crete as his homeland).[64] This, as I noted above, is a dual-processing task, as the listener switches his attention back and forth, almost automatically, between the false tale and the 'real story'.[65] But the poet has added a further complication: the inclusion of Odysseus as a player in two of these tales. It is this complication, which Martin West describes as a 'bamboozling',[66] which forces us, his audience, to work quite hard at this point of the story, in order to resolve the impossible relationship between the teller of the false tale and a particular member of his cast, and, relying on ToM, to guess at the motives of the poet in creating this tangle.

By concentrating on this problem, overlaying the disguised Odysseus on the real Odysseus, and separating both figures from the Odysseus of the false tales, we are able to unravel the poet's 'brain-twister'. As we work through this conundrum, we become acutely aware, fully conscious, of our own processing mechanisms. On resolving this puzzle, and on identifying, at least to our own satisfaction, the poet's motivations for including it, we have a flush of pleasure. This pleasure, Reuven Tsur argues, arises from the confirmation, the reassurance, that these mechanisms of ours are working efficiently.[67]

So, to return to the point at which I began: we now have a working hypothesis from cognitive studies that illuminates the mental activities not only of characters engaging in 'mind-reading' within one of the earliest tales to have survived to us but also of the poem's audience members, as they too rely on ToM to fill 'gaps' and process this complex narrative. In the account above of the work that audience members must undertake to resolve the problem-solving exercises that the narrative poses, we find support for the claim that such creative 'work' leads to greater engagement with the narrative and, ultimately, greater pleasure and greater confidence.[68] In trying to identify the poet's aims as he creates each false tale, I suggest that his goal is not, as West describes it, to bamboozle—to baffle and confuse—his external audience. Indeed, it is vital for the success of his performance that we keep up with his tale. But as I, on behalf of all the *Odyssey*-poet's audiences over the centuries, model a response in cognitive terms to the poet's string of false tales in all their complexity, I conclude that his long-term goal is to leave his listeners and readers deeply impressed, even occasionally dazzled, with the brilliance of his storytelling prowess—and with the agility of his mind.[69]

Notes

1 I offer my thanks to audiences in Canberra and Leiden, and to the reviewers of this written version, for their very helpful comments on earlier versions of this chapter.
2 See, for example, Schank (1979) 278–291; Polanyi (1979) 211–213.
3 I have adapted these 'maxims' from Grice's 'cooperative principle': on this see Grice (1975) 45–50. Although developed for conversational situations, these maxims—quantity, quality, relation, and manner—might also describe effective storytelling, whatever the medium.
4 A principal contributor to this discussion has been Wolfgang Iser, who insisted that we should analyse our own involvement with a text if we are to understand what it is about. For a valuable summary of Iser's thinking, see Holub (1984) 82–106, at 106.
5 Holub (1984) 92–96.
6 For early work of this kind, see, for example, Schank and Abelson (1977); and, more recently, Gerrig and Egidi (2003).
7 Carlson, Koenig, and Harms (2013) 4.394.
8 By inferring motives and intentions we come to understand the chain of cause and effect that underpins any sequence of events: see Schank and Abelson (1977), ch. 2.
9 The classic account of this distinctive motif and of its contribution to the plot of the *Odyssey* is Murnaghan (1987); see also Bowie (2013) 6–15.
10 Walczyk, Harris, Duck, and Mulay (2014); and, for background, see Gombos (2006) 198–199.
11 Walczyk et al. (2014) 24.
12 Talwar and Lee (2008) 876–879.
13 For an important early study on working memory see Baddeley (1986); see now Smith and Kosslyn (2007) chs 5 ('Encoding and Retrieval from Long-Term Memory') and 6 ('Working Memory'). On working memory as a short-term storage site in which 'complex cognitive activities that require the integration, coordination, and manipulation of multiple bits of mentally represented information' are made possible, see 249. For these activities in the context of serious lies, see Walczyk et al. (2014) 24.
14 For functions of ToM in the context of lying, see Walczyk et al. (2014) 30.
15 On executive attention, see Smith and Kosslyn (2007), ch. 7 ('Executive Processes'), esp. at 289–291; Walczyk et al. (2014) 31; Talwar and Lee (2008) 877–878.
16 Walczyk et al. (2014) 30; Debey, De Houwer, and Verschuere (2014) 325.
17 On everyday deception, see Walczyk et al. (2014) 32–33. On cognitive load, see also Debey et al. (2014) 326. The heaviest load occurs when speakers find themselves in unfamiliar, surprising, or complex social contexts: having made the decision that a lie is the appropriate response, they are obliged to work harder to infer what their audiences might expect or wish to hear (Walczyk et al. [2014] 31–33).
18 Walczyk et al. (2014) 31.
19 For Odysseus' lies: to Athena (13.256–286); to Eumaeus (14.192–359 and 468–506); to Antinous (17.415–444); to Amphinomus (18.125–150); to Melantho (19.71–88); to Penelope (19.165–202, 221–248, 262–307, 336–342); to Laertes (24.258–279, 303–314).
20 I suggest that the lie Odysseus tells to Alcinous at 7.303–307 is, by contrast, a white lie. But the external audience, in this case too, must engage in dual-processing in order to detect the lie and observe its construction.

21 For discussions of Odysseus' lies, see Trahman (1952); Walcot (1977); Haft (1984); Emlyn-Jones (1986); Pratt (1993) ch. 2.
22 Herman's italics: Herman (2003) 169. Zunshine goes further: she argues, in her discussion of fiction, that narratives not only 'rely' on our readiness to keep track of individuals' thoughts, desires, and feelings (that is, we are expected to work) but also 'manipulate' and 'titillate' it; and that it is our ToM that makes literature as we know it possible: Zunshine (2006) 3–10, at 5.
23 For an important application of ToM to the action within the storyworld of the Homeric epics, see Scodel (2014) 55–74. Scodel describes ToM as 'a distinct and significant topic in narrative studies' (56).
24 Jones (1988) 88; de Jong (2001) 246.
25 For discussion of emotion regulation from a cognitive perspective, see Koole (2009) 4–10.
26 The other point of the narrative at which Odysseus loses his self-control, again critically, is at 23.181–204, in response to being tested by his wife. In reading this later scene we conclude that the force of Odysseus' indignation is too immediate and too strong for the inhibitory processes of his central executive.
27 Note how the poet conceives of this operation of mental restraint: the hero, as it were, snatches at the truth before it escapes his lips.
28 See also Pratt (1993) 90–91.
29 On this point see Trahman (1952) 35–36.
30 The audience will be predisposed to accepting and even admiring Odysseus' claimed identity: see Haft (1984) 294. On the link with Idomeneus, Haft (at 292–299) argues that the poet has used the Iliadic bond between Idomeneus and Meriones as a framework for the Cretan lies he composes for Odysseus; and that Meriones therefore serves as the model for the first-person hero of Odysseus' Cretan tales.
31 The Phoenicians have a bad reputation: cf. Eumaeus' story at 15.415–484. But, on this occasion, these 'Phoenicians' have, uncharacteristically, respected Odysseus' property.
32 See Haft (1984) 292.
33 This is not an isolated instance in the epics of 'modelling': see Wyatt (1988) 289–297; Pratt (1993) 72.
34 Pratt (1993) 72.
35 Turner (2003); see also Fauconnier and Turner (1998). For useful background on dual attention, especially when the two tasks involved overlap in significant details (as in this case), see Neisser (1976) ch. 5 ('Attention and the Problem of Capacity'); Smith and Kosslyn (2007) 42.
36 The processing of a joke requires us to shift our 'mental set' from perspective to perspective, from plane to plane, in order to 'read' a given narrative from the point of view of different participants, or in different contexts. On shifting mental sets, essential, for example, to the interpretation of jokes, see Tsur (1989) 247–248; for a slightly different account of the role of blending, see Fludernik (2015) 155–175.
37 Indeed, as Joshua Landy comments, '[i]n a way . . . *all* fictions put us in a divided state of mind'. He refers to our awareness that what we are hearing or what we are reading is not real: Landy (2015) 572.
38 The first tale to Eumaeus includes elements from Odysseus' false tale to Athena: association with *Crete, status* as the son of a wealthy man; participation in the *Troy-expedition*, having been asked by the Cretans to lead their contingent along with *Idomeneus*; and involvement with a tricky *Phoenician*. He mentions his arrival in Thesprotia, where he heard news of the Odysseus who had been there also and that he was about to set off for home. It is this element only that Eumaeus doubts: see, for example, 14.378–387.
39 For background to the tale see Bowie (2013) 223.
40 On the 'presumptuous' nature of the tale, see de Jong (2001) 360.
41 And, indeed, he reduces the cognitive load on his external audience as well.
42 Indeed, Eumaeus at 508 describes the story, the αἶνος, as an excellent 'story with a hidden meaning' (de Jong [2001] 360).
43 The cloak-tale qualifies as a 'serious lie'; it was 'devised' by Odysseus/the beggar as a *test* of Eumaeus' loyalty to his master; it was not, however, intended to cheat him of his possessions: cf. Pratt (1993) 89.
44 Despite the fact that Eumaeus is said to be loyal to his master (13.404–406), Odysseus proceeds as Athena had instructed him, testing him also, and cautiously maintaining his disguise in the palace until the time is right to punish the suitors.
45 Jones (1988) 136; de Jong (2001) 360.
46 Cf. Odysseus' other Cretan lies, especially his elaborate lie to Penelope, discussed below.
47 That is, we use ToM to read Odysseus' words as *mock* exasperation: the hero knows that it is proper to ask about a stranger's identity; and, we guess, he knows from experience that his wife is cautious: she will not be satisfied with a superficial or dismissive reply. As we shall see, Penelope will demonstrate this

same trait again, as she refuses to recognize the stranger as her husband until she is thoroughly persuaded (23.205–230).

48 For some discussion of some of these features see de Jong (2001) 468–469. When the beggar spoke with Eumaeus he concocted a 'mixed career' that was not dissimilar to Eumaeus' own. Now, in speaking with Penelope, he claims a royal lineage that will win her confidence: Russo in Russo, Fernández-Galiano, and Heubeck (1992) 85–86.

49 On the possible significance of the name, see Rutherford (1992) 161–162. Odysseus will take another name (Eperitos) in his final lie, to Laertes: 24.304–306.

50 The beggar must tell a story that will convince Penelope that he is a reliable witness, someone whom she might take into her confidence.

51 As experienced also by Menelaus, who was driven to Crete (3.286–303) at this same point. Odysseus is, however, unaware of the Menelaus-story. But we, the audience, recognize the parallel.

52 A skill much admired by Aristotle, *Poetics* 1460a18–26.

53 See also Heitman (2008) 43–49, esp. at 48.

54 Note the hero's use of the adverb νημερτεώς (269), truly, infallibly, as he introduces his claims about Odysseus: this term has appeared as a thread through the Ithaca-narrative: see especially 17.549, 556, 561.

55 This false tale, like an exemplary false tale, includes elements of the truth: the loss of his men, his encounter with the Phaeacians, the gifts that they gave him (273–282).

56 Rutherford (1992) 163 suggests that the hero stresses Aethon's hospitality in order to emphasize his own virtue and the obligation on Penelope to befriend her visitor as her husband would have done (315–316).

57 On the exquisite irony of the scene, as Penelope weeps for her husband, who was seated beside her, in the careful precision of ἑὸν ἄνδρα παρήμενον, at 209, and the pathos of Odysseus, who felt pity for his wife, ἑὴν . . . γυναῖκα, at 210, to whom he could not betray his identity, see de Jong (2001) 470; Russo (1992) 87–88.

58 After the lesson he learned as he left the Cyclops' cave, Odysseus' central executive continues to maintain careful control of his responses.

59 One omission from the false tales under discussion above is the lie Odysseus tells his father Laertes, at 24.258–279, 303–314. Although the hero is sorely tempted, when he encounters his father, to embrace him and identify himself, he falls back on his earlier strategy of the false tale. This lie is not told in a high-stakes context, although Odysseus nevertheless forms a plan to test him (232–240). How to interpret this? On the one hand, Odysseus' behaviour may be an example of 'duping delight': the pleasurable anticipation of lying successfully: on the positive feelings that lying can arouse, see Ekman and Frank (1993) 184–200, at 194–195. But, as soon as Odysseus sees the misery his false tale has caused (315–317), he experiences a harsh moment of shock (δριμὺ μένος, 319) and abandons his disguise (321–322). On the other hand, we may be observing a learned response. It may be that as an outcome of the unhappy finale of the Polyphemus-episode of *Od.* 9, the hero has schooled himself to deceive; even in the presence of his father he feels that he cannot betray himself.

60 Cf. Trahman (1952); Haft (1984); Pratt (1993) 85–93.

61 See especially Tsagalis (2012).

62 Cf. Walcot (1977) 1–19; Emlyn-Jones (1986) 1–10; Pratt (1993) 56–63.

63 Cf. the response of the audience to Odysseus' own long tale at 11.336–341 and at 13.1–2. For commentary see, for example, Rabel (2002) 87; Minchin (2007) 242–243.

64 It is clear that the poet intends that comparison should take place concurrently, since there is no time in an oral performance for any other option. For commentary on dual attention, see above.

65 There is a cognitive burden associated with this task: see Smith and Kosslyn (2007) 303–308, at 304 on switching cost. This dual-processing exercise is, I suggest, possible thanks to our 'training' in everyday cognitive processing. In the real world, for example, we are almost unfailingly able to follow what is happening when we are aware that a friend or a partner is telling a lie.

66 West (2014) 14.

67 Tsur (1989) 247. Tsur refers here specifically to the 'adaptation mechanisms' that allow a listener to find humour in a joke. I suggest that, for a knowing audience, the processing of a complex lie requires a similar ability. See also Zunshine (2006) 16–22.

68 On this, see Zunshine (2006) 20.

69 This too the poet models for us in the responses of the Phaeacians to Odysseus' own tale: his listeners were 'held in enchantment' (κηληθμῷ, 11.334, 13.2) by his telling.

References

Baddeley, A., *Working Memory*, Oxford, Clarendon Press, 1986.

Bowie, A., *Homer: Odyssey Books XIII and XIV*, Cambridge, Cambridge University Press, 2013.

Carlson, S.M., M.A. Koenig, and M.B. Harms, 'Theory of Mind', *Wiley Interdisciplinary Reviews: Cognitive Science* 4 (2013) 391–402.

Debey, E., J. De Houwer, and B. Verschuere, 'Lying Relies on the Truth', *Cognition* 132 (2014) 324–334.

de Jong, I., *A Narratological Commentary on the Odyssey*, Cambridge, Cambridge University Press, 2001.

Ekman, P. and M. Frank, 'Lies That Fail', in M. Lewis and C. Saarni, ed., *Lying and Deception in Everyday Life*, New York and London, The Guilford Press, 1993, 184–200.

Emlyn-Jones, C., 'True and Lying Tales in the *Odyssey*', *Greece and Rome* 33 (1986) 1–10.

Fauconnier, G. and M. Turner, 'Conceptual Integration Networks', *Cognitive Science* 22 (1998) 133–187.

Fludernik, M., 'Blending in Cartoons', in L. Zunshine, ed., *The Oxford Handbook of Cognitive Literary Studies*, Oxford, Oxford University Press, 2015, 155–175.

Gerrig, R. and G. Egidi, 'Cognitive Psychological Foundations of Narrative Experiences', in D. Herman, ed., *Narrative Theory and the Cognitive Sciences*, Stanford, CA, CSLI Publications, 2003, 33–55.

Gombos, V., 'The Cognition of Deception: The Role of Executive Processes in Producing Lies', *Genetic, Social, and General Psychology Monographs* 132 (2006) 197–214.

Grice, H., 'Logic and Conversation', in P. Cole and J. Morgan, eds, *Speech Acts: Syntax and Semantics*, vol. 3, New York, Academic Press, 1975, 41–58.

Haft, A., 'Odysseus, Idomeneus and Meriones: The Cretan Lies of Odyssey 13–19', *The Classical Journal* 79 (1984) 289–306.

Heitman, R., *Taking Her Seriously: Penelope and the Plot of Homer's Odyssey*, Ann Arbor, MI, University of Michigan Press, 2008.

Herman, D., 'Stories as a Tool for Thinking', in D. Herman, ed., *Narrative Theory and the Cognitive Sciences*, Stanford, CA, CSLI Publications, 2003, 163–192.

Holub, R., *Reception Theory: A Critical Introduction*, London and New York, Methuen, 1984.

Jones, P., *Homer's Odyssey: A Companion to the Translation of Richmond Lattimore*, Bristol, Bristol Classical Press, 1988.

Koole, S., 'The Psychology of Emotion Regulation: An Integrative Review', *Cognition and Emotion* 23 (2009) 4–41.

Landy, J., 'Mental Callisthenics and Self-Reflexive Fiction', in L. Zunshine, ed., *The Oxford Handbook of Cognitive Literary Studies*, Oxford, Oxford University Press, 2015, 559–580.

Minchin, E., *Homeric Voices: Discourse, Memory, Gender*, Oxford, Oxford University Press, 2007.

Murnaghan, S., *Disguise and Recognition in the Odyssey*, Princeton, NJ, Princeton University Press, 1987.

Neisser, U., *Cognition and Reality: Principles and Implications of Cognitive Psychology*, San Francisco, CA, W. H. Freeman, 1976.

Polanyi, L., 'So What's the Point?' *Semiotica* 25 (1979) 207–241.

Pratt, L., *Lying and Poetry from Homer to Pindar: Falsehood and Deception in Archaic Greek Poetics*, Ann Arbor, MI, University of Michigan Press, 1993.

Rabel, R., 'Interruption in the Odyssey', *Colby Quarterly* 38 (2002) 77–93.

Russo, J., M. Fernández-Galiano, and A. Heubeck, *A Commentary on Homer's Odyssey*, Oxford, Clarendon Press, vol. 3, 1992.

Rutherford, R., *Homer: Odyssey XIX and XX*, Cambridge, Cambridge University Press, 1992.

Schank, R., 'Interestingness: Controlling Inferences', *Artificial Intelligence* 12 (1979) 273–297.

Schank, R. and R. Abelson, *Scripts, Plans, Goals and Understanding: An Inquiry into Human Knowledge Structures*, Hillsdale, NJ, Lawrence Erlbaum Associates, 1977.

Scodel, R., 'Narrative Focus and Elusive Thought in Homer', in D. Cairns and R. Scodel, eds, *Defining Greek Narrative*, Edinburgh Leventis Studies, 7, Edinburgh, Edinburgh University Press, 2014, 55–74.

Smith, E. and S. Kosslyn, *Cognitive Psychology: Mind and Brain*, Upper Saddle River, NJ, Pearson Education, 2007.

Talwar, V. and K. Lee, 'Social and Cognitive Correlates of Children's Lying Behavior', *Child Development* 79.4 (2008) 866–881.

Trahman, C., 'Odysseus' Lies (*Odyssey*, Books 13–19)', *Phoenix* 6 (1952) 31–43.

Tsagalis, C., 'Deauthorizing the Epic Cycle: Odysseus' False Tale to Eumaeus (*Od.* 14.199–359)', in F. Montanari, A. Rengakos, and C. Tsagalis, eds, *Homeric Contexts: Neoanalysis and the Interpretation of Oral Poetry*, Berlin, De Gruyter, 2012, 309–345.

Tsur, R., 'Horror Jokes, Black Humour, and Cognitive Poetics', *Humor* 2–3 (1989) 243–255.
Turner, M., 'Double-Scope Stories' in D. Herman, ed., *Narrative Theory and the Cognitive Sciences*, Stanford, CA, CSLI Publications, 2003, 117–142.
Walcot, P., 'Odysseus and the Art of Lying', *Ancient Society* 8 (1977) 1–19.
Walczyk, J., L. Harris, T. Duck, and D. Mulay, 'A Social-Cognitive Framework for Understanding Serious Lies: Activation—Decision—Construction—Action Theory', *New Ideas in Psychology* 34 (2014) 22–36.
West, M., *The Making of the Odyssey*, Oxford, Oxford University Press, 2014.
Wyatt, W., 'Homer in Performance: *Iliad* 1.348–427', *Classical Journal* 83 (1988) 289–297.
Zunshine, L., *Why We Read Fiction: Theory of Mind and the Novel*, Columbus, OH, Ohio State University Press, 2006.

7
The forbidden fruit of compression in Homer

Anna Bonifazi

This piece is about the cognitive phenomenon of compression, one of the governing principles of conceptual integration. Both conceptual integration (or blending) and compression have been explored first by Fauconnier and Turner (1998, 2000). A key notion for both ideas is that of mental space, an "array of related mental elements that a person activates simultaneously" (Turner 2015: 212), such as the life of a certain bishop. When we conceptually integrate (or blend) two or more mental spaces, we map selected elements of an individual mental space onto corresponding elements of another mental space, and the resulting blended space develops new emergent meaning. To continue the example of the bishop: a piece of artwork may prompt us to blend the space "life of a certain bishop" with the space "life of a certain saint," and the resulting blend "a bishop saint" carries new meaning—let us think of a miter through which the saint can perform miracles, or of an extraordinary generosity toward the poor.

One of the principles of conceptual integration is that our mind compresses fundamental, "vital" relations between the multiple spaces in questions ("to achieve compressed blended spaces," Turner 2007: 382). In *The Way We Think* (2002: 92–102) Fauconnier and Turner list the following vital relations: Change, Identity, Time, Space, Cause-effect, Part-whole, Representation, Role, Analogy, Disanalogy, Property, Similarity, Category, Intentionality, and Uniqueness. Our ultimate comprehension of a bishop saint is achieved through the compression of the relation of role: we assign to that bishop the compressed role of a saint-and-bishop.

Artistic representations embodying compression across ages and cultures are consciously crafted, and deliberately meaningful. However, human beings take advantage of this cognitive phenomenon many times a day and on many different occasions. Compression is mostly realized unconsciously, and perceiving its effects does not require conscious effort by the mind. The present chapter, therefore, deals with an unconscious and pervasive phenomenon as revealed through conscious and deliberate human artwork. Since nonverbal art somehow seems to illustrate the power of compression in a more straightforward way than verbal art, I will first illustrate the abovementioned notions by pointing to three examples taken from sculpture, painting, and architecture. Next I will switch to poetic language, by commenting on two extended passages in the *Iliad*, and on the Odyssean treatment of a few narrative elements. The piece will conclude with reflections on the consequences of acknowledging compression in Homer in terms of interpretation and interpretive debates.

Compression: examples from nonverbal art

The core piece of sculpture in one of the seven Chapels of the apse in the Cologne Cathedral is the tomb of Archbishop Philipp I von Heinsberg (ca. 1130–1191). The funerary monument, built in ca. 1300, displays a castle in miniature (provided with walls, towers, merlons, and gates) whose interior part constitutes the coffin of the archbishop: a human-scale entrance to a building (perhaps a church) placed on the horizontal level, with the body of the archbishop laying down and occupying the entire space of the entrance, his head leaning on two pillows. Several elements of this complex sculpture would deserve attention; however, the details that are most relevant to the present piece concentrate on the body of the archbishop. He is fully dressed as archbishop, with open eyes; he holds tight a miter in the right hand, and a book in the left hand (Figure 7.1).

Let us consider for a moment the aspect of time; that is, the time relation that is established between different elements of this representation. As we look at this funerary monument we are invited to integrate the life, the death, and the afterlife of the archbishop. He is laying down horizontally and leaning the head on two pillows, suggesting that he is dead. Moreover, he apparently still has force in his hands, as he is tightly holding the miter and the book, and his eyes are open; this suggests that he is still alive. Finally, he is gazing at pictorial representations of St. Valentine and St. Vitus alive (chronologically preceding him by several centuries, as both died in the third century AD); this suggests that sometime after his death he is contemplating two saints. This piece of art makes us grasp at once what the tomb means (to the sculptor, and in the physical context of that chapel) in a temporal perspective. This happens because we perceive, in the sculpture, a blend and compression of time among other blends and compressions.

Let me reiterate that a blend is a mental space emerging from the integration of other mental spaces.[1] In the case of the tomb of Archbishop Philipp I von Heinsberg, the blend refers to an overarching organizing frame underlying all the mental spaces involved; I call this frame *someone's life trajectory in a religious perspective.*[2]

Figure 7.1 The archbishop's tomb, Cologne, Germany

Photo courtesy of the author

We blend or integrate the mental space "life of archbishop X" (input space 1) and the mental space "death of archbishop X" (input space 2). The result is a blended space that receives projections from elements of each input space. These elements are counterparts. For instance, the open eyes in the input space "the life of archbishop X" have their counterpart in the closed eyes of the input space "death of archbishop X." In the blended space, the open eyes are projected, but they are also attached to a laying body gazing at two saints—hence the new meaning, the attitude of Archbishop Philipp I von Heinsberg in his afterlife. It is impossible to derive the sense of eternal worship of the two saints from either of the inputs; the sense of eternity emerges only in the blend. If, by watching the sculpture, we had access only to the life of Philipp I von Heinsberg, we could not know what would have happened after his death in the sculptor's imagination. But in the blend the temporal dimension extends beyond his mortal life. In other words, in this blend the conceptual relation of time is compressed.[3]

Our mind connects things that can only be explained by decompressing them. Our general perceptual experience of a dead body includes our sensing of lifeless muscles and lifeless eyes. But if, due to rigor mortis, the dead body is positioned in such a way that the hands keep grasping an object, and the eyes are kept open, viewers are invited to associate this posture with an imaginary after-death stage characterized by some life-like bodily state. The sculptural representation of what happens after someone's death takes the viewers to a place in an imaginary realm that is inspired by, but constitutionally beyond, our perceptual experience.[4] Turner expresses the tension between the limits of sensory input and the potential of our imagination in these terms:

> From the point of view of evolution, to confuse things that should be kept distinct is like plucking forbidden fruit: we should not do it, on pain of death, quite literally, but also on pain of insanity. Yet, amazingly . . . we pluck that forbidden fruit. We put together what should be kept distinct.
>
> *(Turner 2006b: 109)*

Let us now illustrate compressions of time, place, role, and representation by means of another nonverbal example, which I take from Turner (2006b). In canonical pictorial representations of the Annunciation, Mary faces the Angel while holding a book opened to the narrative of the Annunciation. Turner adds that in one of them (by Rogier van der Weyden, 1440[5]) a medallion on the bed depicts Jesus' Resurrection. The scholar comments: "The painting in this way gives us a compression of eternity or at least, from not yet born to being raised from the dead, all in one momentary scene" (Turner 2006b: 110). This compression of time matches a compression of place. The resurrection—we learn from the Scriptures—takes places far from Nazareth, in the small town of the Annunciation—and the reading of print books including the narrative of the Annunciation can take place anywhere the book in question is available. Yet all these places are compressed into one room, attesting to multiple related events. Moreover, I see a compression of role. Mary is depicted in her own ad hoc role of future mother of Jesus as well as in a less specific role of reader/user of the Sacred Scriptures. The two different roles are compressed into one, which exists only in the blend: the reader of God's will is made to coincide with the ideal disciple, who carries out God's will. Finally, I address compression at the level of representation: the pictorial representation focuses on Mary's assent, an act that occurs at a particular point in time, but the projection of reading alludes to the possibility to remember, re-read, and re-celebrate that act forever and ever (or at least up to the age of the painting).

The third example is inspired by Ferrari's conceptual metaphor reading of ancient Greek sanctuaries (Ferrari 2006). The "integration of the divinity into the polis" (2006: 234) and the divinity's relationship to the citizens are conceptualized in political terms through architectural spaces: precincts, temples, altars, and the *temenos* itself (the site of the sanctuary). Select correspondences show this political understanding: for example, the god corresponds to a human ruler, with rights and responsibilities; the temple corresponds to a royal palace; the sacred area of the sanctuary corresponds to the god's portion of the polis' territory; the tithe claimed by a god (e.g. 10 percent of spoils after a war) corresponds to taxation. I add that all these conceptual mappings suggest the principle "achieve compressed blended spaces": the god's share in the citizens' life is compressed into actual meters of land; the revelation of god's superior rank is compressed into the architectural features of a temple.

Skeptical readers at this point might contend that the conceptual complexity of a Renaissance Christian painting or of an ancient Greek temple is not necessarily an example of a cognitive behavior that is diachronically and cross-culturally persistent. Such an objection can be allayed by considering two aspects. First, human beings seem to be able to represent conceptual integration in art at least since the time of the Löwenmensch (the "lion-man"), apparently the oldest extant portable art object—a small ivory statue dating back to 40,000 years ago.[6] Second, there is a fundamental (perhaps existential) cognitive goal that compression achieves: to reduce to human scale concepts that are beyond human understanding (Fauconnier and Turner 2002: 322–324, 346; Turner 2006b: 110; 2015: 213–214). Compressions allow us to "obtain global insight" and "go from many to one" (Fauconnier and Turner 2002: 323).

Achieving human scale is characteristic of religious practices. Back to Rogier van der Weyden's Annunciation: a unified visual field (and organizing frame) simplifies chronological consequences whose explanations would otherwise require major discontinuities and much more nonverbal information to process, at the risk of confusing people and weakening the message(s). The same holds for Greek sanctuaries: everyday experiences such as walking on a fenced lawn or putting food on a table simplify (and embody) the spiritual process of contact (movement, taste, smell) with a deity.

Compression of time, place, role, identity, and representation in the *Iliad*: the case of Helen's identification of heroes, and Priam's journey

Just like nonverbal artworks, linguistic products may reveal and exploit cognitive compression to obtain global insight: Fauconnier and Turner point out that compression allows humans to "simultaneously control long diffuse chains of logical reasoning, and . . . grasp the global meanings of such chains" (Fauconnier and Turner 2000: 283). Language does that through words, phrases, and entire narrative units whose lexical, semantic, pragmatic, and cognitive features invite us to pluck the forbidden fruit, to put together elements that should be kept distinct. The Homeric epic allows for that, I submit, because of the intrinsically compressed narrative space represented by the genre (on which more later).

The first example reminds me of Rogier van der Weyden's Annunciation on a conceptual and cognitive level. *Iliad* 3.121–124 tells us that Iris, messenger of the gods in disguise, reaches Helen to let her know of the imminent duel between her husband Menelaus and Paris. She leaves the hall immediately and moves toward the Scaean Gates accompanied by two maidens (3.141–145). What has struck readers for centuries are two aspects of the story that are not explicitly connected to the basic action (that is, Iris summoning Helen because of the duel, and Helen reaching the tower to watch the duel). The first aspect is the rather detailed description of what Helen is occupying herself with when Iris arrives (*Iliad* 3.125–128); the second aspect is the traditional motif of the *teichoskopia* (literally "view from the walls"); that is, Helen's identification of enemy warriors on the battlefield from the Trojan walls—a higher vantage point—at the request of Priam (*Iliad* 3.161–244).[7]

Iris finds Helen working on a great robe, depicting Trojans and Achaeans at war:

> [. . .] ἣ δὲ μέγαν ἱστὸν ὕφαινε
> δίπλακα πορφυρέην, πολέας δ' ἐνέπασσεν ἀέθλους
> Τρώων θ' ἱπποδάμων καὶ Ἀχαιῶν χαλκοχιτώνων,
> οὕς ἑθεν εἵνεκ' ἔπασχον ὑπ' Ἄρηος παλαμάων·.
>
> (Iliad *3.125–128)*

> [. . .] she was weaving a great web,
> a red folding robe, and working into it the numerous struggles
> of Trojans, breakers of horses, and bronze-armoured Achaians,
> struggles that they endured for her sake at the hands of the war god.[8]

When Helen reaches the tower, Priam calls her and asks her to identify a few Achaean heroes on the battlefield, although Priam's request represents an anachronism, as Kirk, among others, remarks (1985: 286 and 288): on the tenth year of the war the king of the Trojans hardly needs any introductions of the main heroes among the enemies. This is the text of the first two identifications:

> οὗτός γ' Ἀτρεΐδης εὐρὺ κρείων Ἀγαμέμνων,
> ἀμφότερον βασιλεύς τ' ἀγαθὸς κρατερός τ' αἰχμητής·
> δαὴρ αὖτ' ἐμὸς ἔσκε κυνώπιδος, εἴ ποτ' ἔην γε.
>
> (Iliad *3.178–180)*

> That man is Atreus' son Agamemnon, widely powerful,
> at the same time a good king and a strong spearfighter,
> once my kinsman, slut that I am. Did this ever happen?

> οὗτος δ' αὖ Λαερτιάδης πολύμητις Ὀδυσσεύς,
> ὃς τράφη ἐν δήμῳ Ἰθάκης κραναῆς περ ἐούσης
> εἰδὼς παντοίους τε δόλους καὶ μήδεα πυκνά.
>
> (Iliad *3.200–202)*

> This one is Laertes' son, resourceful Odysseus,
> who grew up in the country, rough though it be, of Ithaka,
> to know every manner of shiftiness and country counsels.

The syntactic, semantic, and pragmatic shape of Helen's utterances has made scholars suggest different communicative contexts than the spoken, online, eye-witnessed identifications from the wall. Already the scholia point out their epigrammatic character.[9] Elmer (2005) not only summarizes the main literature on the topic, but he delves into the catalogic and inscriptional character of those lines. They are supposed to be part of a sequence of parallel identifications (catalogic poetry). Moreover, the concise form featuring the deictic marker οὗτος at the beginning lends itself to a piece of writing that is attached to an object external to the text. The crucial point of the article is the connection between those words and the web Helen is weaving as she is summoned. Ancient and modern commentators do recognize in the description of Helen's folded robe themes that represent the *Iliad* itself (Elmer 2005: 23–24).[10] Elmer goes a step further:

> Iris . . . calls Helen to witness what she had already visualized and pictured on her web. When she mounts the Trojan walls, she hardly needs to look out in order to identify the figures; she can simply "read," or pronounce, the captions she might have applied to her figural representations.
>
> *(Elmer 2005: 25)*

These captions could have been woven into the tapestry (ἐνέπασσεν, 126) and "appended to an image of the Achaeans encamped on the plain of Troy" (Elmer 2005: 1). Iris' words about what Helen may spot from the walls, after all, are ambiguous enough (see Elmer 2005: 24–25 for discussion about θέσκελα ἔργα and the idea of crafted images):

> δεῦρ' ἴθι νύμφα φίλη, ἵνα θέσκελα ἔργα ἴδηαι
> Τρώων θ' ἱπποδάμων καὶ Ἀχαιῶν χαλκοχιτώνων
>
> (Iliad *3.130–131; 131 = 127)*

> Come here with me, dear girl, to behold the marvelous things done
> by Trojans, breakers of horses, and bronze-armoured Achaians

The last point that is relevant to my cognitive reading of compression is the following: Elmer and many other scholars (see e.g. Pantelia 1993; Clayton 1994; Nieto 2008; Karanika 2014; Bottino forthcoming) consider Helen's tapestry a silent and nonverbal equivalent to making poetry (weaving being a well-established metaphor for composing poems).

Let me apply a cognitive reading of compression to all of this in a schematic way. In input space 1 Helen sees the Greek heroes on the battlefield, and tells Priam who they are (*teichoskopia*).

In input space 2 Helen is weaving a tapestry that represents Greek and Trojan heroes, and (Elmer's suggestion) what she pronounces are the words of captions woven in the tapestry.

In input space 3 someone utters epigrammatic lines that have been composed after the death of the Greek heroes being mentioned (epigrammatic poetry thus being temporally unrelated to the moment of war, just like epic poetry).

The Homeric passage on the whole suggests a blend: upon Priam's invitation, Helen presents the Greek heroes participating in the war of Troy by pronouncing captions related to the tapestry she is weaving *and* by performing epigrams composed sometime after the war. By doing so, she plays a blended role: she is not only the female character that acts in the story, she also embodies any poet/performer that celebrates those individuals post mortem. Helen can fulfill this blended role because she is weaving: as a woman she cannot give voice to the commemoration of heroic deeds, but she can nonverbally represent the *kleos* of heroes on elaborate tapestries. She cannot utter any catalogic poetry publicly, but in the blend she can, and she does it.

This blend therefore compresses time (the time of the war and posthumous time converge upon the moment of the *teichoskopia*), place (both the spectacle from the walls and the battlefield on tapestry-size are compressed representations of the war space—or war "theater," to use Clay's term, in Clay 2011), role (Helen-the-epigrammatic poet compresses Helen-Priam's daughter-in-law, Helen-the-weaver, and any poet/performer), and finally representation: the blend compresses the epic hexameters incorporated in the *Iliad*, the epigrammatic tradition, and the inscriptional tradition.

Weaving, like performing/singing, is a powerful medium that grants perennial glory to those who are depicted. Weaving projects the weaver into a larger-than-life dimension of tradition.[11] This is what links, in my view, the Helen passage in Homeric poetry and Rogier van der Weyden's Annunciation. Just as Helen embodies any poet/performer that gives *kleos* to the heroes, Mary, by holding the sacred book, embodies any reader that at any subsequent point in time re-visualizes and brings forward the memory of the scene of the Annunciation. In the former case, the epic genre metapoetically mirrors epigrams and captions; in the latter case the painting mirrors the act of reading. We may call this a compression of medium—as a subtype of the compression of representation.

The second case in point is the Homeric episode of Priam's journey to the tent of Achilles, in the last book of the *Iliad* (24.160–698). Once Achilles has killed and disfigured Hector, Priam visits Achilles to ransom the corpse of his son. Priam reaches the Greek hero and offers him several gifts. After an intense verbal exchange between the two of them, Priam's request is satisfied, and the corpse of Hector is carried home for the funerary procedures.

Let us focus on the first section of this episode; that is, the outbound travel of the Trojan king from the Trojan palace to the Achaean camp (*Iliad* 24.160–472). We can think of this scene as an input space that connects to the previous content. Hector has been killed and dragged for twelve days in the dust (*Iliad* 22.361–515, and 24.15–22), and Zeus has sent Iris to Priam to suggest he ransom the corpse (*Iliad* 24.143–159). The content that follows is even more straightforwardly connected: on the way back to Troy, Priam's daughter (and Hector's sister) Cassandra spotlights the chariot dragging the corpse from the citadel, and everyone starts to mourn and weep over Hector; several ritual laments follow (*Iliad* 24.699–776).

The input space "Priam's Journey" contains the travel of Priam alive to the tent of Achilles alive. It involves roles (e.g. Priam is the main traveler, but others co-travel with him), relations between participants (e.g. the relatives are worried about Priam's departure), actions (e.g. Priam reaches his destination), and props (e.g. he brings along a chariot to transport Hector's body).

This narrative primarily compresses (the conceptual relation of) place. The language of this episode offers phrases and terms that are either ambiguous—that is, suitable for other kinds of travel as well—or incongruous—that is, they do not fit input space 1. The ambiguities and incongruities point to different but coherent sets of features that relate, in my reading, to two different input spaces. I identify input space 2 with the travel of Priam to death, to the house of Hades (thus a metaphorical travel), and input space 3 with the travel of Priam alive to the underworld, where he meets dead Achilles (experience of catabasis, or journey to the land of the dead, for initiation purposes—it includes questioning ghosts[12]).

First let me summarize the numerous and rich elements that secondary literature acknowledges as belonging to a narrative of Priam descending to the underworld.[13] When the messenger of the gods, Iris, reaches Priam, she finds that his relatives are weeping and mourning (24.160–166): she finds γόος "lament[ing] over someone's death" (160); Priam is in the middle (ἐν μέσσοισι, 162), covered by his coat (ἐν χλαίνῃ κεκαλυμμένος, 163), with dung on the head and neck (κόπρος ἔην κεφαλῇ τε καὶ αὐχένι, 164) as a sign of grief. As Hecuba hears that Priam is going to visit Achilles, before any reply she wails aloud (κώκυσε, 200; this verb and the noun κωκυτός in archaic and classical Greek connote primarily the funerary cry of women[14]). Iris exhorts the old man not to fear death (μὴ δέ τί τοι θάνατος μελέτω φρεσὶ μηδέ τι τάρβος, "death and fear be no worry in your thoughts," 181), which evokes "take heart"-exhortations found in epitaphs (Herrero de Jáuregui 2011: 51). Priam wishes to die: αὐτὰρ ἔγωγε / πρὶν ἀλαπαζομένην τε πόλιν κεραϊζομένην τε / ὀφθαλμοῖσιν ἰδεῖν βαίην δόμον Ἄϊδος εἴσω, "as for me, before I see with my eyes the city destroyed and plundered, may I go to the house of Hades," 244–246). At the moment of the actual departure, relatives and friends mourn Priam (ὀλοφυρόμενοι, 328, a verb commonly related to someone's death) as if he is going to death (θανατόνδε, 328, a term occurring only on the occasion of Patroclus' and Hector's deaths, *Iliad* 16.693 and 22.297 respectively[15]). Those who escort Priam move downward (κατέβαν, 329) until they reach the plain (πεδίον δ' ἀφίκοντο, 329). Then, as they pass by the tomb of Ilus (349), they let the mules and the horses drink at the river (στῆσαν ἄρ' ἡμιόνους τε καὶ ἵππους ὄφρα πίοιεν / ἐν ποταμῷ, 350–351). At that point darkness comes (ἐπὶ κνέφας ἤλυθε γαῖαν, "dusk was descending on the earth," 351). The tomb of Ilus, the coming of night, and the river represent boundary markers.[16] At that point Hermes appears. Zeus sends Hermes Psychopomp (see πομπός at 153 and 182) to accompany Priam (333–348), and to make him invisible (337; see also 477). By the way, Zeus calls Priam's destination Πηλεΐωναδε "the place of Peleus' son," 338, which reflects an anomalous use of the suffix –δε to refer to Achilles' tent.[17] Scholars do not comment further, but I submit that the term might refer to some cult place where Achilles is worshipped.

Priam's reaction to the herald's words announcing the presence of a possibly menacing person (Hermes in fact) consists in physical and emotional panic evoking the terror of a catabatic experience more than fear for an enemy (δείδιε δ' αἰνῶς, / ὀρθαὶ δὲ τρίχες ἔσταν ἐνὶ γναμπτοῖσι μέλεσσι, / στῆ δὲ ταφών, "he was badly / frightened, and the hairs stood up all over his gnarled body, / and he stood staring," 358–360). The narration of Hermes and Priam's arrival at the Achaean camp focuses on a few abnormal characteristics of Achilles' tent: it is high (449), roofed with a thatch of meadows (451), surrounded by a large court (αὐλή, 452[18]); most of all it has a bolted door (453–454), which suggests the representation of a gate to the underworld.[19] It is exactly at that point that Hermes leaves Priam alone, and goes back to Mount Olympus (468). After the meeting between Priam and Achilles, Hermes ponders how Priam might escape the "holy gate-wardens" (ἱεροὺς πυλαωρούς, 557–558); Nagler (1974: 185) connects these figures to the sacred keepers traditionally standing at the entrance of the abode of the dead. During the meeting, Priam begs Achilles to let him live and see the light of the sun (ἐπεί με πρῶτον ἔασας αὐτόν τε ζώειν καὶ ὁρᾶν φάος ἠελίοιο, 681), which ambiguously refers to Priam surviving Achilles' fury as well as Priam surviving the dark trip to Hades.

There probably are more lexical, visual, or metaphorical allusions to a descent to the underworld in this excerpt. The ones I mentioned, however, sufficiently show that in secondary literature the interpretations beyond the "logic one" coincide with tracking the linguistic signs of other simultaneous travels.

Let me sketch the input spaces and the related cross-space mappings by unpacking the blend prompted by Homeric language. The "generic space," which includes what the inputs have in common, is quite articulate, as the inputs have a lot in common: the travel is downward, dangerous, done in darkness, and features the crossing of a river as well as the escort of Hermes.

Input space 1 (the literal travel within the land of the living) includes the following elements, which have their counterparts in the other input spaces (that is, the travel to the underworld and the catabasis):

- the kinsmen weep because Hector is dead, and it is as if Priam is going to death as well—CLASH with 2 (see below about clashes)
- the route is from the Trojan citadel down to the plain—CLASH with 2 and 3
- the danger is that Achilles might kill Priam—CLASH with 2 and 3
- the travel takes place at dusk and lasts one night—CLASH with 2
- Hermes escorts and protects Priam
- Priam is and remains alive—CLASH with 2
- Achilles is and remains alive—CLASH with 3
- Achilles' tent is a regular war tent of a military chief—CLASH with 3
- the river Scamander works as a liminal place on the way to the Achaean camp—CLASH with 2 and 3
- Priam comes back home safely—CLASH with 2

Input space 2 (the metaphorical travel to death) includes the following counterparts:

- the kinsmen do not weep so much because Hector is dead, but because Priam is dead or is going to die soon—CLASH with 1 and 3
- the route is from the land of the living down to the land of the dead—CLASH with 1
- the danger consists in the general possibility of suffering while dying
- the travel takes place in the dark
- Hermes escorts dead Priam to the house of Hades—CLASH with 1 and 3
- Priam dies, and does not meet anybody in particular—CLASH with 1 and 3
- the entrance to the underworld is a bolted door—symbol of a sacred venue—CLASH with 1
- the river Scamander works as an infernal river; crossing it marks the transition to the land of the dead—CLASH with 1
- Priam does not come back at all—CLASH with 1 and 3

Input space 3 (the literal and metaphorical experience of catabasis) includes the following counterparts:

- the kinsmen weep because it is as if Priam is going to death—CLASH with 2
- the route is from the land of the living down to the land of the dead—CLASH with 1
- the danger consists in the terror that the descent to the underworld provides
- the travel takes place at night
- Hermes escorts and protects living Priam to the land of the dead—CLASH with 2
- Priam is and remains alive—CLASH with 2
- Achilles is dead in the land of the dead—CLASH with 1 and 2 (in 2 Achilles is not relevant)
- dead Achilles is worshipped in ad hoc shelter and an ad hoc court—symbols of a sacred venue—CLASH with 1
- the river Scamander works as an infernal river; crossing it marks the transition to the land of the dead—CLASH with 1
- Priam, after experiencing the catabasis, comes back to the land of the living—CLASH with 2

The blend is ultimately what the Homeric poem submits to us; that is, what the actual language with its ambiguities and incongruities allows us to approach. The new emergent structure specific to the blend[20] includes these aspects:

- Priam is alive, near to actual death; he is experiencing catabasis, which requires heroism, and he is also a dead hero
- Achilles is alive and can give orders, but he is also a dead hero, and a cult hero
- both can talk and have a meal together even if they are dead, or, perhaps, precisely because they are in an afterlife stage they enjoy a special meal and a special conversation
- Achilles' place is a tent, a shrine, and the underworld
- there is and there isn't a way back.

In this blend I see compression of time (the nocturnal visit to the enemy's tent compresses Priam's lifetime), of place (the route compresses physical and metaphorical routes; Achilles' place compresses the war tent, the posthumous shrine, and the gate to the underworld), and of role (Hector's father is also Priam-the-initiate; Achilles alive and Achilles dead points to a larger-than-life hero blessed by *kleos*, and of representation, to the extent that we take some catabatic allusions as belonging to *catabasis* as a narrative genre). What is particularly interesting in this blend is the presence of several clashing elements at the level of specific mappings, which I indicated above. The numbers refer to the entire input spaces, but the clashes themselves relate only to the corresponding elements present in the mapping. For example, dead Priam in input space 2 is not supposed to talk and eat, as he does in input spaces 1 and 3; however, in the blend he is such a superhuman figure that he can go through the entire experience of the meeting with his own flesh and senses. The same holds for Achilles. A glimpse of the superhuman or afterlife-hero state of both is provided by the text as the two wonder at each other (*Iliad* 24.629–632).[21] "[F]ar from blocking the construction of the network . . . clashes offer challenges to the imagination; indeed, the resulting blends can be highly creative" (Fauconnier and Turner 2002: 131). Oddities in the blend clearly reflect the conceptual integration of multiple input spaces.

About compression in the *Odyssey*, and micro-level compression

Dealing with blends and compressions in literary language means seeing that individual words and phrases express multiple input spaces and cross-space mappings. The way in which we modern readers connect Homeric words and phrases to our mental networks results in patterns of inferences—for example, we understand "swift-foot Achilles" by inferring that Achilles excels in moving fast, and by inferring that a moving fast soldier has better chances to chase and kill enemies. More importantly, our patterns of inferences share at least several elements with the patterns of inferences that composer(s) and performers of the Homeric epic invited throughout the centuries.[22] For example, as we read that Penelope meets Odysseus disguised as a beggar after twenty years of absence, we infer, from our general knowledge of how a disguise works, that Penelope is not supposed to recognize him, or not by default. However, here and there the Homeric language seems to convey that she senses his true identity. We get this from the innumerable inferences we draw about relevant words and phrases on the basis of the co-occurring text, our general knowledge of the world, and our knowledge of the *Odyssey* (the text itself and secondary literature about it).

While reading the dialogue between the two mythical figures in book 19, we can access not only input space 1, Penelope talking to a beggar, and input space 2, Penelope talking to Odysseus; we can also access input space 3, Penelope-as-"weaving," talking to one of the subjects of epic (based on linguistic information about weaving and Penelope, and about the metaphorical connotations of weaving[23]), and input space 4, Penelope as weeping/singing a lament over the death of her husband (see *Odyssey* 19.213, 251, 515–523, and Levaniouk 1999). The list of input spaces could certainly continue.[24] The point is that the Homeric language invites us to integrate all these spaces into one blend, simplified in the compressed form of a conversation, and providing new meaning that the individual input spaces do not convey. Penelope becomes a larger-than-life being that is able to openly recognize her interlocutor, to not recognize him, to mourn him, and to have with him a rich conversation about life transitions and rituals. Likewise, Odysseus becomes a larger-than-life being that is able to reveal himself while sticking to his beggar identity, and to assert the global or cosmic significance of his homecoming.

This is quite different from dramatic irony: in dramatic irony the unawareness of characters is the substance of the conflict between their knowledge and our (omniscient) knowledge. Here, conversely, the dialogue between the two at times shows unmistakable awareness, and displays contents that quite disconcert our search for objective coherence (they are forbidden contents, a forbidden fruit!)—Odysseus should not be so explicit; Penelope should not show full awareness so early. Yet the text does incorporate incoherent features, and we do pluck the forbidden fruit as we connect what should be kept distinct.

The blend that I suggest features compression of identity, role, time, and place, and it achieves human scale by letting the speaking turns of that conversation generate multiple interconnected mental spaces pertaining to the laws of nature, memories, stories, symbols, cult beliefs, initiation aspects, rituals, and poetic and metapoetic procedures. The unified blend of a conversation incorporating all of that—I am ready to submit—makes the listeners remember the contents more effectively.[25]

In the light of this, the so-called "Penelope question" (see esp. Doherty 1995: 31–63) can be said to arise from an unnecessary and limitative presupposition. Since the eighteenth century onward (interestingly, not before), philologists wonder at which point exactly Penelope recognizes Odysseus, as the poet remains silent about it. This question is relevant if we presuppose that the *Odyssey* is a linear story about real-life persons that share with us limited knowledge, and one irreversible life span. The same question becomes conversely irrelevant if we take the language seriously in its ambiguous and incongruous details, and we let ourselves get involved in the spaces and blends that it offers to us, about Penelope and Odysseus, as well as about many other characters. Being heroic/epic persons is not quite the same as being us. The Homeric poetry represents a macro-compression of heroic and epic essence reduced to human-scale figures who think, talk, depart, return, and weep. Yet it projects their deeds into the larger-than-life dimension of *kleos*, which is what epic is for.

One of the jobs of secondary literature is to decompress/unpack multiple blends and multiple compressions. In this way teachers and students can disentangle meaning complexities, and can follow their individual components—individual mental spaces and individual organizing frames.

In this respect, the study of nonverbal communication—that is, what objects, gestures, locations, and visual landmarks signify and communicate in the two poems—could be strengthened at the theoretical level if investigations aimed at decompressing/unpacking blends and compressions. I advance that many of the compressions of vital relations that we process as we read the *Iliad* and the *Odyssey* are facilitated by the description of nonverbal items co-occurring with speech or third-person narration.

With reference to the *Odyssey*, an episode that demonstrates this point through the incorporation of a number of odd and puzzling details is the encounter between Odysseus and Eumaeus in book 14. In Bonifazi (2012: 69–125) I analyze several ambiguous or incongruous wordings in the speeches of the two characters as well as in third-person narration. In order to illustrate that these wordings make sense in multiple communicative situations, I adopt the pragmatic notion of "layering" from Clark (1996: 354–365; see Bonifazi 2012: 78–83). "Layering" denotes our capability, in literary as well as non-literary texts, to relate words to different "staged" situations simultaneously. I regard conceptual blending as the cognitive operation that allows layering to work. Our mind integrates concepts by mapping mental spaces onto other mental spaces; we map elements of situation A onto analogous elements of situation B (and potentially onto further elements of situation C) without losing the discrete sense of the different situations; we just enjoy the simultaneous integration. Those who listen to or read the encounter between Odysseus-beggar and his faithful swineherd are prompted to refer words and nonverbal signs of communication to different layers. They can do so because they organize the frames the words allude to, and map them onto different but somehow related frames. The epic language of *Odyssey* 14 seems to allude to some esoteric significance not only through words, but also (and perhaps primarily) through objects and gestures. For example, the lexical marker for piglets at lines 73 and 81 (χοιροί) evokes ritual sacrifices (especially within Demeter cult) deviating from a secular and humble encounter.[26] Anomalous components of one of the sacrifices being performed let us infer that the swineherd is perfectly aware of the presence of his master. And so on. As we process epic contents about what people say as well as contents about where they talk, when, and by doing what, the blends of Homeric poetry may concern objects and gestures even more than what characters say to each other. Consider, for example, the simple gesture of pouring wine that Eumaeus performs at *Odyssey* 14.447–448 (σπείσας δ' αἴθοπα οἶνον 'Οδυσσῆϊ πτολιπόρθῳ / ἐν χείρεσσιν ἔθηκεν): it is left unspecified to whom the libation is offered, which lends itself to multiple interpretations. The epic language fosters our processing of a blended gesture thanks to the ambiguous dative "to Odysseus sacker of cities": in one mental space the swineherd gives the cup to the beggar; in another space the swineherd offers the libation to Odysseus *ipse*; in the blended space the swineherd does both. An even more pervasive example may be the co-existence, in the mind of performers and audiences from book 13 to book 22, of Odysseus-the-beggar and Odysseus-Odysseus. Making sense of nonverbal items in the *Odyssey* (but also in other literature) may constitute an optimal way to explore and to understand the blends involved. After all, language is only one of the human dimensions in which we can see conceptual blending operating. "Conceptual blending operates . . . in all the areas of thought and action that distinguish human beings from members of other species" (Turner 2007: 383).

Odysseus himself is the man of many blends. His actions, movements, and all the nonverbal signs connected to him define his character as an amalgam of several input spaces derived from the continuum of traditional tales about him as well as from the specific organizing frames of the *Odyssey* that we have. Who he is and what he represents is much more than what the input spaces project in each of the linguistic excerpts concerning him.

In this chapter, I devote attention to compressions in blends that are rather complex and involve a great amount of linguistic material. This does not mean, however, that all linguistic blends operate like that. First of all, Homeric poetry is a blend in itself because it is fictive communication, in that it incorporates "disparate acts of communication, or even non-communicative events, into one single piece of discourse that never took place"; as Pagán Cánovas and Turner explain, "fictive communication is an extremely productive blending pattern combining several generic integration templates" (2015: 3 and 30 respectively). Second, blends in language (literary and non-literary) can be manifested in much smaller discourse units, such as sentences and even noun phrases.

Let us mention one of the uncountable possibilities of compressed blends realized in Homer on the micro-level of language. Back to *Iliad* 3, while summoning Helen, Iris states that at an earlier time Trojans and Achaeans were fighting hard with each other, but at present they stand still—the war momentarily stopped. The Greek wording for "hard fight" is πολύδακρυν Ἄρηα, literally "Ares of many tears," 132. This simple noun phrase represents a blend, and it compresses cause-effect, time, role, and identity. The compression of cause-effect results from the attribution "of many tears" to Ares: the god that causes war is defined by means of the effects of war. The compression of time derives from the inclusion of both the moment of the presence/action of Ares, and the inevitably subsequent moment of grief over those harmed by Ares. Moreover, the war agent (Ares) is characterized through what happens to the war patient (the people who weep because their relatives or friends have died)—this is a compression of roles. Finally, Ares as the impersonation of war reduces to individual human scale the multiple, diverse, and collective aspects of war. Explaining this blend and these compressions requires some time, but comprehending the phrase is immediate. This is how conceptual integration works: it is a basic mental operation, quick and efficient, that allows us to understand complex concepts "in terms close to human experience" (Dancygier and Sweetser 2014: 94).

Conclusion: what it means to acknowledge compression in literary criticism

Conceptual integration, and therefore compression, occurs in all human thought, action, and language. Art is no exception. In nonverbal pieces of art blending achieves human-scale compressions of several kinds, involving several vital relations. Verbal pieces of art such as Homeric poetry do the same, for the same fundamental purpose; that is, to make concepts that exceed human comprehension graspable.

A blend, "if it is a successful one, can be unpacked to access the network of connections that make it meaningful" (Pagán Cánovas and Turner 2015: 3). This chapter suggests and performs a few possible ways to unpack Homeric blends: to look for ambiguous and incongruous elements of episodes, and to pay attention to the representation of nonverbal items (gestures, objects). It also flags the uncountable blends that the language reveals on the micro-level.

Our brain operates conceptual integration and compressions *also* when we read Homer and any other literary text. Acknowledging that brings about some substantial methodological consequences. Much of the secondary literature that we read and publish is engaged with explanations based on inferences and patterns of inferences that we draw on the basis of sophisticated and yet natural and mundane cognitive procedures. Blends and compressions in blends are part of these cognitive procedures, and they directly impact our interpretation of passages or episodes. They sometimes turn out to be at the core of debates in literary criticisms concerning allusions to potentially multiple occasions or identities behind a certain wording. Among uncountable examples, let me recall the vocative Αἰνέα in Pindar, *Olympian* 6.88 ("Aineas, urge the companions to celebrate loudly [κελαδῆσαι] first Hera Parthenia"), which has prompted several mental spaces in the mind of commentators; they deal with different scenarios, including Aineas as a soloist sent to a certain city for the performance of an ad hoc hymn (Heath 1988: 191 and Lefkowitz 1991: 191–201); Aineas as χορηγός of the coral performance of *Olympian* 6 including a section on Hera Parthenia (Mullen 1982: 84; Carey 1991: 195); Aineas not as a person but a personification of *ainos*, "praise" (Too 1991: 263).[27] What I see behind these multiple explanations is the common cognitive operation of compressing the idea of addressing some entity, and relating the addressed entity to the singing "I" (single or collective) through the references to celebrating loudly and praising.

If we accept this, then we see that textual interpretations are influenced by cultural, ideological, and aesthetic factors, but they may or may not take into account natural and daily cognitive procedures. To keep the focus on Homer, several interpretations of the Homeric text presuppose logical linearity, coherence, and unity of time, place, and person,[28] and are certainly affected by cultural, ideological, and aesthetic reasons, but they do not consider conceptual integration and its effects. Priam should not be alive and dead at the same time; Odysseus should not utter Εὔμαιε. Conceptual integration, conversely, makes us pluck the forbidden fruit; we do put together things that we could keep distinct.[29]

I have been considering in particular Helen's complex action of showing heroes to Priam, Priam's articulated journey(s), the exchange between Odysseus and Penelope in *Odyssey* 19, a special libation offered by Eumaeus in *Odyssey* 14, and the nonverbal co-existence of the beggar and Odysseus *ipse*, but the list of Homeric passages prompting compressions of various kinds could be endless. Uncountable tokens could be analyzed in other literary genres as well. The overarching methodological point of this chapter is to provide a theoretical (cognitively oriented) reason for the interpretations of passages that involve the layering of incoherent themes and thoughts. If the linguistic input gives rise to such layering, it means that it encapsulates some conceptual compression of vital relations, and each different interpretation represents a decompression of them. The forbidden fruit is what the language actually allows to be plucked.

My last thoughts concern human-scale concepts. Logical and coherent connections are an important component of a process that cognitive scientists call organizing frames, but organizing frames are—especially in literary texts—exploited to manipulate scripts, and to create meaningful clashes. Poets and writers, who are human beings like us, blend elements of organizing frames to exploit (consciously or not) the potential of blending and compressions.

The overarching goal (in the case of the epic, a genre-related goal) could be to reduce to human scale concepts that are larger than life.

Larger-than-life concepts (metaphorically and non-metaphorically) in fact underlie several examples mentioned in this chapter. These concepts in my view are the cultural, ideological, and aesthetic basis of a number of blends in the Homeric language. Also in this respect, nonverbal art can guide us, as it conveys cognitive complexity more directly than words (think of *eidola* depicted over stelae); perhaps this is one of the reasons why research on that is momentarily more advanced.[30] I posit that the Homeric blends dealing with life and afterlife are largely inspired by nonverbal blends. The compressions that I have shown are not an abstract mental exercise that the composers of the *Iliad* and the *Odyssey* devised for us. Life and afterlife quintessentially characterize heroes and heroic tales *because* they characterize hero cult and actual worship practices, of which heroic tales are fundamental components.[31] In other words, the reality of dead heroes and the narration of living heroes, however culturally, ideologically, and aesthetically mediated, are blended—and very much compressed—in nonverbal experiences of the there-and-then users of epic. Unlike those users, we modern philologists miss the experience of these nonverbal blends and compressions. But we can still at least allow verbal compressions to do what they are supposed to do.

Acknowledgement

Warm thanks in particular to Peter Meineck for his insightful comments on an earlier version of this paper.

Notes

1 For a concise theoretical overview of blending with reference to viewpoint in Homer, see Bonifazi 2018.
2 The organizing frame for blends and for individual mental spaces is "a frame that specifies the nature of the relevant activity, events, and participants. An abstract frame like *competition* is not an organizing frame, because it does not specify a cognitively representable type of activity and event structure" (Fauconnier and Turner 2002: 123).
3 Our focus with regard to the Cologne sculpture is on time, but we could also take into account, for example, the compression of the relation part-whole, and of the relation of disanalogy. The image of Archbishop Philipp I von Heinsberg dressed as archbishop and holding the miter and the book compresses the whole professional life of Philipp I von Heinsberg—his duties, his travels, the decisions that he took, etc. Moreover, the position of the body compresses the disanalogy between Philipp I von Heinsberg alive and Philipp I von Heinsberg dead.
4 On the storage and reactivation of perceptual symbols through sensory-motor mechanisms, see in particular Barsalou 1999.
5 See www.wikiart.org/en/rogier-van-der-weyden/the-annunciation-1440.
6 On display in Ulm (Southern Germany) at the Ulmer Museum: see www.loewenmensch.de/lion_man.html.
7 On the *teichoskopia* as a misplaced catalogue of heroes, see e.g. Tsagarakis 1982. For a recent review of the puzzling aspects of *Iliad* 3 in connection to Iliadic themes, and the analysis of further thematic substrata dealing with wedding poetry, see Karanika 2013.
8 All translations of Homer are by Lattimore (1951).
9 See scholia T 3.178 (ἐπιγραμματικῶς); AbT 1.29d referring to 3.179 as τὸ ἐπίγραμμα; AbT 3.200–202 (τὸ ἐπίγραμμα). On epigrammatic lines in Homer in general, see especially Vox 1975.
10 Nagy (2009: II§373) calls "the essence of pattern-weaving" "an overall metaphor for Homeric narrative."
11 It is well known that the wording of *Iliad* 3.125–126 strongly resembles *Iliad* 22.440–441; Andromache hears the funerary cry about Hector as she is weaving a double-folded robe embroidered with flowers. For a comparison of the two situations, see, e.g., Pantelia 1993: 495–497; for a connection between these

two situations and the purple *peplos* that was pattern-woven for presentation to Athena on the occasion of the grand procession of the Great Panathenaia, see Nagy 2009: II10§§373–406.

12 See e.g. Clark 1979 and Radcliffe 2004.

13 The following list is based on Nagler 1974: 184–185, Wathelet 1988, Mackie 1999: 488–491, and Herrero de Jáuregui 2011. Further references ranging from 1950 to 1984 about the catabatic features of this trip are recorded in Mackie 1999: 488n10.

14 See, e.g., *Iliad* 24.703; *Odyssey* 8.527; Aeschylus *Libation Baearers* 150; Euripides *Medea* 1177.

15 This remark occurs already in Mcleod 1982: 115. I follow West's and others' edition including θανατόνδε; Allen has θάνατον δὲ.

16 See Mackie 1999: 488 and 491. I would add the mules as a symbol of transition and initiation also in Pindar (see Garner 1992).

17 See Herrero de Jáuregui 2011: 47: the suffix in "to [the dwelling of] Achilles" may have prompted an association with its more familiar use with the proper name Hades ("to [the dwelling of] Hades").

18 See Bonifazi 2012: 108–110 about αὐλή designating various kinds of spaces.

19 See Herrero 2011: 46; Mackie 1999: 490; Stanley 1993: 393n15, and on the gates of Hades in general, see Sourvinou-Inwood 1995: 64–65.

20 On the notion of emergent structure in conceptual integration, see especially Fauconnier 2005.

21 See, e.g., Slatkin 2007: 27–28 about the θάμβος effect.

22 See e.g. Minchin 2001 on the Homeric use of cognitive scripts in some typical scenes.

23 For elaborate discussion of weaving Penelope and the construction of epic poetry, see especially Bottino forthcoming.

24 To my knowledge Levaniouk 2011 performs the deepest analysis of cross-myths, cross-figures, and even cross-age matters relating to Penelope and Odysseus and interwoven in their dialogue.

25 Let us think of the successfulness of dialogues in didactic poetry and in philosophy: both genres need to compress substantial amounts of cultural data spanning time, locations, and practices, and aim at transmitting knowledge.

26 See Bonifazi 2012: 95–96.

27 More details and interpretations are reported in Bonifazi 2001: 133–139.

28 Such interpretations are scattered throughout the secondary literature; see, e.g., Heubeck et al. 1988–1992, Vol. II: 224 about *Odyssey* 14.440 (Εὔμαιε): Odysseus should not explicitly mention the name of the swineherd.

29 In this sense this chapter fully supports Turner's basic claim (Turner 1996) that our human mind is a literary mind.

30 An illuminating work relevant to that is Stec and Sweetser 2013 on religious spaces across cultures.

31 Literature on this topic exceeds the size of this chapter. I owe my awareness about that to the works of Gregory Nagy, which ultimately investigate blends and compressions within and outside Homer, and within and outside verbal art. See, among many others, Nagy 1999, 2003, and 2015.

References

Barsalou, L. W. 1999. "Perceptual symbol systems." *Behavioral and Brain Sciences* 22: 577–660.

Bonifazi, A. 2001. *Mescolare un cratere di canti. Pragmatica della poesia epinicia in Pindaro*. Alessandria: Edizioni dell'Orso.

Bonifazi, A. 2012. *Homer's Versicolored Fabric: The Evocative Power of Ancient Greek Epic Word-Making*. Washington D.C. and Cambridge, MA: Harvard University Press.

Bonifazi, A. 2018. "Embedded focalization and free indirect speech in Homer as viewpoint blending." In *Homer in Performance: Rhapsodes, Characters, and Narrators*, J. L. Ready and C. Tsagalis (eds.), 230–254. Austin, TX: University of Texas Press.

Bottino, A. P. Forthcoming. "The *Phâros* of Laertes: Weaving the fabric of epic." ms.

Carey, C. 1991. "The victory ode in performance: The case for the chorus." *Classical Philology* 86: 192–200.

Clark, H. H. 1996. *Using Language*. Cambridge: Cambridge University Press.

Clark, R. J. 1979. *Catabasis: Vergil and the Wisdom-Tradition*. Amsterdam: Grüner.

Clay, J. Strauss. 2011. *Homer's Trojan Theater: Space, Vision, and Memory in the Iliad*. Cambridge: Cambridge University Press.

Clayton, B. 1994. *A Penelopean Poetics: Reweaving the Feminine in Homer's* Odyssey. Lanham, MD: Lexington Books.

Dancygier, B. and E. Sweetser, 2014. *Figurative Language*. Cambridge: Cambridge University Press.

Doherty, L. E. 1995. *Siren Songs: Gender, Audiences, and Narrators in the* Odyssey. Ann Arbor, MI: University of Michigan Press.
Elmer, D. F. 2005. "Helen Epigrammatopoios." *Classical Antiquity* 24: 1–39.
Fauconnier, G. 2005. "Compression and emergent structure." *Language and Linguistics* 6: 523–538.
Fauconnier, G. and M. Turner, 1998. "Principles of Conceptual Integration." In *Discourse and Cognition*, J.-P. Koenig (ed.), 269–283. Stanford, CA: Center for the Study of Language and Information (CSLI) [distributed by Cambridge University Press].
Fauconnier, G. and M. Turner, 2000. "Compression and global insight." *Cognitive Linguistics* 11: 283–304.
Fauconnier, G. and M. Turner, 2002. *The Way We Think: Conceptual Blending and the Mind's Hidden Complexities*. New York: Basic Books.
Ferrari, G. 2006. "Architectural space as metaphor in the Greek sanctuary." In *The Artful Mind: Cognitive Science and the Riddle of Human Creativity*, M. Turner (ed.), 225–240. Oxford: Oxford University Press.
Garner, R. 1992. "Mules, mysteries and song in Pindar's *Olympian 6*." *Classical Antiquity* 11: 45–67.
Heath, M. 1988. "Receiving the κῶμος: Context and performance of Epinician." *American Journal of Philology* 109: 180–195.
Herrero de Jáuregui, M. 2011. "Priam's Catabasis: Traces of the epic journey to Hades in *Iliad* 24." *Transactions of the American Philological Association* 141: 37–68.
Karanika, A. 2013. "Wedding and performance in Homer: A view in the 'Teichoskopia'." *Trends in Classics* 5: 208–233.
Karanika, A. 2014. *Voices at Work: Women, Performance, and Labor in Ancient Greece*. Baltimore, MD: Johns Hopkins University Press.
Kirk, G. S. 1985. *The Iliad: A Commentary. Volume I: Books 1–4*. Cambridge: Cambridge University Press.
Lattimore, R. 1951. *The* Iliad *of Homer*. Chicago, IL: University of Chicago Press.
Lefkowitz, M. R. 1991. *The First-Person Fiction. Pindar's Poetic "I"*. Oxford: Clarendon.
Levaniouk, O. 1999. "Penelope and the Pênelops." In *Nine Essays on Homer*, M. Carlisle and O. Levaniouk (eds.), 95–136. Lanham, MD: Rowman and Littlefield.
Levaniouk, O. 2011. *Eve of the Festival: Making Myth in* Odyssey *19*. Washington D.C. and Cambridge, MA: Harvard University Press.
Mackie, C. J. 1999. "Scamander and the rivers of Hades in Homer," *The American Journal of Philology* 120: 485–501.
Mcleod, C. W. 1982. *Iliad Book 24*. Cambridge: Cambridge University Press.
Minchin, E. 2001. *Homer and the Resources of Memory: Some Applications of Cognitive Theory to the* Iliad *and the* Odyssey. Oxford and New York: Oxford University Press.
Mullen, W. 1982. *Choreia: Pindar and Dance*. Princeton, NJ: Princeton University Press.
Nagler, M. N. 1974. *Spontaneity and Tradition: A Study in the Oral Art of Homer*. Berkeley, CA: University of California Press.
Nagy, G. 1999. *The Best of the Achaeans: Concepts of the Hero in Archaic Greek Poetry* (2nd ed.). Baltimore, MD: Johns Hopkins University Press.
Nagy, G. 2006. "The epic hero" (2nd ed.). Online publication available at http://chs.harvard.edu/CHS/article/display/1302.
Nagy, G. 2009. *Homer the Preclassic*. Online publication available at http://chs.harvard.edu/CHS/article/display/4377.
Nagy, G. 2015. *Masterpieces of Metonymy: From Ancient Greek Times to Now*. http://nrs.harvard.edu/urn-3:hul.ebook:CHS_Nagy.Masterpieces_of_Metonymy.2015.
Nieto Hernández, P. 2008. "Penelope's absent song." *Phoenix* 62: 39–62.
Pagán Cánovas, C. and M. B. Turner, 2015. "Generic integration templates for fictive communication." Available at SSRN: https://ssrn.com/abstract=2694704 or http://dx.doi.org/10.2139/ssrn.2694704.
Pantelia, M. C. 1993. "Spinning and weaving: Ideas of domestic order in Homer." *The American Journal of Philology* 114: 493–501.
Radcliffe, G. E. 2004. *Myths of the Underworld Journey: Plato, Aristophanes, and the 'Orphic' Gold Tablets*. Cambridge: Cambridge University Press.
Slatkin, L. 2007. "Notes on tragic visualizing in the Iliad." In *Visualizing the Tragic: Essays in Honour of Froma Zeitlin*, C. Kraus, S. Goldhill, H. P. Foley, and J. Elsner (eds.), 19–34. Oxford and New York.
Sourvinou-Inwood, C. 1995. *'Reading' Greek Death*. Oxford: Clarendon.
Stanley, K. 1993. *The Shield of Homer: Narrative Structure in the* Iliad. Princeton, NJ: Princeton University Press.

Stec, K. and E. Sweetser, 2013. "Borobudur and Chartres: Religious spaces as performative real-space blends." In *Sensuous Cognition: Explorations into Human Sentience: Imagination, (E)motion and Perception*, R. Caballero and J. E. Díaz Vera (eds.), 265–291. Berlin: De Gruyter.
Too, Y. L. 1991. "Ἥρα Παρθενία and poetic self-reference in Pindar Olympian 6.87–90." *Hermes* 119: 257–264.
Tsagarakis, O. 1982. "The Teichoskopia cannot belong in the beginning of the Trojan War." *Quaderni Urbinati di Cultura Classica* 12: 61–72.
Turner, M. 1996. *The Literary Mind*. New York: Oxford University Press.
Turner, M. 2006a. "Compression and representation." *Language and Literature* 15: 17–27.
Turner, M. 2006b. "Compression." In *The Artful Mind: Cognitive Science and the Riddle of Human Creativity*, M. Turner (ed.), 93–113. Oxford: Oxford University Press.
Turner, M. 2007. "Conceptual integration." In *The Oxford Handbook of Cognitive Linguistics*, D. Geeraerts and H. Cuyckens (eds.), 377–393. New York: Oxford University Press.
Turner, M. 2015. "Blending in language and communication." In *Handbook of Cognitive Linguistics*, E. Dabrowska and D. Divjak (eds.), 211–232. Berlin: De Gruyter.
Vox, O. 1975. "Epigrammi in Omero." *Belfagor* 30: 67–70.
Wathelet, P. 1988. "Priam aux Enfers, ou Le retour du corps d'Hector." *Les Études Classiques* 56: 321–335.

8

Human cognition and narrative closure

The *Odyssey's* open-end[1]

Joel P. Christensen

Although there has been much work on the psychological relevance of Homeric language and on cognitive aspects of memory and identity in Homer (e.g. Minchin 2001 and 2007),[2] literary approaches informed by cognitive science may still help us see both the *Iliad* and the *Odyssey* in new ways. In particular, the application of cognitive science to Homeric poetry promises to help modern readers appreciate how deeply—even if only implicitly—ancient Greeks understood human mental functions, thus providing another perspective on interactions between poem and audience during repeated performances. Collective narratives which developed over time—like those represented by orally derived epic—reflect the ways in which storytelling emerges from and reinforces patterns inscribed in previous narratives and social roles. A range of modern cognitive theories support the centrality of storytelling to shared culture: intersubjectivity, for example, emphasizes how our collective everyday life is built from worlds shared with others through common narratives (see especially in Zlatev et al. 2008).[3] The theory of extended mind, too, helps us to understand how complex mental functions may rely in part on our environment and engagement with others (see Clark and Chalmers 1998). And, as David Hutto (2007) argues, such narrative engagement channels 'folk psychology', echoing the process whereby children develop the ability to attribute mental states to others by hearing and telling stories.[4]

The broad cognitive frameworks for narrative I have just mentioned should encourage us to reconsider representations of minds in Homeric epic and potential interactions between such representations and the minds of epic's audiences. As a step in this direction, this chapter explores some ways in which modern cognitive frameworks help us understand both the problematic ending of the *Odyssey* and, through an exegesis of this end, aspects of literary closure in general. I will start by describing a range of cognitive and literary frameworks for human responses to closure. Then, I will survey signs of closure in the *Odyssey*, some interpretive difficulties of effecting closure in the tale, and the epic's economy of pleasure. I will close by focusing on three representations of problematic closure deployed in *Odyssey* 24 to argue both that the epic anticipates its audiences' responses to complete and incomplete knowledge and that a cognitive-literary approach facilitates a useful reinterpretation of the poem. The epic depicts pleasure coming from narratives that resolve and grief issuing from incomplete endings. In doing so, it provides an anticipative framework for its own ending: inexplicable or problematic endings, rather than being mere causes of grief, present opportunities for agency and redefinition.

Storytelling, human cognition, and closure

As I will discuss below, the Homeric *Odyssey* has a complex end with a history of fraught interpretations. I assume that the strangeness of the epic's end is purposeful and that it derives in part from a basic understanding of the relationship between the human mind and narrative closure. Before adumbrating the basic outlines of the epic's ends and its interpretive difficulties, it will be useful to survey some cognitive theories about the importance and function of storytelling.

There are a range of approaches which emphasize the relationship between narrative and cognition in human minds. For instance, Mark Turner has argued that imagining a narrative *sequence* is a "fundamental instrument of thought" (1996, 4) and, further, that utilizing some narratives as paradigmatic—patterns that guide the way we interpret the world—is "indispensable to human cognition" (5). The experience *of* narrative and, in turn, its generation is essential to cognitive development. From the beginning of our experience of the world, we internalize causality: For example, pre-linguistic infants come to expect outcomes (objects falling when pushed, for example) and express surprise at encountering something non-causal (Bruner 1986, 18). Charles Fernyhough describes well the emergence of a toddler's capacity to remember events along with the ability to tell stories with the self at center (2012, 17).[5] And these early stories—as well as the memories they generate—necessarily entail a sense of cause and effect which may be *imposed* on observed phenomena by human agency.

There is, then, a gap between what happens in the external world and the story our brains develop to explain observed events. As Turner argues, the human mind converts things that happen (an event-story) into *acts* (action-stories) that necessitate agents and objects, which in turn require responsibility and blame; as a result, human audiences will tend to reject narratives that fail to adhere to previous narrative experience. The importance of narrative sequence is implicit in the argument of Aristotle's *Poetics* when he insists that a plot must have a beginning, middle, and end (*Poetics*, 1450b-1451a). For Aristotle, this imitation of events as humans experience them conveys pleasure. Aristotle does not explain why a plot that is causally connected with a clear end is the best one.[6] But if we follow his argument throughout the *Poetics*, a compelling plot—namely one that adheres to audience expectations about causal relations—is more effective in conveying verisimilitude, and therefore most successful in provoking an emotional reaction when the expected pattern is *excepted*, generating surprise and wonder. Aristotle calls this cleansing *catharsis*—and scholars like Martha Nussbaum have argued that catharsis is both emotional and intellectual: that the experience of identifying with a mimetic narrative forces the audience into a "clarification concerning who we are".[7]

So, Aristotle makes it clear that stories with endings that *seem* causally determined facilitate our enjoyment of and identification with a story (along with a clarification of identity). As Andy Clark (2015) proposes in his theory of 'predictive processing', our brains are constantly engaged in a process of predicting outcomes and mapping actions based on a store of prior experiences.[8] These predictive processes govern basic motor functions as well as some higher-order behaviors. While we share such cognitive apparatus with other animals, Clark argues that what makes human cognition different is our ability to shape and rely on our environments and the extended cognitive field provided by human culture and language (14–16).[9] As a result, then, of such basic cognitive 'wiring', outcomes that defy our predictions can cause surprise, if not discomfort.

We can experience a gnawing frustration that comes from not knowing how a story ends; this feeling comes equally from the experience of fiction and from the suspense of real life. The studies I have mentioned so far help to indicate the fundamental level at which this type of emotional feedback is generated. While narrative logic dictates that everything which begins has an end, this logic is a human *need* with philosophical and neurobiological motivations: not knowing how things turn out causes us nearly existential pain. As Clark summarizes from a perspective of perception and neurological function, our cognitive function "depends crucially upon cancelling out sensory prediction error" (2013, 7); breaks in patterns or unexpected outcomes are "explained away" through a higher-order cognitive process. From a literary standpoint, Johnathan Gottschall proposes that our "mind is allergic to uncertainty, randomness, and coincidence. It is addicted to meaning. If the storytelling mind cannot find meaningful patterns in the world it will try to impose them" (2012, 103).

We might be tempted to describe such a neural 'programming' as a narrative causality bias—but it is one that we come to almost by necessity in response to external stimuli. Psychologists like Jerome Bruner have argued that the thing we describe as the *Self* is a type of evolving narrative 'text' written in relation to our experiences in the world (1986, 130). Evolutionary biologist Edmund Wilson takes this further to assert that our minds themselves consist almost entirely of storytelling.[10] Years of working on the human brain and reflecting on these studies have led Joseph LeDoux to conclude "that consciousness is an interpreter of experience, a means by which we develop a self-story that we use to understand those motivations and actions that arise from non-conscious processes in our brains" (2015, 5–6). The species-level advantage of this narrative capacity is enormous; but it also makes us vulnerable to narrative 'blips'. The human tendency toward confabulation is a good illustration of this. And this tendency may be neurologically determined. In a series of experiments on patients with a separation in the hemispheres of their brains, Michael Gazzaniga and Joseph LeDoux provided an image to the right hemisphere which could not be detected or contextualized by the left. LeDoux summarizes:

> For example, in one study we simultaneously showed the patient's left hemisphere a chicken claw and the right hemisphere a snow scene. The patient's left hand then selected a picture of a shovel. When the patient was asked why he made this choice, his left hemisphere (the speaking hemisphere) responded that it saw a chicken and you need a shovel to clean out the chicken shed. The left hemisphere thus used the information it had available to construct a reality that matched the two pieces of information available: it saw a picture of a chicken and it saw its hand selecting a shovel.
>
> *(LeDoux 2015, 10)*

To put this in an Aristotelian frame: the subject receives images that have no actual causal connection and creates a narrative to connect them. Not only are our brains wired in such a way as to invent a story "rather than leave something unexplained" (Gottschall 2012, 99), but the type of story we create tends to have agents and objects whose arrangement is based on individual experience and cultural beliefs. Our brains use our senses to gather data from the world, but they are economical in doing this: we are constantly filling in details that are not there and completing stories. When our brains encounter surprise or outcomes contrary to our expectations, such a "sensory prediction error" (see Clark 2015 cited above) can produce cognitive dissonance. This is why we get so disturbed when we don't know how something ends: most of the time, our brains are picking out endings and reconstructing narratives that would likely lead to them.

The human desire to bring a narrative to its end comes as no surprise from a literary perspective either—drawing on psychoanalysis, critics like Peter Brooks (1992, 102–103) see the death instinct operative in both the pleasure derived from moving toward a text's end and a concomitant impulse to delay that end or to repeatedly return through the same movement. This is confirmed too by psychologists who study the effect of death-anxiety on human behavior:

narrative functions both to safeguard us against mortal fear and to give us meaning despite it.[11] Critics have explored and plotted different generic and cultural attitudes toward those narratives that effect closure and those that appear to remain open. Indeed, many readers have seen a tension or *dialectic* between openness and closure as an indication of sophisticated narrative, of what some might call literature. But much of this leads us down interpretive paths that are both closed and endless—as the late classicist Don Fowler writes, "whether we look for closure or aperture *or a dialectic between* them in a text is a function of our own presuppositions, not of anything 'objective' about the text" (Fowler 1997, 5).

Ending the poem and the economy of pleasure

What the *Odyssey* has to say about the emotional effect of storytelling and the way it shapes teller and audiences derives in part, I believe, from human sensed experience, both of the pleasure of anticipating a tale's end and of the pain of not knowing its conclusion. The epic plays with conventions of story-ending and closure; the complexity of this 'play' both relies upon generic expectations about closure (in a literary sense) and also reflects an understanding of the emotional impact of narrative endings. *Odyssey* 24 offers at least seven closural moments, including: the second underworld scene, the third retelling of the story of Laertes' shroud, the last reunion of the homecoming (Odysseus and Laertes), the split debate among the suitors, followed by the death of Eupeithes, an adjudicating conversation between Zeus and Athena, and the imposed amnesty (*Eklesis*).[12] These moments are entwined; but they also address specific themes and plots explored within the epic.

How the *Odyssey* ends has prevented many readers from appreciating its effects. Since at least the fourth century bce there have been questions about its conclusions—Hellenistic scholars saw resolution in the amorous reunion of husband and wife (23.293–296), while Aristotle saw the epic's *telos* or completion in the payback of the suitors.[13] Eustathius, however, complained that these interpreters "cut off critical parts of the *Odyssey*, such as the reunion of Odysseus and Laertes and many other amazing things" (Comm. ad *Od.*, II.308). And, although book 24 is now accepted as essential to the whole, it has been called "ugly" (Bakker 2013, 129) and "lame, hasty, awkward, abrupt" (Wender 1978, 63).

In order to appreciate the interpretive challenges posed by these closural moments briefly surveyed above, we need to trace the threads of the epic tapestry back to when Odysseus and Eumaios converse after dining (15.398–402):

> Let us delight in one another's gruesome pains while we drink and dine in my home, remembering. A man may delight later on in his pains when he has suffered many and gone through much.[14]
>
> νῶϊ δ' ἐνὶ κλισίῃ πίνοντέ τε δαινυμένω τε
> κήδεσιν ἀλλήλων τερπώμεθα λευγαλέοισι
> μνωομένω· μετὰ γάρ τε καὶ ἄλγεσι τέρπεται ἀνήρ,
> ὅς τις δὴ μάλα πολλὰ πάθῃ καὶ πόλλ' ἐπαληθῇ

When understood within its broader thematic framework, this passage marks the epic's programmatic engagement with expectations concerning and emotional responses to closure. Eumaios supports his injunction to find pleasure in pain with something of a proverb—he moves from a proposition about their individual situation to a comment on universal human conditions. Pleasure for Eumaios and, as implied by his gnomic statement, for others too, emerges in part from a narrative that is definitively in the past. Elsewhere, the epic seems preoccupied with the relationship between narrative and pleasure. But it explores this against a general sensual backdrop: the poem's narrative presents an economy of pleasure where gods and heroes alike derive enjoyment from feasting (e.g. 1.25, 1.422, 4.27), conversation (4.239), athletic competition (4.626 and 17.168), and sex (5.227).[15] Narratives which bring pleasure often involve Odysseus: Helen invites Telemachus and Menelaos to take pleasure in stories about Odysseus (4.238–241). The Phaeacians enjoy the story of the quarrel of Odysseus and Achilles (8.90–92; Odysseus weeps). Odysseus takes pleasure in the story of Hephaestus (8.367–369), later crying at the story of the Trojan horse (8.521–522). Odysseus, in his own account, takes a dangerous pleasure in the song of the Sirens (12.51–54). This basic pattern is interesting for the range of pleasure experienced (which is largely understandable) and the pain felt by Odysseus. Amid these moments, Menelaos "delights his mind with grief sometimes" (ἄλλοτε μέν τε γόῳ φρένα τέρπομαι, 4.102) and Penelope likewise describes her days as pleasured by grieving and lamenting (ἤματα μὲν γὰρ τέρπομ' ὀδυρομένη γοόωσα, 19.513–514). These final two passages help frame the aforementioned examples of narrative enjoyment, if we consider Eumaios' words about taking pleasure in grief more carefully. A man takes pleasure (τέρπεται) *afterwards* (μετά) when he has *finished enduring* pain (ἐπαληθῇ). The presence of the adverb and the aspectual distinction between the verbs of pleasure and suffering give this passage its force and help to guide us to a deeper understanding of the poem's reflections on narrative and closure.

It is not just a narrative's embedding in a previous time that makes it pleasurable, but rather its *completion*. At a very basic level, I suggest, the *Odyssey* tells us that pleasure comes from a narrative that *ends*. Even if you have suffered, you can experience pleasure from a tale that has an ending. It is no accident that in books 1, 4, and 8 we find characters confessing to almost paralyzing grief over a tale whose end is unknown—in fact, over the unknown end of the tale that is being told. This theme is not just about the experience of the *Odyssey*'s characters; it also anticipates audience experience of the *Odyssey* and problematic expectations for the resolution of its plot.

Signaling closure

The final book of the *Odyssey* is in part about how to end a poem. This process of its closure, moreover, itself capitalizes upon both the narrative it has developed prior to this book and an implicit understanding of human psychology. As I explore above, psychological and literary perspectives help to explain why audiences desire to know how a story ends but are troubled by the fact that it must end. I will turn now to explore three signs of closure that emerge in book 24 to show the extent to which the book may prompt similar reflections on closure. First, I will discuss a formulaic pattern that originates in book 1, then the repeated story of Penelope's shroud for Laertes, and, finally, the conversation between Achilles and Agamemnon at the beginning of book 24.

Zeus' opening comments in the *Odyssey* ("Mortals! They are always blaming the gods and saying that evil comes from us when they themselves suffer pain beyond their lot [ὑπὲρ μόρον ἄλγε' ἔχουσιν] because of their own recklessness [σφῇσιν ἀτασθαλίῃσιν]") offer a paradigmatic lesson;[16] but they also inspire the construction of a narrative in response: the event of suffering initiates an action-story, relating misery to a complex web of human and divine agency. When characters feel pleasure or pain at the telling of a tale, the external audience is cued to seek out its cause.

Storytelling's effects, then, are at issue early in the poem—indeed, Zeus is responding to one narrative, the story of Aigisthus and Orestes, and expressing frustration over the fact that human beings fail to learn the extent of their own responsibility. For Zeus, the story of Aigisthus presents a paradigmatic problem; but it is a repeated pattern. His frustration derives from suffering the same experience again and again. Different emotional effects from storytelling are central to the conversation between Penelope and Telemachus in book 1 (337–344):

> Phemios, you know many other spells for mortals, the deeds of men and gods, the things singers make famous while they remain here singing as these men sit drinking their wine in silence. Stop this grievous song: it wears always on the heart in my chest and unforgettable grief [*penthos alaston*] has come over me especially. Always remembering that sort of man, I long for my husband whose fame [*kleos*] spreads wide through Greece and Argos.
>
> Φήμιε, πολλὰ γὰρ ἄλλα βροτῶν θελκτήρια οἶδας
> ἔργ' ἀνδρῶν τε θεῶν τε, τά τε κλείουσιν ἀοιδοί·
> τῶν ἕν γέ σφιν ἄειδε παρήμενος, οἱ δὲ σιωπῇ
> οἶνον πινόντων ταύτης δ' ἀποπαύε' ἀοιδῆς
> λυγρῆς, ἥ τέ μοι αἰὲν ἐνὶ στήθεσσι φίλον κῆρ
> τείρει, ἐπεί με μάλιστα καθίκετο **πένθος ἄλαστον**.
> τοίην γὰρ κεφαλὴν ποθέω μεμνημένη αἰεὶ
> ἀνδρός, τοῦ κλέος εὐρὺ καθ' Ἑλλάδα καὶ μέσον Ἄργος.

Penelope asks for the suitors to sing a different song, because this one causes her *penthos alaston*, ceaseless pain. Note how she limits the function of storytelling for this audience: she characterizes the song as *bewitchment* [θελκτήρια], entertainment. There is a gap here in the way the internal audiences respond to the same narrative: the suitors' entertainment is Penelope's pain. This contrast also bears fruit in its psychological reflections. The cause of the difference in their responses becomes clearer when Telemachus speaks (346–355):

> My mother, why do you begrudge the singer to delight wherever his mind leads him. Singers aren't to blame but Zeus is who allots to each of mortal men however he wishes. There's nothing wrong with him singing the terrible fate of the Danaans, for men make more famous the song which comes most recently to their ears. Let your heart and mind be bold enough to listen: Odysseus wasn't the only man who lost his homecoming day in Troy, many other men died too.

Telemachus acknowledges that the story is the "terrible fate of the Danaans" but endorses the telling of it for two reasons: first, the most entertaining tale is the most recent one—and no story is more current than this. And, second, the grief is not only hers: other men died during the return home. The difference in the responses of Penelope and Telemachus is easy to explain from an emotional perspective. Penelope is emotionally connected to the absence of her husband—and its effects—in a way Telemachus cannot be. Even though Odysseus' delayed return has had a negative material impact on Telemachus, his father is still just a *story* to him. Accordingly, his reception of the tale relativizes it: Odysseus is one of many. But what often goes unnoticed is that Telemachus and the suitors are able to derive pleasure because they have implicitly provided a different end to the story: They think that Odysseus is dead.[17]

There is additional thematic relevance beyond the divergent emotional responses. When Telemachus makes it to Sparta in book 4, Menelaos confesses to indulging in grief as he thinks back to his companions, the war, and the terrible returns home (4.104–112):

> Often, while grieving and mourning everyone and while sitting in my home I sometimes delight my mind with lamentation; and other times I stop, since my fill for shrill lament is fast-coming. I don't grieve so much for all the others when I mourn as for one who troubles my sleep and my food as I remember him—since no one of the Achaeans toiled and achieved as much as Odysseus did. I will always feel grief for him and my woe [*akhos*] for him is always unforgettable [*alaston*] because he has been gone so long and we do not know if he is alive or dead. So, too, must they mourn for him, I imagine, elderly Laertes, prudent Penelope and Telemachus, the child he left just born in their household.

Here again we find an expression of emotion similar to Penelope's. In contrast to Telemachus, Menelaos makes Odysseus exceptional; he singles him out from the many others and imagines the response of those bereft of him. The diction ties his response to Penelope's too: He isolates Odysseus for causing him grief, here too described as *alaston*.[18] Menelaos provides another clue to the difference between Telemachus' response to the homecomings and Penelope's. The unrelenting grief marked by the adjective *alastos* is steeped in uncertainty or a lack of resolution: Penelope does not know if Odysseus is alive or dead. Telemachus' comments make it clear that he does not feel similarly because to his mind the story is over. That Telemachus has written an end to his father's tale is implied in later books. During a conversation with Nestor, in fact, he declares Odysseus dead and incapable of achieving a "true return" (κείνῳ δ' οὐκέτι νόστος ἐτήτυμος, 3.241). Later when he asks Menelaus for some fame of his father (εἴ τινά μοι κληηδόνα πατρὸς ἐνίσποις, 4.316), he attempts to put a limit on the tale by framing this request as a wish to be informed of the "grievous ruin" of that man (κείνου λυγρὸν ὄλεθρον ἐνισπεῖν, 4.323).

This dynamic between the pleasurable pain from a sad tale that has an end and the destructive grief from the unresolved narrative appears elsewhere in the *Odyssey*. As I anticipated in discussing Eumaios' invitation to Odysseus to indulge in telling each other sad tales, the fact that the tales are complete and behind them is crucial for their pleasure-content—a positive cognitive feedback loop from telling a complete tale. Stories that have been told and are over are narrative experiences that communicate who the characters are—the stories that we tell each other of our pasts establish identities for our present.

When stories are incomplete and cannot be told, the epic marks them out as having a detrimental effect. Eumaios, in a slightly earlier scene, reveals that he too mourns without ceasing (*alaston*) for Telemachus because he does not know if the boy is alive or dead and it causes grief because he is helpless to effect any change at all (14.174–190). Eumaios draws a direct connection between hearing about his guest's suffering and knowing him (ἀλλ' ἄγε μοι σύ, γεραιέ, τὰ σ' αὐτοῦ κήδε' ἐνίσπες / καί μοι τοῦτ' ἀγόρευσον ἐτήτυμον, ὄφρ' ἐὺ εἰδῶ·). The adjective *alastos*, also applied to Odysseus' absence earlier by Penelope and Menelaos, has etymological associations to explain its use. I have translated it in different ways (e.g. "ceaseless" or "inescapable"), but its etymology points most clearly to a root that might render it "unforgettable"—marking something that cannot be forgotten no matter how hard one tries.[19] The adjective *alastos* is applied to grief and grieving (with three different lexical roots: *penthos*, *akhos*, and *oduromai*) over events that have no clear ending; they are "unforgettable" in the sense of inescapable. These associations are metanarrative and metacognitive: such grief is connected both to the incompleteness of the *Odyssey* and to our desire to turn events into action-narratives that have clear outcomes. Just as the poem's external audience witnesses its internal audiences struggling with the incompleteness of tales, the external audience experiences strains of similar frustration: in one part, in sympathy for the characters themselves; and in another, in response to a growing need to know the end of the tale itself.

A final occurrence of the same adjective helps to support these last assertions. During the Ithacan assembly on the deaths of the suitors, Eupeithes stands to speak about what the aggrieved families should do.[20] Before he speaks, the narrative describes him (24.428–430):

> Among them then Eupeithes stood and spoke, for unforgettable grief [*alaston . . . penthos*] filled his thoughts over his son Antinoos whom Odysseus killed first.
>
> τοῖσιν δ' Εὐπείθης ἀνά θ' ἵστατο καὶ μετέειπε·
> παιδὸς γάρ οἱ **ἄλαστον ἐνὶ φρεσὶ πένθος** ἔκειτο,
> 'Αντινόου, τὸν πρῶτον ἐνήρατο δῖος 'Οδυσσεύς·

This moment departs from the earlier examples in significant ways. First, the narrative describes Eupeithes as suffering "unforgettable grief"—in the earlier passages, it is the characters themselves who use the phrase. Second, Eupeithes is speaking of someone who is already dead and whom they have just buried. This second feature challenges the claim that there is an essential lack of closure to grief described with the adjective *alastos*. Eupeithes explains the source of his grief in the speech that follows (24.428–438):

> Friends, this man has accomplished a 'great' deed for the Achaeans: He led many fine men away on his ships—then he lost the ships, and he lost the men. And, once he returned, he killed those who were best of the Kephallenians. But come, let us go, before that man flies off to Pylos or shining Elis where the Epeians rule. Otherwise we will be ashamed forever. This will be an object of reproach even for men to come to learn, if we do not pay back the murders of our relatives and sons. It cannot be sweet to my mind at least to live like this. But instead, I would rather perish immediately and dwell with the dead. But, let's go so that those men don't cross to the mainland first.
>
> ὦ φίλοι, ἦ μέγα ἔργον ἀνὴρ ὅδε μήσατ' 'Αχαιούς·
> τοὺς μὲν σὺν νήεσσιν ἄγων πολέας τε καὶ ἐσθλοὺς
> ὤλεσε μὲν νῆας γλαφυράς, ἀπὸ δ' ὤλεσε λαούς,
> τοὺς δ' ἐλθὼν ἔκτεινε Κεφαλλήνων ὄχ' ἀρίστους.
> ἀλλ' ἄγετε, πρὶν τοῦτον ἢ ἐς Πύλον ὦκα ἱκέσθαι
> ἢ καὶ ἐς Ἦλιδα δῖαν, ὅθι κρατέουσιν 'Επειοί,
> ἴομεν· ἦ καὶ ἔπειτα κατηφέες ἐσσόμεθ' αἰεί.
> λώβη γὰρ τάδε γ' ἐστὶ καὶ ἐσσομένοισι πυθέσθαι,
> εἰ δὴ μὴ παίδων τε κασιγνήτων τε φονῆας
> τεισόμεθ'· οὐκ ἂν ἐμοί γε μετὰ φρεσὶν ἡδὺ γένοιτο
> ζωέμεν, ἀλλὰ τάχιστα θανὼν φθιμένοισι μετείην.
> ἀλλ' ἴομεν, μὴ φθέωσι περαιωθέντες ἐκεῖνοι.

The grief attributed to Eupeithes expands our understanding of what qualifies as *alaston*. As Eupeithes explains, the source of his pain is a different type of incomplete story: he—and the other families—are due to pay back Odysseus for the deaths of so many Ithakans or else suffer reproach in the future. So strong is this impulse that Eupeithes would rather die than go on living without completing this obligation. Penelope and Menelaos powerlessly wait for resolution to their anxiety about Odysseus while Eupeithes suffers because he believes he must act to continue the story and write a new ending for his son.

Two additional features make the attribution of *alaston penthos* to Eupeithes more clearly motivated and, indeed, especially powerful in the context of the epic tradition. The first is etymological: already by the time of Homer, the related noun *alastor* for avenger was active—the man who seeks and exacts vengeance is one who by nature cannot or will not forget.[21] In addition, the passage draws additional force from its resonance with repeated descriptions of parents' *unforgettable grief*. In the *Iliad*, Thetis, carrying unforgettable sorrow in her heart, has *penthos alaston* in her thoughts as she grieves for Achilles, whom, though still alive, will die. She is powerless to act to save him (**πένθος ἄλαστον** ἔχουσα μετὰ φρεσίν· οἶδα καὶ αὐτός· *Il.* 24.105). In the Homeric *Hymn to Aphrodite*, the Trojan king Tros feels unforgettable grief when his son Ganymede disappears ("Unforgettable grief overtook Tros' mind because he did not know where the divine wind had taken his dear son [Ganymede]. He mourned him thereafter continually every day"; Τρῶα δὲ πένθος ἄλαστον ἔχε φρένας, οὐδέ τι ᾔδει / ὅππῃ οἱ φίλον υἱὸν ἀνήρπασε θέσπις ἄελλα· / τὸν δὴ ἔπειτα γόασκε διαμπερὲς ἤματα πάντα, 207–209). And, in Hesiod's *Theogony*, Rhea has the same emotional response when Kronos eats her children ("[Kronos] did not keep a blind watch, but he noticed the children [being born] and ate them up. And unforgettable grief took Rhea"; τῷ ὅ γ' ἄρ' οὐκ ἀλαοσκοπιὴν ἔχεν, ἀλλὰ δοκεύων / παῖδας ἑοὺς κατέπινε· Ῥέην δ' ἔχε πένθος ἄλαστον, 466–467). In each case, the inescapable emotion comes from a situation outside the character's control, is related to a strong emotional bond, and happens at a time of paralysis or inaction. In these examples, the formulaic invocation of grief depends upon a narrative pattern: the pain issues from a lack of resolution, from not knowing whether one thing or another has happened, and not being able to do anything about it.

Here, Homeric formulaic language and its marked deployment echo what modern studies have demonstrated regarding the impact of a certain type of grief. Psychological research has identified unresolved grief—sometimes called "ambiguous loss" (see Boss 1999)—as a special category with symptoms similar to "anxiety, depression, and somatic illnesses" (Boss 1999, 10). The uncertainty or lack of resolution can cause inaction and undermine confidence in the self and the world (ibid., 107). Indeed, additional studies have shown that individuals who are coping with "complicated grief" demonstrate a diminished capacity for attention and compromised cognitive functions (Hall et al. 2014). This is caused in part by a dysfunctional return to the cause of uncertainty, "a repetitive loop of intense yearning and longing that becomes the major focus of their lives".[22] The courses of treatment effective for unresolved grief have significant implications for my arguments about the emotional impact of narrative in the *Odyssey*. Pharmacological interventions have been shown to be of limited efficacy. Instead, storytelling and long-term psychotherapy have proved to provide the only durative relief.[23]

Without the intervention of another narrative, people who suffer from unresolved grief remain like the parents in Homeric poetry mentioned above—paralyzed in an obsessive cycle of reflection. At times, however, their mental apparatus provides a response from prior experience, from the store of narrative detail that helps to structure their thoughts and guide their actions. When the pattern of unforgettable pain is applied to Eupeithes—and the end of the *Odyssey*—it induces a variation that imperils the survival of Odysseus and threatens a different end to this tale. Each of the strands of the epic's narrative is clipped or tied off, but Eupeithes has a claim to action that cannot be ignored. His *unforgettable grief* elicits a paradigmatic pattern: children are open-ended tales, narratives that go on after the end of the parent. For this Homeric father, a dead child is certainly not a closed tale; revenge is not just the quickest resolution, it is the primary and conventional end to this tale. In short, Eupeithes addresses his own endless grief with a narrative solution: he plans to write the ending of the tale by killing Odysseus. In this, he is applying a paradigmatic solution. Odysseus has killed his son; Odysseus needs to die.

The sign of the shroud

Understanding the cognitive basis of our relationship to narrative also entails appreciating possible maladaptive effects. Eupeithes' sorrow and his compulsion to write an ending to his story, then, advances a thematic treatment of the relationship between the closure of a narrative and the experience of grief. The implication is that this is a Homeric dramatization of the psychological impact of narrative which we now know has specific cognitive aspects. The depth of the epic's interest in this theme emerges throughout book 24. Eupeithes' speech at its end, which imposes a conclusion on a story in progress, corresponds well to a conversation that begins the book in the Nekyia where a dead suitor tries to re-order the past. The desire to end a tale is prefigured through an examination of causality and completion.

The longest detail of Amphimedon's story of the death of the suitors recounts Penelope's betrayed trick of promising to marry after the completion of her work (24.125–155). At first, this story has a clear function: it is a famous illustration of Penelope's guile that shows how similar to her husband she is, and it characterizes the suitors' frustration by explaining how she resisted them for so long. The interpretive challenge is that this is actually the *third* time in the epic it has been described. In book 2, Antinoos tells the tale to the assembled Ithacans (2.93–110). And then in book 19, Penelope narrates the tale to a disguised Odysseus (19.137–161). Repetition allows the scene to function as a metanarrative reflection of the Homeric art: the shroud stands as a metaphor for the completion of stories in general. Indeed, this metaphorical meaning may have been charged in Greek culture and in early hexameter poetry especially where covering garments function widely as metaphors and metonymies for death, as Douglas Cairns argues (2016, 8–9). While the language has only minor divergences, there are details added with each retelling.[24] Most important among these differences is that in the final description the completion of the fabric is followed by the return of Odysseus and the completion of the *Odyssey*, to a point (24.139–146):

> So she was completing it, even though she was unwilling, she finished it, under force. When she showed us the robe she wove on the great loom after she washed it, it shone like the sun or the moon. And then a wicked god brought Odysseus from somewhere from the farthest part of the country, where the swineherd lives. That's where godly Odysseus' dear son came home too from sandy Pylos, sailing with his black ship. The two of them came to the famous city, devising an evil death for the suitors—well, Odysseus came later, it was Telemachus who led him there first.
>
> ὣς τὸ μὲν ἐξετέλεσσε καὶ οὐκ ἐθέλουσ', ὑπ' ἀνάγκης.
> εὖθ' ἡ φᾶρος ἔδειξεν, ὑφήνασα μέγαν ἱστόν,
> πλύνασ', ἠελίῳ ἐναλίγκιον ἠὲ σελήνῃ,
> καὶ τότε δή ῥ' Ὀδυσῆα κακός ποθεν ἤγαγε δαίμων
> ἀγροῦ ἐπ' ἐσχατιήν, ὅθι δώματα ναῖε συβώτης.
> ἔνθ' ἦλθεν φίλος υἱὸς Ὀδυσσῆος θείοιο,
> ἐκ Πύλου ἠμαθόεντος ἰὼν σὺν νηῒ μελαίνῃ·
> τὼ δὲ μνηστῆρσιν θάνατον κακὸν ἀρτύναντε
> ἵκοντο προτὶ ἄστυ περικλυτόν, ἦ τοι Ὀδυσσεὺς
> ὕστερος, αὐτὰρ Τηλέμαχος πρόσθ' ἡγεμόνευε.

By narrating the events in this way, Amphimedon appears to be translating them into an 'action-story' in which there is a causal relationship between the completion of the weaving and the completion of the *Odyssey*'s narrative(s). He and the suitors are not responsible for their deaths;

but rather a god and Odysseus. This passage elides the events of the epic itself where there is no clear collusion between Penelope and Odysseus. Not only does it appear to deprive Penelope of some agency and correctly identify some divine agency, but it also depicts Amphimedon as blaming the gods for his own suffering (confirming Zeus' lament discussed above). But, in addition, this passage features a Homeric *character*-mind trying to make sense of a series of events and attributing agency and causality in different directions: now Penelope, now Odysseus and a god, now Telemachus. In his questing to tell a story that makes sense, Antinoos actually weaves together a fairly 'true' picture of agency and causality.[25] But he also imposes an interpretation on the tale—that Odysseus and Penelope colluded ahead of time—which the external audience knows did not clearly happen. The epic, then, presents a character intentionally re-reading his own experiences in a way that allows him to make sense of the world he inhabits even as the audience of the poem is engaged in a similar process.

At the beginning of book 24, then, the *Odyssey* re-centers problems of how we interpret the tales we hear and the impact of our own expectations and needs on the way we retell our stories. The metapoetic and metacognitive nature of his closing gesture is reinforced by the act of weaving the shroud and the object itself. As many have observed, weaving is often a metaphor not just for intelligence but for poetic composition in Greek culture and others.[26] In the *Iliad*, Helen weaves a *pharos* that depicts "The many struggles of the horse-taming Trojans and the bronze-girded Achaeans / All the things they had suffered for her at Ares' hands" (*Il.* 3.121–128). An ancient scholar recognized in Helen's weaving an embedded metaphor for Homer's own art, which he calls "a worthy archetype for his own poetry" (ἀξιόχρεων ἀρχέτυπον ἀνέπλασεν ὁ ποιητὴς τῆς ἰδίας ποιήσεως, Schol. bT ad Il. 3.126–127). If we pursue the relevance of the weaving metaphor to the *Odyssey*, we find that it continues to shape itself around the relationship between audience desire and narrative closure. Weaving appears throughout the poem, but its decoration goes undescribed. Helen gives Telemachus a garment to give to his future wife (*Od.* 15.123–130). Calypso (5.62) and Circe (10.222) weave while singing. Nausicaa leaves a robe for Odysseus (6.214) which Arete recognizes because she made it (7.234–235). We even hear that the Naiads who live on the shore of Ithaca weave "sea-purple garments, wondrous to see" (13.108), but we never *see* them. The lack of description might be less confounding if Penelope's delaying were not understood as equivalent to the delaying narrative strategies of the *Odyssey*.[27] But few commentators have worried about what might be pictured on the finely woven cloth. Barbara Clayton writes:

> Homer's audience would have assumed an implicit narrative component in Penelope's web, perhaps that she is depicting the heroic deeds of Laertes . . . I do not think that Homer's silence on this point represents the omission of an unimportant detail. I would argue instead that Homer deliberately leaves the narrative content of the web within the realm of potentiality. And this aspect of potentiality in turn complements the fact that Penelope's web is potentially never complete.
>
> *(2004, 34)*

The undescribed content of the shroud is a metaphor for the unbounded and complete nature of the *Odyssey* itself. It simultaneously responds both to our reluctance to end a tale and our need to do so. Its completion, coterminous with Odysseus' return, seals its connection with that narrative—especially considering that, like the tale of the epic, the shroud was woven and unwoven before it was finally 'made'. But the refusal to provide an image on the shroud—or to describe the image that is there—leaves narrative work to the audience itself.[28] Amphimedon emerges in the poem as a stand-in for someone who 'writes' his own story on the blank surface of the shroud. But since he is an observer and a participant in the narrative, his retelling of the tale engages with the themes I discussed above regarding the way Homeric characters process and act upon narratives. Even in the underworld, Amphimedon reconsiders and retells his story in an attempt to take control of it. He shows us how he completed it by placing himself in a perspective where he was the victim of unexpected collusion. In doing so, he models (mis-) reading for the narrative's audiences and, further, contributes to the epic's presentation of the impact of narrative on human life. The blankness of the shroud leaves the narrative work to us and we are compelled to finish its tale. We have to tell that story just as, I propose, we have to imagine what happens after the end of the *Odyssey*.

The end of a poem

While the epic dramatizes our need to provide an end to a tale through the depiction of Eupeithes' death and an exploration of the theme of unforgettable grief, its deployment of the shroud motif applies an implicit understanding of the human narrative mind by creating a puzzle for its audiences. Our modern understanding of cognitive science can help us understand the effect this has *outside the poem*. The unexplained here becomes not a cause of grief but instead an opportunity for agency.

I opened this chapter by discussing some of the thematic and compositional problems of the *Odyssey*'s final book. One advantage for using a cognitive approach to the *Odyssey* is that it helps us to acknowledge that it is within the epic's range of narrative strategies *not* to make sense, or to challenge our expectations for narrative in part because these very expectations are based on *other* narratives. The device of raveling and unraveling narrative becomes a powerful sign of indeterminacy and control. Book 24 returns to this image and features someone whose life (and story) is over, encouraging the audience to think of the shroud and its attendant interpretive issues as the story ends.

Earlier I emphasized the powerful desire on the part of the audience to hear—if not produce—the end of the tale. This desire is connected to our causal sense of narrative, embedded on a cognitive level and present in a compulsion to bring stories—even our own—to completion. This perspective makes me hear the repeated report of the shroud's completion differently: "so she was completing it, though unwilling, under compulsion" (ὣς τὸ μὲν ἐξετέλεσσε καὶ οὐκ ἐθέλουσ', ὑπ' ἀνάγκης). *Anangkê* in Greek poetry can signal physical force or threat (as it does here) but it can be generalized as an externally imposed compulsion, something like fate. Thus, this line signals that the sign of the shroud is also in part about the necessary completion of a thing, be it a garment, a poem, or even a life itself.

In its retelling and redeployment as a sign of closure, the shroud is not altogether the most complicated motif: it is delivered by a dead man among the dead. Book 24, then, starts with a moment of narrative surplus modeling how part of its own tale is received by audiences who have no further tales to live. But before Amphimedon speaks, we get to eavesdrop on a conversation between Agamemnon and Achilles (24.93–98):

> So you, when you died, didn't lose your name, but your fame will always be noble among all men, Achilles. But what consolation is this for me when I ran the war? Zeus devised ruinous pain for me in my homecoming at the hands of Aigisthos and my destructive wife.
>
> ὣς σὺ μὲν οὐδὲ θανὼν ὄνομ' ὤλεσας, ἀλλά τοι αἰεὶ
> πάντας ἐπ' ἀνθρώπους κλέος ἔσσεται ἐσθλόν, 'Αχιλλεῦ·
> αὐτὰρ ἐμοὶ τί τόδ' ἦδος, ἐπεὶ πόλεμον τολύπευσα;
> ἐν νόστῳ γάρ μοι Ζεὺς μήσατο λυγρὸν ὄλεθρον
> Αἰγίσθου ὑπὸ χερσὶ καὶ οὐλομένης ἀλόχοιο.

Agamemnon says this to Achilles after the latter has sympathized with him that he did not receive the glorious burial he deserved after fighting at Troy (24.22–34). Before the central players of the *Iliad* hear Odysseus' tale, they appear to contemplate their own status as objects of fame. Agamemnon, in fact, reflects upon the very tale Zeus contemplated earlier. Here, the Aigisthos model is set next to the theme of glorious death. But before some type of synkrisis or comparison can be completed, their reminiscence is interrupted by the arrival of Amphimedon who tells the tale of the shroud and ends by lamenting that the suitors are not yet buried. Agamemnon responds only to the part about Penelope (192–202):

> Blessed child of Laertes, much-devising Odysseus, you really secured a wife with magnificent virtue! That's how good the brains are for blameless Penelope, Ikarios' daughter, how well she remembered Odysseus, her wedded husband. The fame of her virtue will never perish, and the gods will craft a pleasing song of mindful Penelope for mortals on the earth. This is not the way for Tyndareos' daughter. She devised wicked deeds and since she killed her wedded husband, a hateful song will be hers among men, she will attract harsh speech to the race of women, even for those who are good.
>
> ὄλβιε Λαέρταο πάϊ, πολυμήχαν' Ὀδυσσεῦ,
> ἦ ἄρα σὺν μεγάλῃ ἀρετῇ ἐκτήσω ἄκοιτιν·
> ὡσ ἀγαθαὶ φρένες ἦσαν ἀμύμονι Πηνελοπείῃ,
> κούρῃ Ἰκαρίου, ὡσ εὖ μέμνητ' Ὀδυσῆος,
> ἀνδρὸσ κουριδίου. τῶ οἱ κλέος οὔ ποτ' ὀλεῖται
> ἧσ ἀρετῆς, τεύξουσι δ' ἐπιχθονίοισιν ἀοιδὴν
> ἀθάνατοι χαρίεσσαν ἐχέφρονι Πηνελοπείῃ,
> οὐχ ὡς Τυνδαρέου κούρη κακὰ μήσατο ἔργα,
> κουρίδιον κτείνασα πόσιν, στυγερὴ δέ τ' ἀοιδὴ
> ἔσσετ' ἐπ' ἀνθρώπους, χαλεπὴν δέ τε φῆμιν ὀπάσσει
> θηλυτέρῃσι γυναιξί, καὶ ἥ κ' εὐεργὸς ἔῃσιν.

In this response Agamemnon ruminates on *kleos* and the reception of narrative. And, through this speech and the one cited before, the epic also expresses a concern about how narrative is used. Agamemnon praises Penelope for her intelligence and loyalty but only as a transition into lamenting Klytemnestra again: he claims that her ill-fame is so powerful that it will negatively affect future responses to women regardless of their behavior. It is easy to identify the negative expectancy Agamemnon's prediction effects. Additionally, we witness Agamemnon selectively interpreting the tale: he does not celebrate Odysseus' accomplishment in the slaughter, but emphasizes instead the part of the tale that resonates (or 'dissonates') most strongly with his own. Women will suffer suspicious rumor (*khalepên phêmin*) because of the hateful song told of Agamemnon's wife despite the availability of Penelope's positive tale. This too echoes cognitive understandings of narrative: Agamemnon, like Amphimedon and even Eupeithes, imposes an ending (or interpretation) on the story he hears. Like Agamemnon, we respond to narratives that resonate with what we already know. We apply paradigmatic tales even when they may not apply and this can distract us from seeing how the pattern might not apply.

The forced end of the *Odyssey* runs the risk of giving its audiences something unresolvable and unforgettable, of leaving us with *penthos alaston*, which disrupts our expectations. Such disruption produces, as discussed above, a cognitive dissonance—this dissonance can in turn provide a motivation to resolve it by seeking out a different explanation or integrating information into a new pattern.[29] Thus, the epic's ending turns into a moment of what we might call *aporia* or pathlessness in a Platonic dialogue. In recent work, Laura Candiotto (2015) argues that emotion and reason collaborate cognitively in the aporetic state. Candiotto draws on cognitive studies that examine "epistemic emotions" to argue that the aporetic state entails a "transformative process that allows us to find, within negativity itself, the key to imagine an otherness" (242). Further, she suggests that the emotional shame and the cognitive field required to achieve this state are dependent upon group intellectual and emotional work, what some have called the extended mind and the extended emotions.[30] From the perspective of performance, we would here understand the instrumental nature of the audience in shaping its own and each other's perception of and response to the narrative. Except, in this mutual reshaping, the audience itself is remade as well.

The *Odyssey*'s final book begins with concerns about the reception of narrative and reflects upon the difficult challenges of bringing a tale to its end. The conversations of the dead also prompt audiences to think about what kind of tale this is: Achilles received immortal fame for dying in war; Agamemnon became part of a negative homecoming paradigm with his wife. The open-end remains for Odysseus' own story. By retelling Odysseus' *killing of the suitors* within parenthetical considerations for different types of fame, the epic makes clear that it is really Odysseus' story that is at issue here and, further, that the final details added in this story transform what kind of story it has become. The end of the *Odyssey* acknowledges, then, the paradigmatic forces of audience expectation and cognitive dissonance exerted on its closure which motivate the compulsion to bring the story to an end.

A longer exploration of this theme would include a fuller explication of the *eklêsis*, the suddenly *imposed* end of the *Odyssey*. But this chapter's comments have helped to frame the effect of such immediate and 'false' closure on its audience(s). If the epic leaves us with *unforgettable* grief, it comes from the pain of not having our questions answered and knowing they likely will not be. These moves are critical of a simplistic and overly paradigmatic use of myth and they echo both what Aristotle says about the importance of predictability and surprise in effective narratives and what modern science has told us about the human mind: that we desire clear causality and closure so much that we will fabricate it if necessary. The grief that the epic's players experience at not knowing Odysseus' fate echoes the real-life pain of not knowing a loved one's fate, not understanding how to live (or act) after a momentous event, or the anxiety of an unknown death that awaits everyone. Such abruptness and lack of closure play upon our reflexive desire to know a story's end and demand that its audiences consider what *other* kinds of endings might be possible.

Notes

1 Versions of this chapter were presented at the 2016 CAMWS Annual Meeting in Williamsburg, VA, New York University, and Harvard University. Much gratitude is due to helpful comments from audiences there and to Peter Meineck for insightful and helpful comments on an earlier draft.
2 For overviews of Homeric psychology, see Harrison 1960; Russo and Simon 1968; and Russo 2012.
3 Palmer (2010) calls "intermental thought" a type of "extended cognition or intersubjectivity" that characterizes the dynamic relationship between external and internal functions of minds (39–41). For a theory of mind related to this, see Zunshine 2006, 6–8.
4 For "folk psychology", cf. Bruner 1986, 6; and White 2007, 102–106.
5 For the evolutionary development of the human capacity for narrative, see Gottschalk 2012, 26–31; cf. Dennet 2017, 177–204; cf. Logan 2007, 41–58 for the emergence of language and the theory of extended mind. Churchland (2013, 204–205) argues against the proposal that language makes consciousness possible.
6 For a recent analysis of Aristotle's plots, see Meineck 2017, 30–51. Meineck draws on the theory of predictive processing which proposes that our brains are always predicting outcomes for given

circumstances based on prior models or patterns. Surprise creates a 'sensory error' that sharpens and unsettles cognition. Cf. Clark 2015.

7 See Nussbaum 1986, 390–391; contra Nuttall 1996, 10–16.

8 See also Clark 2013 and the integration of his theories in Meineck 2017.

9 For the importance of human language in the development of consciousness, see above, note 5. There is some debate about the extent to which cognitive function and narrative can be universalized. For an overview of recent debates and a nuanced presentation of the relationship between core cognitive operations and cultural variations, see Senzaki et al. 2014. Cf. Kaplan et al. 2017.

10 Cf. Le Hunte and Golembiewski 2014, 75: "Thanks to storytelling, evolution can take place in a single lifetime. You don't need to die of thirst to realize that going into the desert without water is a bad idea."

11 See Solomon, Greenberg, and Pyszczynski 2015, 80–82 for the connection between mortal anxiety and both cultural discourse and individual narratives.

12 In addition to these signs of closure, the promise of an inland journey foretold by Teiresias' prophecy in book 11 adds another layer of indeterminacy to how the epic ends. On this see especially Peradotto 1990, 75–78 and Purves 2010, 77–89.

13 Schol. in *Od.* 23.296 HMQ list this as "the end [*péras*] of the *Odyssey*" whereas Schol. in *Od.* 23.296 M.V. Vind. 133 attest this as "the end [*télos*] of the *Odyssey*." For an overview of the epic's end, see Bertman 1968; Moulton 1974; Wender 1978.

14 All translations are my own.

15 See also discus-throwing (4.624 and 17.167–169), watching the beggar Iros fight Odysseus (18.36–39), banqueting (17.604–606), and revelry (18.304–306).

16 For Zeus' lines as programmatic see Adkins 1960, 19–20; cf. Marks 2008, 22–23. Contra Van der Valk 1949, 243; Clay 1983.

17 For character belief about Odysseus' death, see Barker and Christensen 2015, 94–95; for Telemachus' desire, see Murnaghan 2002.

18 For a short analysis of the adjective *alaston*, see Barker and Christensen 2015, 94–96. Cf. Loraux 2006 (cited below, note 19).

19 Cf. Slatkin 2011, 95–96 for Thetis; cf. Marks 2008, 67–68. ἄλαστος: *alastos*, likely from the root **lath–*, 'escape memory'; cf. Gk. λανθάνω ('escape notice'); ἀληθής ('true'); λήθη ('forgetfulness'). It is also realted to ἀλάστωρ (*alástôr*, 'avenger') as 'one who does not forget'. See Chaintraine s.v. ἀλάστωρ.

20 See Loraux 2006, 156–161 for a discussion of the poetics of "mourning that cannot be forgotten".

21 See above, note 12.

22 Zisook and Shear 2009, 69. They go on to describe a "maladaptive" excess of avoidance and obsession leading to social isolation.

23 See Boss 1999, 129; Zisook and Shear 2009, 70–71 for "complicated grief treatment" which "combines cognitive behavioral techniques with aspects of interpersonal psycho-therapy and motivational interviewing". Cf. Shear et al. 2016.

24 For the differences, see Lowenstam 2000.

25 See Lowenstam 2000, 339–341.

26 On weaving and female fame, cf. Mueller 2010. Murnaghan 1987, 95–96: Penelope is also a weaver of plots. On weaving in the *Odyssey* and *mêtis*, see Slatkin 2011, 234–237; Clayton 2004, *passim*.

27 Cf. Austin 1975, 253; Peradotto 1990, 83–84. For the possibility that in other traditions of Odysseus' return home Laertes and Penelope were colluding, see Haller 2013.

28 See, again, Clayton 2004, 38.

29 See Harmon Jones et al. 2015 for the relationship between cognitive discrepancy and dissonance reduction.

30 For a succinct articulation extended mind theory—which posits that other people and the environment function as an essential part of the functioning of human minds—see Clark and Chalmers 1998. Cf. the longer exploration in Logan 2007. For intermental thought and the importance of 'social minds' for understanding how fiction works, see Palmer 2010, 240–245.

References

Adkins, A. W. H. 1960. *Merit and Responsibility: A Study in Greek Values*. Oxford.

Austin, Norman. 1975. *Archery at the Dark of the Moon: Poetic Problems in Homer's* Odyssey. Berkeley, CA.

Bakker, Egbert. 2013. *The Meaning of Meat and the Structure of the* Odyssey. Cambridge.

Barker, Elton T. E. and Christensen, Joel P. 2015. "Odysseus's Nostos and the *Odyssey*'s Nostoi." In G. Scafoglio, *Studies on the Epic Cycle*. Rome: 85–110.
Bertman, Stephen S. 1968. "Structural Symmetry at the End of the *Odyssey*." *GRBS* 9: 115–223.
Boss, Pauline. 1999. *Ambiguous Loss: Learning to Live with Unresolved Grief.* Cambridge.
Brooks, Peter. 1992. *Reading for the Plot: Design and Intention in Narrative.* Cambridge.
Bruner, Jerome. 1986. *Actual Minds, Possible Worlds.* Cambridge.
Cairns, Douglas. 2016. "Mind, Body, and Metaphor in Ancient Greek Concepts of Emotion." *L'Atelier du centre de recherche historique* 16: 1–18.
Candiotto, Laura. 2015. "Aporetic State and Extended Emotions: The Shameful Recognition of Contradictions in the Socratic Elenchus." *Ethics and Politics* 17: 233–248.
Churchland, Patricia. 2013. *Touching a Nerve: Our Brains, Our Selves.* New York.
Clark, Andy. 2013. "Whatever Next? Predictive Brains, Situated Agents, and the Future of Cognitive Science." *Behavioral and Brain Sciences* 36: 181–204.
Clark, Andy. 2015. "Embodied Prediction." *Open MIND* 7: 1–21.
Clark, Andy and Chalmers, David J. 1998. "The Extended Mind." *Analysis* 58: 7–19.
Clay, Jenny Strauss. 1983. *The Wrath of Athena: Gods and Men in the* Odyssey. Princeton, NJ.
Clayton, Barbara. 2004. *A Penelopean Poetics: Reweaving the Feminine in Homer's* Odyssey. Lanham, MD.
Dennet, Dale C. 2017. *From Bacteria to Bach and Back: The Evolution of Minds.* New York.
Fernyhough, Charles. 2012. *Pieces of Light.* New York.
Fowler, Don. 1989. "First Thoughts on Closure: Problems and Prospects." *MD* 22: 75–122.
Fowler, Don. 1997. "Second Thoughts on Closure." In D. H. Roberts, F. M. Dunn, and D. Fowler (eds.) *Classical Closure: Reading the End in Greek and Latin Literature.* Princeton, NJ: 3–22.
Gazzaniga, Michael S. and LeDoux, Joseph. 1978. *The Integrated Mind.* New York.
Gottschall, Jonathan. 2012. *The Storytelling Animal: How Stories Make Us Human.* Boston, MA.
Hall, Charles A. et al. 2014. "Cognitive Functioning in Complicated Grief." *Journal of Psychiatric Research* 58: 20–25.
Haller, Benjamin. 2013. "Dolios in *Odyssey* 4 and 24: Penelope's Plotting and Alternative Versions of Odysseus' *Nostos*." *TAPA* 143: 263–292.
Harmon-Jones, Eddie, Harmon-Jones, Cindy and Levy, Nicholas. 2015. "An Action-Based Model of Cognitive-Dissonance Processes." *Current Directions in Psychological Science* 24: 184–189.
Harrison, E. L. 1960. "Notes on Homeric Psychology." *Phoenix* 14: 63–80.
Hutto, David. 2007. *Folk Psychological Narratives.* Cambridge.
Kaplan, Jonas T. et al. 2017. "Processing Narratives Concerning Protected Values: A Cross-Cultural Investigation of Neural Correlates." *Cerebral Cortex* 27.2: 1428–1438.
LeDoux, Joseph. 2002. *The Synaptic Self: How Our Brains Become Who We Are.* New York.
LeDoux, Joseph. 2015. "Feelings: What Are They and How Does the Brain Make Them." *Daedalus* 144.
Le Hunte, Bern and Golembiewski, Jan A. 2014. "Stories Have the Power to Save Us: A Neurological Framework for the Imperative to Tell Stories." *Arts and Social Sciences Journal* 5.2: 73–76.
Logan, Robert K. 2007. *The Extended Mind: The Emergence of Language, the Human Mind and Culture.* Toronto.
Loraux, Nicole. 2006. *The Divided City: On Memory and Forgetting in Ancient Athens.* Corinne Pache and Jeff Fort, trans. New York.
Marks, Jim. 2008. *Zeus in the* Odyssey. Washington, D.C.
Meineck, Peter. 2017. *Theatrocracy: Greek Drama, Cognition, and the Imperative for Theatre.* Routledge.
Minchin, Elizabeth. 2001. *Homer and the Resources of Memory: Some Applications of Cognitive Theory to the Iliad and the* Odyssey. Oxford.
Minchin, Elizabeth. 2007. *Homeric Voices: Discourse, Memory, Gender.* Oxford.
Moulton, Carroll. 1974. "The End of the *Odyssey*." *GRBS* 15: 153–169.
Mueller, Melissa. 2010. "Helen's Hands: Weaving for Kleos in the *Odyssey*." *Helios* 37: 1–21.
Murnaghan, Sheila. 1987. *Disguise and Recognition in the* Odyssey. Princeton, NJ.
Murnaghan, Sheila. 2002. "The Trials of Telemachus: Who Was the *Odyssey* Meant For?" *Arethusa* 35: 133–153.
Nussbaum, Martha. 1986. *The Fragility of Goodness: Luck and Ethics in Greek Tragedy and Philosophy.* Cambridge.
Palmer, Alan. 2010. *Social Minds in the Novel.* Columbus, OH.
Peradotto, John. 1990. *Man in the Middle Voice: Name and Narration in the* Odyssey. Princeton, NJ.
Purves, A. C. 2010. *Space and Time in Ancient Greek Narrative.* Cambridge.

Russo, Joseph. 2012. "Re-thinking Homeric Psychology: Snell, Dodds and their Critics." *QUCC* 101: 11–28.
Russo, Joseph and Bennett, Simon. 1968. "Homeric Psychology and the Oral Epic Tradition." *Journal of the History of Ideas* 28: 43–58.
Senzaki, Sawa, Masuda, Takahiko, and Ishii, Keiko. 2014. "When Is Perception Top-Down and When Is It Not? Culture, Narrative, and Attention." *Cognitive Science* 38.7: 1493–1506.
Shear, M. K., Reynolds, C. F., Simon, N. M., Zisook, S., Wang, Y., Mauro, C., Duan, N., Lebowitz, B., and Skritskaya N. 2016. "Optimizing Treatment of Complicated Grief: A Randomized Clinical Trial." *JAMA Psychiatry* 73: 685–694.
Slatkin, Laura. 2011. "Genre and Generation in the Odyssey." In *The Power of Thetis and Selected Essays. Hellenic studies 16.* Washington, D.C.: 157–166.
Solomon, Sheldon, Greenberg, Jeff and Pyszczynski, Tom. 2015. *The Worm at the Core: On the Role of Death in Life*. London.
Stanford, W. B. 1965. "The Ending of the *Odyssey*: An Ethical Approach." *Hermathena* 100: 5–20.
Turner, Mark. 1996. *The Literary Mind: The Origins of Thought and Language*. Oxford.
Van Der Valk, Marchinus. 1949. *Textual Criticism of the Odyssey*. Leiden.
Wender, Dorothea. 1978. *The Last Scenes of the Odyssey*. Leiden.
White, Michael. 2007. *Maps of Narrative Practice*. New York.
Willcock, M. M. 1964. "Mythological Paradeigmata in the *Iliad*." *CQ* 14: 141–151.
Wilson, Donna F. 2002. *Ransom, Revenge, and Heroic Identity in the* Iliad. Cambridge.
Wilson, Edmund. 2014. "On Free Will and How the Brain Is Like a Colony of Ants." *Harper's* September 2014, 49–52.
Zisook, S. and Shear, K. 2009. "Grief and Bereavement: What Psychiatrists Need to Know." *World Psychiatry* 8: 67–74.
Zlatev, Jordan et al. 2008. *The Shared Mind: Perspectives on Intersubjectivity*. New York.
Zunshine, Lisa. 2006. *Why We Read Fiction: Theory of Mind and the Novel*. Columbus, OH.

9

"I'll imitate Helen"!

Troubling text-worlds and schemas in Aristophanes' *Thesmophoriazusae*

Antonis Tsakmakis

This chapter deals with one of the earliest passages in western literature which problematizes the consciousness of the multifaceted relationship between reality and fictionality: the parody of Euripides' *Helen* in Aristophanes' *Thesmophoriazusae* (846–928). In this parody two remarkable features produce comic effects: the abusive treatment of the Euripidean original (a standard practice in literary parody), but also the failure of the internal audience to acknowledge it. Critylla, the woman who is on stage during the performance, cannot apprehend the world of the performed tragedy in its own right and distinguish it from her own comic world. In consequence, the performance gives rise to a series of misunderstandings which not only expose several conventions pertaining to fiction and role-playing, but also call for a study of the cognitive responses to phenomena such as drama, literature and genre.

Scholarship on *Thesmophoriazusae* has especially focused on questions of gender and identity, as well as on the blurring of genre boundaries, themes that are of salient importance in the comedy.[1] From the perspective of cognitive poetics, these aspects can be fruitfully discussed as examples of conceptual blending on the grounds of the theory of mental spaces.[2] In this chapter the discussion of particular aspects of the *Helen*-scene will be primarily based in Text World Theory, which is compatible with blending theory and provides a tool for a more detailed analysis of textual elements that are employed in the scene and their effects on the construction of different conceptual frames.[3]

Worlds in the mind: Text World Theory

Text World Theory was introduced by Werth (1994, 1999) and further developed by Gavins (2007). It seeks to understand 'the precise structure and cognitive effects of individual mental representations'[4] that are triggered by discourse. Text World Theory hypothesizes the existence of mental representations of 'worlds' as they become manifest in discourse. Text-worlds are communication events, and as such they include the participants of the communication and their discourse; besides, the context which is suggested as relevant for the discourse to become fully intelligible is also part of a text-world. In the words of Peter Stockwell, '[a] world is a language event involving at least two participants, and is the rich and densely textured real-life representation of the combination of text and context'.[5]

A text-world is constructed in analogy to the real world, as the real world informs our experience of life and communication, but each text-world remains a language event, i.e. it is by no means bound to a particular reality external to the mind. It is only discourse that stimulates the building of text-worlds in the mind. Text World Theory distinguishes between the discourse-world, at the primary level of communication (a world 'prototypically involving face-to-face discourse participants, such as two speakers in a conversation, or a letter-writer and receiver, or an author and reader'[6]) and text-worlds embedded within the discourse ('fictional spaces distinct from the discourse-world',[7] distinguished from it by an ontological divide). The discourse-world consists of the representation of the event of communication itself, and therefore it is a mental and textual representation of the real-life experience of being engaged in a communicative exchange. On the other hand, embedded text-worlds (Werth had used the term sub-worlds which was later abandoned) can be representations of past events in the mind, as well as any other type of knowledge and belief, conceptual structures triggered by modalized propositions dealing with non-real situations (probabilities, possibilities, improbabilities and impossibilities), hypothetical or imaginary assumptions etc.

In every communication common-ground assumptions are constantly retrieved, and therefore become part of the discourse-world: contextual factors (perceptions, previous knowledge, expectations, culturally determined behavioral rules and acknowledgments, ideological assumptions etc.) become relevant or are negotiated in the course of the communication event. According to Gavins, Text World Theory provides a framework within which to analyze discourse which 'is fully sensitive to all the situational, social, historical and psychological factors which play a crucial role in our cognition of language'.[8] The relevance of contextual knowledge depends wholly on the text; hence Text World Theory resides on the principle of text-drivenness.[9]

A thorough adaptation of Text World Theory for the analysis of dramatic performance is still desired.[10] Nevertheless, a text-based approach of dramatic texts is undoubtedly possible. While it mainly consists of the analysis of discourse, it can also take into consideration those elements of the performance which can be inferred from the text. Students of Greek and Roman drama are familiar with the limitations of such an enterprise, which does not enable us to come closer to the visual (and other) experience of ancient audiences; given these insuperable obstacles, Text World Theory nonetheless provides a tool which can be helpful in revealing some subtleties of dramatic texts and improving our appreciation of their art.

Waiting for Euripides

Let us now turn to the *Thesmophoriazusae*. In this play, the women of Athens are plotting against Euripides because he draws, as they believe, a negative image of them in his tragedies. According to the comic plot, they use the opportunity of their gathering for the celebration of the Thesmophoria – a women's festival – in order to condemn the tragedian. A kinsman of Euripides volunteers to defend him. Disguised as a woman, he infiltrates the festival and addresses the female assembly in a speech which irritates the audience. After the revelation of his sex, he is put under arrest. The incident has been reported to the authorities and a request to appear has been extended to the *prytanis*. The kinsman seeks Euripides' help, hoping that the poet will keep to his word of honor and appear to his aid. A first attempt to induce his intervention, inspired from Euripides' tragedy *Palamedes* of 415 BC,[11] has failed, and now, after the Parabasis (785–845), the kinsman and his guard Critylla are on stage. The hero still wears female garments and reflects on his desperate situation:

ἰλλὸς γεγένημαι προσδοκῶν· ὁ δ᾽ οὐδέπω.
τί δῆτ᾽ ἂν εἴη τοὐμποδών; οὐκ ἔσθ᾽ ὅπως
οὐ τὸν Παλαμήδην ψυχρὸν ὄντ᾽ αἰσχύνεται.
τῷ δῆτ᾽ ἂν αὐτὸν προσαγαγοίμην δράματι;

I've gone cross-eyed with looking out for him, and still no sign of him.
What can be holding him up? It can only be
That he's ashamed of *Palamedes* because it was such a bore.
What play *can* I use to entice him here?[12]

(846–849)

The transitory passage (which also includes the two subsequent verses that announce the parody) raises a series of overarching issues that have to be dealt with at length. Therefore, we'll devote to it a detailed discussion, while the parody-scene will be later examined as a whole in a more comprehensive way.

The kinsman's discourse is part of the text-world which is constructed by *Thesmophoriazusae* (we can call this world the base text-world of the comedy, in order to emphasize the hierarchical distinction between this world, the primary level at which the comic plot evolves and further text-worlds constructed in the characters' discourse). Even though in drama the characters engage in direct verbal communication, so that their situation is very similar to a discourse-world, there is always a marked distance from the real world which inhibits any such impression. Discourse in drama is not processed in the same way as if spectators were eavesdropping on a real conversation. The characters are usually fictive persons in a different setting and time, and they retrieve background knowledge that is irrelevant to the situation of the spectators, and they almost always ignore what is going on in the world of the spectators. In classical theater, the physical position of a restricted acting area 'below' the spectators, as well as visual elements that highlighted 'otherness', such as the masks and costumes, together with the often 'remote' settings – both temporally and, usually, spatially – upheld the spectators' awareness that this is a different world than their own discourse-world. Verse and stylized diction also contributed to the same effect.[13]

These four lines (846–849) are not addressed to any particular addressee on stage. Spectators who are familiar with conventions pertaining to drama process them as a monologue. Contextual knowledge also suggests that monologues of this kind are usually employed to express the speaker's thoughts. Thus, we can assert that monologic discourse shapes a new text-world, while at the same time it re-enforces the spectator's feeling of immersion in a state of direct communication with the speaker as the latter's sole addressee (notwithstanding the fact that the gap that separates them can never be crossed). New elements can be introduced into it, which do not have any impact on the discourse-world, since, at first sight, the presence of Critylla can be ignored. The opening line ('I've gone cross-eyed with looking out for him') reveals the speaker's psychological state: he is upset at Euripides' failure to appear. This feeling is verbalized through a metaphorical expression which, taken literally, would direct the addressees' attention to the speaker's external appearance – to an eye-squint gesture. Observing such a gesture in the theater would be practically impossible for several reasons, such as the distance of the spectators from the acting area and the mask. In consequence, spectators are rather invited to *imagine* a facial expression, perhaps supported through the actor's gesture, which is interpreted as an external sign for a feeling of physical exhaustion and despair of the still-waiting man. As the present scene has been separated from the initial attempt to summon Euripides by the Parabasis, and a certain time interval has elapsed, spectators are supposed to be able to share with the kinsman the experience of waiting, whereas they do not share his emotive state. The second half-line 'and still no sign of him' brings a confirmation of these inferences; it creates a negative text-world, which is an alternative conceptualization of the base text-world.

The question 'What can be holding him up?' cues the construction of a new text-word with Euripides being at some other place, allegedly trying to approach, with an unspecified obstacle preventing him from appearing. As the speaker's monologue evolves, a hypothesis about this obstacle is articulated as a further embedded text-world consisting in a mental state attributed to Euripides: 'That he's ashamed of *Palamedes* because it was such a bore'. This joke is only intelligible if spectators relate it *not* to the comic character Euripides, but to his counterpart in the discourse-world.[14] The more the kinsman profiles himself as an expert on Euripidean tragedy, the nearer he comes to the discourse-world. Nevertheless, despite the mutual approach of hero and spectators triggered by textual elements, the ontological divide never ceases to exist;[15] on the contrary, it is a necessary condition for a comic effect to be generated. Once the divide disappears and audiences engage in serious discussion of the kinsman's opinions, the source of amusement vanishes; inversely, the divide guarantees that whatever is claimed within the fiction is not subject to refutation. On the other hand, a devastating judgment on Euripides' art is not a source of laughter for fictional characters. The comic effect is produced by the discrepancy between discourse that takes place on the level of the text-world and contextual knowledge that is retrieved while this discourse is processed by participants of the discourse-world.

The supposed reason for Euripides' feeling of shame, namely that his recent *Palamedes* was, from an aesthetic point of view, disappointing, expands on the spectators' knowledge of the real Euripides' work and/or career. We cannot specify the point of this verdict with precision (if there was any, and the kinsman is not trying to find an excuse for the failure of his initial attempt), because the play is lost, but there is a wide range of possibilities: the predicate may allude to specific features of the tragedy; or to possible reactions to the play (Euripides' trilogy was not victorious, but this, again, was not unusual); or make fun of the fashion of contemporary literary criticism; it can also reflect naïve spectators' embarrassment, caused by elements of the performance, especially novelties, that remained incomprehensible to them; at any rate, the judgment, which is comically attributed to Euripides himself, also exploits the inherently subjective character of aesthetic response to any form of art.[16] Whatever the case, the utterance does not necessarily imply Aristophanes' own opinion about the play.

The effect of the kinsman's statement, and more generally the broad question of whether and when Aristophanic satire is 'serious', can be approached through blending theory. The statement exploits the dynamism of an emergent structure which is the result of blending between two distinct mental spaces as inputs. The work of Euripides, especially its most outstanding and remarkable traits, is the first input space, and the distorted image of comic satire the second. As much as the satire can be unmistakably associated with real Euripides, the mapping of a new space is generated; it results in a new, blended representation of 'Euripides', which is neither the historical person nor the dramatic character; it is informed by both. According to Fauconnier and Turner, '[b]lending can compose elements from the input spaces to provide relations that do not exist in the separate inputs' (2002, 48). In the stage of completion, additional structure is brought to the blend. Euripides is the author of *Palamedes* but he is also in a way responsible for the failure of the kinsman's trick in the comedy. The typical structure of comedy and the expectations concerning the behavior of a comic character provide an additional frame. This frame is integrated in the blend in the stage of elaboration, which allows the blend to run dynamically: now, things can happen that are not present in the inputs. The women's charges are a characteristic example: the comic plot would be unintelligible without any knowledge of Euripidean tragedy; on the other hand, the motif of Euripides' misogyny is for the first time introduced in the *Thesmophoriazusae* as a fiction. The blend consists in presenting Euripides as having at least all the common characteristics of the real and the comic Euripides, but the new, blended image of Euripides acquires a life of its own. This is the way Aristophanes coined a comic version of Euripides and influenced subsequent views on the poet, including the later biographical tradition.

The next line (849), 'What play *can* I use to entice him here?', is a function-advancing proposition which constructs a modal text-world, as the speaker explores possibilities to satisfy his wish to attract the poet. The term 'play' (δράματι), however, alerts the spectators to the possible creation of a further text-world as it introduces the idea of applying a new, still unspecified, theatrical parody, following the parody of *Palamedes* shortly before. The diction of the passage (the reference to an 'obstacle' in 847 and the instrumental dative in 849) suggests that a practical, utilitarian logic governs the base text-world (the fact that not only the instrument but also the obstacle emanate from Euripides' art is a further source of fun for the spectators). This logic is inherently opposed to notions and expectations related to aesthetic experience and aesthetic values, which are prevalent in discourse about theater in the discourse-world, and in the blends which are partly based in this world as an input.

The imitation of the new *Helen* and some ambiguities

Contextual knowledge from the discourse-world will prove indispensable for the audience to process the announcement of the next rescue, too. In lines 850–851 the kinsman reveals his new plan of action: ἐγᾧδα· τὴν καινὴν Ἑλένην μιμήσομαι. πάντως ὑπάρχει μοι γυναικεία στολή ('I know; I'll act his new Helen. I've got the woman's costume already, anyway'). While the reference to the costume refers to the base text-world and draws attention to the kinsman's disguise, which is visually perceptible by the audience, the future tense expresses an intention, i.e. it creates a new text-world, which involves an image of the speaker 'imitating' a certain 'new Helen' in the very near future. The speaker's words are ambiguous in two points: Helen is both a Euripidean play and a person (a dramatic character and a mythical personage), while imitation refers both to behavior which copies the behavior of another person, and, as a technical term, to dramatic role-playing. According to the precise meaning we give to these terms, different text-worlds will be constructed. The key to solving these ambiguities is the highlighted function of the 'female dress', which is mentioned thereafter – it points toward role-playing. It follows by inference that Helen is a dramatic role, and imitation is playacting.[17] Costume, or more generally external appearance, is thus suggested as the foundation of theatrical performance. The visible object (the garment) is explicitly treated as a contact point between the two text-worlds. In consequence, it is construed as a blend: at the same time visible and symbolic. But it also has a pre-history, it has been there for a while. It proved useless within the base text-world after the hero's disguise had been revealed; now it becomes again an instrument which will enable him to assume a different identity. The double function of the female costume as an instrument of deceptive disguise and as a dramatic accessory further enriches the blend. Finally, discourse about costume as an instrument of change can only be processed against the background of what it conceals or changes – in this case, a male body. The consequences of this implicit evocation of the male body as a constant background for the conceptualization of drama in Greek antiquity are evident, as under every costume, male or female, a male body could be imagined and *Thesmophoriazusae* foregrounds this idea.[18]

The importance of imitation in the present context cannot be overstated, but the kinsman's words do not specify the precise character of the intended imitation. Earlier he had merely employed an idea and performed an action (a trick from Euripides' *Palamedes*, which, in the Euripidean play, was not performed on stage; it is comedy that visualized the reported action), but the customary meaning of *mimesis* in a theatrical context is the imitation of a person. In the prologue of the play, Agathon had exposed his theory of drama (149–156), and μίμησις played a key role in it. According to Agathon, in order to be creative, a poet has to assimilate himself to characters who have a different 'nature' (φύσις). This is attained through imitation, and disguise was an effective means to reach this goal. More specifically, cross-dressing allowed a male poet to write good plays about women.[19] So far mimesis concerns the imitation of a person's character and the way it is manifested. The verb μιμήσομαι in the kinsman's announcement evokes Agathon's theory of μίμησις for an additional reason: it was Agathon who had borrowed the female costume for the kinsman. A further dimension will be added to this understanding of imitation. As soon comes out, however, the imitation of a heroine takes place in the context of an entire play about her. The kinsman seeks a re-performance of a distorted and abridged version of Euripides' *Helen*. Spectators are invited to use their knowledge about theater and classify what they attend as a performance within a performance, a processing that requires the construction of blended spaces. On the level of the base text-world, an opposition between tragedy and comedy is avoided, unlike other Aristophanic plays that foreground it – even in the Parabasis. This is important: discourse on this level ignores generic distinctions; it concerns drama and performance as a whole.

Agathon's theory does not explicitly refer to any effects imitation can have on the mental state of an author who composes a drama about different characters. But poetic theories from the last decades of the fifth century BC to Aristotle assumed that the *ethos* of the figures who were presented in poetry or on stage had an impact on the *ethos* of the audience. On the other hand, Aristotle believed that the character of poetry depended on the *ethos* of the poet, a theory which is not very far from Agathon's expositions in *Thesmophoriazusae*. We can thus safely infer that from the last decades of the fifth century BC onward poetic theories assumed that ethical qualities could be transferred through poetic mimesis from poets to poetry, from poetry to (actors and) audiences.[20] Against the background of such views, Agathon's theory in Aristophanes' comedy appears as a superficial doctrine which undermines essentialist assumptions (the view which assigns priority to nature and advocates a steadiness of personality and the self), with naïve constructivism, which admits the possibility of character change to the degree that assimilation of foreign traits is an ordinary process.[21] In contrast to Agathon, who believes in the ability of costume and appearance to effect a change in a person's inner state, the kinsman's concerns are far more trivial. The purpose of disguise is to change the view that others maintain about a person.

There is a final ambiguity regarding the adjective which modifies the name of the play. Spectators knew that *Helen* has been produced recently, yet the meaning of the adjective καινήν ('new') in the present context is far from clear. According to the principle of minimal departure (see above n. 14), it has to be related to the discourse-world, but it can be understood in a twofold way. At first sight its meaning is temporal: it specifies the play as a recent one. Yet, as the parody itself later suggests, another possibility exists as well, namely that the play was 'innovative'. In Aristophanic comedy the adjective is usually understood in a positive sense, but the rules of tragedy are different. Again, the kinsman's words trigger a blended image of Euripides with wide-ranging consequences.[22]

Critylla intervenes in a way that makes evident that, against our expectations, she *did* listen to the kinsman's monologue. This was the woman's first direct address to the captive.

> τί αὖ σὺ κυρκανᾷς; τί κοικύλλεις ἔχων;
> πικρὰν Ἑλένην ὄψει τάχ', εἰ μὴ κοσμίως
> ἕξεις, ἕως ἂν τῶν πρυτάνεών τις φανῇ.

> What are you cooking up now? What do you keep staring this way and that for?
> I'll give you a Helen or two in a minute, if you don't behave yourself properly
> until one of the Prytaneis turns up.
>
> *(851–854)*

Critylla is clearly unhappy with her own duties as a guard and looks forward to the arrival of a state official – the verbalization of the expectation causes a world-switch, an implicit threat for the kinsman, as the arrival of a prytanis would put an end to any hope for fleeing. Critylla is interested in keeping the captive under discipline until then, and the time interval up to this moment is conceptualized mentally in a hypothetical, negative text-world. The apodosis consists in a threat expressed by a visual metaphor which makes an ironic play with the name Helen. The metaphor relies on a familiar idiom of the Greek language (the ironical repetition of a noun used by a previous speaker, with an inappropriate, unexpected modifier, in a way that suggests an opposite meaning). Thus, a 'bitter Helen' as the object of the kinsman's perception becomes a metaphor for an unpleasant experience, a punishment; this is one more embedded text-world which is created in Critylla's discourse. In this text-world, however, Critylla's words acquire a meaning which extends beyond her own intentions. For spectators who import contextual knowledge from the discourse-world, apart from linguistic irony, the polysemy of the name Helen amplifies the function-advancing properties of the proposition. As a consequence, Helen matters, regardless of which future projections of the text-world will eventually prevail. As it turns out, the performance of *Helen* will be not completed and proves ineffective.

On the whole, visual elements abound in the three lines spoken by the woman. Her first question uses a visual metaphor;[23] the second refers directly to the kinsman's appearance (as it deals with the way he is looking around, it corresponds to his own reference to his 'squinted' eyes); then the woman uses the metaphor of seeing to denote her confidence in what she is expecting as the certain outcome of the kinsman's attempt (for her, this metaphorical image expresses a meaning that is much more true than any real but deceptive spectacle); finally, the impending arrival of a prytanis is verbalized as 'appearance', 'coming into sight'.

A further characteristic of the woman's discourse is that she is absent from the constructed text-worlds; she never claims a role as an agent in them. Similarly, she will avoid foregrounding herself in her discourse throughout the following scene.[24] Perhaps this style is in alignment with expectations of women's discourse in the public sphere in classical Greek society. Be that as it may, it also foreshadows her primary role of addressee rather than the enactor involved in the embedded text-world during the following performance.

Regarding the content of the exchange, it is not clear that the kinsman's words made sense for Critylla. The precise nature of the text-world he is constructing seems to remain obscure for her. In her reaction she picks up the world-builder (Helen), but refers only vaguely to the instantiation of the text-world. At any rate, her response to the name Helen does not indicate that she associates it with a mythological or dramatic figure. This impression will be corroborated by her interventions in the course of the parody. Thus, the spectators can infer the existence of a knowledge gap between her and the kinsman, and between her and themselves. While the kinsman imports more context than expected – and even when this is not expected – Critylla seems to lack the experiences which would allow her to do so.

The parody

From 855 the kinsman embarks on his performance. He first speaks the opening three lines of the Euripidean play, following the original faithfully till the middle of the third line, where the text is comically distorted:

> Νείλου μὲν αἵδε καλλιπάρθενοι ῥοαί,
> ὃς ἀντὶ δίας ψακάδος Αἰγύπτου πέδον
> λευκῆς νοτίζει μελανοσυρμαῖον λεών.

> This is the beauteous maiden stream of Nile,
> Who takes the place of heaven's showers, and waters
> Egypt's white plains and swarthy laxative-takers.
> *(855–857)*

The original verses established in the tragedy the physical setting of the fictive text-world in Egypt. In Text World Theory all elements which identify the spatial setting of the action and the participants involved in it are world-building elements, i.e. they have a key function in the construction of a text-world.[25] Despite Critylla's interventions, the kinsman will go on reproducing more or less faithfully lines from the Euripidean prologue in which further world-builders will be accumulated such as: 'I come from Sparta, my father is Tyndareos' (859–860); 'My name is Helen' (862); 'I was the cause for the Trojan War' (864–865); 'I am now in Egypt. My husband is Menelaus' (866–867). World-building elements also include expressions about a relation between persons; such relations, especially family relations, also play a cardinal role in the passage.[26]

All the kinsman's claims are contested or directly refuted by Critylla,[27] who, however, does not seem to treat them as a coherent world-building project. She responds to every statement 'locally'; for her, the kinsman is a villain who lies, is mistaken, or is deceiving through cross-dressing. Still, she relates everything to her own text-world; she cannot acknowledge that the kinsman is playing theater and that his utterances presuppose a different context than she assumes.[28] In her comic text-world there is no tragedy and no theater, and she naturally judges the kinsman by these standards. Her theatrical inexperience is emphasized by the fact that she is not entirely enclosed in the base text-world of the comic fiction. Occasionally, she imports contextual information from the discourse-world (she is supposed to know the kinsman's father by name, presumably a real Athenian, perhaps the father of an existing relative of Euripides, or – a quite fascinating hypothesis – the actor's father). But this emblematic ability of comedy to access the context of the discourse-world contrasts sharply with the woman's ignorance of theater. On the other hand, her naïve reaction foregrounds the absurdity of the performance. Spectators realize both her right to remain stable in the context of her comic world and the kinsman's right to exploit the real world. Nevertheless, his treatment of theater is problematic. The tragic original is being heavily abused. Selections, distortions, abridgement, modification and mixture with comic diction at first sight reduce it merely to a source of world-builders, carefully selected for a practical purpose. Presumably, however, the acting style, the gesture and the voice of the kinsman already created the effect of a parody of tragic performance. For the spectators, these elements sufficed to evoke the contextual knowledge of theater, tragedy, Euripides' plays and, more specifically, *Helen*. Thus, in the mind of the spectators the kinsman's performance was progressively framed in conceptual structures already available to them; this process results in a blend: unlike Critylla, the spectators progressively develop the impression that they attend a performance which is embedded in the comedy. The kinsman as Helen belongs to a new text-world.

The establishment of this text-world is secured when function-advancing elements are introduced, i.e. when actions and events take place within the text-world.[29] By the end of the prologue, a function-advancing element is contained a negative text-world which is constructed by the kinsman's discourse: ('but my unhappy husband, my Menelaus, still he comes not hither', 866–867). A new function-advancing process is inferred from the modal text-world constructed by an indirect wish of the speaker to put an end to her life ('Why then do I yet live?', 868), to be immediately replaced by an expression of hope and a prayer (869–870). Finally, Euripides enters using tragic lines from *Helen*. His appearance adds to the performance a hitherto missing element: dialogue. This development is crucial for the stabilization of the conceptual structure of theatrical performance. In addition, it justifies optimism, given that the *Palamedes* failed because

Euripides did not appear. For Critylla who attends the first exchanges between the complotting actors, there is still space to negotiate a common text-world with the stranger, but she soon gives up, when he realizes that he is also 'insane'. The dispute is also about visible objects, such as the scene-building, which in the base text-world is the Thesmophorion and in *Helen* becomes a palace, or the altar in the orchestra, which in *Helen* is called a tomb. A reference to the kinsman's costume (890) also recalls by association the dispute about the kinsman's gender and identity. Material objects cut across text-worlds and are used as an implicit reminder of the divide between conceptual spaces and physical space.

As the negotiation evolves, Critylla is directly invited to participate in the performance. A role is explicitly offered to her when the kinsman introduces her to Euripides as Theonoe – 'This is Theonoe, daughter of Proteus' (897), but she refuses to access the new conceptual structure by assuming this name. Instead, she insists on attempting to frame every utterance within the context of the base text-world. As she is not familiar with theater, she cannot construe the text-world of theater as a blended world, one that would allow her to maintain her identity in her world, while participating in a different role.[30] In principle she is right, because the two men intended to attract her completely into the performance, in order to switch off her adherence to the base text-world. This is in contrast to what happens with audience participation in modern immersive theater. In this scene, theater threatens to become a substitute for reality – therefore, she steadfastly keeps her identity, as she also refuses to acknowledge the alleged (tragic) identity of the two male characters. She is equally unwilling to participate in their fiction and to behave like a spectator. She does not even open herself to the play, by acknowledging that the two men's discourse may have sense in a way she still has to discover. She neither has the knowledge that might enable her to accommodate the kinsman's discourse against the contextual framework of a theatrical performance, nor is she able to process it in a creative way. This might have led her to a new construal, a conceptual structure that would fit the requirements of this play in a coherent way, a 'learning'-process, like that of a spectator who is for the first time exposed to the experience of theater.

For their part, the spectators may have realized that Euripides uses phrases that in Euripides' tragedy were attributed to Teucros, the first stranger in *Helen* to arrive in Egypt.[31] Such discrepancies only confirm the power of the conceptual frame. Although the performed play, strictly speaking, is not Euripides' *Helen*, in a sense it *is*. The identity of the play is a conceptual construct which is not harmed by individual deviations. The frame is not a sum of particular traits; once a mental construct has been established as relevant context, no single discrepancies can delete it. Against this background, however, some problematic aspects of the performance become visible. The ceaseless interruptions by Critylla not only disrupt the integrity of the performance, but also highlight the lack of an audience on the level of the base text-world, and this indicates that the two men are not interested in the performance *per se*. Consequently, they don't need and they don't respect the entire contextual framework they borrow from the discourse-world. The fragmented and shattered text they recite now appears as a pretext, very much like the female costume which could not any more convince anybody that the kinsman is not a male. They don't need the conventions of drama; they only need the Euripidean play as a convenient conceptual structure which makes the fulfillment of their aims possible. The two men act for a purpose which has nothing to do with the embedded text-world of their performance. They are not really interested in keeping the different text-worlds apart. On the contrary, their goal is practical and egoistic and their whole performance is a function-advancing device on the level of the base text-world. The excessive presence of world-builders in the first part of the parody contrasts with the obvious priority of the intended outcome, the kinsman's escape. Euripides' play and whatever is metonymically associated with it proves empty and meaningless; the play is just a routine that has to be hurried through, with no significance for its own sake. In this sense, Critylla's rejection of it seems justified.

After Critylla has revealed her name and demos of origin (899), a fact that suggests her own claim to import contextual elements from the discourse-world, she remains silent for a while and the embedded play seems to rapidly evolve toward the intended finale. At the peak of the dramatic plot, a recognition-scene is enacted. It becomes clear that tragedy is reduced to a sequence of actions and events that leads to an already known end. Helen and, metonymically, tragedy are reduced to a standardized plot structure. The two men enact the complete sequence to conjure the desired finale. However, the performance has to be terminated too early, as the arrival of a state official and of a Skythian Archer (responsible for public order in classical Athens) is announced (922–923). Euripides flees and the kinsman remains helpless at the disposal of his opponents. Euripidean tragedy has once more proved useless for him.

Escaping by playing theater is no doubt an unusual idea. If there is a schema of escaping – a familiar conceptual structure which consists of a set of interrelated actions and situations (involving an initial state of resignation with the conditions of detention, communication with potential assistants, taking notice of the conditions of custody, identification of weaknesses, precautions, acquisition of some necessary tools, keeping them in a hidden place, deciding about the right moment, and finally performing the necessary acts to materialize the plan, evading the attention of opponents, running away etc.) – then staging Euripides certainly is not part of this schema. By assigning to Euripidean tragedy a place in the escape-schema, or more generally to a problem-solving situation, the kinsman does not make any significant contribution to the spectators' conceptualization of real-life situations. But he certainly draws attention to the innovative and unconventional elements of some recent Euripidean tragedies, suggesting the idea of raising the plot structure of *Helen* to the status of a literary schema.

In parallel, the absurdity of replacing a world schema by a literary schema may have pointed to the absurdity of literary theories that trivialized poetry by trying to establish its practical value. In the *Frogs* Aristophanes will satirize such theories, which suggests their expansion in the last decades of the fifth century BC. A significant contribution of Aristophanes' parody of *Helen* is that it highlights the importance of contextual frames in communication, literature and drama. The kinsman's attempt to perform a tragedy in *Thesmophoriazusae* and Critylla's exposure of it as nonsense makes a valid point against any attempt to implant fragments of a text-world created in discourse into an irrelevant context.

Concluding remarks

The parody of Euripides' *Helen* in Aristophanes' *Thesmophoriazusae* raises a series of questions that touch upon important aspects of drama, literature and literary theory: the role played by knowledge and experience for the appreciation of literary works and dramatic performance; the status and validity of genre conventions; the impact of art on reality and, inversely, its dependence on reality; and so on. Old Comedy incessantly blurs the boundaries between fiction and reality, and, hence, prompts the audience to retrieve a large amount of knowledge and experience from outside the text in order to adequately process it. *Thesmophoriazusae* not only presents artful examples of these features but also cues self-conscious confrontation with relevant questions from a theoretical point of view. Text World Theory with its sensitivity for context proved an outstandingly fruitful framework for a discussion of the subtleties which produce the comic effect of the scene under examination and illuminate its importance for Aristophanic poetics.

The construction of embedded text-worlds in this scene from the second part of the comedy is exploited as a source of misunderstandings between comic characters. Situated between scenes of direct confrontation and physical violence, the parodies of *Palamedes* and *Helen* are attempts to apply literary experience in order to solve practical problems. While the first scene did not involve any participants apart from the comic hero who performed a trick inspired by Euripidean tragedy, the second consists in a performance of a condensed and distorted version of an entire Euripidean play before a female spectator. The female spectator's lack of contextual knowledge leads to a series of conflicts that could be analyzed in terms of divergent cognitive responses to drama and literature. The characters of *Thesmophoriazusae* disagree on the fundamental pre-requisites of theater: on the conditions which make a text-world appear as a discourse-world. It is through this mechanism that the play prompts the spectators' awareness of theatrical convention and prevents them from being 'lost' in either text-world.

Thesmophoriazusae explores the ontological divide which separates the text-world from the discourse-world and does not allow us to take statements made in comedy at face value and treat comedy as 'serious'. It is the same divide that calls for a rejection of poetic theories and literary critics that seek to distill from literature practical lessons for real life. The effort to make tragedy 'useful' in the scene we discussed (perhaps in alignment with contemporary views about literature) leads to a comic debacle which is illustrated on the comic stage.

If, however, comedy exposes and rebuts attempts to establish direct links between the discourse-world and text-worlds constructed in theatrical performance, the study of cognitive processing of the text-world as it is triggered by Aristophanes' text may provide important indications about the addressees of Old Comedy. Critylla's complete ignorance of Euripides' tragedy and dramatic performance makes better sense if women had no first-hand experience of dramatic performances. This estimation fits in well with the central theme of the play, namely that women condemn Euripides based on hearsay from their husbands who are coming home from the theater, and even adds evidence to the much-debated question of whether most Athenian women attended the theater.[32]

Notes

1 Cf. Zeitlin (1981); Duncan (2006) 32–57; Platter (2007) 143–175.
2 See Fauconnier and Turner (2002).
3 Text World Theory and mental space theory provide alternative models for the description of similar processes; cf. Dancygier (2012) 36: 'The primary difference is possibly that the text world theory is much more text-driven than the mental spaces-and-blending framework, while also being more focused on the analysis of discourse, rather than the elucidation of linguistic concepts'. For an application of blending theory in the context of Text World Theory see Gavins (2007) 146–164. For an analysis of the *Palemedes*-parody of *Thesmophoriazusae* based in mental space theory see Tsakmakis (forthcoming).
4 Gavins (2007) 10.
5 Stockwell (2002) 136.
6 Stockwell (2002) 136.
7 Gibbons (2016) 72.
8 Gavins (2007) 9.
9 Gavins (2007) 29.
10 For some recent approaches see Cruickshank and Lahey (2010); Gibbons (2016). The former deals with the reading of dramatic texts, the latter with immersive theater.
11 Traditionally *Thesmophoriazusae* has been dated to 411. For an alternative suggestion (Lenaea of 410) see Tsakmakis (2012).
12 All translations (and emphasis) are from Sommerstein (1994).
13 We avoid the term staged world which was introduced by Cruickshank and Lahey (2010), because it suggests a more radical differentiation from the embedded text-worlds within the drama; besides, it may have misleading connotations when it is used in the context of reading and not for the experience of a performance.
14 This is guaranteed by the so-called principle of minimal departure: 'In the case of literary texts, precisely where along this scale of accessibility a particular world is positioned can affect how readers respond to its content. In line with cognitive approaches to discourse-comprehension, however, possible worlds

theorists recognise that in all cases readers will begin processing a text with the assumption that its textual world has an identity with the actual world until they are presented with information to the contrary' (Gavins 2007, 12).

15 A cognitive poetics approach could also account for the so-called dramatic illusion, including its alleged 'breaking' in comedy; see for opposite views Sifakis (1971) 7–14, and Bain (1977) 3–7.

16 Although Euripides' shame is sufficiently motivated through the context, it is not a typical psychological phenomenon for a comic hero. In case this dissonance with the spectators' expectations is realized, a search for an appropriate context which can provide a frame that renders the statement meaningful is cued; the knowledge of Euripidean tragedy, as a relevant background, provides the key: the kinsman's explanation makes more sense against the background of Euripides' well-known inclination for psychological realism.

17 Up to this moment the dress served as an instrument of deception (to conceal the kinsman's sex and identity); now it will be used a conventional theatrical device. Thus, its significance is altered following the world-switch. However, on the visual level it remains the same. Problems arising from the 'sameness' of the spatial setting will be a source of misunderstandings in the course of the 'performance'.

18 For comedy (and especially this comedy) the ultimate proof of gender identity is the physical body. The kinsman's identity had to be disclosed through the exposure of his genitals (643–648); but also Agathon, despite his cross-dressing, is identified by the kinsman as a male because his gown has the smell of a penis (254).

19 In the words of Austin and Olson (2004) 106, 'the primary focalizers are female'.

20 Agathon's theory is the earliest text dealing with the process of writing as imitation. Gorgias, a source earlier than *Thesmophoriazusae*, referred to the effect of drama on the spectators as 'deception' (ἀπάτη), and he also described the impact of λόγος in general on the soul as a process of replacing one belief or opinion (δόξα) by another. On a possible application of similar theories on actors and acting see Tsakmakis (1997).

21 Cf. Duncan (2006) 32: 'At the same time that he destabilizes boundaries and seems to point toward the idea of identity as constructed, however, he also insists on a kind of essentialism'.

22 Although Euripides was not the first to challenge the traditional version of the Helen-myth (Stesichorus had done this almost two centuries earlier), it is not necessary to look for further alternative explanations: the notion of eccentricity is sufficiently justified here despite the primacy of Stesichorus. See also Austin and Olson (2004) *ad loc.* who suggest: 'there is probably also an allusion to the allegedly novel type of female heroine featured in the play'.

23 The questions that open Critylla's intervention are rhetorical; in the first question she used a dead metaphor ('stirring up' for what she perceives as disorderly behavior). The image of mixing various materials together is ultimately a visual one, and is used as a metaphor for her perception of the hero's words and behavior as confusing. In contrast to the captive, the woman perceives the present state of affairs as (a re-establishment of) order and any attempt to change it as disorder.

24 She will only tell her name to correct the kinsman who tries to involve her in the fiction (898), and express personal opinion in 920, but even there she is not foregrounded as a syntactical subject (μοι δοκεῖς).

25 Gavins (2007) 36–38.

26 Family relations: πατήρ (861), πόσις (866), παιδί (891), κασιγνήτῳ (900), πόσιν (901), δάμαρτος (912); relation to a place which functions as a deictic center: πατρίς (859), ξένους (872), ξένε (882), ξένη (890, 896), ξένον (892), ἐπιχωρία (907). Even gender is treated as controversial (862: 'Becoming a woman again, are you, before you've even paid the penalty for your last female impersonation?').

27 After an embarrassed exclamation ('You're a rogue, you are, by Hecate the Bringer of Light', 858) as a response to the identification of the setting as 'Egypt', the woman energetically contests all further references to persons and places which threaten to establish a new reality (e.g. kinsman: 'My native land is not unknown to fame; 'Tis Sparta, and Tyndareos is my father' – Critylla: 'He's your father, you scum? More likely Phrynondas is!' 859–861).

28 The comic potential of a dispute over text-builders will be amply exploited throughout the whole parody, as the mock-tragic verses are overloaded with deictic elements (demonstrative pronouns and adverbs, proper names, first and second-person pronouns and verb forms); they force Critylla to use similar language in her replies, as she has to repulse the kinsman's (and later Menelaus') assertions about the comic text-world. Demonstrative pronouns in vv. 855–919: αἵδε (855), ἐνθάδε (866), τῶνδε (870), τάδε (874), τούτῳ (879), τουτογί (880), ἐκεῖνος (861), τόδε (886), τάσδε (889), οὗτος (893), αὕτη (897), τουτί (904); proper names or derivatives: Νείλου (855), Αἰγύπτου (856), Αἴγυπτον (878), Σπάρτη, Τυνδάρως (860), Ἑλένη (862), Σκαμανδρίοις (864), Μενέλεως (867), Πρωτέως (874, 891, 897), Πρωτέας (876, 883), Πρωτεύς (881), Θεσμοφόριον (880), Φρυνώνδας (861), Θεονόη (897), Κρίτυλλα, Ἀντιθέου, Γαργηττόθεν (898), Μενέλεων, Τροίᾳ (901), Ἑλληνίς (907, 908), Ἑλένη (909), Μενελάῳ (910), Τυνδάρειον, Σπάρτην (919); first-person pronouns: ἐμοί (859), ἐμέ (864, 906, 918), ἐγώ (866, 910), ἐμήν (869, 918), ἐμόν (895, 901), με

(904, 913 twice, 915 twice); second-person pronouns: σοί (860), σύ (865, 899, 906, 918), σε (896, 903, 906, 909, 910, 915, 917), σῷ (900), σόν (908), σῆς (912); first-person verb forms: εἰμί (866), ζῶ (867), εἰσεκέλσαμεν (877), πεπλώκαμεν (878), καθήμεθα (886), βιάζομαι (890), γαμοῦμαι (900), αἰσχύνομαι (903), εἰσορῶ (905), θέλω (908), εἶδον (909), κύσω (915); second-person verb forms εἶ (858, 899, 905, 906, 907), γίγνει (862), ὤφελες (865), πείθει (879), ναυτιᾷς (882), ἐρωτᾷς (884), ἐξόλοιο, ἐξολεῖ (887), τολμᾷς (888), θάσσεις (889), ἐξαπατᾷς (892), βάυζε (895), βούλει, λέγε (899), εἶπας, στρέψον (902), ἔγνως (911), λαβέ (913 twice), περίβαλε (914), φέρε (915), ἄπαγε (915 four times), κλαύσετε (916). On the importance of names in *Thesmophoriazusae* see also the remarks of Zeitlin (1981).

29 Cf. Gavins (2007) 56: 'Set against the deictic background constructed by the world-building elements of the text, function advancing propositions can be seen in many ways to be the items which propel a discourse forwards'.

30 Cf. Fauconnier and Turner (2002) 66: 'Drama performances are deliberate blends of a living person with an identity. They give us a living person in one input and a different living person, an actor, in another. The person on stage is a blend of these two'.

31 Commentators failed to single out the implications of this trick. Indeed, spectators who still had a clear memory from *Helen* may have been duped: they must have mistaken Menelaus for Teucros, until he behaves as Helen's husband. Even after the beginning of the recognition, his utterances are ambiguous, until his identity is explicitly confirmed (910–911).

32 For details on this debate see Roselli 2011.

References

Austin, C. and Olson, D. (2004), *Aristophanes'* Thesmophoriazusae: *Edited with Introduction and Commentary*, Oxford.

Bain, D. (1977), *Actors and Audience: A Study of Asides and Related Conventions in Greek Drama*, Oxford.

Cruickshank, T. and Lahey, E. (2010), 'Building the Stages of Drama: Towards a Text World Theory Account of Dramatic Play-Texts', *Journal of Literary Semantics* 30.1: 67–91.

Dancygier, B. (2012), *The Language of Stories: A Cognitive Approach*, Cambridge.

Duncan, A. (2006), *Performance and Identity in the Classical World*, Cambridge.

Fauconnier, G. and Turner, M. (2002), *The Way We Think: Conceptual Blending and the Mind's Hidden Complexities*, New York.

Gavins, J. (2007), *Text World Theory: An Introduction*, Edinburgh.

Gibbons, A. (2016), 'Building Hollywood in Paddington: Text World Theory, Immersive Theatre, and Punchdrunk's *The Drowned Man*', in: J. Gavins and E. Lahey (eds.), *World Building: Discourse in the Mind*, London and New York, 71–89.

Platter, C. (2007), *Aristophanes and the Carnival of Genres*, Baltimore, MD.

Rabinowitz, P.J. (2015), 'Toward a Narratology of Cognitive Flavor', in: L. Zunshine (ed.), *The Oxford Handbook of Cognitive Literary Studies*, Oxford, 85–103.

Roselli, D.K. (2011), *Theater of the People, Spectators and Society in Ancient Athens*, Austin, TX.

Sifakis, G.M. (1971), *Parabasis and Animal Choruses*, London.

Sommerstein, A.H. (1994), *Aristophanes'* Thesmophoriazusae, Warminster.

Stockwell P. (2002), *Cognitive Poetics: An Introduction*, London and New York.

Tsakmakis, A. (1997), 'Nikandros, Schauspieler des Aristophanes? (Diogenes von Babylon bei Philodemos, ΠΕΡΙ ΜΟΥΣΙΚΗΣ)', *Zeitschrift für Papyrologie und Epigraphik* 119: 7–12.

Tsakmakis, A. (2012), 'Persians, Oligarchs and Festivals: The Date of *Lysistrata* and *Thesmophoriazusae*', in: A. Markantonatos and B. Zimmermann (eds.), *Crisis on Stage: Tragedy and Comedy in Late Fifth-Century Athens*, Berlin and Boston, MA, 291–302.

Tsakmakis, A. (forthcoming), 'Aristophanic Parody and the Invention of the Euripidean Style', in: A. Vatri and T. Liao (eds.), *Style in Discourse*, Berlin and Boston, MA (*Trends in Classics Suppl.*).

Werth, P. (1994), 'Extended Metaphor: A Text World Account', *Language and Literature* 3.2: 79–103.

Werth, P. (1999), *Text Worlds: Representing Conceptual Space in Discourse*, London.

Zeitlin, F.I. (1981), 'Travesties of Gender and Genre in Aristophanes' *Thesmophoriazusae*', in: H. Foley (ed.), *Reflections of Women in Antiquity*, London, 169–217.

Zimmermann, B. (2014), 'Aristophanes', in: M. Fontaine and A.C. Scafuro (eds.), *The Oxford Handbook of Greek and Roman Comedy*, Oxford, 132–159.

10

The body-*as*-metaphor in Latin literature[1]

Jennifer Devereaux

How involved is the body in structuring classical Latin texts? In this chapter, I focus on ubiquitous processes of figuration that attest to the body playing an important role in the performance of knowledge and the expression of sentiment in classical Latin texts. The examples provided suggest that the emergence of complex social concepts took place through gestalt structures that treat the body *as* metaphor and thereby transcend the linguistic level of lexical semantics.

In identifying the body *as* metaphor and using the term 'body-*as*-metaphor', I do not treat the body as 'merely a construct'. Rather, I treat is as a whole that serves as a source and basis for metaphors that structure other (esp. non-physical) concepts. This involves, I argue, pre-linguistic invocations of the body-*as*-metaphor which I call *enactive analogies*. I chose this term to qualify the body-*as*-metaphor because pre-linguistic information resides in the body but is expressed through the integration of internal and external worlds by Roman authors – situating readers at the threshold of undecidability between objective and subjective experience. The structuring of emotional content in this way maps neatly onto Varela, Thompson and Rosch's (1991, revised 2016) suggestion that cognition is enactive – meaning, in the simplest possible terms, that there are no internal symbolic representations to divide from the pre-given external features of the world. In accordance with the embodied position that knowledge of all manner of phenomena, including abstract concepts, emerges through bodily engagement with the world, they posit that embodied experience of the environment *is* cognition. As I will demonstrate, *enactive analogies* are body-based knowledge structures that interact with and are related to embodied metaphors, which are derived from bodily experience of the environment, making them identifiable in ancient texts. They extend Cognitive Metaphor Theory (CMT) (Lakoff and Johnson 1980) and treat the body *as* metaphor, the body-*as*-metaphor as transhistorical and literature as *intercorporeal*. I build upon CMT in this way in order to:

1) Identify a basic organizing principle based on a well-defined paradigm
2) Provide a schematic account of emotion based upon consistencies across texts
3) Identify and quantify sentiment
4) Provide a comparative model

By identifying paradigms, correlating emotional schemas, and measuring and comparing sentiment, I will demonstrate how a self-reflexive network of reciprocal sympathies creates a dialogue for communicating norms and values. By drawing attention to a tendency of human expression to (re)instantiate the social world through the enactment of bodily experience, I hope to encourage readers to consider how deeply embedded is bodily knowledge in Latin texts. My aim is to open up the discussion of the body's rhetorical and transmissive role in narrative construction, not resolve it.

In this direction, I will begin by defining *enactive analogies* as gestalt structures that treat the body *as* metaphor. Demonstrating that abstract meaning emerges from these underlying structures, I ground a 'cognitive poetics' in the analysis of specific textual features and provide a theoretical basis for how those features deliver meaning. That theoretical basis and the apparent paradox of pre-linguistic discourse are explained and addressed through a three-pronged discussion of the embodiment hypothesis, embodied semantics, and Vittorio Gallese and Vittoria Cuccio's notion of *intercorporeality*. Throughout the discussion, I propose several examples of a pre-linguistic paradigm that illustrates the ubiquitous nature of the body-*as*-metaphor, before providing a case study of Apuleius' Cupid and Psyche, which includes a comparative model. Through this four-part study, I explore associations between sexual attraction and bodily experiences of inebriation to forward the suggestion that bringing our bodies to the texts is essential to a more complete understanding of ancient culture and the transmission of its various elements across texts and contexts.

What is an enactive analogy?

A detailed explanatory model will follow, but let's begin with an example of what an *enactive analogy* 'looks like'. Consider a fragment of Pacuvius cited most fully by Cicero at *De Orat.* 3.157:

> . . . inhorrescit mare,
> tenebrae conduplicantur, noctisque et nimbum occaecat nigror,
> flamma inter nubis coruscat, caelum tonitru contremit,
> grando mixta imbri largifluo subita praecipitans cadit,
> undique omnes venti erumpunt, saevi exsistunt turbines,
> fervit aestu pelagus . . .
>
> . . . a shivering takes the sea,
> Darkness is doubled, and the murk of night
> And storm clouds blinds the sight, flame 'mid the clouds
> Quivers, the heavens shudder with thunderclaps,
> A sudden hail with bounteous rain commingled
> Falls headlong, all the winds from every quarter
> Burst forth, and savage whirlwinds rise; the sea
> Surges and boils . . .
>
> *(Rackham 1948)*

In this passage, the metaphors that convey emotion are crafted through the movement and bodily experiences expressed by the verbs. *Inhorresco* denotes bristling and trembling and is thus similar in meaning to the Greek term φρίκη. As Douglas Cairns observes, φρίκη denotes the physical symptoms of shivering, shuddering and horripilation (goosebumps), which state in general, whether or not it is used metonymously to stand for the emotion it expresses, 'is particularly associated with automatic responses to sudden visual or auditory stimuli' (Cairns 2017). *Inhorresco* is similarly characterized by Seneca (*Ep.* 57.4: *inhorrescet ad subita*) and is associated with fear by Horace (*Od.* 1.23), who refers to a deer trembling (*pavidam*) with fear (*metu*), which is caused by the woods metaphorically shivering (*inhorruit*) with sounds that cause the heart and knees to tremble (*et corde et genibus tremit*). According to Cicero, *terror* was used to refer to autonomic responses, like changes in complexion related to blood-flow and the trembling of the body, *timor* involved the approach of frightful things and *pavor*, which literally means 'a trembling', refers to the 'movement' of *mens* or *sapientia* away from its place in the chest, which is suggested by Cicero to refer to the escape of breath from the body (*Tusc.* 4.19 citing Ennius). Internal and external shivering mirror one another, revealing a feedback loop between brain, body and world. That is to say, internal mental 'movement' (*pavor*) mirrors external physical movement (*inhorrere*), becoming a constitutive part of it, which is understood in terms of autonomic processes (*inhorrere*) that manifest in the visible body (*tremere*).

The correspondence between inner and outer worlds that Horace's use of *inhorruit* facilitates is seemingly present in the Pacuvian passage as well. The episode is identified by Cicero (*Div.* 1.24) as pertaining to a storm that destroyed much of the Greek fleet after the sack of Troy (*Od.* 1.326–7, 3.130–85, see Mankin 2011: 244), the embodied experience of which is seemingly externalized into the environment. That is to say, the sea is embodied in order to externalize the autonomic response (i.e. trembling and horripilation) of the sailors to the sudden storm (Cf. Cic. *Fin.* 5.31: *quin etiam 'ferae', inquit Pacuvius, 'quibus abest ad praecavendum intellegendi astutia', iniecto terrore mortis 'horrescunt'*). Trembling is amplified by *coruscat* and *contremit*, both of which carry a sense of 'shaking' or 'trembling'. Together with this, winds 'break out' (*erumpunt*) and the sea 'boils' with heat (*fervit aestu pelagus*). These verbs (*erumpunt* and *fervit*) contribute further to the externalization of embodied experience, as they play into a ubiquitous cognitive metaphor that construes intense emotion as hot liquid in a container from which it is prone to burst – that is, become externalized through expression (for Latin, see Riggsby 2015: 114; for general discussion, see Gibbs 2017: 118; Sanford and Emmott 2012: 60; Kövecses 2002; Lakoff 1987). This metaphor is derived from the embodied experience (i.e. the phenomenology) of passionate emotions, like the skin becoming flushed, and breath, tears and sounds seeming to 'burst' from the body. This element of embodiment tacitly associates an intense emotion with the trembling and bristling characteristic of an automatic response to frightening stimuli. One can thus imagine that the sailors' embodied emotional experience is externalized into the environment by Pacuvius with verbs that enact a coherent and familiar experience. The body provides evidence for the veracity of an emotional experience by acting as the source domain for metaphor – the sense made of the environment is full-bodied. The body-*as*-metaphor acting in this way to create meaning is what I refer to as an *enactive analogy*. The next three sections on embodiment, intercorporeality and embodied semantics will detail how we are able to identify the body-*as*-metaphor when reading Latin texts.

The embodiment hypothesis

The ineffaceable feedback loop that exists between mind, body and world, as well as the meaning-making potential of the body, is of central concern to theories of embodied cognition. These theories suggest that certain aspects of cognition are dependent upon or influenced by the body beyond the brain. Proponents of embodied cognition theorize that thoughts, feelings and behaviours are grounded in sensory experiences and bodily states (for review see Barsalou 2008; Spellman and Schnall 2009). A general contention of the embodiment paradigm is that mental processes of all kinds involve simulations of body-related perceptions and actions. The term 'simulation' in this context refers to the reactivation or reuse of the neural networks that encode our physical experiences in the world. Such reuse has been explained in a variety of different, not mutually exclusive, ways (for summary and discussion see Meier, Schnall, Schwarz and Bargh 2012). One prevalent theory is that we 'evolved from creatures whose neural resources were devoted primarily to perceptual and motoric processing' (Wilson 2002: 625); another that complex mental processes reuse evolutionarily older programs (Anderson 2010). Others emphasize developmental processes and suggest that our early experiences with the physical world (e.g., moving around in space) structure our later understanding and representation of more abstract concepts (e.g., likes and dislikes), a process referred to as scaffolding (Williams, Huang and Bargh 2009). This last theory, which can be traced back to Jean Piaget's work on developmental psychology (1970), suggests that as brain-body organisms we rely on sensory information to understand and explain not only our physical and mental experiences, but also our social worlds. It is this last theory of embodiment that will guide our reading, so I will supply a brief illustration of what such a perspective entails.

Enactive analogies can be quite slight – a singular movement – and yet they enable us to navigate complex networks of meaning. Consider, for example, an emotional concept derived from the experience of being stationary and stretching upwards. It is a gesture that originates in infancy, reflecting simultaneously the rejection of a current circumstance and a desire for social reassurance and/or cooperation:

The infant in Figure 10.1 is reaching both *away* from the floor and *towards* his mother, so we can describe the picture by saying 'the baby wants off the floor' or 'the baby wants his mother', or both, depending on our perspective on the situation.

Let's now look at the myth of Daphne and Apollo. In the myth, as told by Ovid, Daphne, desperate to escape the amorous advances of Apollo, is transformed into a tree by her father, a river god. Her transformation (*Met.* 1.549–50), which occurs at the height of her desperation, is described as her feet sticking to the ground (*haeret*) and her arms growing skyward (*crescunt*). Let's think about this with the help of Bernini's 1625 sculpture (Figure 10.2):

Daphne, not entirely unlike the infant, is seeking the help of her parent, but is conceived by Bernini not as stretching *towards* her father, but as stretching *away* from Apollo, although *both* are true in the realm of bodily analogy. The ease with which the same bodily movement and position can accommodate and emphasize a variety of perspectives is precisely why it can so easily enact meaning and scaffold social knowledge.

Figure 10.1 Infant reaching away from social isolation and towards cooperation[2]

Figure 10.2 Bernini's Daphne reaching away from social cooperation and towards isolation[3]

Intercorporeality

The ability for the body to scaffold social knowledge is attributed to what Vittorio Gallese and Vittoria Cuccio (2015) call *intercorporeality*. They propose that traces of bodily experience in the brain are accessed both when we visually perceive and when we imagine other bodies, and that the reuse of mental states or processes is the foundation from which shared meaning arises. Our bodily experiences thus constitute the primordial source of knowledge that we have of others (cf. Gallese 2007). Gallese (2016) further argues that mundane bodily experience, especially kinetic movement, makes it possible to connect the common bodily experiences of the pre-linguistic sphere to the linguistic one (see also Gallese and Lakoff 2005; Glenberg and Gallese 2012). In other words, language encoded with sensorimotor information is meaningful and facilitates intersubjectivity, which refers to 'the common sense, shared meanings constructed by people in their interactions with each other and used as an everyday resource to interpret the meaning of elements of social and cultural life' (Seale 2004).

To illustrate intercorporeality as a way of reading, let's return to Daphne transforming into a tree at *Met.* 1.549–50: *in ramos bracchia crescunt, pes modo tam velox pigris radicibus haeret* (her arms grow up into branches, and her active feet as clinging roots fasten to the ground), and consider with it the anguished love of Echo for Narcissus later in the text (*Met.* 3.393).

Echo's sexual desire for Narcissus similarly 'adheres' and 'grows' upwards (*haeret amor crescitque*). The desire to escape from sex – encapsulated by Daphne's bodily movement – in turn affects the conceptualization of Echo's longing for Narcissus, which in turn effects a reconceptualization of Daphne's experience by conflating aversion with desire through a self-reflexive network of reciprocal sympathies that illustrates the potential for bodily movement to create a dialogue for communicating norms and values. From this vantage point we might even imagine the combined movements of Daphne and Echo to metamorphose (e.g.) Sappho, who conceived of her love object 'loving against her will' (1.24: φιλήϲει κωὐκ ἐθέλοιϲα).

Embodied semantics

In terms of the communicative potentialities of the body, the seminal discussion is Lakoff and Johnson's *Metaphors We Live By* (1980). Lakoff and Johnson advanced the hypothesis that we use what we know about our physical experience to provide understanding of countless other subjects by analogy. This is why, for example, one might be 'quick on one's feet', where physical speed is mapped to intellectual facility, or have a 'warm' relationship with someone, where degree of temperature is mapped to intensity of emotion. The study of metaphors like this in literature is by now standard practice, and the embodied perspective has led to the valuable discovery of diachronically consistent conceptual networks. For example, thoughts were conceived as having tangible qualities, like shape and weight, and the mind was construed as haptic, taking part in activities like grasping (Short 2012). These examples represent embodied conceptual metaphors that are identifiable in a variety of modern languages as well, which demonstrates that an embodied perspective can account for figurative practices in ancient authors, as well as our ability to understand them despite gulfs of time and culture (for additional examples see also Short 2008, 2013 and Chapter 4 in this volume; Fedriani 2016). While the interpretation of bodily movements and activities may vary, that which is being interpreted is relatively constant, which allows for a certain degree of communicative continuity across time and space, even if it does not necessitate constraints upon cross-cultural and historical variation. By virtue of its consistency of form and its consistent subjection to the laws of physics, the body is fundamentally *transhistorical*. The role of the body in meaning-making is thus critical to scholars concerned with social knowledge and how it is structured across contexts. After all, how can we responsibly receive literature, much less treat it critically, if we allow any aspect of the body and its relation to the social world to remain an unanalysed element of not only our study but also our practice? The body is a *sine qua non* in terms of meaning-making and represents a basic way in which humans conceive of their own behaviour and psychology – but I am not suggesting it is the *only* way. I focus on bodily experience simply because it is an extraordinarily common source of meaning – so much so that it is easily taken for granted, which is problematic, because to conceive of signification as 'the act of a radically disembodied consciousness, or rather, the act that radically disembodies that consciousness' is to utterly lose track of how culture and its related knowledge structures emerge (Butler 2011: 176 in reference to Sartre and de Beauvoir).

As previously stated, this chapter builds upon and extends the hypotheses of embodied semantics to consider the role of the body in structuring abstract thought in terms that are slightly less representational than those generally explored with CMT. It takes interest in the role of the body in analogically structuring the abstract socioemotional knowledge that exists beneath the surface of texts. We will next continue to explore the structure of this knowledge and the intuitive role of the body in meaning-making through the same paradigm that structures Daphne and Echo's experiences above. The *avoidance-approach* paradigm states, intuitively, that we tend to avoid (either literally in the physical world or figuratively in the social world) what we view negatively, and approach (either literally in the physical world or figuratively in the social world) what we view positively. Some researchers have suggested that *avoidance-approach* is a fundamental and basic distinction 'that should be construed as the foundation on which other motivational distinctions rest' (Elliot and Covington 2001). The *avoidance-approach* paradigm, perhaps because it is fundamental to all manner of reasoning and behaviours, is a full-body response, meaning that the body itself *is* the response – hence the body-*as*-metaphor emerging from it.

Indeed, we are increasingly coming to understand the diverse ways in which the body *as a whole* acts as a communication channel. For example, multiple studies from various disciplines have shown that bodily expressions are as powerful as facial expressions in conveying emotions (see Kleinsmith and Bianchi-Berthouze 2013 for summary), and technologies encountered by the average person on a daily basis afford multimodal interactions in which bodily expressions assume an important role that goes beyond that of gesture (Ibid.). Take as examples the game Tetris, which requires physical perspective manipulation in order to perform optimally at an otherwise more generally 'mental' task (Wilson 2002), and whole-body computer games in which bodily movement is not only a means to control the interaction between users and games, but is also a way to capture and affect their own emotional and cognitive performances (Ibid.).

The cognitive performance of readers is also likely influenced by the bodily movements that ground gestalt structures in Latin literature (see Devereaux 2016). The Neural Theory of Language (NTL) offers explanation for this, arguing that thought is physical and reasoning is the (re)activation of certain neuronal groups that capture bodily experience (see Feldman 2006 and Lakoff 2008). According to NTL, all mental simulation is embodied insofar as imagining an action uses the same neural substrate as performing that same action (Lakoff 2008). For example, imagining the performance of physical exercise results in several bodily parameters behaving as if the actions were actually being executed (Decety et al. 1991; Clark et al. 2014). Other studies have suggested that reading or listening to verbs or sentences about action automatically produces motor impulses corresponding to that action (Klepp et al. 2014; Raposo et al. 2009; Pulvermüller 2005, 2008; Buccino et al. 2005). Reading sentences involving giving or receiving objects, for example, was found to modulate motor system activity as measured in the hand (Glenberg et al. 2008; Aziz-Zadeh et al. 2006), and seeing a picture depicting someone kicking or simply hearing the word 'kick' was shown to activate areas of the brain responsible for that particular motor movement (Tettamanti et al. 2005; Hauk et al. 2004; Kable et al. 2002). Evidence even suggests that the understanding of the most conventional metaphoric action retains a link to sensory-motor systems involved in action performance (see Desai et al. 2011). In this vein, Raymond Gibbs (2006) suggests that our ability to make sense of embodied metaphors resides in the automatic construction of a simulation whereby we imagine performing the bodily actions referred to in these expressions. Citing the regularity with which emotional experience in particular is construed as a process of moving through 'affective space', Gibbs argues that one reason why people can so readily interpret such metaphors is because metaphoric language rooted in bodily movement allows people to imaginatively recreate those movements during their ordinary use of language. NTL thus suggests a tremendous influence on the part of bodily experience on the development of patterns of inference and the creation of abstract meaning.[4]

We can see this unfold by continuing to trace the stretching *away*/stretching *towards* paradigm. We see, for example, Creusa's bodily movement used to demonstrate existing social bonds and/or willing bonds of affection in Virgil's *Aeneid* 2.673–4: *Ecce autem complexa pedes in limine coniunx haerebat, parvumque patri tendebat Iulum.* Similar bodily movement is also present in Homer's *Iliad* (1.500–1): καί ῥα πάροιθ' αὐτοῖο καθέζετο, καὶ λάβε γούνων σκαιῇ, δεξιτερῇ δ' ἄρ' ὑπ' ἀνθερεῶνος ἑλοῦσα. Although 'sitting' and 'taking' are less vivid in terms of movement, it is unambiguous that Thetis must stretch her arm upwards as she sits at Zeus' feet. She is stationary and stretching *towards* him. Given Zeus' one-time courtship of Thetis, affection surely colours this supplicatory action. Taking notice of the pre-linguistic architecture does not preclude a meaningful correlation between the *Aeneid* passage and, say, the famous scene between Andromache and Hector at the end of *Iliad* 6, but it adds emotional dimension that enables us to see a deeply entrenched similarity between the concepts of Aeneas and Zeus which is quietly instantiated through the movement and positioning of female bodies, a feature of the texts that is grounded in the pre-linguistic, socially structured unconscious.

On the opposite end of the emotional spectrum we find grief, as in *Aeneid* 10.844–5: *et ambas ad caelum tendit palmas et corpore inhaeret*, where the same embodied paradigm is used in a different context. Mezentius, mourning the death of Lausus, clinging to the corpse, stretches his hands up to the sky, a bodily position that closely resembles that of Daphne and the infant pictured above. In stretching towards the sky he is also stretching *away* from the corpse and the lost social bond it represents. 'Love' was for women, it would seem, conceived of in terms of bodily movement that 'felt' not unlike 'grief' for men. This of course ties into the general 'marriage is (like) death' notion encapsulated (e.g.) by the myth of Persephone and Hades. Through these simple examples we thus get a sense of the generative matrix sustained by bodily schemas that function as epistemological paradigms.

What we see in Latin, then, is emotional states of love and grief conceived as a mixture of (if not a conflict between) avoidance and approach intuitions (cf. the paradigmatic Catullan *odi et amo*). In terms of what we have discussed so far, this bodily dynamic might be (re)visualized like this (Figure 10.3).

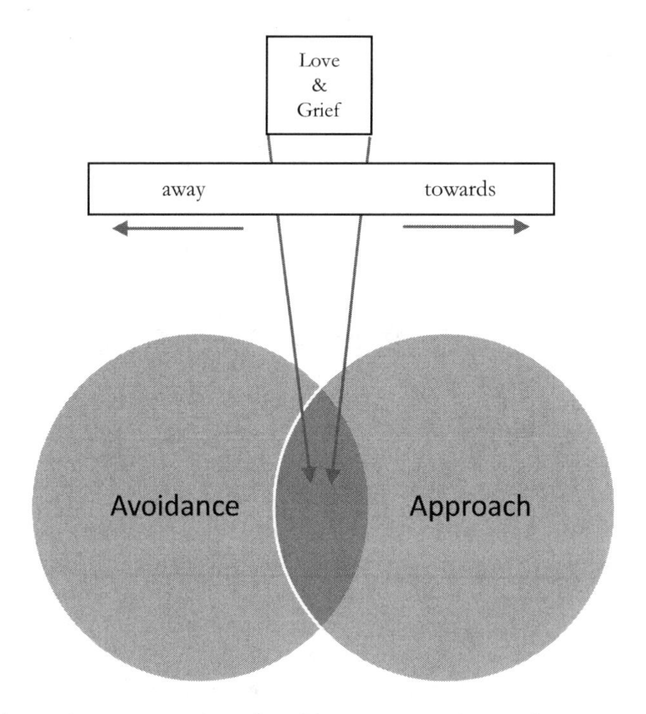

Figure 10.3 Schematic representation of avoidance-approach paradigm

A case study

This case study explores bodily experience as a coherence device that structures sentiment around the paradigmatic *avoidance-approach* schema traced through a number of poetic texts above. In his *Metamorphoses* (5.21), Apuleius describes Psyche, the lover of Eros, as follows:

> . . . aestu pelagi simile maerendo fluctuat . . . titubat multisque calamitatis suae distrahitur affectibus. Festinat differt, audit trepidat, diffidit irascitur; et, quod est ultimum, in eodum corpore odit bestiam, diligit maritum.
>
> . . . in her grief she ebbed and flowed like the billows of the sea . . . she wavered irresolutely, torn apart by the many emotions raised by her dilemma. She felt haste **(approach)** and procrastination **(avoidance)**, daring **(approach)** and fear **(avoidance)**, despair **(avoidance)** and anger **(approach)**; and worst of all, in the same body she loathed the beast but loved the husband.
>
> *(Hanson 1989)*[5]

Apuleius' conceptualization of Psyche's mental activity being moved in opposite directions foreshadows the later positioning of her paradigmatic body (grabbing hold of and dangling from Cupid's leg as he flies away: 5.24) and in so doing seems to treat in psychological terms the physical science of attraction outlined by Lucretius (4.1192–1208):

> Nec mulier semper ficto suspirat amore
> quae conplexa viri corpus cum corpore iungit
> et tenet adsuctis umectans oscula labris;
> nam facit ex animo saepe et, communia quaerens
> gaudia sollicitat spatium decurrere amoris.
> nec ratione alia volucres armenta feraeque
> et pecudes et equae maribus subsidere possent,
> si non, ipsa quod illarum subat ardet abundans
> natura et Venerem salientum laeta retractat.
> nonne vides etiam quos mutua saepe voluptas
> vinxit, ut in vinclis communibus excrucientur?
> in triviis cum saepe canes, discedere aventes,
> divorsi cupide summis ex viribus **tendunt**,
> quom interea validis Veneris compagibus **haerent**.
> quod facerent numquam, nisi mutua gaudia nossent,
> quae lacere in fraudem possent vinctosque tenere.
> quare etiam atque etiam, ut dico, est communis voluptas.
>
> Nor does a woman always feign the passion which makes her sigh, when she embraces her mate joining to body, and holds his lips in a long kiss, moistening them with her own. For she often does it from the heart, and seeking mutual joys rouses him to run the full course in the lists of love. Nor otherwise could birds or cattle, wild beasts or sheep or mares submit to the male, were it not that their own nature, overflowing, is on heat and burning, and they thrust gladly against the penis of the mounting male. Do you not see also, when mutual pleasure has enchained a pair, how they are often tormented in their common chains? For often dogs at the crossways, desiring to part, pull hard in different directions with all their strength, when all the while they are held fast in the strong couplings of

> Venus. But this they would never do, unless they both felt these joys which were enough to lure them into the trap and to hold them enchained. Therefore again and again I say, the pleasure is for both.
>
> *(Rouse and Smith 1992)*

Seeking a pleasurable bond while straining against it engages with the same epistemological paradigm in which Daphne and Echo, and even Thetis and Creusa, are enmeshed, all of whom engage with the *avoidance-approach* schema, embodying emotions pertaining to social bonds, including those that could be forged 'against one's will'.

If we truly seek to understand the way in which systemic or pervasive political and cultural structures are enacted and reproduced through individual acts and practices, and how the analysis of ostensibly personal situations is clarified through situating the issues in a broader and shared cultural context (cf. Butler 1997), we must consider the role of bodily movement and positioning in ancient texts, as the textual body actively constructs gender through specific corporeal acts. The simple positioning of the female body in relation to various male figures in fact demonstrates that female embodied experience, as we can identify it in ancient texts, is an historical situation that develops through the constant engagement of that body in certain cultural and historical possibilities that are continually realized (cf. Merleau-Ponty 1962; de Beauvoir 1974; Butler 1997). Those possibilities were of course realized by male authors, which we should bear in mind as recipients of the texts, not least of all because the movements and positions of textual bodies bear similarities to performative acts within theatrical contexts, as we are intercorporeal readers. That is to say, we, as historically contingent viewers, effortlessly and even unconsciously 'perform' the embodied text, and are thus subject to the properties of authority that emerge from it within our own readerly contexts.

Moving forward, we will pay attention to descriptive generalizations and authoritative expressions of bodily knowledge that coalesce around the *avoidance-approach* schema and more importantly to the properties of authority that emerge from them.

We first note that it is with the bidirectionality of the *avoidance-approach* paradigm engaged that Psyche visually perceives Cupid for the first time (5.22):

> Videt capitis aurei genialem caesariem ambrosia temulentam, cervices lacteas genasque purpureas pererrantes crinium globos decoriter impeditos, alios antependulos, alios retropendulos, quorum splendore nimio fulgurante iam et ipsum lumen lucernae vacillabat. Per umeros volatalis dei pinnae roscidae micanti flore candicant, et quamvis alis quiescentibus extimae plumulae tenellae ac delicatae tremule resultantes inquieta lasciviunt.
>
> On his golden head she saw glorious hair drunk on ambrosia: wandering over his milky neck and rosy cheeks were the neatly shackled ringlets of his locks, some prettily swaying in front, others swaying behind; the lightning of their great brilliance made the lamp's light flicker. Along the shoulders of the winged god white feathers glistened like flowers in the morning dew; and although his wings were at rest, soft and delicate little plumes along their edges quivered restlessly in wanton play.
>
> *(Translation adapted from Hanson 1989)*

While it would be simple and even valid enough to read the passage and the rebuke of Psyche that follows as the censure of the female appropriation of a masculine subject position – that is, acting as subject rather than object of the gaze – I think there is more going on here. Note that the tottering back and forth (*titubat*) of Psyche's mind is transmuted into Cupid's appearance: *pererrantes, antependulos, retropendulos, vacillabat. Perrerantes* and (*lumen lucernae*) *vacillabat* extend to the language of mistake-making and the disorderly movement associated with it (cf. Short 2013). With this transmogrification, Psyche essentially self-objectivizes, taking a third-person perspective on her internal experience. By transmuting a subjective experience into the objective world – making it visible – Apuleius shows his 'expertise' on the subject at hand, which first becomes clear in the description of Cupid's wings which, when taken together with *temulentam*, is deeply reminiscent of Ovidian didacticism:

> Vinaque cum bibulas sparsere Cupidinis alas,
> Permanet et capto stat gravis ille loco.
> Ille quidem pennas velociter excutit udas:
> Sed tamen et spargi pectus amore nocet.
> Vina parant animos faciuntque caloribus aptos:
> Cura fugit multo diluiturque mero.
> Tunc veniunt risus, tum pauper cornua sumit,
> Tum dolor et curae rugaque frontis abit.
> Tunc aperit mentes aevo rarissima nostro
> Simplicitas, artes excutiente deo.
> Illic saepe animos iuvenum rapuere puellae,
> Et Venus in vinis ignis in igne fuit.
> Hic tu fallaci nimium ne crede lucernae:
> Iudicio formae noxque merumque nocent.
>
> And when wine has sprinkled Cupid's thirsty wings,
> He abides and stands overburdened, where he has taken his place.
> He indeed quickly shakes out his dripping plumes;
> Yet it hurts even to be sprinkled on the breast with love.
> Wine fashions minds apt for passions:
> Care flees and is drowned in much wine.
> Then laughter comes and even the poor find vigour,
> Then sorrow and care and the wrinkles of the brow depart.
> Then simplicity, most rare in our age, lays bare the mind,
> With the god shaking off knowledge.
> At such time often have women bewitched the minds of men,
> And Venus in the wine has been fire in fire.
> Do not overly trust the treacherous lamp at such a time;
> Darkness and drink impair your judgment of beauty.
>
> (*Ars Amatoria 1.233–46; translation adapted from Mozley 1979*)

Cupid's wings are synonymous with the internal seat of affection (*sparsere Cupidinis alas . . . spargi pectus*), a conceptualization of attraction that may go back to Plato's winged soul in the Phaedrus (251a–c). When the soul sprouts feathers at the sight of beauty, Plato says, there is a shivering and something of fear (πρῶτον μὲν ἔφριξε καί τι τῶν τότε ὑπῆλθεν αὐτὸν δειμάτων), which likely descends from a similar psychosomatic experience as that which informs the Aeschylean notion of δεῖμα fluttering around the heart (*Ag.*975–7). The release of adrenaline is common to fear and sexual arousal, as is also an increased or 'fluttering' heartrate. Intense emotion could indeed cause one's heart to beat in such a way as to seem as though a bird's wings were beating in one's chest. In Ovid we have a similar enaction of the heart 'fluttering' (*excutit udas . . . artes excutiente deo*), which represents an interoceptive metaphor: the heart flutters in the chest like wet wings. Engagement with the metaphor portrays Psyche very much like Ovid's inebriated would-be lover, whose mind is like hers: multiform (*vina parant animos faciuntque caloribus*), and subject to perception that is not to be trusted (*hic tu fallaci nimium ne crede lucernae*). Apuleius thus seemingly incorporates male 'expertise' on psychosomatic experience, codes the knowledge as feminine and thereby reifies his subject and her circumstances. Initially dominant, like the female rulers of elegy, Psyche ultimately comes to resemble the passive elegist, who is subject to the torments of abandonment (Gold 2012). This suggests that Apuleius constructs female identity in accordance with a prior authoritative work concerned with internal masculine experience, which is woven into Apuleius' Psyche, whose externalization of internal experience ultimately denies her an internal locus of control.

The externalization of the interoceptive metaphor evokes the body as a viewable object that is subject to judgement. This effect is heightened by the inebriated body-*as*-metaphor. In his description of Cupid's appearance, Apuleius, who himself crafts invective around the common connection made between wine and sex when telling of the cruel and licentious Miller's wife (*Met.* 9.14; cf. Bauman 192), invokes the language of Pliny's quasi-scientific description of the drunken body in *NH.* 14.28:

> hinc pallor et genae pendulae, oculorum ulcera, tremulae manus effundentes plena vasa, quae sit poena praesens furiales somni et inquies nocturna, praemiumque summum ebrietatis libido portentosa ac iucundum nefas.
>
> Tippling brings a pale face and hanging cheeks, sore eyes, shaky hands that spill the contents of vessels when they are full, the condign punishment of haunted sleep and restless nights, and the crowning reward of drunkenness: monstrous lust and forbidden delights.
>
> *(translation adapted from Rackham 1968)*

Pliny	**Apuleius**
pendulae	antependulos; retropendulos
tremulae	tremule
inquies	inquieta
ebrietatus	temulentam
iucundum nefas	lasciviunt

The seemingly artistic play of referring to Cupid's hair as 'drunk' quietly interjects that Psyche's mental state is *unhealthy*. Suggestive of this are the playful feathers on Cupid's wings, which seem to mirror the 'wicked thoughts' (*iucundum nefas/lasciviunt*) of their viewer. Underlining the externalization of female interiority is the coherence the scene has to Stoic conceptualizations of vision (for Apuleius' familiarity see (e.g.) *Ap.* 15.33). According to the Stoics, vision involves the mind (*hêgemonikon*) being imprinted or reconfigured through unbroken physical contact with an external object. The Stoic mind

> is like an octopus, each sense a tentacle grasping the external world, reshaping itself and the world accordingly. Vision is a physical connection between seer and seen, operating – with the usual Stoic fondness for paradox – like the walking stick of a blind man, through two-way transmittal of tensile motion. Mind and world communicate as if along the threads of a spiderweb.
>
> *(see Habinek 2011: 64–5)*

It is indeed the schemas that accompany the *avoidance-approach* paradigm that communicate a large sum of information, not the least of which being that the passage accommodates the Stoic worldview that 'we gradually align ourselves with material reality to the extent that we and it become indistinguishable' (Habinek 2011: 71). There is thus a suggestion that Apuleius is also performing Plinian knowledge. Both Pliny and Apuleius use the overlapping terms in a physical description of a very particular state of being, with Cupid's naughty feathers nodding to the moral authority of the Plinian passage, and a Stoic inflection triggering a blend of Psyche's perception with the drunken body. Add to this that the narrator, who acts as a vehicle for Apuleius' performance, is a female who is likely drunk (*Met.* 4.7) and thus embodies the physical state embedded in the narrative. We then see that only an intercorporeal understanding of the text allows readers to contemplate the fullness of Apuleian expression, demonstrating the value of attention to pre-linguistic affective scaffolding and the mutual resonance of bodies across texts and contexts.

To briefly recap, the semantics of the transhistorical body-*as*-metaphor have helped us to identify the *avoidance-approach* paradigm at work structuring the Cupid and Psyche episode, which has enabled us to situate it within a particularized cluster of knowledge pertaining to the Roman social world and the emotional regime underpinning it – from Lucretius' paradigmatic discussion of sex and Pliny's physical description of drunkenness to Ovid's interoceptive wings and the Stoic notion of vision as a mutually constitutive process. Can it also quantify sentiment and draw meaningful distinctions between the compositional structure of otherwise similar texts? It would seem so. The Apuleius passage is morally inflected in a quantifiable way. Emphasis is placed on downward movement in the description of Psyche's psychosomatic experience at the beginning of the passage (*de*territa, *de*fecta, *de*sedit, *de*lapsum, *de*fecta).

> At vero Psyche tanto aspectu deterrita et impos animi, marcido pallore defecta tremensque desedit in imos poplites et ferrum quaerit abscondere, sed in suo pectore. Quod profecto fecisset, nisi ferrum timore tanti flagitii manibus temerariis delapsum evolasset. Iamque lassa, salute defecta, dum saepius divini vultus intuetur pulchritudinem, recreatur animi.
>
> But Psyche was terrified at this marvellous sight and put out of her mind; overcome with the pallor of exhaustion she sank faint and trembling to her knees. She tried to hide the weapon – in her own heart. And she would certainly have done so, had not the blade slipped down and flown away from her reckless hands in its horror of so atrocious a deed. She was now weary and overcome by the sense of being safe, but as she gazed repeatedly at the beauty of that divine countenance her spirit began to revive.
>
> *(Hanson 1989)*

In Latin, as in other languages, movement that is downward, frequently expressed by the prefix *de-*, has a negative force in many conceptualizations. '*De*mens' is to be mentally unstable, the harshest form of banishment is *de*portatio, to lose one's claim in court is to have one's hand knocked down ('*de*pellere'), etc. (on Latin spatialized readings of *de* see Short 2012; on orientational experiential metaphors see Fedriani 2016). At least partial explanation for this tendency is found in modern evidence for a link between affective judgements and physical metaphors. In *Philosophy in the Flesh* (1999), Lakoff and Johnson contend that metaphor allows people to think abstractly because it links abstract concepts (e.g., affect) to concrete sensory experiences, so abstract thought is not simply aided by physical metaphors; it is actually constituted in metaphorical terms. Bodily knowledge that 'down is bad', which derives from early bodily experiences like falling down, is captured by the iterated prefix, generating an opportunity for audiences to intuitively collectivize their evaluative response to the narrative.[6] What that response might be is also imparted by the prefix *de-*, which in addition to downward movement also denotes movement *away* (*depellere* is an especially clear example of this, meaning both 'to drive *away*' and 'to cast *down*'). *Abscondere*, too, plays on the iteration of the *de-* prefix, helping not only to create an opportunity for collective sentiment to unfold over time, but also to provide a structure for that sentiment by way of the *avoidance-approach* paradigm, which is punctuated by Psyche, whose body ends up in exaggerated *avoidance-approach* position (dangling from the leg of Cupid), ultimately falling *down* and *away* (5.24: *delabitur solo*). With this additional information about the sentiment that qualifies emergent authority in mind, let's now turn to a comparative model.

A comparative model

We find the love-is-drunkenness embodied metaphor directly represented by Catullus, who uses the term *pueri ebrii ocelli* (45.11):

> Acmen Septimius suos amores
> tenens in gremio 'mea,' inquit, 'Acme,
> ni te perdite amo atque amare porro
> omnes sum adsidue paratus annos,
> quantum qui pote plurimum perire,
> solus in Libya Indiave tosta
> caesio veniam obvius leoni.'
> hoc ut dixit, Amor sinistra ut ante
> dextra sternuit adprobationem.
> at Acme leviter caput reflectens
> et dulcis pueri ebrios ocellos
> illo purpureo ore suaviata
> 'sic,' inquit, 'mea vita, Septimille,
> huic uni domino usque serviamus,
> ut multo mihi maior acriorque
> ignis mollibus ardet in medullis.'
> hoc ut dixit, Amor sinistra ut ante
> dextra sternuit adprobationem.
> nunc ab auspicio bono profecti
> mutuis animis amant amantur.
> unam Septimius misellus Acmen
> mavult quam Syrias Britanniasque:
> uno in Septimio fidelis Acme
> facit delicias libidinesque.
> quis ullos homines beatiores
> vidit, quis Venerem auspicatiorem?
>
> Holding his girlfriend Acme close upon his
> lap, Septimius said: 'My darling Acme,
> if I don't love you madly, if I'm not quite,
> quite resolved to be constant all my lifetime,
> insurpassably, desperately devoted,
> in far Libya or burning India may I

meet up, solo, with a green-eyed lion!'
At these words, Love leftward as beforehand
rightward sneezed his approbation. Then sweet
Acme, gently tilting back her head and
with those rich red lips bestowing kisses
on her darling boy's besotted eyes, said:
'Thus, Septimius, *thus*, my life, my precious,
may we serve this single lord forever,
while more strongly and fiercely day by day this
hot flame blazes through my melting marrow.'
At these words, Love leftward as beforehand
rightward sneezed his approbation. Now from
this auspicious omen setting out, they
give and receive true love with equal passion.
Poor Septimius now rates Acme over
all the hoopla of Syria and Britain;
with Septimius only, faithful Acme
runs the gamut of all delights and pleasures.
Who, pray, ever saw two or more triumphant
lovers, who a Venus more auspicious?

(Green 2005)

Here, *sinistra* and *dextra* engage the bi-directional movement associated with the schema, with motion *towards* emphasized by *(ad)probatio*. Beyond engagement with the *avoidance-approach* schema no other element of the gestalt found in Apuleius is engaged. Catullus' inebriation metaphor instead engages with the social-warmth schema (*tosta . . . ignis . . . ardet . . . amant amantur*). The social-warmth schema accounts for why a modern reader might describe a close relationship as 'warm', or a sexually appealing person as 'hot'. Physical proximity modulates bodily temperature – (physical) proximity 'is' (social) closeness, and closeness enables the perception of (bodily-*as*-social) warmth. Heat as an element of sexual attraction, which is clearly relevant in Catullus, derives from the social-warmth bodily schema (for interest see Williams, Huang and Bargh 2009). So the love-as-drunkenness metaphor, when applied to male emotional experience by Catullus, is associated with social warmth and movement *towards*, while it is incorporated by Apuleius to identify female emotional experience with bodily instability and movement *away* and *down*.

We can visualize the schematic distribution of bodily knowledge in the texts through a simple representation of the data, which can be generated through straightforward analysis. That is to say: we can count the words that fall into the same category of conceptual metaphor within a given block of text. In Figure 10.4 below, words that engage with the metaphor 'love is inebriation' in terms of bodily experience (interoceptive and proprioceptive), spatialized affect (towards, away, down) and the embodied social concept of 'warmth' are counted. Engagement with these schemas is represented in terms of approximate percentages, with the first and last schematic lexeme marking the text on either side of the text being counted, from which pronouns, determiners and conjunctions are excluded (total words: Catullus: 36 / Ovid: 59 / Apuleius 5.22: 78). This can of course be done by the expert reader. For this chapter, however, I had the help of TACIT, a text analysis, crawling and interpretive tool developed by the University of Southern California's Computational Social Sciences Laboratory,

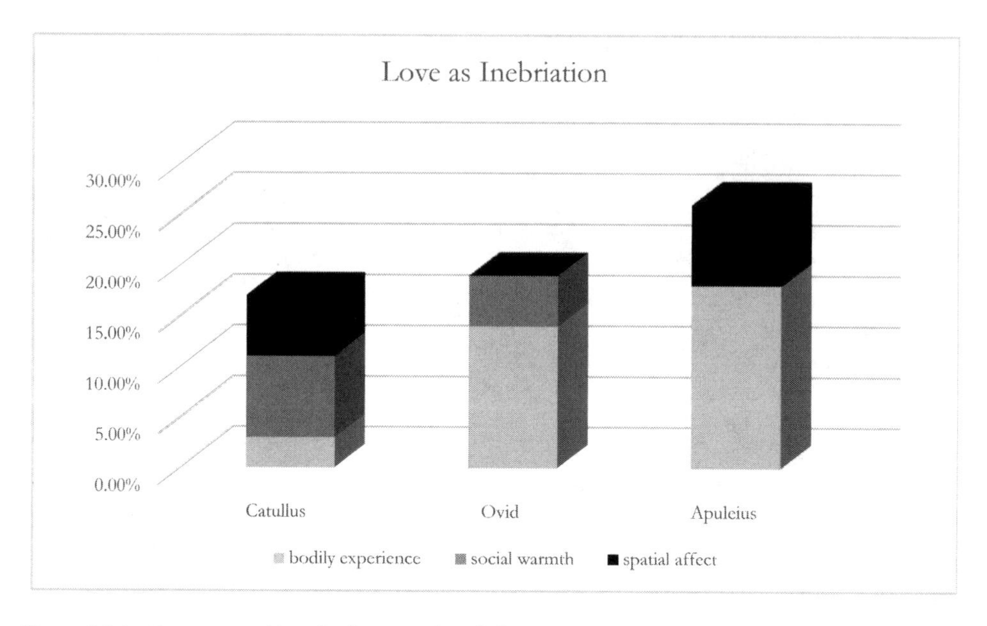

Figure 10.4 Representation of schema-related data across texts

which was quickly able to locate the conceptual overlaps that occur in the texts. The reader should thus note that this limited case study is strictly for illustration and is not meant to be definitive. These passages were selected for their marked similarities or differences in terms of schematic content. I am not suggesting they have a priority amongst passages that explicitly or implicitly associate love with drunkenness (see e.g. Pl. *Aul.* 4.10.19–21, Verg. *Aen.* 1.749, Sen. *Ep.* 105.6, Sen. *Med.* 69, etc.), merely that they illustrate similarities and differences well, and therefore provide a useful illustration of schematic textual features and how they might be interpreted.

This graphic representation of schematic content, by visualizing social concepts and values in terms of the bodily experiences underpinning the texts, helps to illustrate how enactive analogies operate in literature. That is to say, Figure 10.4 demonstrates that bodily schemas generate expressions of value that scaffold the propositional content of the texts. The emotion associated with the inebriation metaphor for Ovid's male audience was an internal experience rather than an externalized one. Spatial affect is absent for Ovid's male subject, and emphasized for the self-objectifying female. Social warmth is present in both narratives that engage the metaphor in the context of male emotional experience, but it is absent in the narrative that engages the metaphor to construct female emotional experience.[7] So, love-as-drunkenness was not constructed as a universal experience, but rather as a gendered one. Male emotion radiates from an internal locus of control, while female emotion is self-objectified and associated with movement that is downward and away. Socially structured understandings of subjective experience thus emerge from the body itself, providing something along the lines of what Pierre Bourdieu called *le sens du jeu* – a feel for the game. At this level of pre-linguistic expression, we can both more deeply appreciate the ways in which ideology and identity, as well as topic and genre, interact in ancient literature, and begin to interrogate the role bodily schemas might have in (re)producing sentiment and culture.

Conclusion

Although the degree to which Apuleius may have assumed knowledge on the part of his readers of the material he seemingly draws from is a vexed question, it is, I should think, reasonable to expect, given the popularity of Stoicism, as well as the volume of discussion about inebriation in the ancient world, that many readers of both genders would have recognized his expertise (cf. Bowditch 2005: 283). But even if they didn't – which we cannot know – it is still of value to identify the influence of these various bodies of knowledge, all of which are structured around the body, on the structure of the text. The *enacted analogy* developed through the *amor*-as-drunken-body metaphor demonstrates how, by expanding upon Cognitive Metaphor Theory, we can observe the transhistorical body (re)create culture through the scaffolding of affect and knowledge about the self and others. It provides readers with a full-bodied account of how a segment of text conveys meaning, showing that the body-*as*-metaphor is essential to Roman processes of figuration, which, like ours, are *intercorporeal.* Attention to *intercorporeality* can thus complement any number of other ways of reading texts. It provides a model for organizing knowledge that is structured around the body, saving us from reducing passages like these to decorative metaphors, conventional language, literary practices, etc., failing to recognize the significant role of the body in socially structuring the cognitive unconscious, which we inhabit from within our own embodied readerly context.

Notes

1 Tremendous thanks to Peter Meineck and William Short, both of whom have offered expert guidance, charitable insights and tireless support. Thanks also to Douglas Cairns for his comments and patience with me through myriad growing pains. And, as always, heartfelt gratitude to Thomas Habinek for shepherding and inspiring me in so many ways.

2 www.gettyimages.co.uk/detail/photo/crying-baby-reaching-for-mother-royalty-free-image/74010161.

3 https://commons.wikimedia.org/wiki/Category:Apollo_and_Daphne_(Bernini)#/media/File:Apollo_and_Daphne_by_Bernini_(Galleria_Borghese).jpg.

4 While perhaps not informed by the same scholarship, the connection between body and meaning has not been lost on philologists, who have already begun to explore the relationship between bodily movement and emotion in Roman texts (see e.g. Zanobi 2008; Kubiak 1981). Nor was this lost on ancient authors. Aristotle, for example, recognized the importance of movement to crafting metaphor (*Rhet.* 3.11.1:1411b), and both Cicero and Seneca regularly identify various mental states as *motus animi*, movement of (some embodied aspect of) mind.

5 On anger as approach oriented see (e.g.) Harmon-Jones et al. 2008.

6 Such negative judgement of metaphorical drunkenness is consistent with male perceptions of female behaviour in the Roman world. Authors frequently tell their readers that excessive wine-drinking on the part of women is associated with adultery (see Edwards 1993 who cites examples from Dion.Hal. 2.25.5–7; Val.Max. 2.1.5; Gel. 10.23). Apuleius himself, drawing on a long-standing association between wine, adultery and murder by poisoning (see Bauman 1992), crafts invective around the connection between wine and sex when telling of the Miller's wife (*Met.*9.14), a cruel and licentious woman who killed her husband with witchcraft. Such associations underpin Valerius Maximus' telling of Egnatius Mecennius, who, without criticism, beat his own wife to death with a club for drinking, on which Maximus remarks: *et sane quaecumque femina vini usum immoderate appetit, omnibus et virtutibus ianuam claudit et delictis aperit* ('and surely it is the case that any female who seeks to drink wine without moderation closes the door to every virtue and opens it to every vice', 6.3.9. See also Pliny *N.H.* 14.89). The imperilment of virtue by way of wine also led Augustus, when he exiled his daughter, Julia, first to Pandateria and then to Rhegium, to deny her wine and the ability to see any man without his permission (Bauman 1992). Even in a ritual context, drinking was thought to remove women from their proper role and behaviours (E.g. *AUC* 39.8.5–8).

7 I considered whether to include *lumen* and/or *lucerna* as metonymies that could fit into the social-warmth schema, and ultimately decided against it at this stage. If one were to include either term, then

there would seem to be an element of social warmth in the Cupid and Psyche episode, but because Cupid is physically injured by hot oil from the lamp – a manmade object – the association would not appear to be with social warmth in the positive sense for which I measured. Because light is not generally associated with temperature, because heat as an element of metaphor and analogy is not limited to expressions of social closeness, and because the introduction of manmade objects raises a host of questions beyond the scope of this chapter, I excluded both terms from the schema count at this time.

References

Anderson, M. 2010. Neural reuse: A fundamental organizational principle of the brain. *Behavioral and Brain Sciences*, 33, 245–66.

Aziz-Zadeh, L., Wilson, S., Rizzolatti, G., and Iacoboni, M. 2006. Congruent embodied representations for visually presented actions and linguistic phrases describing actions. *Current Biology*, 16(18), 1818–23. doi: 10.1016/j.cub.2006.07.060.

Barsalou, L. W. 2008. Grounded cognition. *Annual Review of Psychology*, 59, 617–45.

Bauman, R. A. 1992. *Women and Politics in Ancient Rome*. Routledge.

Bowditch, P. L. 2005. Hermeneutic uncertainty and the feminine in Ovid's Ars Amatoria: The Procris and Cephalus digression, in Acona, R. and Greene, E. (eds.), *Gendered Dynamics in Latin Love Poetry*. Arethusa Books.

Buccino, G., Riggio, L., Melli, G., Binkofski, F., Gallese, V., and Rizzolatti, G. 2005. Listening to action related sentences modulates the activity of the motor system: A combined TMS and behavioral study. *Cognitive Brain Research*, 24(3), 355–63. doi: 10.1016/j.cogbrainres.2005.02.020.

Butler, J. 1997. Performative acts and gender constitution: An essay on phenomenology and feminist theory, in Conboy, K., Medina, N. and Stanbury, S. (eds.), *Writing on the Body: Female Embodiment and Feminist Theory*. Columbia University Press.

Butler, J. 2011. *Gender Trouble: Feminism and the Subversion of Identity*. Routledge.

Cairns, D. 2017. Horror, pity, and the visual in ancient Greek aesthetics, in Cairns, D. and Nelis D. P. (eds.), *Emotions in the Classical World: Methods, Approaches, and Directions*. HABES 59, Steiner.

Clark, B., Mahato, N., Nakazawa, M., Law, T., and Thomas, J. 2014. The power of the mind: The cortex as a critical determinant of muscle strength/weakness. *Journal of Neurophysiology*, 112(12), 3219–26. doi: 10.1152/jn.00386.2014.

de Beauvoir, S. 1974. *The Second Sex*, trans. H. M. Parshley. Vintage.

Decety, J., Jeannerod, M., Germain, M., and Pastene, J. 1991. Vegetative response during imagined movement is proportional to mental effort. *Behavioral Brain Research*, 42(1), 1–5. doi: 10.1016/S0166–4328(05)80033–6.

Desai, R., Binder, J., Conant, L., Mano, Q., and Seidenberg, M. 2011. The neural career of sensory-motor metaphors. *Journal of Cognitive Neuroscience*, 23(9), 2376–86. doi: 10.1162/jocn.2010.21596.

Devereaux, J. J. 2016. Embodied historiography: Models for reasoning in Tacitus's *Annales*, in W.M. Short (ed.), *Embodiment in Latin Semantics*. John Benjamins. 237–68.

Edwards, C. 1993. *The Politics of Immorality in Ancient Rome*. Cambridge University Press.

Elliot, A. J. and Covington, M. V. 2001. Approach and avoidance motivation. *Educational Psychology Review*, 13(2), 73–92.

Fedriani, C. 2016. Ontological and orientational metaphors in Latin: Evidence from the semantics of feelings and emotions, in Short, W. M. (ed.), *Embodiment in Latin Semantics*. John Benjamins. 115–40.

Feldman, J. 2006. *From Molecule to Metaphor: A Neural Theory of Language*. The MIT Press.

Gallese, V. 2007. Before and below Theory of Mind: Embodied simulation and the neural correlates of social cognition. *Philosophical Transactions of the Royal Society of London* B, 362, 659–69.

Gallese, V. 2016. The multimodal nature of visual perception: Facts and speculations. *Gestalt Theory* 38(2/3), 127–40.

Gallese, V. and Cuccio, V. 2015. The paradigmatic body: Embodied simulation, intersubjectivity and the bodily self, in Metzinger, T. and Windt, J. M. (eds.), *Open MIND*. MIND Group. 1–23.

Gallese, V. and Lakoff, G. 2005. The brain's concepts: The role of the sensory-motor system in conceptual knowledge. *Cognitive Neuropsychology*, 22(3–4), 455–79.

Gibbs, R. W. 2006. Metaphor interpretation as embodied simulation. *Mind and Language*, 21(3), 434–58.

Gibbs, R. W. 2017. *Metaphor Wars: Conceptual Metaphors in Human Life*. Cambridge University Press.

Glenberg, A. and Gallese, V. 2012. Action-based language: A theory of language acquisition production and comprehension. *Cortex*, 48, 905–22.

Glenberg, A., Sato, M., Cattaneo, L., Riggio, L., Palumbo, D., and Buccino, G. 2008. Processing abstract language modulates motor system activity. *Quarterly Journal of Experimental Psychology*, 61(6), 905–19. doi: 10.1080/17470210701625550.

Gold, B. K. ed. 2012. *A Companion to Roman Love Elegy* (Vol. 187). John Wiley & Sons.

Green, P. 2005. *The Poems of Catullus: A Bilingual Edition*. University of California Press.

Habinek, T. 2011. Tentacular mind: Stoicism, neuroscience, and the configurations of physical reality, in Stafford, B. M. (ed.), *A Field Guide to a New Meta-Field: Bridging the Humanities-Neuroscience Divide*. The University of Chicago Press. 64–83.

Hanson, J. A. 1989. *Apuleius Metamorphoses 1–6*. Harvard University Press.

Harmon-Jones, E., Peterson, C., Gable, P. A., and Harmon-Jones, C. 2008. Anger and approach-avoidance motivation, in Elliot, A. J. (ed.), *Handbook of Approach and Avoidance Motivation*. Psychology Press. 399–413.

Hauk, O., Johnsrude, I., and Pulvermüller, F. 2004. Somatotopic representation of action words in human motor and premotor cortex. *Neuron*, 41, 301–7. doi: 10.1016/S0896–6273(03)00838–9.

Kable, J., Lease-Spellmeyer, J., and Chatterjee, A. 2002. Neural substrates of action event knowledge. *Journal of Cognitive Neuroscience*, 14, 795–805. doi: 10.1162/08989290260138681.

Kleinsmith, A. and Bianchi-Berthouze, N. 2013. Affective body expression perception and recognition: A survey. *IEEE Transactions on Affective Computing*, 4(1), 15–33.

Klepp, A., Weissler, H., Niccolai, V., Terhalle, A., Geisler, H., and Schnitzler, A. 2014. Neuromagnetic hand and foot motor sources recruited during action verb processing. *Brain and Language*, 128(1), 41–52. doi: 10.1016/j.bandl.2013.12.001.

Kövecses, Z. 2002. *Metaphor: A Practical Introduction*. Oxford University Press.

Kubiak, D. P. 1981. The Orion episode of Cicero's Aratea. *Classical Journal*, 77, 12–22.

Lakoff, G. 1987. *Women, Fire, and Dangerous Things: What Categories Reveal about the Mind*. Chicago University Press.

Lakoff, G. 2008. The neural theory of metaphor, in Gibbs, R. W. (ed.), *The Cambridge Handbook of Metaphor and Thought*. Cambridge University Press.

Lakoff, G. and Johnson, M. 1980. *Metaphors We Live By*. University of Chicago Press.

Lakoff, G. and Johnson, M. 1999. *Philosophy in the Flesh*. Basic Books.

Mankin, D. ed. 2011. *Cicero De Oratore Book III*. Cambridge University Press.

Mead, G. H. 1934. *Mind, Self and Society*. University of Chicago Press.

Meier, B. P., Schnall, S., Schwarz, N., and Bargh, J. A. 2012. Embodiment in social psychology. *Topics in Cognitive Science*, 4(4), 705–16.

Merleau-Ponty, M. 1962. The body in its sexual being, in *The Phenomenology of Perception*, trans. Colin Smith. Routledge and Kegan Paul.

Piaget, J. 1970. *Structuralism*. Basic Books.

Pulvermüller, Friedemann. 2005. Brain mechanisms linking language and action. *Nature Reviews Neuroscience*, 6(7), 576–82. doi: 10.1038/nrn1706.

Pulvermüller, Friedemann. 2008. Brain embodiment of category-specific semantic memory circuits, in Semin, G. and Smith, E. (eds.), *Embodied Grounding: Social, Cognitive, Affective, and Neuroscientific Approaches*. Cambridge University Press. 71–97. doi: 10.1017/CBO9780511805837.004.

Rackham, H. 1948. *Cicero, De Oratore Book III*. Harvard University Press.

Rackham, H. 1968. *Pliny Natural History 12–16*. Harvard University Press.

Raposo, A., Moss, H., Stamatakis, E., and Tyler, L. 2009. Modulation of motor and premotor cortices by actions, action words and action sentences. *Neuropsychologia*, 47(2), 388–96.

Riggsby, A. M. 2015. Tyrants, fire, and dangerous things, in Williams, G. D. and Volk, K. (eds.), *Roman Reflections: Studies in Latin Philosophy*. Oxford University Press.

Sanford, A. J. and Emmott, C. 2012. *Mind, Brain and Narrative*. Cambridge University Press.

Seale, C. ed. 2004. *Researching Society and Culture*. Sage.

Short, W. M. 2008. Thinking places, placing thoughts: Spatial metaphors of mental activity in Roman culture. *Quaderni del Ramo d'Oro*, 1, 106–29.

Short, W. M. 2012. A Roman folk model of the mind. *Arethusa*, 45(1), 109–47.

Short, W. M. 2013. Getting to the truth: Metaphors of mistakenness in Greek and Latin. *Arion*, 21(2), 111–40.

Spellman, B. A. and Schnall, S. 2009. Embodied rationality. *Queen's Law Review*, 35, 117–64.

Tettamanti, M., Buccino, G., Saccuman, M., Gallese, V., Vittorio, D., Massimo, S., Paola, F., Ferruccio, R., Giacomo, C., Stefano, F. C., and Perani, D. 2005. Listening to action-related sentences activates fronto-parietal motor circuits. *Journal of Cognitive Neuroscience*, 17(2), 273–81. doi: 10.1162/0898929053124965.

Varela, F. J., Thompson, E., and Rosch, E. 2016. *The Embodied Mind: Cognitive Science and Human Experience*. MIT Press.

Williams, L. E., Huang, J. Y., and Bargh, J. A. 2009. The scaffolded mind: Higher mental processes are grounded in early experience of the physical world. *European Journal of Social Psychology*, 39(7), 1257–67.

Wilson, M. 2002. Six views of embodied cognition. *Psychonomic Bulletin and Review*, 9, 625–36.

Zanobi, A. 2008. The influence of pantomime on Seneca's tragedies, in Hall, E. and R. Wyles (eds.), *New Directions in Ancient Pantomime*. Oxford University Press. 227–57.

Part III

Social cognition

11

Group identity and archaic lyric

We-group and out-group in Alcaeus 129

Jessica Romney

Around the beginning of the sixth century BCE in the polis of Mytilene on Lesbos, Pittacus the son of Hyrrhas secured his position as tyrant and thus gave material for a poetic campaign that would paint him as slovenly, power-hungry, and gluttonous. Alcaeus, Pittacus' contemporary and political rival, created a poetic character of Pittacus that made the tyrant into the ultimate out-group to the sympotic group(s) composed of Alcaeus and his *hetairoi* so that Pittacus alone stood for his entire group of supporters. Alcaeus' portrayal of Pittacus ostracized the tyrant from the convivial group of the *symposion* by denying him the sympotic traits of restraint, moderation, and loyalty: the man who ruled Mytilene alone drank alone, opposite in every way to the sympotic and political group to which the poet and his audience belonged. Fragment 129, where Alcaeus gives one of the fullest descriptions of Pittacus and his actions, engages in a process of group identity construction. The fragment characterizes Pittacus in negative, asymptotic terms, and the implicit and explicit oppositions drawn between the in-group (Alcaeus and audience) and out-group (Pittacus and his invisible supporters) characterize the poet's group in turn as well. By attacking his political opponent, Alcaeus constructs a group identity for those listening and attempts to persuade them of its importance to their social identity.

This contribution applies a social identity perspective to the study of ancient identity and analysis of archaic lyric. In the following pages, I first present an overview of the social identity approach, which examines the cognitive and social processes and structures underlying group formation and social identities. I then turn to a case study of social identity construction in the context of the archaic *symposion* through an analysis of Alcaeus 129 and the poet's use of in-group (Us)/out-group (Them) framing, where Alcaeus uses a Them-group (Pittacus) to encourage the coherence of an Us-group (Alcaeus and his audience).

Social identity

Social identity is the conception of self as determined by membership in a group and/or category and involves a strong inter-relational element whereby levels of similarity with other individuals are established; in one succinct definition, social identity is "who we are to one another."[1] Individuals can have multiple social identities that share a common core and which are primed by the social context and group at hand. Yet while identity is often spoken of as a thing, identity theorists using a discourse analysis perspective have shown that "identity" is in fact a public phenomenon, performed in discourse or "language in use" as connected to a community's social structures and institutions.[2] Although social identity is not wholly produced through language use, social discourses and the values, behaviors, and social norms perpetuated through discourse help to shape, maintain, and adapt group identities. "Identity" should thus be thought of as "identification,"[3] as an ongoing process of asserting similarity to the group or category reference point.

Groups are collections of individuals (two or more) who recognize their shared membership in the collective organization and whose existence is recognized by at least one other outside of the group.[4] Psychologically, members of a group relate themselves to it "subjectively for social comparison and the acquisition of norms and values" and thereby privately accept membership in the group; this group membership then influences individual attitudes and behaviors so that there is a leveling effect as group members come to a common consensus on what attitudes, behaviors, and values define the group.[5] There are three criteria for group formation: "identity" (a cognitive criterion whereby individuals have a collective awareness of their shared membership and common identity); "interdependence" (group members should be positively interdependent, whether motivationally, socially, or psychologically); and "social structure" (the gradual stabilization of intragroup relations and the development of a) role and status differentiations and b) social norms and values that regulate the group's beliefs, attitudes, and conduct).[6]

As noted above, this contribution adopts a social identity perspective concerning social groups and their shared identities, combined with a discourse analysis approach to analyze the construction of social identities via literary production (here, sympotic lyric). Focusing on the cognitive processes and structures of group membership alongside the associated social processes and structures, the social identity perspective[7] seeks to explain how "people relate to one another as members of social groups and categories"[8] within the specific social context(s) which give meaning to the cognitive and psychological processes underlying the development of social identities. At the heart of group formation are the cognitive processes whereby an individual's conception of self as a biographical individual is exchanged for a conception of self as an exemplar of the group who is interchangeable with any other group member ("depersonalization"). This process is, according to Turner, the basic cognitive process underlying group phenomena such as cohesion, social stereotyping, in-group/out-group categorizing, coordinated action, and so forth.[9] As individuals tend to view themselves positively, they come to view the social groups on which their sense of self depends positively as well, which leads in turn to in-group bias and social stereotypes.[10]

An in-group is a positive reference group for an individual's social identity, while an out-group is a negative reference group which exists on a spectrum from "not Us but ok" to "not Us and major threat."[11] It is this opposition between the Us and Them that fosters group cohesion and the depersonalization process whereby individuals identify themselves with/as the ideal group member. In the case of we-identities, there are two types: the first, an "internal-we," is based on an assertion of shared characteristics amongst the group members (*we* are all the same, "we-hood"), while the second, the "external-we," takes its sense of we-ness from a contrast with a group of others (*Us* versus Them, "us-hood").[12] The we-group in fragment 129 is an external-we as an in-group Us against the out-group of Pittacus.

A quick note on Pittacus as the out-group in fragment 129: while a group, by definition, consists of multiple individuals, in fragment 129 Pittacus acts alone, though the final stanza may have linked him with Myrsilus, the previous tyrant of Mytilene (v. 28; cf. fr. 70). Yet it is unlikely that Pittacus acted without political support or left the *hetaireia* to which he once belonged and to which Alcaeus still belongs on his own; those who supported him and/or had ties of *philia* and kinship to him would have likely also left with him. The fact that Pittacus became tyrant and held on to the tyranny especially suggests this, as he would have needed allies. His former *hetaireia* would have been a prime place for recruiting them in addition to finding supporters from outside of the *hetaireia*. Pittacus in fragment 129 thus stands as the figurehead of his group, and Alcaeus focuses on him to the point that Pittacus' allies can be forgotten and Pittacus thereby stands as a singular Them on which the we-group (Us) can focus their vitriol.[13]

When we turn to archaic lyric and Alcaeus' poetry, we see how the social groups and categories of "elite/*agathos*," "*hetaireia* against the tyrant," "symposiast," etc. come together to create a group encompassing Alcaeus and his audience whose values and social norms revolve around appropriate sympotic behavior (shared, moderate consumption of wine), political action against the tyrant(s) of Mytilene, and group loyalty. They arise from the social institution of the archaic *symposion* and the political structures of archaic Mytilene, where, as far as can be told, elites operated within power blocks known as *hetaireiai* and competed with one another for political office, sometimes to the point of *stasis*. The reproduction of these values and norms in archaic lyric not only reinforces them by embedding such values in the group's language productions (and thus, every time a poem or theme is repeated, the values and norms are reinforced), but it also maintains their importance for elite groups and the social identities.

The archaic *symposion* was a place of constructive drinking,[14] where the shared consumption of alcohol by a social group makes visible ideal social relationships and the social world in which they exist. For the *symposion*, this social world contains like-minded individuals who all partake in the triad of good birth, education, and economic status, and who are all equal, or are at least notionally so.[15] The speaking rules of the *symposion*—all speak in turn—reinforce the notional equality of the group, while the label of *pistos philos hetairos* ("dear and trusted companion") establishes the principle of group loyalty, and Pittacus, according to Alcaeus, violated both group values in his quest for political power.

The *polis* of Mytilene seems to have been particularly afflicted by political *stasis* in the early archaic period; later writers record a succession of tyrants beginning with the removal of the Penthiliadai in the seventh century: first Melanchrus, then Myrsilus, and finally Pittacus. Myrsilus was likely the tyrant against whom the original oath to free the *demos* in fragment 129 was aimed; Pittacus, who had been a member of the same *hetaireia* as Alcaeus and who participated in this oath to overthrow the tyrant, changed sides and the conspiracy fell apart (fr. 129). Alcaeus and some or all of his *hetairoi* went into exile (possibly several times), while Pittacus married into the Penthiliadai (Alc. 70) and became tyrant himself at some point later (fr. 348). For whatever reason, Pittacus stopped seeing himself as an interchangeable member of his *hetaireia*, which caused a conflict of values when his individual striving for political power clashed with his group's collective attempt for it. His subsequent departure, along with that of any supporters, weakened the social ties amongst the remaining members of the *hetaireia*. Fragment 129, which participates in Alcaeus' rhetorical campaign to blacken Pittacus' name, also serves to strengthen the group's cohesion in the face of Pittacus' departure and in the context of exile by protecting the integrity of the group's shared identity as "we." The out-grouping of Pittacus and his invisible supporters (Them) serves as a catalyst to prompt greater identification on Alcaeus' and his audience's part with the in-group (Us), thereby producing greater cohesion within the group and adherence to the group's social identity.

Fragment 129[16]

]. ρạτα τόδε Λέςβιο]ι
. .]. . . . ς εὔ]δε[ιλον τέμενο]ς μέγα
ξῦ]νον κάτ]. σσα[ν, ἐν δὲ βώ]μοις
ἀθα[νάτων] μακά[ρων ἔθη]καν (4)
κὰ[πωνύ]μασσα[ν ἀντί]αον [Δ]ία,
σὲ δ' Α[ἰολήιαν] κυδα[λιμαν] θ[έο]ν
π ά]ν[των γενέθλαν, τὸν δὲ] τερ[το]ν
τόνδε κεμήλιον] ὠ[ν]ύμα[σσ][α]ν (8)

Ζόννυσον ὠμήσ]ταν. ἄ[γ][ι][τ', εὔνοο]ν
θῦμον ςκέθοντε]ς ἀμμετ[έρα[ς] ἄρας
ἀκούςατ', ἐκ δὲ τῶ]ν[δ]ε μό[χθων
ἀργαλέας τε φύγας ṣ[(12)
τὸν Ὕρραον δὲ παῖδα] πεδελθ[έτω
κήνων Ἐρ[ίνν]υς, ὤς πο]τ' ἀπώμ[νυ]μεν,
τόμοντες ἄ. φ[́.]ν. [ν]
[μηδάμα μηδένα τὼν ἐταίρων, (16)
ἀλλ' ἢ θάνοντες γᾶν ἐπιέμμενοι
κείσεσθ' ὐπ' ἄνδρων οἲ τότ' ἐπικ ̣ ́ ηṇ
ἤπειτα κακκτάνοντες αὔτοις
δᾶμον ὐπὲξ ἀχέων ῥύεςθαι (20)
κήνων ὀ φύσγων οὐ διελέξατο
πρὸς θῦμον, ἀλλὰ βραϊδίως πόσιν
ἔ]μβαις ἐπ' ὀρκίοισι δαπτει
τὰν πόλιν ἄμμι δέδ[.]. . [.]είπαις (24)
οὐ κὰν νόμον [. .]. . ε [ ]. ́ [ ]
γλάυκας α[.]. . [.]. . [
γεγρα. [
Μυρσιλ[(28)
. . .]. [
[ ]
[ ]
.]. . [(32)

. . . This the Lesbians . . . established, a great, visible *temenos* in common, in which they set altars of the blessed immortals (4)
and they named Zeus *Antiaos* and you, glorious Aeolian goddess, the mother of all, and the third, this one they named *Kemelios*, (8)
Dionysus, Eater of Raw Flesh. But come now! Set your *thumos* favorably and hear our prayer. (Save us?) from these trials and painful exile (12)
Let the Erinys of those matters chase down the son of Hyrrhas, since we all once swore, having cut . . . not even one of our companions (16)
but to die and lie clothed in earth at the hands of those men who then . . . or instead to kill them and to save the *dēmos* from its woes. (20)

Of those matters Sausage-Gut had no consideration in his *thumos*, but recklessly
with his feet trampling the oaths he devours our *polis* . . (24)
Not by law . . . grey . . . wrote (?) . . . Myrsilus . . (28)
. . . (32)

Alcaeus 129 and the construction of a we-group

Turning to Alcaeus 129, we can see how the above principles concerning social groups and their identities work in practice in sympotic lyric. Twice in his extant fragments Alcaeus claims that Pittacus had once been considered a *hetairos* but had betrayed his *hetairoi* by joining the tyrant Myrsilus (frr. 70, 129). Fragment 129 addresses the issue of identity under threat from both internal (Pittacus) and external (exile) factors: by leaving the group of *hetairoi*, Pittacus caused internal fissures to the group, and the exile that the we-group experiences in this fragmentation of the group threatens their social identity (poet and audience included). Identity in the Archaic and Classical periods depended on one's place in their *polis*, and exile forcibly removed that anchor.[17] To mitigate the damage posed by these threats, Alcaeus links the we-group in fragment 129 to the larger, historic Lesbian community that founded the *temenos* or shrine in which the fragment is set, while at the same time disassociating the remaining members of the group from Pittacus. As he turns the tyrant into an out-group whose inability to display sympotic restraint reasserts the sympotic and social norms on which the group's integrity depends, the poet seeks to remove any internal threat to the we-group by excising Pittacus from the group's past and turning the internal threat into an external one. Pittacus is not a group member now, and he never was *then*.[18]

Along with the call for the gods "to hear *our* prayer" (v. 10), the verb ἀπώμ[νυ]μεν, "*we* swore" (v. 14), asserts a we-identity on the audience through the associative semantics of the first-person plural, which include the audience/addressees in the verbal and adjectival number along with the speaker.[19] By imposing an affinity between speaker and audience, the first-person plural integrates the latter into the linguistic body of the former and thereby eases the audience's adoption of the speaker's perspective and values.[20] Both uses of the first-person plural in fragment 129 refer to speech acts (praying and swearing), and thus as the poet intones the request "to hear *our* prayer" and asserts that "*we* once swore," the audience joins him in the act of praying and swearing and thereby notionally participates in the entire narrative of fragment 129 and the values structuring it, namely unity in political endeavors and a desire to return to *their polis*.

The opening of a *symposion*, where all involved participated in pouring a series of libations and singing hymns, creates a group through shared action and shared speech; Alcaeus' incorporation of the audience into the acts of praying and swearing through the first-person plural in lines 10 and 14 returns to this initial construction of a unified group through their joint participation in a speech event. As they had earlier opened the *symposion* together, the audience now also pray to return home, for revenge on the son of Hyrrhas, and they too swear to engage in *stasis* by "saving the *demos* from its woes" (v. 20), thereby internalizing the we-identity constructed in the fragment. The we-identity imposed by the speech acts in "*our* prayer" and "*we* swore" aims to embed the audience in the poem's narrative as it asserts a level of sameness between poet and audience, collectively grouped together as "we." It reinforces the assumption that all members of a sympotic group share the same values and desires, and the denigration of Pittacus in line with sympotic values, as we will see below, also contributes to this fiction and the group cohesion based on it. As out-group, Pittacus opposes everything the we-group as in-group is.[21]

While out-groups do not necessarily constitute a major threat, Pittacus as out-group in fragment 129 does. In general, out-groups serve as a foil to the in-group in terms of one or more characteristics that contribute to an Us versus Them mentality. They thus serve, in sympotic poetry, to persuade a poet's audience to identify with the poet's group identity by embodying its opposite and by posing a threat to the sympotic group. These threats can range from an implied opposition to the group (e.g., Archil. 124b) to an explicit opposition between poetic group and out-group (e.g., Sol. 4c);[22] Alcaeus 129 falls into this latter category, with its firm opposition between Pittacus and the we-group who have been harmed by Pittacus' actions. Out-groups are maintained as such through the perpetuation of language or discourses hostile to them by dominant groups,[23] and the ability to label others as members of an "out-group" conveys the social power on the in-group. The venom with which Alcaeus attacks Pittacus as "out-group" in fragment 129 may be due to a struggle for such power, as the poet and his group, who have been made an out-group by Pittacus by merit of their exile from Mytilene, try to wrest back control over who is a political in-group by making Pittacus into an out-group in their poetic productions.

The construction of Pittacus as an out-group begins in line 14 as the poet bids "let an Erinys of those matters chase down the son of Hyrrhas [Pittacus]." The primordial manifestation of justice, the Erinys appears here in her role as the avenger of broken oaths (cf. Hes. *Op*. 803–424), and her presence foreshadows the oath's dissolution even as her position in the line protects the oath which will be re-sworn through the poem's narrative from Pittacus (son of Hyrrhas—Erinys—oath).[24] The stanza, by calling on an Erinys because of a broken oath and only then narrating the swearing of that oath, accuses Pittacus of insincerity, which in turn contributes to the editing of the group's past to protect the efficacy of the we-group in the present: as Pittacus is not part of the group *now*, he must not have been *then*, as evidenced by his disregard of the oath-swearing ritual. The re-swearing of the oath via its narration in line 14 involves only the "we" present at the moment of performance. Pittacus is no longer involved and so the oath is protected and the we-group are no longer oath-breakers,[25] which Pittacus' betrayal had made them by preventing them from accomplishing the oath for whatever reason, for they (re)direct the oath against the current woes afflicting the *demos* (i.e. Pittacus).

By narrating the oath, its contents, and its breaking, Alcaeus retrieves the group's memory of the event and reinforces his version of events along with Pittacus' lack of "true" group membership. As the performance then proceeds, the retrieved memory is (re)consolidated and stored, and every time the poem is performed, a similar process takes place so that the group's collective memory of the event is retrieved, modified to suit the narrative, and then stored. This process of memory reconsolidation[26] informs the group's shared history and its perception of difference between itself as in-group and Pittacus plus supporters as out-group by imputing long-standing incompatibility of values and even hostility toward the in-group on the part of the out-group.

The most explicit out-group characterization of Pittacus, however, comes as the poet names the son of Hyrrhas *Phusgōn*, "Sausage-Gut" (v. 21).[27] This insult draws on a discourse of political consumption that we see in the Homeric epics as well as throughout archaic lyric and which denies the glutton the right to hold political power.[28] It also opposes Pittacus to the *symposion* and its rules of consumption. The *symposion* imposed a "regime of regulation" on its participants,[29] where consumption was to be shared, reciprocal, and in equal amounts. Sympotic consumption, to be considered "moderate," required drinking to pace so that no one consumed more (or less) than the others. And from wine to food to politics: Pittacus' position as tyrant reveals, by this logic, a political appetite that necessitates sole rule and thereby alimentary greed on the tyrant's part. This in turn leads to the oversized paunch suggested by the appellation *Phusgōn*. Pittacus' past actions concerning the oath have already set his interests against those of the we-group, and with the abusive epithet *Phusgōn*, his current actions toward the *polis* continue the opposition between the two and argue for the merit of re-swearing the oath "to save the *demos* from its woes," i.e. Pittacus. Defined as the opposite of Pittacus, the we-group now has a task to complete to reinforce the social cohesion already garnered by the threat Pittacus poses to the group and their long-term social ties as implied by the poem's narrative.[30]

The violent appellation of *Phusgōn* charges that Pittacus has an appetite so extreme that it has affected his bodily appearance, giving him a distended paunch that makes his lack of self-control evident. Even worse, Pittacus acquired this paunch by "wolfing down the *polis*." This statement removes Pittacus from the group of decorous, human consumers (and symposiasts) in two ways. The first comes through δάπτω, "devour, rend, wolf-down," which describes the ravenous consumption of wild animals, as well as destruction wrought by spears and fire; the verb thus has a conceptual frame that denotes the savage rending and tearing of its object.[31] The second removal derives from that which Pittacus devours, namely the *polis* and the body politic. To rend and consume a polis was to rend and consume the demos, so Alcaeus here accuses Pittacus of metaphorical cannibalism, which appears based on the model of Achilles' pronouncement of Agamemnon as a "*demos*-devouring king" (δημοβόρος βασιλεύς [*Il.* 1.231]).[32] As *Phusgōn* and the devourer of the *polis*, Pittacus is as far from moderate sympotic consumption as one can get, and that is the point.[33] Alcaeus not only furthers the out-group characterization of Pittacus through his attack on Pittacus' appetite, but he also reinforces the importance of sympotic consumption and its surrounding values for the we-group who define themselves opposite the tyrant.

The extreme individualization of Pittacus in fragment 129 focuses on the leader at the expense of the followers, exacerbating the threat posed by Pittacus to the we-group. By so doing it also protects the fiction of sameness that holds the group together as established by fragment 129: if only Pittacus left, then there is no need to reevaluate the group's values, since he was never really a member of the group. But if multiple members left with him, then their departure could have ruptured the cohesion of the we-group. By remembering Pittacus so vehemently, Alcaeus and his audience actively forget any supporters.[34] The audience, if they accept this edited recollection of events and players, would then contribute to the process of group identity and cohesion by preserving the fiction of same-mindedness and shared values amongst all group members.

By naming Pittacus *Phusgōn* and devourer of the *polis*, Alcaeus 129 restores the sympotic world Pittacus upturned when he left the group by denying Pittacus trade in the sympotic traits that define the we-group. The insults, created in line with sympotic values and rules, reinforce the importance of those values for the poet and audience, as well as for the wider world of the Greek elite. Pittacus' betrayal thus reconstitutes the sympotic group as the narrative of 129 imposes the rules of sympotic behavior and speech on its audience. Since Pittacus cannot display restraint or loyalty and cannot join in the speech acts of swearing and praying in the fragment's narrative, his departure no longer threatens the efficacy of the we-group, who have been (re)defined in opposition to him.

Conclusions

The process of identity construction is ongoing, and the archaic *symposion* gave performance space to declarations of *agathos*-status and to poems which constructed social identities for their users, connecting them under the headings of *agathos*, symposiast, *hetairos*, hoplite, and so on. The repetition of common themes concerning values and behaviors marks *agathos*-status and its associated identities across sympotic lyric, and the performance of these poems in a social institution was central to elite social life and *agathos*-status, which highlights the values and behaviors of the Greek elite. Sympotic poems such as Alcaeus 129 engaged in an ongoing process of identification between poet and audience and amongst the wider group of *agathoi*, and by understanding how the poems did so, we can then understand the larger processes by which lyric poetry served as social communication and the ways by which the archaic elite presented themselves and others in the *symposion* and constructed their social identities by categorizing in-groups and out-groups.

Alcaeus 129 takes vehement abuse and inserts it into a lyric prayer for revenge so that the narrative marks the poet and his audience as the wronged party in reference to the target (Pittacus) and imposes an in-group/out-group opposition on the parties involved. As the analysis above shows, this opposition depends on the performance context of the *symposion*: Pittacus as out-group cannot possess the traits and values that designate the we-group as proper symposiasts, nor can he participate in the shared speech of sympotic discourse that marks those who share speech and drink as a single, cohesive group bound by a fiction of sameness. The discourses which define Alcaeus' we-group identity insist on the group's adherence to the shared values of loyalty, political action on behalf of the *demos*, and moderate appetite. On their own, these values are commonplace, but applied to the specific scenario of fragment 129—exile and betrayal—they become important parts in the process of identification.

Notes

1 Bethan Benwell and Elizabeth Stokoe, *Discourse and Identity* (Edinburgh: Edinburgh University Press, 2006), 4.
2 See, among others, Jenkins (2004); Benwell and Stokoe (2006); and Hall (2000), 19; Coupland and Jaworski (2009), 11. For an introduction to discourse analysis, see the works collected in Jaworski and Coupland (2006a) as well as Fowler (1981, 1996) and Fairclough (1995, 2003).
3 Following Jenkins (2004), 5.
4 Rupert Brown, *Group Processes*, 2nd ed. (Oxford: Blackwell Publishers, 1988), 3.
5 John C. Turner *et al.*, *Rediscovering the Social Group: A Social Categorization Theory* (Oxford: Basil Blackwell, 1987), 1–2. See also Matelski and Hogg (2015) for an introduction to the social psychological dimension to group processes.
6 Turner *et al.*, *Rediscovering the Social Group*, 19–20.
7 This approach encompasses Social Identity Theory or SIT (Tajfel 1979, 1982; Tajfel and Turner 1979, 1985), Social Categorization Theory or SCT, which is also called the social identity theory of the group (Turner 1982, 1985, among others; Turner *et al.* 1987; Haslam and Reicher 2015), and the Social Identity or SI approach (Abrams 1999). Both SCT and SI developed from SIT; SIT and SCT stress the cognitive processes underlying group identities and categorization while SI focuses on the social structures and process affecting the cognitive processes underlying group formation. For introductions to the social identity perspective see Hogg and Abrams (1999), Operio and Fiske (1999), and Reynolds (2015).
8 Dominic Abrams, "Social Identity, Social Cognition, and the Self: The Flexibility and Stability of Self-Categorization," in *Social Identity and Social Cognition*, ed. Dominic Abrams and Michael A. Hogg (Oxford: Blackwell, 1999), 197.
9 Turner *et al.*, *Rediscovering the Social Group*, 50.
10 This assumption underlies the SIT developed by Tajfel and Turner (n. 7 above).
11 Garret Fagan, *The Lure of the Arena: Social Psychology and the Crowd at the Roman Games* (Cambridge: Cambridge University Press, 2011), 159–60; the examples he gives for a neutral or even positive out-group is Canadians to the American in-group.
12 Mühlhäusler and Harré (1990) employ the terms "internal-we" and "external-we"; "we-hood" and "us-hood" come from Hylland Eriksen (1995).
13 This process of "active forgetting" is discussed further below.
14 O'Connor (2015, 101) applies the concept, which is developed by Douglas (1987), to the *symposium*.
15 This is best seen in the Theognidea, but most sympotic poems assume that the audience group is composed of like-minded individuals who are notionally equal and of similar social background.
16 Text from Hutchinson (2001); translation is my own.
17 Jan Felix Gaertner, "The Discourse of Displacement in Greco-Roman Antiquity," in *Writing Exile: The Discourse of Displacement in Greco-Roman Antiquity and Beyond*, ed. Jan Felix Gaertner (Leiden: Brill, 2007), 11.

18 In the pages that follow, the "Pittacus" whom I discuss is a poetic construct; I am not interested so much in whether the Pittacus of fr. 129 is historically accurate or not, but in how Pittacus is used as an out-group to help cohere the we-group.
19 See Wechsler (2010, 333) further on the associative semantics of the first and second-persons plural.
20 Mühlhäusler and Harré call this the "integrative function" of the first-person plural (1990, 174–5).
21 There is the issue that the out-group of fr. 129 is an out-group of a single individual; see the discussion at the end of this section, which addresses this point.
22 While all genres of sympotic poetry contain instances of out-groups, the strategy is particularly popular in iambic poetry and in poems adopting an iambic stance, where the speaker is the wounded, yet in the right, party.
23 Such as, for example, the perpetuation of racist discourses against ethnic minorities (van Dijk *et al.* 1997).
24 See Konstantinidou on the connection between the Erinys and oaths, which is part of his larger analysis of oaths as conditional self-curses (2014, 9–11; for fr. 129 as an example of such, p. 20 n. 54).
25 Anne Pippin Burnett, *Three Archaic Poets: Archilochus, Alcaeus, Sappho* (London: Duckworth, 1983), 160.
26 See Alberini and LeDoux (2013) for an introduction to memory consolidation, as well as Alberini's work in general.
27 φύσγων [=φύσκων] derives from φύσκη, the large intestine and any sausage or pudding made from stuffing it (LSJ, s.v. φύσκη). "Fatty" is the usual translation for *phusgōn*, which is visually accurate; current English usage of "fatty," however, while damaging to its targets, does not carry the violence of appetite connoted by *phusgōn*. Bachvarova connects Pittacus' swollen gut to his oath-breaking (2007: 185–6, with Near Eastern and Greek parallels).
28 We see this discourse in Achilles' insult of Agamemnon as a "*demos*-devouring king" (*Il.* 1.231) and in the actions of the suitors in the *Odyssey*. It also appears in Solon's poems, the Theognidea, and as a general social commentary throughout lyric, particularly iambus. See Davidson on the politics of appetite in Classical Athens (1997).
29 Sean Corner, "Transcendent Drinking: The Symposium at Sea Reconsidered," *Classical Quarterly* 60 (2010): 357–8.
30 The accomplishment of a joint task is the basis of task cohesion, which has been shown to develop strong, short-term cohesion within a group, even when it is made up of strangers. By assigning a task, Alcaeus increases the group's social cohesion. Weaker social cohesion, which depends on social bonds and trust within a group, survives longer than task cohesion, which depends on the job at hand. For an introduction to group cohesion, see Dion (2000) and Siebhold (2006).
31 LSJ s.v. δάπτω.
32 Franco Ferrari, *Sappho's Gift* (Ann Arbor, MI: Michigan Classical Press, 2010), 91.
33 On the relationship between insults and the social institutions in which they are uttered, see Collins (2004), 233.
34 Flower has identified two tendencies in the practice of "active forgetting": "the urge to remember the villain so that his fate might be a warning to others" and "to obliterate his name and career as if he had never existed" (1998, 180). Alcaeus applies the first to Pittacus, and the second to any supporters Pittacus may have had.

References

Abrams, Dominic. "Social Identity, Social Cognition, and the Self: The Flexibility and Stability of Social-Categorization." In *Social Identity and Social Cognition*, edited by Dominic Abrams and Michael A. Hogg, 197–229. Oxford: Blackwell Books, 1999.

Alberini, Cristina M. and Joseph E. LeDoux. "Memory Reconsolidation." *Current Biology* 23 (2013): 746–50.

Bachvarova, Mary R. "Oath and Allusion in Alcaeus fr. 129." In *Horkos: The Oath in Greek Society*, edited by Alan H. Sommerstein and Judith Fletcher, 179–88. Exeter: Bristol Phoenix Press, 2007.

Benwell, Bethan and Elizabeth Stokoe. *Discourse and Identity*. Edinburgh: Edinburgh University Press, 2006.

Brown, Rupert. *Group Processes*, 2nd edition. Oxford: Blackwell Publishers, 2000.

Burnett, Anne Pippin. *Three Archaic Poets: Archilochus, Alcaeus, Sappho*. London: Duckworth, 1983.

Collins, Derek. *Master of the Game: Competition and Performance in Greek Poetry*. Washington, D.C.: Centre for Hellenic Studies, 2004.

Corner, Sean. "Transcendent Drinking: The Symposium at Sea Reconsidered." *Classical Quarterly* 60 (2010): 352–80.

Coupland, Nikolas and Adam Jaworski. "Social Worlds through Language." In *The New Sociolinguistics Reader*, edited by Nikolas Coupland and Adam Jaworski, 1–22. Basingstoke: Palgrave Macmillan, 2009.

Davidson, James. *Courtesans and Fishcakes: The Consuming Passions of Classical Athens.* London: HarperCollinsPublishers, 1997.

Dion, Kenneth K. "Group Cohesion: From 'Field of Forces' to Multidimensional Construct." *Group Dynamics: Theory, Research, Practice* 4 (2000): 7–26.

Douglas, Mary, ed. *Constructive Drinking: Perspectives on Drink from Anthropology*. Cambridge: Cambridge University Press, 1987.

Fagan, Garrett G. *The Lure of the Arena: Social Psychology and the Crowd at the Roman Games.* Cambridge: Cambridge University Press, 2011.

Fairclough, Norman. *Critical Discourse Analysis: The Critical Study of Language*. London: Longman, 1995.

—. *Analysing Discourse: Textual Analysis for Social Research.* London: Routledge, 2003.

Ferrari, Franco. *Sappho's Gift: The Poet and Her Community*. Translated by Benjamin Acosta-Hughes and Lucia Prauscello. Ann Arbor, MI: Michigan Classical Press, 2010.

Flower, Harriet I. "Rethinking 'Damnatio Memoriae': The Case of Cn. Calpurnius Piso Pater in AD 20." *Classical Antiquity* 17 (1998): 155–87.

Fowler, Roger. *Literature as Social Discourse: The Practice of Linguistic Criticism.* London: Batsford Academic and Educational Ltd., 1981.

—. "On Critical Linguistics." In *Texts and Practices: Readings in Critical Discourse Analysis*, edited by Carmen Rosa Caldas-Coulthard and Malcolm Coulthard, 3–14. London: Routledge, 1996.

Gaertner, Jan Felix. "The Discourse of Displacement in Greco-Roman Antiquity." In *Writing Exile: The Discourse of Displacement in Greco-Roman Antiquity and Beyond*, edited by Jan Felix Gaertner, 1–20. Leiden, Brill, 2007.

Hall, Stuart. "Who Needs Identity?" In *Identity: A Reader*, edited by Peter du Gay, Jessica Evans, and Peter Redmen, 15–30. London: Sage Publications Inc., 2000.

Handler, Richard. "Is 'Identity' a Useful Cross-Cultural Concept?" In *Commemorations: The Politics of National Identity*, edited by John R. Gillis, 27–40. Princeton, NJ: Princeton University Press.

Haslam, S. Alexander and Stephen D. Reicher. "Self-Categorization Theory." In *International Encyclopedia of the Social and Behavioral Sciences*, 2nd edn., edited by James D. Wright. Elsevier, 2015. www.sciencedirect.com/science/referenceworks/9780080970875.

Hogg, Michael A. and Dominic Abrams. "Social Identity and Social Cognition." In *Social Identity and Social Cognition*, edited by Dominic Abrams and Michael A. Hogg, 1–25. Oxford: Blackwell Books, 1999.

Hutchinson, G.O. *Greek Lyric Poetry: A Commentary on Selected Larger Pieces: Alcman, Stesichorus, Sappho, Alcaeus, Ibycus, Anacreon, Simonides, Bacchylides, Pindar, Sophocles, Euripides*. Oxford: Oxford University Press, 2001.

Hylland Eriksen, Thomas. "We and Us: Two Modes of Group Identification." *Journal of Peace Research* 32 (1995): 427–36.

Jaworski, Adam and Nikolas Coupland, eds. *The Discourse Reader*, 2nd edition. Abingdon: Routledge, 2006a.

—. "Introduction: Perspectives on Discourse Analysis." In *The Discourse Reader*, 2nd edition, edited by Adam Jaworski and Nikolas Coupland, 1–37. Abingdon: Routledge, 2006b.

Jenkins, Richard. *Social Identity*, 2nd edition. London: Routledge, 2004.

Konstantinidou, Kyriaki. "Oath and Curse." In *Oaths and Swearing in Ancient Greece*, edited by Alan H. Sommerstein and Isabelle C. Torrence, 6–47. Göttingen: De Gruyter.

Matelski, Monique H. and Michael A. Hogg. "Group Processes, Social Psychology of." In *International Encyclopedia of the Social and Behavioral Sciences*, 2nd edn., edited by James D. Wright. Elsevier, 2015. www.sciencedirect.com/science/referenceworks/9780080970875.

Mühlhäusler, Peter and Rom Harré. *Pronouns and People: The Linguistic Construction of Social and Personal Identity*. Oxford: Basil Blackwell, 1990.

O'Connor, Kaori. *The Never-Ending Feast: The Anthropology and Archaeology of Feasting*. London: Bloomsbury, 2015.

Operio, Don and Susan T. Fiske. "Integrating Social Identity and Social Cognition: A Framework for Bridging Diverse Perspectives." In *Social Identity and Social Cognition*, edited by Dominic Abrams and Michael A. Hogg, 26–54. Oxford: Blackwell Books, 1999.

Reynolds, Katherine J. "Social Identity in Social Psychology." In *International Encyclopedia of the Social and Behavioral Sciences*, 2nd edn., edited by James D. Wright. Elsevier, 2015. www.sciencedirect.com/science/referenceworks/9780080970875.

Siebhold, Guy. "Military Group Cohesion." In *Military Life: The Psychology of Serving in Peace and Combat*, vol. I, edited by T.W. Britt, C.A. Castro, and A.B. Adler, 185–201. Westport, CT: Praeger Security International, 2006.

Tajfel, Henri. *Differentiation between Social Groups: Studies in the Social Psychology of Intergroup Relations*. London: Academic Press, 1979.

—, ed. *Social Identity and Intergroup Relations*. Cambridge: Cambridge University Press, 1982.

Tajfel, Henri and John C. Turner. "An Integrative Theory of Intergroup Conflict." In *The Social Psychology of Intergroup Relations*, edited by W.G. Austin and S. Worchel, 33–47. Monterey, CA: Brooks/Cole, 1979.

—. "The Social Identity Theory of Intergroup Behaviour." In *Psychology of Intergroup Relations*, edited by S. Worchel and W.G. Austin, 7–24. Chicago, IL: Nelson-Hall, 1985.

Turner, John C. "Towards a Cognitive Redefinition of the Social Group." In *Social Identity and Intergroup Relations*, edited by Henri Tajfel, 15–40. Cambridge: Cambridge University Press, 1982.

—. "Social Categorization and the Self-Concept: A Social Cognitive Theory of Group Behaviour." In *Advances in Group Processes*, vol. 2, edited by E.J. Lawler, 77–122. Greenwich, CT: JAI Press, 1985.

Turner, John C., Penelope J. Oakes, Stephen D. Reicher, and Margaret S. Wetherell. *Rediscovering the Social Group: A Social Categorization Theory*. Oxford: Basil Blackwell, 1987.

van Dijk, Teun A., Stella Ting-Toomey, Geneva Smitherman, and Denise Troutman. "Discourse, Ethnicity, Culture and Racism." *Discourse as Social Interaction: Discourse Studies, a Multi-Disciplinary Introduction*, volume 2, edited by Teun A. van Dijk, 144–80. London: SAGE Publications Ltd., 1997.

Wechsler, Stephen. "What 'You' and 'I' Mean to Each Other: Person Indexicals, Self-Ascription, and Theory of Mind." *Language* 86 (2010): 332–65.

12

Plato's dialogically extended cognition

Cognitive transformation as elenctic catharsis

Laura Candiotto

Introduction

In the late 1990s, a new hypothesis for understanding the cognitive system was proposed, leading to a very vivid debate, which is still very alive today. I am referring to the hypothesis of the extended mind (Clark & Chalmers 1998), which suggests that cognition emerges through the brain-body organism's constitutional entanglement with the environment, and that the mind goes beyond the boundaries of the skin and skull. Three models have been developed to grasp this process, namely the original model that was based on the parity principle, for which we should state a functional equivalence between parts of the world and inner states under certain circumstances;[1] the second on cognitive integration, for which cognitive systems operate through integration of neural and bodily functions (Menary 2007); and the third on the social dimension of extension, that investigates those cases in which individual mental states may be partly constituted by the states of other thinkers, and more systematically by "mental institution" (Gallagher & Crisafi 2009; Gallagher 2013: 6–7; Slaby & Gallagher 2015).

Another different approach to cognition was proposed by Edwin Hutchins and colleagues (Hutchins 1995; Hollan *et al.* 2000) at the Department of Cognitive Science in San Diego. In their approach, cognition is distributed among team members pursuing the same task. The distributed cognition approach, which in my opinion is consistent with the third wave of the extended mind hypothesis, is one of the best models for explaining group knowledge. Its explanatory power regarding the emergence of group knowledge, based on the complementarity of functions within a dynamic cognitive system, enables us to conceive the "cognitive transformation" produced through group knowledge. By "cognitive transformation" I mean the process of reshaping the cognitive environments, thanks to the emergence of new and different functions (Klein & Baxter 2006; Menary 2007).[2]

Studies on group mind and group knowledge have recently increased in the fields of social ontology, epistemology and cognitive science.[3] The ancient model of Socratic dialogue is an excellent example of how social interactions shape and vehicle cognition. Thus, it may provide historical and conceptual evidences to the current debate regarding the generation of knowledge in a group. At the same time, the adoption of a cognitive model to further our understanding of Socratic inquiry would be beneficial for Classics as well, in line with the prominent and innovative investigation of the Cognitive Humanities.[4]

In this chapter, I will contextualize the collective cognitive dimension of Socratic dialogue using socially extended mind and distributed cognition to identify the mechanism that leads to cognitive success through the *elenchus* as cross-examination, a procedure that tests out the consistency of an interlocutors' beliefs. Being liberated by those errors of reasoning that affect our way of life is for Socrates the most prominent cognitive success. Success, for the *elenchus*, does not mean the production of knowledge – this will be the aim of the following phase, the maieutic one – but the purification (κατηαρσισ) of false beliefs, understandable as cognitive transformation. I will argue that Socratic *elenchus* is a dialogically distributed cognitively motivational state that leads to a purification of reasoning through cross-examination, acting as cognitive therapy that aims at intellectual and moral enhancement of the interlocutors and the public. By "public", I mean both the internal and external audience of the dialogue. I cannot discuss here the issue regarding the identity of the external public that attended the dialogue's public readings because it discloses the venue of research about the audience of Plato's dialogues as a literary work, which is not pertinent to the aim of this chapter. For Plato, the cognitive procedure that moves from errors to truth leads to moral enhancement as well, since the purification of beliefs implies a transformation of behaviors.

Catharsis will be depicted in its active stance as cognitive transformation from errors to truth. Examining Socratic *elenchus* as dialogically extended cognition, I will thus analyze how Socrates orchestrated the flow of information among the interlocutors in order to reach this cathartic outcome, not only for the direct recipient but to the entire epistemic group. I introduce the thesis that the elenctic *aporia* is the dialogically distributed embodiment of the interlocutors' and the public's motivational state for knowledge. Only if you care for truth, and you are motivated toward it, would you accept to be fully engaged in the Socratic inquiry and, thus, to be refuted. Consequently, motivation for knowledge is an epistemic disposition which spreads over the group, and it enables the function of the elenctic *aporia* as extended catharsis, which is the collective purification of the errors of reasoning through dialogical extended cognition.

First, I will introduce extended and distributed cognition, drawing attention to their explanatory power regarding dialogical interactions. Then, I will analyze the system that makes the *elenchus* extended within the distributed system of Socrates, the interlocutors and the public, depicting in detail the *aporia*'s function. Finally, the public will be understood not only as a recipient of the purification, but also as a resonator of dialogically extended *elenchus*.

1. Socially extended cognition and dynamic cognitive systems

The so-called "third wave" of the extended mind hypothesis aims to illuminate cognitive extension within the social realm, focusing on interactions among subjects, groups and institutions. Notably, social interactions maximize cognition (Gallagher 2013; De Jaegher *et al.* 2010). This not only means that the intersubjective dimension of cognition is the vehicle of the collective production of knowledge, but also that it enables the achievement of such a cognitive success that would not be produced by a single individual. These cognitive successes should be understood as a cognitive transformation (Menary 2010) of the epistemic agents' cognitive environment through social interactions (Kirchhoff 2012), and its function is to reshape the agent's understanding of a certain content. In slogan form this model says that changing your mind belongs to the process of learning, but that you are able to do so only under certain circumstances – as in the social dimension of dialogical inquiry.

Arguably, dialogue should be understood as one of the most prominent cognitive tools (Tylén *et al.* 2010) that permit this kind of social extension. As Clark and Chalmers (1998: 18) have claimed, language may have evolved to enable the extensions of human cognitive resources within actively coupled systems. This means that verbal communication permits a brain-to-brain coupling (Hasson *et al.* 2012) for which the mental processes in one brain are dynamically interrelated with the mental processes of another brain via verbal communication. Following this line of reasoning, language should be understood as the tool that permits humans to create cognitive niches (Clark 2008: 76–81) that support extended cognition.[5] Clark and Chalmers (1998) provide some examples: the couple in a long-term relationship that through dialogical interaction remember some events of their common past, completing each other's phrases, and the waiter who can anticipate the order of a frequent customer, then suggesting the meal (Clark & Chalmers 1998: 17–18). What these examples show is that remembering is a cognitive function activated not only by dialogue but also by a stable interaction over a period of time.[6] In fact, this mutual engagement between agents is essential to dialogue as a vehicle for cognition.[7]

Fusaroli and colleagues (2013, 2014) have provided some empirical evidence about dialogical interaction as a vehicle for cognitive extension and the shared roles individuals have in the construction of concepts. Understanding language as a social activity,[8] Fusaroli and colleagues (2013, 2014) have analyzed language as what enables skillful intersubjective engagement, demonstrating the functional linguistic coordination that belongs to dialogical interactions as what permits one to perceive a mutual understanding among the dialogue's partners, both in verbal and nonverbal communication. Joint linguistic actions and the communicative interactive alignment enables the right endeavor to produce some socially extended mental states (cf. also Gangopadhyay 2011: 381), since they ground the cognitive coupling in the wider and vital dimension of the social interactions. Some empirical evidence has been provided about this, such as the integration of sensory information between individuals in the collective process of dialogical decision making, where it is exactly linguistic dialogue that constitutes the coupling link for the dialogical dyad (Bahrami *et al.* 2010); but also about more basic coordinative mechanisms in play as social interactions, such as the one for which it is possible to measure sensitivity to others' perspectives and engagement (De Jaegher *et al.* 2010), or to make a meaning thanks to a process of contextually sensitive reciprocal adaptation (Tylén *et al.* 2013). These results are very important for my thesis since they show that certain epistemic functions may be performed better by a pair or a group than the best individual through dialogue. Of course, there are certain requirements that should be satisfied to get a skillful coordination beneficial for socially extended cognition – as the "glue and trust" criteria (Clark 2010: 46) brought to the social realm of interactions, or the joint commitment[9] for group knowledge (Candiotto 2017), but this does not invalidate the thesis. On the contrary, and specifically for our topic, it asks to detect the characteristics that make Socratic *elenchus* a relevant case of socially extended cognition. As Fusaroli *et al.* (2014: 35) have argued, not all the kinds of dialogical interactions are cases of extended cognition, but only the one regulated by skillfulness. In the next sections I will explain why we should individuate in Socrates' expertise and know-how what assures *elenchus* of being a successful case of dialogically extended cognition.

The distributed cognition approach proficiently explains the machinery through which an intersubjective system, such as dialogue, can organize information derived from different sources and, thus, produce new knowledge. According to the model of distributed cognition, a cognitive transformation emerges from the coordination/cooperation among group members performing various functions to attain a joint goal. Just think about your own experience as a student, and in particular when a cooperative group learning task has been assigned to you. How much did you learn from discussing the topics with your classmates, especially from ones that disagreed with you? And how much does collaboration count not only for reaching a mutual understanding, but also for revising your position and learning something new? Do you remember how important it was that different abilities and skills were available to your group for achieving the collective goal?

Education and learning practices are gold mines for finding examples of cognitive transformation, but the key insight provided by the application of the distributed cognition hypothesis to group knowledge is that certain cognitive phenomena – as aporetic states in the Socratic inquiry – are better understood as distributed processes through which the computation, traditionally ascribed to the individual mind understood as an inner processor, is disseminated within the interactions among the different agents that are part of this decentralized system. Assuming this framework, dialogue, as a device for group cognition, appears to be a dynamic system, rather than only a sum of the mental states possessed by the single individuals. This means that the cognitive processes at work within a dialogical system go beyond the processes of the individual, emerging from the interactions that take place among epistemic agents. Through the complementarity of functions, skillful dialogue acts to process the information and produce cognitive transformation, which is one of the most important goals of the epistemic group agent.

2. Socratic dialogue as a distributed cognitive system

My hypothesis is that social interaction is a necessary component for the arising of some specific mental states – such as the aporetic ones, and the Socratic *elenchus* is a good case study for proving it. Socrates, the interlocutors and the public are members of the epistemic group that uses elenctic inquiry as a tool for acquiring knowledge. It is not that an individual cannot think on her own. It is only that she needs someone to test if what she conceives is consistent. As Socrates claimed in the *Protagoras* (348c5–d5), one may achieve an understanding alone but, after that, he should discover somebody to whom he can show it off, and who can corroborate it; or, as in the *Gorgias* (506a), when Socrates asks his interlocutors to refute him; or when dialogue facilitates agreement about the things each interlocutor conceptualizes in his own mind (*Soph.* 218c).

Even the most solitary inner reasoning is understood by Plato as inner dialogue, or as a debate among the different parts of the soul. In the same lines of the *Protagoras*, Socrates said that what pushes him to converse is to examine the puzzles that occur to himself. Consequently, Socratic *elenchus* may be understood as the externalization of Socrates' inner reasoning, conceived of as problem solving.[10] Inner doubt motivates one to search out someone with whom to test their assumptions and beliefs.[11] Alternatively, doubt will arise within the dialogical interaction, naturally or induced by Socrates. This is not a contradiction because it depicts the very nature of reasoning as problem solving. It is a procedure that may include many puzzles placed at different steps, and that should comprise elements of surprise as well, as related to the detecting of errors and *aporiai*.[12] This means that one specific doubt may motivate the dialogical inquiry but other doubts may arise within the process. Moreover, this point stresses the importance of framing single dialogical interactions within a broader framework of reiterated dialogues.[13] One single session of dialogue is not enough to grasp the truth together. We need more sessions to further the inquiry. Socrates believes that we should converse as often as we find the right interlocutor (*Prot.* 348 d6).

We can also conjecture that the inner doubt that pushes Socrates to search out someone with whom to discuss it comes from inner self-reflection about the understanding achieved from the previous dialogical inquiry. However, the key idea is that *elenchus* requires at least two people, or two inner debaters, to reach some outcomes as to the resolution of the problem. What is crucial within dialogical inquiry is the complementarity between the partners, in particular among their cognitive functions, which produces a new cognitive achievement. The intuition, namely that social interactions nurture our quest for truth – even in the negative side of the recognition of our faults – grounds dialogical practice and it will be developed by Plato in the philosophical method of *dialectic*, understood as the refined procedure for the examination of one's assumptions and concept-formation. Thus, the dialogical dimension of inner reasoning and of social interactions seems a clear indication of the social dimension of knowledge. Despite mental internalism as a feature of Socratic thought, Plato's epistemology may be considered an ancestor of externalism, because dialogue stems not only from social activity, but also from the very nature of reasoning.[14]

Following this line of thought, dialogue seems not only a contextual factor for cognition, but also that which enables specific mental states, such as aporetic ones. The *elenchus*, as the procedure that tests out the consistency of the interlocutors' opinions, requires dialogical interactions for detecting the unjustified assumptions that may underlie the beliefs. If, as two epistemic agents confront some task, let's say to resolve epistemic doubt, they jointly perform cognitive processes P and P^* in such a way that, taken separately, P or P^* would not be functionally sufficient to perform the task, we will have no problems recognizing the joint functionality of the two processes within one individual cognitive system. Thus for the parity principle we should state that these processes are socially distributed over a single cognitive system distributed in two epistemic agents. The task, namely the resolution of epistemic doubt, will be recognized as a newly emergent property of the epistemic dyad as a distributed cognitive system.[15]

It follows that the interlocutors check one another's judgments through questions and answers, in order to face the aporetic states and move beyond them. Thus, "we-reasoning" is motivated by Socrates' epistemic doubt, recognized as the best medicament toward unjustified beliefs, and, in the meantime, it spreads these antecedent states to the interlocutors. After that, it will lead the procedure to the next step, the maieutic one, namely the one through which the interlocutors will give birth to the forgotten knowledge.[16] This means that the shaking of the interlocutor's confidence in what he believes, combined with the enhancement of the desire to go beyond the *aporia*, will lead the process to the multi-agent final state of generation of knowledge, as error correction.

Arguably, skepticism does not have the last word in this procedure: not only because the recognition of errors has a positive value of its own, but also because the desire to know brings the *aporia* to the maieutic step of the procedure. Therefore, if the aporetic states are motivated by the desire for knowledge, they enable the achievement of the cognitive transformation into knowledge. This transformation is the basis of maieutic as generation of knowledge.

Analyzing the elenctic procedure, we find an important difference from the distributed model. For distributed cognition, the cognitive processes are decentralized, meaning that there are no directors that orchestrate the flux of information. Here, on the contrary, it is Socrates who puts in place this complex process.[17] Socrates induces the aporetic states to the interlocutors and, thanks to the public that act as resonators of the aporetic state, he achieves catharsis through extended *elenchus*, as I will explain in the next sections. Socrates is like the torpedo fish that perplexes whoever touches him (*Men.* 80b), or like the horsefly that stings the citizens of Athens with his questioning (*Apol.* 29c–30c; 30e–32a). The starting point for the elenctic procedure belongs to Socrates, the one that orchestrates the machinery, in order to extend the aporetic state to the interlocutors and the public through questioning. That is why I think that the model of extended cognition has a specific role to play here as well: Socrates is like the brain that extends into the environment to realize those mental processes that require an external vehicle to be performed. Socrates does it to distribute *aporia* among the interlocutors and achieve their cognitive transformation. Thus, the theory of distributed cognition explains the Platonic account of dialogue as a dynamic system that enables the emergence of group knowledge.

3. Elenctic catharsis, extended

If it is clear that aporetic states imply an inner conflict in search of a solution, we need to understand the cognitive structure that enables its contagion. I think it is exactly the social dimension of Socratic inquiry that permits it, and thus the actions performed by Socrates to achieve the refutation of the interlocutor (as questions, criticism toward the provided definitions, analysis of examples, etc.) should be understood as vehicles for the distribution of the aporetic state.

As the Eleatic Stranger states in the *Sophist* (230b–e), refutation is a form of purification (κάθαρσις). The cleansing works as self-transformation: it is a cognitive and emotional acknowledgment of errors that pushes the cognitive transformation. Through cognitive transformation, the agent is able to transform her behavior as well. For Socratic intellectualism, true thinking is the basis for right action.[18]

I conceive catharsis as the outcome produced by the dialogically extended *elenchus*. It points to the fact that *aporia* extends our cognition to attain a transformation of character. Moreover, it illuminates another crucial element of the process. The aporetic state involves the psychological recognition of the errors – specifically through shame and anger[19] – integrated within reasoning. *Aporiai* are felt by the agent as distressful emotions that make the character in need of a solution, and thus prone to accept the purification's process. Aporetic states are those radical states whereby epistemic doubt transforms into the recognition of contradictions. Polus, for example, says that Gorgias, feeling ashamed to admit that he did not know what justice is, has fallen into contradiction (*Gorg.* 461c). The inner and extended debates produce a kind of conflict that also implies a feeling of inward anger, which leads to catharsis as the release from this conflictual state. For example, in the *Laches* (194a–c, 197c 5–7), Laches is angry with himself because he is not able to explain what he has in mind. And, thus, in the lines of the *Sophist* that we are discussing, it is said that the refuted is angry with himself for the contradiction embedded in his opinion (230b) (Candiotto 2018). If we consider the agonistic values of Greek society (Dover 1974), we could grasp how revolutionary this idea may be. In fact, Socrates lights the fire of conflict thanks to the elenctic procedure, in order to produce the inward anger that, in cooperation with shame, will bring the interlocutor to the purification. Therefore, the conflict functionally works for dialogically extended cognition understood as a cooperative and distributed activity. The conflict brought by the *aporia* is not for its own sake, but it is used by Socrates as a tool for purification. This means that the aporetic state is employed as a cognitively motivational state for inducing cognitive transformation.

Shame, as recognition of inferiority and, thus, related to the need of restoration of honor, aims at prompting the desire to accept the purification and to go beyond the *aporia*. This active stance of catharsis is conceived within the procedure of *elenchus* as maieutic: purification makes the interlocutor available to further the discussion with Socrates to grasp the truth, and to change his lifestyle as well. Specifically, the Socratic refutation plays on the feeling of shame that obliges him to admit his own ignorance and to become devoted to the discovery of truth. Thus, the dialogically extended *elenchus* is not only a test of truth,[20] but also a device that brings the refuted to the next step of the procedure, which is the eliciting of knowledge.

As a consequence, *elenchus* functions as an extended cognitive therapy that uses the integrated function of emotions and cognition within dialogue to obtain purification and liberation from cognitive mistakes.[21] What makes catharsis unique for distributed and extended cognition approaches is that it is a mental function that can only be achieved through the socially extended *elenchus*, i.e. something involved with yet external to our brain. The mind needs the socially extended *elenchus* to perform cognitive purification. In this way, dialogue achieves what individual thinking cannot. This cognitive purification is the basis for new belief-building as a maieutic process. The core hypothesis of this approach is that if we free interlocutors from cognitive mistakes, they will be able to change their behavior as well.[22]

The questions asked by Socrates to make the interlocutors aware of their errors are moved by the desire for their enhancement, broadly understood. What is at work here is not just Socrates' self-centered motivation, i.e. finding someone to test out his assumptions. While unearthing certain beliefs of the interlocutors, Socrates acts as a healer moved by an empathic response toward the suffering caused by fallacies. Consequently, Socrates converses with the interlocutors to free them from these errors, understood as diseases. In this case, the *elenchus* is the remedy against illnesses of reason, and thus it should be understood as a virtuous practice to perform through dialogue. As I have already mentioned, in the Noble Sophistry passage of the *Sophist*, the *elenchus* is explicitly recognized as a remedy. It is like a purgative that extirpates the impediments rooted in the soul that do not allow it to generate knowledge.[23] In the *Gorgias* (476d–479c), the medicament is understood as a punishment (κόλασις), since Socrates think that his interlocutors are culpable for their wrong behavior.[24] I should underline that, in the *Gorgias*, Socrates was conversing with some rhetoricians and politicians and, thus, we can suppose that Plato needed to strongly criticize their work, pointing to their wrong behavior. In this case, the *elenchus* acquires the value of political criticism. This recognition does not falsify my argument. *Elenchus* is a cognitive therapy in this case too, and it reaches purification through punishment. Punishment plays an important function here, in that it can be assimilated, in more positive terms, to the shame induced by Socrates as physician of the soul. At the same time, here the healing function of the *elenchus* is directed to the public, and to their false beliefs, perceived as illness, as I am going to explain in the next section. Especially in the case of Callicles, it clearly appears that Socrates is not so naive as to think he could take care of his illness through *elenchus*. Therefore, my hypothesis is that, in such cases, his procedure is directed toward other patients, such as the public that is assisting the dialogue.

4. Socially extended catharsis

Just as in the tragedy where the public is emotionally released by the vision of the tragic plot,[25] the public's false beliefs are purified by the elenctic procedure in these specific cases of refutation. The Socratic therapy is not a corrective for the interlocutors only. Its cognitive purification from false beliefs is extended to the public too. This means that the extended dimension of the *elenchus* does not include the interlocutors only, but the public as well (Candiotto 2015, 2018), conferring a cultural and political valence to the extension,[26] as the third wave of the extended cognition has pointed out.

The proper context to grasp the functioning of *elenchus* is that of medicine and, thus, of healing. As it is a common experience, we accept to be treated unless it causes pain because we know it will be for the sake of health. In the same way, Socrates expects that his audience will accept in good spirit his treatment, knowing that it is for the best. It is said, for example, that the refuted grows angry with himself, but gentle toward others (*Soph.* 230b, πρὸς δὲ τοὺς ἄλλους ἡμεροῦνται; the verb alludes to making a wild beast tame), especially toward the doctor that is taking care of his disease. Moreover, as Socrates says in *Gorgias* 458a, he feels pleasure (ἡδέως) at being refuted if he says anything untrue, and he refutes with pleasure anyone else who might speak untruly. In the *Meno* (84b), Socrates says that by causing the interlocutors to doubt (i.e. giving him the torpedo fish's shock), he has not done him any harm. Instead, he has certainly given him some assistance in finding out the truth of the matter. The reason Socrates feels pleasure during a painful treatment is that he perceives refutation as a great benefit. Consequently, pleasure is a reaction that takes place after a significant moral and epistemic transformation achieved through the elenctic treatment.[27]

The public should be educated to accept purgation. There are cases in which counterproductive reactions to *elenchus* happen (*Gorg.* 457d–e). These reactions are affected by context. For example, the partisan public that shares beliefs with Callicles or Trasimachus may be outraged to see their fellow refuted. Nevertheless, I think that elenctic performance serves the function of convincing the public of their own mistakes too and, thus, Socrates aims to produce in them a cognitive state that is different from the one experienced by the interlocutor that does not accept purgation. The participation of the public does not only mean that it could identify with the interlocutor, mirroring[28] his emotional state, but also that, in certain cases, it should learn to think differently from him. It is exactly this explicit concern toward the partisan public and, thus, the necessity to make them change their minds that should have motivated Plato to build a device[29] with more secure outcomes. That is why I think that Plato has ascribed to the public the function to make the refutation resonate (Candiotto 2015).[30] Ideally, if the public would experience disgust toward the interlocutor that does not accept the refutation, he will perform a complementary function to the one of Socrates, maximizing the pressure toward the interlocutor to change his mind.

The public is both the recipient of purification and the performer of a different and complementary function in the process. That's why dialogue, as a dynamic distributed system, should include the public among its members. The public does not only see the representation, mirroring the states of the interlocutors, but also enhances the pressure toward the refuted, in order to encourage him to accept the purification. Socrates orchestrates this distributed system to ensure that the flux of information he sends to the interlocutors and the public through the experience of the aporetic state may be enhanced, exactly as the feedback loop "promiscuously crisscrosses the boundaries of brain, body and world" (Clark 2008: xxvi).

As I have already mentioned, the idea is that a cognitive transformation is achievable through the extended *elenchus* as a therapeutic procedure, and that a transformation of thinking implies a transformation of behavior. The difference here is that Socrates does not simply address questions to the interlocutor or make him aware of the contradictions embedded in his reasoning in order to reach this positive outcome. Socrates uses the public as an extended vehicle for the recognition of the interlocutors' mistakes that, eventually, may be the same embraced by the public.

Moreover, if one accepts that a function of the Socratic dialogues is to promote the identification of the public with Socrates (Kahn 1996), one can then grasp the political valence embedded in the socially extended catharsis as the one that educates people to philosophy. If the interlocutors and the public would be available to recognize Socrates as their healer, then they would pursue a philosophical life, devoted to truth. In this regard, the transformation of character pertains mostly to the development of those intellectual virtues that lead the epistemic agent to knowledge.

To summarize this point: social reasoning, and the *elenchus* in particular, functions as medication not only regarding the direct interlocutor, but also in cases in which the catharsis is extended to the public. This outcome is made possible by the distributed cognitive system exploited by Socrates, who uses dialogue to orchestrate the processes necessary to cognitive transformation-as-purification.

5. Conclusion

The core idea here is that *elenchus* is a very powerful cognitive medication, and that it functions thanks to well-defined dialogically extended dynamics. Someone that possesses the experience to administer the purge is required. That someone is Socrates, identified here as a physician caring for truth. Socrates is aware that we can give birth to false beliefs and, thus, we need others to test them. Not with just anyone, but with the right interlocutors – the friends. Thus, Socrates needs the interlocutors as well: only through the "we-reasoning", which is the socially extended dimension of dialogue, can he really test out his/their assumptions. Moreover, just by testing their assumptions, he can purify them, or, at least, know if they are incurable. Thus, Socratic *elenchus*, as the most prominent tool of dialogically extended and distributed cognition, not only involves interaction among the interlocutors; it also involves the audience. The audience not only receives the purification, but also enacts a crucial role for maximizing the purification outcome, namely to make the refutation resonate.

In this chapter, I have looked to theories of distributed cognition to emphasize the value of Socratic dialogue both as a learning practice and as a cognitive therapy. Catharsis is the prized goal of *elenchus*. Socrates uses the *elenchus* to achieve a cognitive transformation of the interlocutors and the public. Contrary to purely internal processing, these dialogical manipulations of others' mental states allow for a maximized purification. I argued that the *aporetic* state is the dialogically extended dimension of *elenchus*, not just in the interlocutor's mental state. Specifically, I emphasized how the aporetic states, performed within the social dimension of dialogical inquiry, enable the transformation of unquestioned false beliefs into knowledge, opening up the maieutic phase of the procedure. The most effective form of beliefs' purification as cognitive therapy is accessible in a dialogic context, understood here as extended dialogical cognition, which is pursued through *elenchus*.

More work should be done in this exciting field of the cognitive humanities. In that vein, I hope to have properly shown that the extended and distributed models for cognition are very proficient in explaining the functioning of the Socratic dialogue. In particular, my contention is that both the elenctic *aporia* and the cathartic outcome are extended cognitive states that confer to Plato's epistemology a very new flavor, despite the traditional internalism. Additionally, I think that the Socratic account on dialogically extended cognition proves very fruitful for contemporary debates. It depicts dialogically extended cognition as a form of therapy, conferring to it a particular ethical meaning that unlocks the practical outcomes of our epistemic practices.

As is well known, Ludwig Wittgenstein's philosophy was also a kind of therapy. Just like Socrates, the Austrian-British philosopher pointed out new ways of looking at language, specifically through his use of it. May the cognitive humanities inspire new investigations on this fascinating topic.

Notes

1 "If, as we confront some task, a part of the world functions as a process which, were it done in the head, we would have no hesitation of recognizing as part of the cognitive process, then part of the world is (so we claim) part of the cognitive process" (Clark & Chalmers 1998: 8).

2 This diachronic account of knowledge-building differs from the standard storehouse account, for which knowledge is a collection of information stored by our brain. Moreover, it plays a significant role in the second wave of the extended mind hypothesis, having been modeled by Richard Menary and Michael Kirchhoff as the result of the integrated functionality of mind and environment in the development of a cultural extended expertise, the so-called "extended expertise" (Menary & Kirchhoff 2014). What I want to highlight here, adopting the distributed account of cognition, is that this cognitive transformation may be achieved through social interactions, and specifically having philosophical dialogues as vehicles.

3 Deborah Tollefsen's recent book on group minds (Tollefsen 2015) is a very useful and clear introduction to the current research on collective cognitive agency in the philosophy of mind and cognitive science. Goldman and Blanchard 2016 is an excellent introduction for those interested in group knowledge and we-intentionality within the field of social epistemology.

4 Let me emphasize points that seem crucial for the development of the cognitive humanities. First, the recent studies on cognitive science may be very helpful to enlighten some features of the ancient models of mind and knowledge which, without them, risk being unrecognized. Second, the ancient models may provide some intuitions to the contemporary debate that, if well developed, may support new venues of inquiry.

5 In transposing work in evolutionary niche construction theory (Odling-Smee *et al.* 2003) to theory of mind, Clark has introduced the notion of "cognitive niche", meaning the continuous and dynamic reconstruction of the cognitive environment, through recursive developmental interactions between the cognitive agent and her environment, to fulfill the cognitive agent's needs. Sterelny (2012) has applied this theory to human social life, arguing for the mutual and cooperative production and transmission of information among humans from generation to generation.

6 Some readers may find an intriguing parallel with Plato's theory of *anamnesis* through dialogue in this idea, for example in the *Meno* 81b–85d. Taking the different metaphysical assumptions, it enlightens the procedural role of dialogue for memory processing.

7 These cases may serve as examples for a "dialogical" predictive processing too. For Clark's theory of active predictive processing, in fact, when we deal with incoming sensory data, the brain uses top-down processing to make predictions about what is being experienced. This may well explain why the waiter can anticipate the customer, or the ability of one of the members of the couple to "predict" what her/his partner would say, but also many other cases of group brain-storming and dialogical inquiry. Further researches should be done regarding this point, but let me just emphasize that linguistically enabled social interactions should play a prominent role in the explanation of our social cognitive abilities through predictive processing. Moreover, this model may play a significant role in the understanding of the Classics as well, as the recent book written by Peter Meineck (2018) excellently testifies.

8 They claim that Clark (2008) overshadows the processes that make language a social activity, focusing more on the code-like products of linguistic activities. They take language as "an activity that allows us to coordinate actions, perceptions and attitudes, share experiences and plans, and to construct and maintain complex social relations on different time scales" (Fusaroli *et al.* 2014: 33), and they bring this conception to work for the extended cognition hypothesis. Thus, their model seems to me consistent with the third wave of the extended mind, for which cognition is spread over social interactions – made of gestures, affective resonance and joint actions – and not only mental states.

9 For an introduction to the notion of commitment, and to the psychology of those commitments that are jointly undertaken, cf. Michael *et al.* 2016.

10 As has been underlined by Rudebusch (2005: 197), Socrates is proposing problem solving as a partnership between friends who seek to investigate together the nature of human excellence. For the meaning of *elenchus* as problem solving, cf. Politis 2006. In particular, the author underlines the function of the *zetetic aporia*, explaining that solving particular *aporiai* is part of the search for knowledge. This idea, developed by Aristotle (i.e. *Met.* B1. 995a34–b1), not only has had great luck in the history of Western philosophy – just think about Descartes' radical doubt – but it is also consistent with the empirical evidence on the structure of reasoning pursued by cognitive science.

11 For example, in *Charmides* 166d, Socrates stressed his motivation for the inquiry, underlining that he was examining the argument mainly for his own sake. Immediately after, he added that he was doing it for the benefit of his friends too. I think that, without denying Socrates' irony, especially toward those interlocutors that he needs to refute in public, we should recognize his inner motivation as epistemic doubt.

12 This is very similar to what happens in theater, where surprise is functional to the cognitive transformations (as catharsis) of the audience. Notably, Meineck (2018: 31–38) understands this element of surprise as belonging to a process of predictive processing and of probabilistic thinking (Wohl 2014). The similitude that I am here establishing between Socratic dialogues and theater is not extrinsic or accidental, not only because both have catharsis as one of their main functions, or because dialogues are new types of drama (Charalabopoulos 2012) – as many studies devoted to Plato's literary writing have argued, but also because the same cognitive process should be understood – in Plato's terms – as inner dialogue (Plato, *Soph.* 263e), an inner drama for resolving the inner conflict among the three parts of the soul (Plato, *Rep.* IV, 440c–441c)

13 For Plato, differently from the historical Socrates, the reiterated dialogues will be embedded in the shared life of a philosophical community. Cfr. *Theaetetus* (150d–e) and the *Letter Seventh* (34 c4–d2).

14 I am not discussing the innatism/empiricism debate here, but I am stressing that the procedure that elicits knowledge should be understood as dialogically extended.

15 Theiner *et al.* (2010), and then Szanto (2013), have developed the social parity principle, which functions in a quite similar way to what I am depicting here. The authors states that as we agree to recognize as cognitive this procedure if internalized, we should do the same if the cognitive processes are externalized. Here, nevertheless, I do not want to reduce what is achieved within a cognitive social interaction to what usually happens within the internal debate. On the contrary, I want to stress – and that is why I introduce the emergentist notion of cognition that belongs to the distributed model – that what is realized in the social realm is something that cannot be achieved in solitude.

16 In the *Theaetetus*, Socrates as a midwife brings out the knowledge from the soul of his interlocutor who is in labour: "those who associate with me, although at first some of them seem very ignorant, yet, as our acquaintance advances, all of them to whom the god is gracious make wonderful progress, not only in their own opinion, but in that of others as well. And it is clear that they do this, not because they have ever learned anything from me, but because they have found in themselves many fair things and have brought them forth. But the delivery is due to the god and me . . . Now those who associate with me are in this matter also like women in childbirth; they are in pain and are full of trouble night and day, much more than are the women; and my art can arouse this pain and cause it to cease" (Plato, *Theaet.*, 150d–151a, tr. Fowler). Plato uses the image of "giving birth to a forgotten knowledge" for depicting the process of knowledge as remembering (αναμνεσισ), as his solution to the paradox of learning (Plato, *Men.* 80d–e). As has been underlined by Sarah Kofman (1983: 53), within a discussion of Detienne and Vernant's work on cunning intelligence (Detienne & Vernant 1974), the troubles produced by Socrates' aporetic states are not like the ones of the sophists that just immobilize the interlocutors, but they awaken the desire to give birth. Notably, for the Socratic method this awakening is realized through dialogue. Later on, in the *Phaedo* for example (Plat., *Phaed.* 73d5–74a1), other sources will serve this function, as images and visual perceptions.

17 I am referring to the fact that it is Socrates who orchestrates the process, and it is thanks to his know-how that it may be recognized as a skillful dialogical interaction. In fact, Socrates possesses a procedural knowledge: he can orchestrate the dialogue because, as his mother, he uses the maieutic method (Plato, *Theaet.* 149a–151d) and, thus, he knows how to manage the pain of the delivery without losing the orientation provided by the understanding. However, this does not mean that he already knows the content toward which he would push the interlocutors but, on the contrary, he wishes to search for it with the interlocutors. Moreover, the philosophical ignorance is different from the ignorance of the common people, since it implies the metacognitive awareness of ignorance. In Socrates' view, this awareness is crucial for going beyond ignorance by the practice of epistemic doubt.

18 As a side note, I would mention that this point enhances the value of the elenctic practice beyond epistemology, stressing Plato's ethical and political aims.

19 It is interesting to notice that shame is depicted by Plato in its bodily expression too, as blushing. For example, in the *Republic* 350d1–3, Thrasymachus blushes during the refutation. On Thrasymachus' shame, see Pilote 2010.

20 There is a vivid debate about what the *elenchus* may establish – truth or coherence only. I think, in agreement with Vlastos, that Socrates – notably in the *Gorgias* – is seeking the truth, not simply the validity of the arguments. One of the reasons for thinking that derives from the realistic stance of Plato's epistemology. Nevertheless, it is not said that Socrates can succeed at finding what he declares to look for through the *elenchus*, and maybe Plato has invented the dialectics exactly to provide a more logical structure. For further details on the debate, cf. Scaltsas 1989.

21 Jonathan Lear has worked extensively on Socratic irony as therapy, mostly within the psychoanalytic approach. For example, Lear (2013, 2014) has argued for a creative and life-changing use of irony, depicting the related positive outcomes of shame. This point is important for my analysis because shame is exactly the feeling that took place within the refutation and that enables the purification. For Renaud (2002), this is the moral valence of the *elenchus*: to exhort the interlocutors to change their life. In the *Lysis*, in particular, the author finds the exhortatory valence of the *elenchus* coupled with *psychagogia*. For the psychoanalyst Francois Roustang (2009), Socrates is not only the first philosopher, but also the first therapist, since he knew how to use some cognitive techniques in order to change the mind and the life of his interlocutors. Notably, this idea has been brought by Lacan (1991) as well, but with a different meaning. In fact, Lacan, discussing Plato's *Symposium*, has stressed

the role performed by desire in relation to both the agent and the truth. However, if in the first seminaries Lacan has recognized Socrates as the first psychotherapist, later on he would depict his *atopia* as the death drive. By the way, what I am analyzing here is the fundamental relation between thinking, emotions and behavior that underlies the function of *elenchus* as purification. Following this road, the *elenchus* appears as cognitive therapy because it aims to change the interlocutors' style of life by identifying and changing wrong thinking. For cognitive therapy, cf. Beck 1975; Ellis & Harper 1975.

22 Gregory Vlastos (1983: 37) has notably recognized the existential dimension of the Socratic *elenchus*. The *elenchus*, in fact, examines both propositions and lives.

23 "the purifier of the soul is conscious that his patient will receive no benefit from the application of knowledge until he is refuted, and from refutation learns modesty; he must be purged of his prejudices first and made to think that he knows only what he knows, and no more" (Plato, *Soph.* 230c–d).

24 Explaining the procedure through which injustice is caused by psychological disorder, Shaw (2015) has argued that, in Socrates' opinion, the true judge will use refutations as punishments. This point is very significant because it underlines the judgmental structure of *elenchus*, which aims for the moral purification of the interlocutors. As I am arguing in this chapter, this procedure is similar to the one performed by doctors in order to free patients from illness. This point is explicit in the *Gorgias*, and not by chance in my opinion. In fact, Socrates' interlocutors were exactly those leaders that should be judged, i.e. refuted by the *elenchus*, and purified by the philosophical use of the medical catharsis. Larivée (2007) has depicted this method as homeopathic, since it aims to heal from the illness produced by the sophistry and heuristic through the use of a method that uses many violent strategies similar to those of the sophists. I think that this idea is very insightful and permits us to grasp not only the conflictual tone of the *elenchus* in the *Gorgias*, but also the commonalities between Socrates and the sophists. However, we should confer to Socrates' strategies a very positive value, the one of cognitive and moral enhancement, despite the apparent common features with the sophistry.

25 Aristotle in the *Poetics* (1449b21–28) explains that catharsis describes the effect of tragedy on the audience.

26 By "cultural and political valence" I mean the positive outcomes for society they derive from the healing procedure directed to the public through the extended *elenchus*. Healing the public for Plato means also healing Athenian citizens, establishing a new *paideia* (Candiotto 2017). Regarding the production of knowledge in the public sphere, cf. Farenga 2014.

27 Jessica Moss (2005) has defended Socrates' accusation of using shame instead of reasoning, arguing for the moral valence of shame. In general agreement with the author, my aim here is to explain why shame may be a moral emotion, and I am doing it stressing his cleansing value within the *elenchus*. Accordingly, shame and reasoning are not antagonistic, but they are integrated within the cognitive process of purification.

28 Peter Meineck (2011) has explained this crucial element for the engagement of the spectators in the performance through the theory of the "mirror neurons". In particular he has showed how much the use of masks enhances the process. I think that this pattern may be applied to the Socratic *elenchus* as well, if we would agree that it is a performance that involves the public. Nevertheless, I think that it may be an important difference: Socrates does not look for an empathic responsiveness only. He aims to make the public disagree with the interlocutors, especially in those dialogues where the interlocutors are the incurables.

29 Much important research has been done over recent years to explain the aims of Plato's writing and literary style. Cf. Gill 2006; Rowe 2007; Bonazzi *et al.* 2009. On the relation between writing and orality, in particular for the aporetic dialogues, cf. Erler 1987.

30 There are cases, as in the *Protagoras* (358c), where the public positively mirrors the cognitive states of Socrates' interlocutor. The shame felt by Protagoras is partaken by the whole of the company. Here, again, the internal public belongs to Protagoras' entourage, but, contrary to the *Gorgias* case I mentioned before, it accepts the shame, as the interlocutor does.

References

Bahrami, B. *et al.* (2010) "Optimally interacting minds", *Science* 329 (5995): 1081–1085.

Beck, A.T. (1975) *Cognitive therapy and the emotional disorders*, Madison, CT: International Universities Press.

Bonazzi, M., Dorion, L.-A., Hatano, N., Notomi, T., and Van Ackeren, M. (2009) "Socratic Dialogues", *Plato, The Electronic Journal of the International Plato Society*, https://gramata.univ-paris1.fr/Plato/article88.html.

Candiotto, L. (2015) "Aporetic state and extended emotions: The shameful recognition of contradictions in the Socratic elenchus", *Ethics & Politics* XVII, special issue: "The Legacy of Bernard Williams's Shame and Necessity", ed. A. Fussi, 233–224.
— (2017) "Boosting cooperation: The beneficial function of positive emotions in dialogical inquiry", *Humana Mente. Journal of Philosophical Studies* 33, special issue: "The learning brain and the classroom", eds. A. Tillas and B. Kaldis, 59–82.
— (2018) "Purification through emotions: The role of shame in Plato's *Sophist* 230b4–e5", *Educational Philosophy and Theory* 50 (6-7), special issue: "Bildung and paideia: Philosophical models of education", eds. J. Dillon and M. L. Zovko, 576–585. DOI: 10.1080/00131857.2017.1373338.
Charalabopoulos, N. (2012) *Platonic drama and its ancient reception*, Cambridge and New York: Cambridge University Press.
Clark, A. (2008) *Supersizing the mind: Embodiment, action, and cognitive extension*, Oxford: Oxford University Press.
— (2010) "Memento's revenge: The extended mind, extended", in R. Menary (ed.) *The extended mind*, Cambridge, MA: MIT Press, pp. 43–66.
Clark, A., and Chalmers, D. (1998) "The extended mind", *Analysis* 58: 10–23.
De Jaegher, H., Di Paolo, E., and Gallagher, S. (2010) "Can social interaction constitute social cognition?", *Trends in Cognitive Science* 14/10: 441–447.
Detienne, M., and Vernant, J.-P. (1974) *Les ruses de l'intelligence. La mètis des Grecs*, Paris: Flammarion.
Dover, K. (1974) *Greek popular morality in the time of Plato and Aristotle*, Berkeley, CA: University of California Press.
Ellis, A., and Harper, R. (1975) *A guide to rational living*, Chatsworth, CA: Wilshire Book Company.
Erler, M. (1987) *Der Sinn der Aporien in den Dialogen Platons. Ü bungsstü cke zurAnleitung im philosophischen Denken*, Berlin: de Gruyter.
Farenga, V. (2014) "Open and speak your mind", in V. Wohl (ed.) *Probabilities, hypotheticals, and counterfactuals in ancient Greek thought*, Cambridge: Cambridge University Press, pp. 84–100.
Fusaroli, R., Gangopadhyay, N., and Tylén, K. (2014) "The dialogically extended mind: Language as skilful intersubjective engagement", *Cognitive Systems Research* 29/30: 31–39.
Fusaroli, R., Raczaszek-Leonardi, J., and Tylén, K. (2013) "Dialogue as interpersonal synergy", *New Ideas in Psychology* 32: 147–157.
Gallagher, S. (2013) "The socially extended mind", *Cognitive System Research* 25/26: 4–12.
Gallagher, S., and Crisafi, A. (2009) "Mental institutions", *Topoi* 28(1): 45–51.
Gangopadhyay, N. (2011) "The extended mind: Born to be wild? A lesson from action-understanding", *Phenomenology and the Cognitive Science* 10: 377–397.
Gill, C. (2006) "Le dialogue platonicien", in L. Brisson and F. Fronterotta (eds.) *Lire Platon*, Paris: Presses universitaires de France, pp. 53–75.
Goldman, A., and Blanchard, T. (2016) "Social epistemology", in E. N. Zalta (ed.) *The Stanford encyclopedia of philosophy*, https://plato.stanford.edu/archives/win2016/entries/epistemology-social.
Hasson, U., Ghazanfar A., Galantucci, B., Garrod, S., and Keysers, C. (2012) "Brain-to-brain coupling: A mechanism for creating and sharing a social world", *Trends in Cognitive Sciences* 16(2): 114–121.
Hollan, J., Hutchins, E., and Kirsh, D. (2000) "Distributed cognition: Toward a new foundation for human-computer interaction research", *ACM Transactions on Computer-Human Interaction* 7(2): 174–196.
Hutchins, E. (1995) *Cognition in the wild*, Cambridge, MA: The MIT Press.
Kahn, C. H. (1996) *Plato and the Socratic dialogue: The philosophical use of a literary form*, New York: Cambridge University Press.
Kirchhoff, M. D. (2012) "Extended cognition and fixed properties: Steps to a third-wave version of extended cognition", *Phenomenology and the Cognitive Sciences* 11(2): 287–308.
Klein, G., and Baxter, H. C. (2006) "Cognitive transformation theory: Contrasting cognitive and behavioral learning", *Interservice/Industry Training, Simulation, and Education Conference (I/ITSEC)*, paper n. 2500.
Kofman, S. (1983) *Comment s'en sortir?*, Paris: Galilée.
Lacan, J. (1991) *Le Séminaire. Livre VIII. Le Transfert, 1960-1961*, texte établi par Jacques-Alain Miller, Paris: Seuil.
Larivée A. (2007) "Socrate et sa méthode de soin homéopatique dans le *Gorgias*", in M. Erler and L. Brisson (eds.) *Gorgias-Menon: Selected papers from the seventh Symposium Platonicum*, Sankt Augustin: Academia Verlag, pp. 317–324.
Lear, J. (2013) "The ironic creativity of Socratic doubt", *MLN* 128: 1001–1018.

— (2014) *A case for irony*, Cambridge, MA: Harvard University Press.
Meineck, P. (2011) "The neuroscience of the tragic mask", *Arion: A Journal of Humanities and the Classics* 19(1): 113–158.
— (2018) *Theatrocracy: Greek Drama, Cognition, and the Imperative for Theatre*, London and New York: Routledge.
Menary, R. (2007) *Cognitive integration: Mind and body unbounded*, Basingstoke: Palgrave Macmillan.
— (2010) "Dimensions of mind", *Phenomenology and the Cognitive Sciences* 9(4): 561–578.
Menary, R., and Kirchhoff M. (2014) "Cognitive transformations and extended expertise", *Educational Philosophy and Theory* 46(6): 610–623.
Michael, J., Sebanz, N., and Knoblich, G. (2016) "The sense of commitment: A minimal account", *Frontiers in Psychology* 6: article 1968.
Moss, J. (2005) "Shame, pleasure, and the divided soul", *Oxford Studies in Ancient Philosophy* 29: 137–170.
Odling-Smee, J., Laland, K. N., and Feldman, M. W. (2003). *Niche construction: The neglected process in evolution*, Princeton, NJ: Princeton University Press.
Pilote, G. (2010) "Honte et réfutation chez Platon", *Phares* 10, http://revuephares.com/parutions/volume-10-2010.
Politis, V. (2006) "*Aporia* and searching in the early Plato", in J. Lindsay and V. Karasmanis (eds.) *Remembering Socrates: Philosophical essays*, Oxford and New York: Clarendon Press, pp. 88–109.
Renaud, F. (2002) "Humbling as upbringing: The ethical dimension of the elenchus in the *Lysis*", in G. A. Scott (ed.) *Does Socrates have a method? Rethinking the elenchus in Plato's dialogues*, University Park, PA: Pennsylvania State University Press.
Roustang, F. (2009) *Le secret du Socrate pour changer la vie*, Paris: Odile Jacob.
Rowe, C. (2007) *Plato and the art of philosophical writing*, Cambridge: Cambridge University Press.
Rudebusch, G. (2005) "Socratic love", in S. Ahbel-Rappe and R. Kamtekar (eds.) *A companion to Socrates*, New York: Wiley-Blackwell, pp. 186–199.
Scaltsas, T. (1989) "Socratic moral realism: An alternative justification", *Oxford Studies in Ancient Philosophy* VII: 129–150.
Shaw, J. C. (2015) "Punishment and psychology in Plato's *Gorgias*", *The Journal for Ancient Greek Political Thought* 32: 75–95.
Slaby, J., and Gallagher, S. (2015) "Critical neuroscience and socially extended minds". *Theory Culture & Society* 32(1): 33–59.
Sterelny, K. (2012) *The evolved apprentice: How evolution made humans unique*, Cambridge, MA: MIT Press.
Szanto, T. (2013) "Shared extended minds: Towards a socio-integrationist account", in D. Moyal-Sharrock, A. Coliva, and V. A. Munz (eds.) *Mind, language and action: Proceedings of the 36th International Wittgenstein Symposium*, Kirchberg am Wechsel, pp. 400–402.
Theiner, G., Allen, C., and Goldstone, R. L. (2010) "Recognizing group cognition", *Cognitive Systems Research* 11(4): 378–395.
Tollefsen, D. (2015) *Groups as agents*, Cambridge: Polity.
Tylén, K., Fusaroli, R., Bundgaard, P. F., and Østergaard, S. (2013) "Making sense together: A dynamical account of linguistic meaning making", *Semiotica* 194: 39–62.
Tylén, K., Weed, E., Wallentin, M., Roepstorff, A., and Frith, C. D. (2010) "Language as a tool for interacting minds", *Mind & Language* 25(1): 3–29.
Vlastos, G. (1983) "The Socratic *elenchus*", *Oxford Studies in Ancient Philosophy* 1: 27–58.
Wohl, V., ed. (2014) *Probabilities, hypotheticals, and counterfactuals in ancient Greek thought*, Cambridge: Cambridge University Press.

13

Cognitive dissonance, defeat, and the divinization of Demetrius Poliorcetes in early Hellenistic Athens

Thomas R. Martin

"But then the gods of the heavens brought this very calamity upon you . . ."
(Achilles speaking to Priam in Homer, Iliad *24.547)*

The concept of cognitive dissonance explains consequences of the catastrophic defeat of Athens in 322 BCE in the Lamian War that deeply affected the long-term political and religious history of Athens, the most spectacular being the divinization in 307 of Demetrius Poliorcetes.[1] This argument reflects the importance of the Cognitive Science of Religion (CSR) for the study of ancient religion; CSR offers new insights relevant not only to methodological discussions but also to the interpretation of specific phenomena such as, in the case under investigation here, the development and persistence in the Hellenistic period of religious beliefs about the divinization of human beings with the status of savior gods.[2]

As Jennifer Larson (2016: 379–380) explains in a recent study on understanding ancient religion, "[CSR maintains that] . . . religious beliefs arise in connection with the normal operation of human cognitive structures shaped by natural selection." As she also says, CSR does not constitute the whole of an explanatory model for religion: "[it] does not explain the variables that occur within the parameters set up by our mental tools; these are the result of the historical contingencies of culture, physical environment and individual agency."[3] Nevertheless, CSR does reveal that "mental tools generate content which is relatively invariable across cultures."[4] In fact, as Beck (2006: 88-101) argues, CSR shows that "we may safely assume that we form our representations of supernatural beings, to all intents and purposes, just as the ancients did." In sum, it seems to me legitimate and constructive to apply the theory of cognitive dissonance and CSR to Greek antiquity.[5]

The theory of cognitive dissonance got its start with the colorful case study published in 1956 by Leon Festinger and his collaborators entitled *When Prophecy Fails: A Social and Psychological Study of a Modern Group that Predicted the Destruction of the World*. Based on direct observation of a mid-twentieth-century apocalyptic group called the Seekers, the researchers investigated the psychological consequences for the group's members that ensued after the failure of the prediction by the group's prophetic leader that a cataclysm would soon strike the world, but that the Seekers would be saved by extraterrestrial aliens taking them away from our planet in a spaceship to safety. Most of the members responded to the severe cognitive dissonance they experienced from the incontrovertible disconfirmation of the prophecy, to which they had been intensely committed, by revising their beliefs and increasing their proselytism.

In 1957 Festinger then published a scholarly monograph that went well beyond the cognitive consequences of failed prophecy to discuss cognitive dissonance in human experience in general and the results of controlled experiments testing the theory.[6] In sum, he argued that cognitive dissonance is a reaction of mental discomfort in the face of inconsistency of knowledge, belief, expectation, etc.[7] The following excerpts outline Festinger's theory in his own words (1956: 25-26; 1957: 3, 9-11):[8]

> Dissonance and consonance are relations among cognitions—that is, among opinions, beliefs, knowledge of the environment, and knowledge of one's own actions and feelings. Two opinions, or beliefs, or items of knowledge are *dissonant* with each other if they do not fit together—that is, if they are inconsistent, or if, considering only the particular two items, one does not follow from the other . . . Dissonance produces discomfort, and correspondingly, there will arise pressures to reduce or eliminate the dissonance. Attempts to reduce dissonance represent the observable manifestations that dissonance exists. Such attempts may take any or all of three forms. The person may try to change one or more of the beliefs or opinions or behaviors involved in the dissonance; to acquire new information or beliefs that will increase the existing consonance and thus cause the total dissonance to be reduced; or to forget or reduce the importance of those cognitions that are in a dissonant relationship. If any of the attempts are to be successful, they must meet with support from either the physical or social environment . . . The magnitude of the dissonance will, of course, depend on the importance of the belief to the individual and on the magnitude of his preparatory activity . . . [T]he major point to be made is that *the reality which impinges on a person will exert pressures in the direction of bringing the appropriate cognitive elements into correspondence with that reality* [Festinger's italics].[9]

A host of scholars have critiqued and revised the theory over the past sixty years, but its basic schema continues to be seen as "a reliable phenomenon" (Cooper 2007: 6), which experimental science suggests originated deep in our evolutionary past (Egan *et al.* 2007). In his extended review of the decades of psychological research on the theory, Joel Cooper concisely formulates the contemporary understanding of that schema: "The state of cognitive dissonance occurs when people believe that two of their psychological representations are inconsistent with each other" (2007: 29).

Festinger's theory of cognitive dissonance therefore provides a well-tested approach for understanding the psychological aftermath of the defeat and catastrophic punishments inflicted on the Athenians in 322 by the Macedonian commander Antipater at the conclusion of the Lamian War. Modern scholarship's tendency, at least until recently, to see Philip II of Macedon's victory at Chaeronea in 338 as the decisive turning point in the erosion of Greek freedom has obscured the reality that the events of 322 and their enduring consequences added up to the worst disaster in the history of the Athenian *polis*.[10] That reality was, as Borza (1999: 59) epigrammatically observes, that the Lamian War, not Chaeronea, "changed everything."[11]

The situation in 322 was an unprecedented catastrophe for Athens because, first, the Athenians' foreign conqueror imposed a foreign military garrison on their *polis* on the twentieth day of the month Boedromion; second, he abolished their democracy;[12] third, he installed an oligarchy based on wealth; fourth, he disenfranchised well more than half the citizen body;[13] and fifth, he displaced thousands and thousands of citizens far away from home in a new settlement in Thrace, there surrounded by an often hostile non-Greek population.[14]

The specific reference to the twentieth day of the Athenian month Boedromion (Plutarch, *Phocion* 28) makes clear that Athens' punishment began on the same day of the year that the ancestors of the Athenians of 322 had received direct help from the gods in their glorious victory at the Battle of Salamis in 480 during the invasion of the Persian king Xerxes. As Herodotus describes (8.65), the gods had signaled that they would assist Athens in securing its salvation/preservation (*soteria*) from foreign attack on that fateful day.[15] So, the Athenians in 322 were asking themselves in dismay and confusion, why had this horrible calamity struck them on the twentieth of Boedromion?

To make matters even more disconcerting for the Athenians, not only had they not committed any impiety that would have justified their awful punishment at the conclusion of the Lamian War in 322, they had overtly been fighting for the sake of the traditional gods.[16] This view of the Athenians' motivation in the war is explicitly documented in an officially commissioned public speech delivered by the orator Hyperides at Athens on a solemn national occasion commemorating the war dead during the late winter/early spring of 323/322.[17] His *Funeral Oration* dramatically (its language and verbal images are extraordinary—see, for example, the image of rebirth at section 28)[18] insists that the Athenians are winning the war against Antipater; the oration in fact amounts to a triumphalist prophecy of victory.

Hyperides emphasizes that Athenian troops fought and died in this war against the Macedonians not only in service to the political goal of preserving the freedom of the Greeks against tyrants and kings, but also to defend the worship of the traditional gods against those seeking to worship a living human being as a god—meaning the recent movement to divinize Alexander the Great.[19] The speech was reportedly very well received by the citizens of Athens.[20] In fact, as Plutarch reports, they celebrated their wartime success in the service of freedom and piety by joyfully expressing thanks to the gods through festivals and sacrifices (*Phocion* 23.5-6).

Shockingly to these same Athenians, however, only a few months after Hyperides' exultant oration and their joyous celebrations, they suffered their disastrous defeat and punishment. What made this situation completely traumatic was that many Athenians—surely the great majority of citizens, who favored democracy instead of oligarchy—concluded that this multi-faceted punishment, of a ferocity that they had never before experienced, *had been overseen by the gods*. This conclusion is reported by Plutarch in his biography of the Athenian leader Phocion (*Phocion* 28.2-3):[21]

> The very timing added not a little to their suffering, for the garrison was installed on the twentieth of Boedromion, with the Mysteries being celebrated, on the day that they send Iacchus from the city to Eleusis, so that with the initiation rites being thrown into confusion, many people analyzed the more ancient actions of the divine compared with the recent ones. For in the past, there had occurred mystic apparitions and voices during their best good fortunes, to the consternation and astonishment of their enemies, but now during these same religious ceremonies, *the gods were overseeing [or, managing] the worst sufferings of Greece* [my italics].

So, the Athenians' dismay arose especially from their belief that the gods had not just neglected to notice, or passively permitted, the worst sufferings in their history, but indeed that the gods had been *responsible* for these events. Plutarch's Greek makes this point clear. The crucial word in Greek is ἐπισκοπεῖν; numerous other passages show that this word means "to oversee as managers or supervisors" and can apply to the gods.[22] Unfortunately, modern English translations mislead with versions that say the gods were simply unconcerned, observing Athens' sufferings with an Epicurean dispassion. Waterfield (2016: 129) renders the key phrase as "the gods were looking down impassively," while Scott-Kilvert and Duff (2011) have "the gods apparently looked down unmoved."[23]

The consequences of this divine management, as the Athenians perceived it, were for most of the citizens, excepting perhaps the rich social elite, both horrific and—this is the crucial point—bewildering.[24] Therefore, the defeat and punishment of Athens in 322 became for the majority what in cognitive dissonance theory is called a focal or generative cognition (Harmon-Jones and Harmon-Jones 2015: 8). Unmistakable evidence from reality generated this cognition, but it was dissonant with the Athenians' long-held cognition about the nature of their protective relationship with the gods. That is, the Athenians had a traditional belief to which they were strongly committed, namely that there were patron gods favoring Athens, and furthermore that the Athenians would experience salvation/preservation (*soteria*) so long as they met their responsibility to their commitment to proper worship of the gods (*eusebeia*) and did not anger the gods by committing impiety.

In traditional Greek religion, "Human suffering in the form of plagues, droughts, or military defeat was *prima facie* evidence for sin and divine wrath" (Attridge 2009: 75; cf. Harris 2001: 137); that is, calamity is comprehensible. In this case, however, the Athenians' belief that they had been completely pious in fighting the Lamian War meant they could find no plausible explanation for why the gods had inflicted the evil that their enduring misery represented. Their dissonant cognitions anticipated what Bulloch (1984: 229) identifies as the religious implications of the poems of the early Hellenistic author Callimachus questioning the comprehensibility of strongly held religious beliefs of his time: "the orderliness assumed by traditional religion" seemed "illusory."

Consequently, in the Athenians' minds, the gods' role in overseeing the unprecedented and undeserved punishment of Athens created a dissonant cognition that its citizens could not ignore. Indeed, their resulting cognitive dissonance was compelling because their traditional belief in the salvific role of their patron gods was crucial for them, yet they now confronted incontrovertible proof that they had failed to receive this expected divine salvation/preservation in 322, despite the strong demonstration of their national piety (*eusebeia*) so forcefully described by Hyperides. CSR confirms the inevitability of this dissonance: "having extraordinary powers does not absolve gods of the necessity of fair play. We expect our gods to follow the rules of social exchange" (Tremlin 2006: 118-119). In short, the Athenians' situation was overwhelmingly inexplicable.[25]

This inexplicability made the Athenians' cognition concerning the fact of their dire situation after the Lamian War disturbingly dissonant with their cognition concerning their traditional national political theology.[26] The level of their cognitive dissonance was extreme because they had been so strongly committed to their traditional belief about the role of their protective patron deities; now, they found themselves no longer able to rely on their long-held religious beliefs as support in coping with disaster.[27]

Furthermore, their dissonant cognitions were motivational because they were relevant to one another by having prior motives associated with them. To illustrate what is meant by prior motives, there is the often-cited example of cognitive dissonance being aroused when a person who smokes tobacco is confronted with reports that smoking is dangerous to human health:

> The cognition corresponding to smoking ("I am smoking") and the cognition corresponding to reading about lung cancer and smoking ("Smoking causes death from lung cancer") are relevant to one another because obvious motives interact, namely, "gain pleasure from smoking" and "avoid painful death," in addition to many other possible motives.
>
> *(Wicklund and Brehm 1976: 255-256)*[28]

In the Athenians' case, one strong prior motive was associated with their cognition that they had patron gods: they would be protected in an uncertain and dangerous cosmos. Another prior motive stemmed from their confidence that they would receive salvation in 322 in return for their piety in fighting the Lamian War to protect the honors of the gods, as Hyperides had proclaimed they had done. But the incontrovertible evidence of their sufferings in 322 and thereafter put these cognitions into dissonance, providing strong motivation for resolution.

Like the Jews in ancient Israel during and after crises with Assyria and Persia from the eighth century BCE on, like the followers of Jesus in the first century CE, like the followers of William Miller and his prediction of the Second Advent of Jesus in the 1840s, and like the Christians in the Confederacy during and after the US Civil War in that same century, the late-fourth-century BCE Athenians had suffered an undeniable disconfirmation of a deeply held belief that they had a special relationship with the divine.[29] This relationship, to reiterate what they fervently believed, provided them with safety and gave them a special identity in the world.

When confronted with a completely unexpected contradiction of this foundational cognition, the Athenians experienced a high level of dissonance because, to use Festinger's terminology, they had experienced disconfirmation of a belief to which they were deeply committed as the result of "exposure to information" (cf. Wicklund and Brehm 1976: 8-9). In fact, the defeated and punished Athenians had been exposed to "incontestable evidence" (Wicklund and Brehm 1976: 297), which produced a strong cognitive response: since "historical events . . . cannot be changed . . . cognitions concerning them will therefore be highly resistant to change" (Wicklund and Brehm 1976: 3-4). Accordingly, the Athenians could not avoid experiencing a seemingly crushing magnitude of dissonance.

Research on cognitive dissonance shows, as mentioned above, that such a high level of dissonance in turn generates a strongly felt drive to reduce, and if possible resolve, the dissonance: "Festinger concluded that . . . because of the inconsistency between your expectation and your experience . . . people who are in the throes of inconsistency in their social life are *driven* to resolve that inconsistency" (Cooper 2007: 2-3). The drive to reduce and relieve dissonance is so pronounced that "it has behavioral properties in common with other motivational states, especially the so-called homeostatic drives [physiological processes outside of an individual's conscious control]" (Wicklund and Brehm 1976: 107, 252). Plutarch's report of what Athenians were saying in response to the catastrophe of 322 reveals that they were experiencing just this sort of strongly felt manifestation of the drive state of dissonance.

So, how did the Athenians react to this extraordinary psychological pressure? The undeniable nature of the Athenians' current reality meant that they emphatically did not have the option of "not thinking about the dissonant relations" (Wicklund and Brehm 1976: 138).[30] The option most helpful to the Athenians in their highly stressful circumstances was to add consonant cognitions "organized around the cognition that is most resistant to change" (Wicklund and Brehm 1976: 5, 124-135).[31] In their case, the cognition most resistant to change was of course the belief that patron gods, when properly and piously worshipped, do provide salvation for the *polis*.

The new consonant cognition that the Athenians adopted to reduce their painful dissonance provides the culmination of my argument: they satisfied their overpowering drive for dissonance reduction by divinizing human beings as a source of national salvation. This new cognition arose as a result of the reality of the liberation of Athens from oligarchy and Macedonian control in 307 by a living and present human being—the Macedonian commander Demetrius the City-Besieger.[32] His liberation in person of Athens—carried out in accordance with the wishes of his father, Antigonus the One-Eyed—proved to be the salvation of the *polis* that had not occurred on the twentieth of Boedromion in 322, or in the fifteen woeful years for Athens since that day so memorable for all the wrong reasons.[33] The intentional actions of Demetrius (and Antigonus) at last lifted the military occupation of Athens, restored its democracy, and re-integrated its disenfranchised and displaced citizens. In addition, the liberators made an enormous gift to Athens of food for the people and timber for building warships.[34] In 307, then, the Athenians discovered that a human being could and would do what previously only traditional gods had been regarded as capable of doing: provide salvation/preservation to the *polis*. The Athenians were jubilant and grateful. Bayliss (2011: 160) eloquently expresses their reaction: "We cannot underestimate the enthusiasm the Athenians must have felt . . . How emotional must they have been when democracy was restored after a decade of political oppression under Demetrius of Phalerum?"

The Athenians' salvation seemed paradoxical, as the ancient historian Diodorus remarks (20.46.3); that is, human beings had now done what patron gods had purposely—but inexplicably—failed to do for a decade and a half. Applying insights from the theory of cognitive dissonance about the ways in which people seek to resolve dissonant cognitions explains why the Athenians in 307 responded to the paradox not only by recognizing their human liberators as kings, but also by divinizing them as savior gods (*theoi soteres*).[35] Divinization was a new cognition for the Athenians, in the sense of that term as applied in the theory of cognitive dissonance. There had certainly been instances of attributing divine attributes to human beings before this date.[36] But divinization of father and son at Athens in 307 was a tipping point in the development of the idea of a savior god incarnate in a human body and present on the earth. Harmon-Jones and Harmon-Jones (2015: 11), in their review of a half-century of research on cognitive dissonance theory discussing the action-based model of dissonance (for which see Harmon-Jones 1999, 2000), comment that "dissonance evokes a negative affective state that signals the organism that something is wrong and motivates the organism to engage in behavior to solve the problem." In the context of the catastrophe at Athens that occurred in 322, that organism was the Athenian people, and the behavior engaged in was to divinize Demetrius and Antigonus following the liberation of 307.

The details of the Athenians' behavior in divinizing human beings in 307 resulted from the learning process that can be generated by cognitive dissonance, as Adcock (2012: 588) explains: new information that contradicts a mental schema (based on existing knowledge) requires mental effort in accommodation and assimilation for the integration of the new knowledge with the existing schema. This particular decision to divinize human beings, taken to reduce the dissonance resulting from the Athenians' defeat in the Lamian War and consequent suffering, had the form that it did because they had experienced a strong motivation to accommodate and assimilate to new information about the nature of the gods, especially as saviors, and about the nature of the divinities' special protective relationship with the *polis* of Athens. As Hardyck and Braden (1962: 137) explain in their summary of Festinger's theory, to reduce dissonance, believers must experience content change by seeking "new information consonant" with their beliefs. So, for the Athenians, their learning process from exposure to new information had generated a new cognition that required their accommodating contradictory notions of the human and the divine, at least insofar as the gods' natures had traditionally been understood in ancient Greek religious belief.[37] In other words, the content of their cognitions about the nature of the divine had changed.

In addition, in the case of the Athenians reacting to the state of their community in a radically refashioned world order (in the eastern Mediterranean region), the resolution of dissonance through content change both depended on and supported the continuance of changes in cognition and in behavior on their part *collectively* (if not, of course, universally; there are always outliers in every complex human situation). As Taylor (2002: 582) observes,

> Festinger formulated the theory of cognitive dissonance precisely in a study of a corporate experience and response thereto, not an individual one. The possibility of a collective and intense experience of dissonance, particularly in a society in which an individual would be so conscious of and sensitive to his or her collective identity, would therefore seem to merit some consideration.[38]

The clear disconfirmation of a cherished religious belief experienced by the Athenians as a group motivated them to generate a new cognition.[39]

CSR explains why their acceptance of a seemingly paradoxical new cognition—a human being as a true savior god—was acceptable, even desirable, for the community as a whole. The idea and process of divinizing a human being fit well with insights from CSR on how people characteristically ascribe properties of various kinds to supernatural beings. Those properties, found in beliefs about gods across time and cultures, include anthropomorphism (Barrett and Keil 1996: 244; Boyer 1996; Berner 2011; Tremlin 2006: 98-100), theory of mind and intentionality (Luhrmann 2011; Goldman 2012), hyperactive agency detection (Barrett 2000; Boyer 2000; Lawson 2006; Tremlin 2006: 75-86; Ma-Kellams 2015), and minimally counterintuitive concepts (Batson 1975; Burris *et al.* 1997; Barrett and Nyhof 2001; Boyer and Ramble 2001; Pyysiäinen 2002, 2009; Atran and Norenzayan 2004; Tremlin 2006: 86-98; Barrett 2008; Baumer and Boyer 2013).[40] Moreover, the concept of divinization fits well both with the theory of the dual-process model of cognition, in which human beings arrive at impressions and conclusions through the complex interaction—or lack of interaction—of two systems of reasoning in the brain, one unconscious or automatic or implicit and the other conscious or controlled or explicit, and also with the evidence for people frequently accepting the coexistence of multiple, or contradictory, mental representations of religious concepts (Barrett 1999: 325-326; Boyer 2001: 78-89; Tremlin 2006: 181-182).[41] Finally, the divinization of human beings also corresponds to the observation that the human mind constructs the special concept of god (Barrett and Keil 1996) on the ontological category of "Person" (see the helpful Table 3.2 in Tremlin 2006: 93).

Perhaps the most revealing of these insights from CSR is that people define gods as counterintuitive intentional agents. That is, gods are on the one hand certainly not like human beings—they are immortal, possess great powers, can be invisible, etc.—but on the other hand they are also like human beings because they have minds, operate with intentionality in analysis and action, and perform as agents capable of producing effects. As Tremlin explains,

> In counterintuitive concepts, properties that, from a cognitive perspective, are actually quite ordinary are matched with the properties that make them extraordinary. This structural combination of ordinary and extraordinary properties is key to the attraction of counterintuitive concepts . . . As a rule, then, counterintuitive concepts contravene one or more intuitive expectations associated with an ontological category while at the same time preserving others.
>
> *(2006: 87, 90, 106)*

The level of counterintuitiveness of the ontological category of gods is, in technical terms, minimal, so that the cognition can be acceptable rather than incomprehensible, as would be the case with the attribution to them of bizarre, unimaginable characteristics seemingly suited only for phantasy (Barrett and Nyhof 2001; Boyer and Ramble 2001).

Furthermore, it is important to recognize that the Athenians' new cognition about the existence of divinized human beings as savior gods did not abolish their long-held belief in the efficacy of *all* of the traditional gods under *all* circumstances. In fact, the Athenians in 307 (or soon thereafter) publicized the consonance of their new cognition with their older one by having the images of Demetrius and Antigonus woven into the robe (*peplos*) presented to Athena on the acropolis in her role as patron deity of the *polis*.[42] There, the new gods Demetrius and Antigonus appeared alongside the old gods Athena and Zeus in a depiction of the salvific power of these traditional divinities—the battle in which they overcame the giants, thereby saving the world.[43]

The language of the so-called Hymn to Demetrius preserved in Athenaeus 6.253d–f (citing the late-fourth/early-third-century historian Duris of Samos), a text composed perhaps some two decades after the original liberation of Athens, further reveals the characteristics of this new cognition about divinized savior gods and how it had been made consonant with the Athenians' traditional cognition about the role of the gods in order to reduce or resolve cognitive dissonance.[44] This poetic song was performed not just publicly by a chorus at Athens but also sung at private social gatherings; the latter point speaks against the dismissal of the text, by Duris and by some modern scholars, as nothing more than official (and cynical) flattery without any significant content deployed in the hopes of pleasing a powerful commander.[45]

The description of Demetrius presented in the verses of the Hymn clearly reveals aspects of the supernatural concept of divinization that served to reduce the Athenians' cognitive dissonance. First, it says that Demetrius comes to Athens with Demeter, revealing that the new cognition has been made consonant with the old one about traditional savior gods, subject to some modifications that are revealed a bit further on in the text. Demetrius himself is hailed as "cheerful (*hilaros*), good-looking (*kalos*), and laughing (*gelōn*), as a god should be."

Hilaros is especially interesting as perhaps the most striking cognition concerning the definition of this new form of divinity. It is a word known mostly from comedy in the classical era that gained wider currency later on; the works of Plutarch provide many instances. It seems predominately to be used to describe human characteristics of good cheer and joy exhibited in both "workaday" situations and celebratory circumstances.[46] Here, it seems to belong to a new category—a divinized god interacting personally with human beings. As for laughter as a property of this new kind of divinity, Plato in the fourth century, though perhaps not Homer earlier on, would have regarded this characteristic as a radical departure from a traditional representation of the gods (*Republic* 3.388e–389a).

Demetrius is also named as the son of Poseidon and Aphrodite; this pedigree connects him with the Athenians' long-standing belief in traditional gods.[47] The next lines reveal revisions in traditional notions of the divine that were required for dissonance reduction: other gods are far off (note the Greek does not say "*the* other gods," meaning all of them, as is sometimes the translation), or do not have ears (meaning they do not respond to prayer and petitions for help), or do not exist, or pay no attention at all to the Athenians.[48] He (Demetrius) is visible, being present, not made of wood or stone, but "true"; the Athenians pray to him, who is "beloved" (*philtatos*) and "in charge" (*kyrios*).

That this new kind of god can and will answer the Athenians' prayer by bringing them salvation/preservation (*soteria*) is the clear expectation. In fact, by alluding with its language not only to Demetrius's great naval triumph in 306 at Salamis on the island of Cyprus, but also to the glorious victory at Salamis near Athens in 480, the hymn emphasized the Athenians' recognition of the salvific power of Demetrius as a divine human being who preserved Athens' freedom and democracy.[49] This connection of the memory of the national salvation of Athens in 480 and the hymn's extolling of Demetrius' power to benefit Athens was in fact explicit because, as pointed out above, Demetrius in 307 had liberated Athens from the loss of freedom and democracy that Antipater had imposed in 307 on the very anniversary of the first Battle of Salamis. Since "dissonance theory predicts that being induced to make a counter-attitudinal statement would lead to attitude change in the direction of the speech" (Cooper 2007: 15), this helps explain the persistence of the cognitive effect testified to by the language of the Hymn to Demetrius and the Athenians' repeating it to themselves in public and in private.

To conclude: the late-fourth-century Athenians faced a religious "belief crisis" caused by "frustrated or disconfirmed beliefs" (Wernik 1975). The theory of cognitive dissonance explains the Athenians' reaction to this crisis by its applicability to social groups who create content change in their beliefs. This phenomenon is perceptively described by Uri Wernik in his study of early Christianity, and it is worthwhile to close my argument by citing his work as a useful analogy to the interpretation of the case of Athens presented here. Wernik specifically applies the theory of cognitive dissonance to the idea that Christ's crucifixion disconfirmed a religious belief, namely that Jesus was the Messiah. He then shows how content change served as a method to resolve dissonance. As Wernik explains (p. 98): "[t]he main research hypothesis . . . is that Jesus' death created in his believers a cognitive dissonance, whose resolution constitutes the Christology of the New Testament." His hypothesis has the following propositions:

1 There was a multi-faceted belief in Jesus by his disciples and followers.
2 There was a commitment to this belief.
3 The belief was disconfirmed by Jesus' crucifixion. Thus, a belief crisis came into existence . . .

Similar points can be applied to the situation of the Athenians resulting from the unexpected and inexplicable (on the traditional content of their beliefs) catastrophe that struck them in 322:

1 Athenians held a multi-faceted belief about the traditional gods and the gods' special protective relationship to the Athenians.
2 The prayers, festivals, and rituals of the Athenian *polis* show the commitment to this belief, strongly emphasized by Hyperides' *Funeral Oration* for the specific occasion of the Lamian War of 323-322.
3 The Athenians' belief was disconfirmed by the disaster of 322 and its long-lasting consequences.

The resolution to the resultant dissonance of the Athenians takes place in 307 with the divinization of Demetrius (and Antigonus, as the father of the present-in-Athens *theos soter*) as an addition to the company of at least some of the traditional gods, thereby changing the content of the Athenians' belief in the nature of gods.

This new cognition, consisting of the content change represented by the divinization of living human beings as savior gods, persisted over time. This was so despite the serious military and political tensions and conflicts that the Athenian *polis* encountered in dealing for the first time with a formerly human but now *theos soter* in the decades between the liberation of Athens in 307 and the end of Demetrius's career in the 280s.[50] Perhaps the most discomforting episode for the Athenians was the defeat of Demetrius and Antigonus by other "successor kings" at the Battle of Ipsus in 301 and the citizens' decision not to receive Demetrius afterwards.[51] The subsequent revival(s) of Demetrius' power compelled further revision by the Athenians of their notion of the nature of a divinized human as savior god, including the realization that their relationship with this new form of salvific divinity could be very painful if they did not consistently respect their obligations as defined by the "human now god."

In this respect, the situation reflected *mutatis mutandis* what they had always known about their obligations to the traditional gods.

Laboratory experiments on cognitive dissonance indicate that it is not surprising that "the longevity of dissonance reduction effects" depends "largely on whether or not the circumstances of dissonance arousal lend themselves to a continued salience" (Wicklund and Brehm 1976: 117-119). Continued salience was certainly the case regarding national salvation/preservation in the religious and political history of Hellenistic Athens in particular and the eastern Mediterranean Hellenistic Greek world in general. Over the long term, this new cognition evolved in diverse directions: soon into cult for the rulers of temporal kingdoms, and later into worship of the ruler of a spiritual kingdom; it competed successfully with Epicurean philosophy as an understanding of the role of gods in human affairs.[52]

On the significance of this evolution, here there is only space to say that I disagree with Parker (1996: 263) concerning whether "watershed" is an apt metaphor for the religious history that followed upon the events of 307 down into the 280s:

> [They], therefore, are not quite the watershed that they may appear. After a period of some confusion, traditional cults and cults of god-kings settled into a mode of co-existence. And it was, in all seeming, through external circumstance, not a development or decline in belief, that the efflorescence of ruler-cult at Athens occurred at the time that it did.

Equally unconvincing to me is the denial of Jon Mikalson (1998: 307) that significant change occurred in Athenian religious beliefs in the early Hellenistic period.[53] More persuasive seems the succinct observation of Françoise Dunand (2010: 262): "The desire to see a god embody himself in living form is undoubtedly one of the new characteristics of religious life in the Hellenistic age." This seems to me a very significant development in Athenian religious history—and beyond—that cognitive dissonance in the context of CSR does much to illuminate.

In a retrospective look at the history of his theory thirty years after its inception, Festinger called for more correlational studies based on archival research that apply sociologically oriented social psychology based not on experimental designs and laboratory results but on historical events:

> One thing that I think has to be done is for more research to go on on dissonance producing situations and dissonance reduction processes as they occur in the "real world" . . . I think we need to find out how dissonance processes and dissonance reducing processes interact in the presence of other things [outside a laboratory setting] that are powerful influences on human behavior and human cognition, and the only way to do that is to do studies in the real world. They're messy and difficult. You don't expect the precision out of these studies that you get in the laboratory. But out of them will emerge more ideas which we can then bring into the laboratory to clarify and help to broaden and enrich the work.
>
> *(Festinger 1999: 384-385; cf. Hood, Jr. 2011: 35-36)*

I suspect that Festinger would agree that CSR opens up new horizons in research in Classics, and that there is much to gain for our field in expanding into the "real world of the past" the kind of research that Festinger wanted to see in his field.[54] For these reasons and more, the history of divinization in pre-Christian antiquity needs to be the subject of an extensive study in the context of CSR. As Pyysiäinen (2012: 16) says, "within CSR, there are only a few studies on the historical spread of particular religious concepts and beliefs (e.g. Beck 2006; Pyysiäinen 2009; Martin and Sørensen 2011)."

Notes

1 I would like to express my warm thanks to Andrea Martin-Nieuwland for guidance in cognitive science and to Peter Meineck for his initiative in supporting the integration of cognitive theory with research in Classics.

2 On CSR, see Guthrie 1980; Lawson and McCauley 1990; Guthrie 1993; Boyer 1994, 2001, 2003; Barrett 2000, 2007, 2011; Ozorak 2005; Lawson 2006; Slone 2006; Tremlin 2006; Boyer and Bergstrom 2008; Pyysiäinen 2012; Parker 2014; Larson 2016: 373-384. For the types of methodological discussion for which CSR could be illuminating, see Scheid 2016: 1-21 and Koortbojian 2013.

3 For diverse perspectives on the nature of religious belief, see Leach 1966; Needham 1972; Gager 1975: 38-39; Sperber 1996; Beit-Hallahmi and Argyle 1997; Pargament 1997: 221-232; Barrett 2000, 2008; Exline 2002: 185-187; Boyer 2003: 123; Exline and Rose 2005: 316-317; Gervais and Henrich 2010; Banerjee and Bloom 2013; Baumard and Boyer 2013; Grubs, Exline, and Campbell 2013; Norenzayan and Gervais 2013; Bae 2016. See Renfrew 1994 and 2007 on "the archaeology of mind" as revealed by cognitive archaeology, especially concerning religion and belief, and on "the prehistory of mind." See Holt 2012: 160-210 on cognitive numismatics and the history of Hellenistic Bactria (Afghanistan). For belief in ancient Greek religion, see Parker 2011: 1-39; Harrison 2015. On belief in contemporary American evangelical religion, see Luhrmann 2012.

4 On culture and cognition, see Marcus and Kitayama 1991; Luhrmann 2011; Fessler and Machery 2012.

5 For arguments in favor of the applicability of the theory of cognitive dissonance across cultures, races, and time, see Hoshino-Brown *et al.* 2005; Kitayama *et al.* 2006; Cooper 2007: 135-156. Nongbri 2013 argues that the concept of religion cannot be applied to pre-modern cultures, or indeed to non-Protestant Christianity. From among many scholarly reviews of his controversial thesis, see the critical response by Frankfurter 2015. See Taylor 2002: 579-584 for discussion of CSR's applicability to antiquity.

6 See also Festinger 1964.

7 In this context, Versnel 1990 studies "the notion of inconsistency as a much neglected or even rejected, though important aspect of history" (p. 1) in the religious mentality of ancient henotheism.

8 For overviews of Festinger's theory, see, among numerous choices, Wicklund and Brehm 1976: 2–10; Darity, Jr., ed. 2008: 599–601; Hood, Jr. 2011: 26–28; Stone 2011: 45–47, 58; Tumminia 2011: 3–4. For extensive reviews of research on the theory, see Cooper 2007; Harmon-Jones and Harmon-Jones 2015.

9 Gawronski 2012 rebuts arguments that deny a drive for cognitive consistency is a core motive.

10 The long tradition of seeing the aftermath of the battle of Chaeronea as the beginning of the end of the freedom and significance of the world of the Greek *polis* stretches, in English scholarship, at least as far back as George Grote, who near the end of his immense and influential *A History of Greece* (12 vols., 1846–1856) remarks "[T]he freedom of Hellas, the life and soul of this history from its commencement, disappeared completely during the first years of Alexander's reign" (vol. 12: 391). More recently, "The End of Greek Freedom" is the title of ch. 10 in Worthington 2013, who there (p. 252) says of the outcome at Chaeronea, "The battle for Greek freedom was over" (they are both therefore echoing Pausanias 1.25.3: "the misfortune at Chaeronea *began* the evil for all the Greeks"). In fact, this defeat did not signal the end of Greek freedom, or generate the evil that was to befall Athens in 322. To the contrary, Athens actually grew significantly richer in the years between the battle of Chaeronea and the Lamian War (Burke 1985; Rhodes 2012: 125; Worthington 2013: 307-308; for general surveys of Athenian history from Chaeronea to the Lamian War, see Will 1983; Wirth 1999; Worthington 2013: 237–344; Harding 2015: 44–46).

11 For the Lamian War (or Hellenic War), see Hyperides, *Funeral Oration*; Diodorus 18.8-18; Plutarch, *Demosthenes* 27-28, *Phocion* 23–28; Pausanias 1.25.3-5; Arrian in F. Jacoby. 1962. *Die Fragmente der griechischen Historiker*. Leiden: Brill = *FGrH*, no. 156 F 1.9, 12 (= Photius, *Bibliotheca* 92, p. 69, lines b16-23, b29-33, ed. I. Bekker. Vol. I. 1824. Berlin: G. E. Reimer); Justin 13.5. For modern studies, see, among many, Westlake 1949; Mossé 1973: 96–101; Tritle 1988: 123–133; Schmitt 1992; Sealey 1993: 215–219; Tracy 1995: 23–29; Habicht 1997: 36–42; Bosworth 2003; Lübke 2007; Romm 2011: 93–155.

12 Cartledge 2016: 217 comments that Antipater's punitive actions in 322 "dealt a mortal blow" to Athenian democracy in the long run; see also Green 1990: 11; Oliver 2003, 2007: 45; Harding 2015: 44–46, 55–58; T. Mitchell 2015: 293–294.

13 For the debate over the total number, which was in any case enormous relative to the size of the citizen body, see Harding 2015: 70–73.

14 Diodorus' positive evaluation of the effects of Antipater's actions (18.18) directly contradicts his own subsequent report (18.66.5) on what the Athenians themselves said: they saw the situation as "slavery for their homeland." Cf. Plutarch, *Phocion* 28, who offers a more realistic evaluation of the dire situation. On the fate of the people displaced to Thrace, see Baynham 2003. Some scholars nevertheless downplay the severity and suffering of the punishment imposed on Athens in 322. See, for example, Habicht 1997: 40-45; Waterfield 2011: 40; Pownall 2016: 48.

15 Plutarch gives this date for the battle of Salamis at *Camillus* 19; cf. Graton and Swerdlow 1988: 18–20, 36–37. For an alternate date, see Frost 1980: 157–159. On the tradition of divine assistance to the Greeks at the battle of Salamis, see Mikalson 2003: 67–85; Boedeker 2007; Petridou 2015: 116–119. On the date of the procession of Iacchus, see Robertson 1998; for an overview of the Mysteries at Eleusis, see Clinton 2010. The significance of the date as an occasion of "salvation" is emphasized by the report that in 325 Alexander's fleet embarked on its dangerous voyage back from India on that very day, following a sacrifice to Zeus the Savior (Arrian, *Indica* 21). For vase paintings from the 320s interpreted as showing the gods supporting the Greeks against the Persians from a "concern for the welfare of Hellas," whether at the time of the Persian Wars or of Alexander the Great, see C. Long 1987: 349–351.

16 On the concept of impiety in Greek religion, see Bowden 2015. For divine punishment of city-states for the impiety of breaking oaths, see Sommerstein and Bayliss 2013: 167–175. Lycurgus 1 (*Against Leocrates*) builds his argument against Leocrates on the theme of divine punishment of impious actions; see Whitehead 2006.

17 For recent studies, see Wirth 2004; Herrman 2009.

18 On this passage, see Hess 1938: 83.

19 On the role of freedom in the Athenians' public commitment to the war, see Ashton 1984: 154; Lehmann 1988; Dmitriev 2011: 107. Alexander's claims to divinity and responses to it are recorded in various sources: Dinarchus, *Against Demosthenes* 94; Hyperides, *Against Demosthenes* 31–32, *Funeral Oration* 21; Polybius 12.12b.3 (= Timaeus *FGrH* 566 F 155); Valerius Maximus 7.2 ext. 13; Plutarch, *Alexander* 28, *Moralia* 180E (cf. 341B), 219E-F, 408B, 842D; Arrian 7.20.1 (= Aristobulus *FGrH* 139 F 55), cf. 7.23.2 on the arrival in Babylon of *theoroi* from Greek city-states in 323 BC; Aelian, *Varia Historia* 2.19, 5.12; Diogenes Laertius 6.63; Athenaeus 6.250F-251C. On this much-discussed topic, see Badian 1981; C. Long 1987: 188-189; Bosworth 1988: 278-290, 1996: 98–132; Klauck 2000: 266–274; Worthington 2001, 2004: 199-206; Fredricksmeyer 2003; Cartledge 2004: 339–340; Dreyer 2009; Anson 2013: 83-120; L. Mitchell 2013; Whitmarsh 2015: 145-148. On the possible existence of a cult at Athens, see Jaschinski 1981: 93-119.

20 [Plutarch], *Lives of the X Orators: Hyperides* 849f; Photius, *Bibliotheca* 266, p.496a. For English translations of these passages, see Roisman and Worthington 2015: 69, 317.

21 οὐ μικρὸν δὲ τῷ πάθει προσέθηκεν ὁ καιρός, εἰκάδι γὰρ ἡ φρουρὰ Βοηδρομιῶνος εἰσήχθη, μυστηρίων ὄντων, ᾗ τὸν Ἴακχον ἐξ ἄστεος Ἐλευσινάδε πέμπουσιν, ὥστε τῆς τελετῆς συγχυθείσης ἀναλογίζεσθαι τοὺς πολλοὺς καὶ τὰ πρεσβύτερα τῶν θείων καὶ τὰ πρόσφατα, πάλαι μὲν γὰρ ἐν τοῖς ἀρίστοις εὐτυχήμασι τὰς μυστικὰς ὄψεις καὶ φωνὰς παραγενέσθαι σὺν ἐκπλήξει καὶ θάμβει τῶν πολεμίων, νῦν δὲ τοῖς αὐτοῖς ἱεροῖς τὰ δυσχερέστατα πάθη τῆς Ἑλλάδος ἐπισκοπεῖν τοὺς θεούς, καὶ καθυβρίζεσθαι τὸν ἁγιώτατον τοῦ χρόνου καὶ ἥδιστον αὐτοῖς, ἐπώνυμον τῶν μεγίστων κακῶν γενόμενον. This text is from Perrin 1919 (Loeb Classical Library, as reproduced in the Perseus Digital Library). It is the same, except for minor variations in punctuation, as the text printed in K. Ziegler. 1993. *Plutarchi Vitae Parallelae*. Vol. II. Fasc. 1. Stuttgart: Teubner.

22 See Martin and Sun 2017. Lycurgus 1.94 (*Against Leocrates*) from 331 BCE documents this meaning unambiguously: "I believe, men (of the jury), that the gods' diligent attention (*epimeleia*) oversees (*episkopein*) all human actions, especially, as is fitting, concerning our piety (*eusebeia*) toward our parents and those who have died, and toward themselves." (Engels 2008 does not discuss this passage.) For earlier instances illustrating the meaning of *episkopein*, see Sophocles, *Antigone* 1136; Euripides, *Iphigenia in Tauris* 1414 and *Phoenician Women* 661; Aristophanes, *Knights* 1173, 1186; from later Greek, see, for example, Josephus, *Jewish Antiquities* 10.50, *Contra Apionum* 2.160-161. Of the more than fifty instances in Plutarch, see *Lycurgus* 14.1, *Numa* 9.4, *Aemilius Paulus* 38.5, *Moralia* (*Quaestiones Conviviales*) 654F.

23 These versions seem aligned with that of the Loeb Classical Library edition by Perrin from a century ago: "the gods looked down with indifference" (1919: 209). Nineteenth-century translations by Langhorne and Langhorne and by Clough are similar. The so-called Dryden translation from the seventeenth century has "at the same season, the Gods themselves stand witnesses of the extreme oppressions of Greece." Neither LSJ9 (with a revised supplement 1996) nor *The Brill Dictionary of Ancient Greek*

(2015) by Franco Montanari includes any definition of the verb reflecting the seemingly Epicurean view anachronistically incorporated into the English translations cited here. Modern French and Italian translators have been more circumspect in their translations; see, for example, Flacelière and Chambry 1976 and Bearzot 1993. The reading *episkopein* bothered the early editors of the text of Plutarch's *Phocion*, who apparently could not accept the word's attribution of responsibility to the gods. They therefore emended it away, despite the absence of any manuscript evidence for a different reading. Ziegler 1932: 57 ruled out this emendation, although he, too, mistranslated *episkopein* as "*ansehen ruhig*."

24 I use "bewildering" to separate my view, which admittedly pertains in this case specifically to religion rather than to philosophy as a whole, from that of A. A. Long 1986: 3, who says "It is difficult to find anything in Hellenistic philosophy which answers clearly to a new sense of bewilderment." Long himself was contrasting his interpretation both with that of Bevan 1913: 32, whose expression "bewildered world" was the source of Long's choice of the term "bewilderment," and also with the view of Edelstein 1966: 13 arguing for a "new consciousness of man's power that arose in the fourth century, the belief in the deification of the human being."

25 The Athenians certainly did not accept what Lefkowitz 2016: 50 concludes was the answer to be found in the tragedies of Euripides: "Mortals cannot expect to receive redemption, salvation, or even assistance from the gods, even when they deserve to receive it." On theodicy in antiquity, see Johnston 2004. On divine envy seen as a source of human misery in this general period, see Whitehead 2009. On the other hand, Leocrates 1.94 (*Against Leocrates*) expresses what was surely the common Athenian opinion: "From [the gods] we have the origin of our being alive and have experienced the greatest goods." Cf. Plato, *Republic* 2.379c–380c. For the centrality in the religious views of modern US teenagers (and probably adults, as well) of the notion that God plays fair with human beings by providing them with "therapeutic benefits," see Smith and Denton 2005: 162-171 on what they label "Moralistic Therapeutic Deism."

26 See Mazlish 1990: 267-268 for "psychic repository" as another possibility for expressing the concept of "national group psychology"; myths are "probably the most fundamental" of the materials forming this repository. On religion in the Athenian *polis* at the supra-individual level, see Versnel 1995; Jameson 1998; Parker 2005: 89-115; Deacy 2010; Blok 2014; Mikalson 2016.

27 On religious belief and psychological coping, see Phillips *et al.* 2004; Pargament *et al.* 2005.

28 Wicklund and Brehm (*loc. cit.*) add that their analysis "has enabled us to elaborate cognitive dissonance in a way not explicitly stated by Festinger (1957) in his theory: *Cognitive dissonance is a general 'motivational state' that occurs when there is some prior motive associated with the cognitions that are dissonant.* Although a dissonant relationship is defined more broadly, we are not confident that other forms of psychological implication would in fact provide the conditions necessary to be arousal of dissonance."

29 On cognitive dissonance and Old Testament prophecy, see Carroll 1979, especially pp. 86-128. On early Christianity, see Gager 1975 (reprinted Horrell 1999: 177-194), especially pp. 37-49 on Festinger's theory of cognitive dissonance, and the critical response by Rodd 1981, who also discusses Carroll 1979 and suggests in conclusion (p. 33 in the Chalcraft 1997 reprint) that "historical sociology is impossible" (Holmberg 1990: 6-17 and 78-81 responds specifically to Rodd); Wernik 1975; Wicklund and Brehm 1976: 302-303; Segal 1990; Fredriksen 1988: 133-135, 168, 184; Horrell 2002; Taylor 1998, 2002. On the Millerites, see Festinger *et al.* 1956: 14-23; Festinger 1957: 248-251; Knight 1993. On the Confederacy, see Silver 1957; Wilson 1980; Beringer *et al.* 1986: 82-107, 268-298, 336-367. Hood, Jr. 2011: 26-28 offers a summary of the application of the theory of cognitive dissonance to Montanists, Millerites, and early Christians. See also Stone 2000 for the varied consequences of failed prophecy in millenarian cults and the followers of prophets. There are, of course, numerous other similar instances of disconfirmation to be found in history.

30 This is one of the possible types of response observed among people experiencing the drive to reduce dissonance. See Festinger 1957: 19-24; Abelson 1959: 344, 354; Batson 1975: 176; Wicklund and Brehm 1976: 124-139; Taylor 2002: 579; Tumminia 2005: 155-156; Stone 2011: 43; Tumminia 2011: 3.

31 Prus 1976: 133 argues that dissonant cognitions are socially constructed: "While the degree of importance attributed to a discrepancy is critical vis-à-vis dissonance-reduction motivation, the intensity of any dissonance is not an intrinsic quality of the discrepancy in question, but is a problematic and negotiable subjective assessment, reflecting one's cultural experiences, specific referent and contact [with] others, and immediate/anticipated interests." Even if this should be true, the intensity of the discrepancy for Athenians was surely of the highest degree. As Festinger *et al.* 1956: 5 remark: "There is usually no mistaking the fact that [the predicted events] did not occur and the believers know that. In other words, the unequivocal disconfirmation does materialize and makes its impact on the believers."

32 Diodorus 20.45-46; Plutarch, *Demetrius* 8–14; cf. Habicht 1997: 67–71.

33 On this period, see Diodorus 18.64–68, 19.78 and Plutarch, *Phocion* 29–38, and, among many modern studies, Mossé 1973: 108–114; Williams 1982; Habicht 1997: 36–66; Tracy 2000; Lape 2004: 40–52; O'Sullivan 2009; Bayliss 2011; Anson 2014: 83-164; Harding 2015: 57–63; Luraghi 2016.

34 Diodorus 20.46.4; Plutarch, *Demetrius* 10.

35 Plutarch, *Demetrius* 9–10; cf. Diodorus 20.46.2. On divinization/deification, see Balsdon 1950; Habicht 1979; Petrovic 2015. For this particular episode, see Billows 1990: 148–150, 235–236; Santi Amantini *et al.* 1995: 331–334; Parker 1996: 258–262. On the phenomenon of introducing new gods, see Garland 1992; Auffarth 1995; Anderson 2015. On Greek savior gods, see Wendland 1904; Bleeker 1963; Larson 2007: 301 (references *s. v.* "savior").

36 For the case of the fifth-century BCE Spartan commander Brasidas being lauded by the citizens of Amphipolis after his death as a *soter*, see Larson 2016: 287-289. Slightly later, the Spartan Lysander became, in Plutarch's words (*Lysander* 18, citing Duris of Samos), "the first Greek for whom the Greek city-states set up altars as for a god and made sacrifices." What this report actually means has been much discussed; see Flower 1988: 129-134. In any case, as with Brasidas, Lysander was dead at this point, not alive and present like Demetrius; this was a highly significant difference, as the language of the Hymn to Demetrius makes clear, as does the appearance of "Manifest" (*Epiphanes*) as an epithet in later ruler cult. For the case of the Spartan Agesilaus turning down such honors (obviously while alive), also see Flower 1988. See above concerning the cult of a divine Alexander in the 320s. On the divine honors previously decreed to Antigonus by the city-state of Scepsis, see Erskine 2014.

37 On the traditional concepts informing the notion of gods in Greek religion, see Henrichs 2010.

38 See Kindt 2012 for the argument that the "theology" of ancient Greek religion is embedded in a symbolic universe involving inherently polyvalent, and therefore necessarily inconsistent, religious symbols, an observation with obvious potential for relevance to the theory of cognitive dissonance as applied to collective ancient Greek religious beliefs.

39 On the disconfirmation of strongly held religious beliefs, especially in groups, see Zygmunt 1972; Wicklund and Brehm 1976: 297–303; Sande and Zanna 1987: 64–66; Goethals 1987; Dein 2001; Dawson 2011: 89; Hood, Jr. 2011; Pyysiäinen 2012: 15, citing Geertz 2010; McKimmie 2015. On dissonance reduction in religious belief disconfirmation, see Burris *et al.* 1997; Dawson 1999: 76.

40 For the anthropomorphic characteristics associated with traditional gods in ancient Greek religion, see Erskine 2010; Petridou 2015: 32-43. On the "cognitive optimum" of intuitive and counterintuitive properties, which is key for cultural transmission, see Sperber and Wilson 1995; Barrett 2000; Boyer 2001: 86; Pyysiäinen 2009: 206; 2012: 13. On the concept of historical agents, see Sørensen 2011: 187–188. On the conceptual overlap of kings and gods in Greek political thought, see Brock 2013: 1–24.

41 The dual-process model is widely used in analyzing human thought, though there is no set of universally agreed-upon terms to describe the differences between the functioning of System 1 and System 2.

42 Diodorus 20.46.2; Plutarch, *Demetrius* 10, 12. On the *peplos* of Athena, see Barber 1992; cf. Håland 2004.

43 Euripides, *Hecuba* 466–469, *Iphigenia in Tauris* 223–224; Plato, *Euthyphro* 6b-c; Suda *s. v.* Πέπλος (Adler π 1006). On the significance of the scene, see Barber 1991: 380.

44 On the Hymn to Demetrius Poliorcetes, its uncertain date, and its religious significance, see Weinreich 1926; Scott 1928; Ehrenberg 1935, 1946: 179–198; Manni 1953: 93–95; Landucci Gattinoni 1981; Marcovich 1988: 8–19; Henrichs 1999: 243–247; Shipley 2000: 160–162; Chaniotis 2003: 431–432; Kolde 2003: 378–392; O'Sullivan 2008a; Chaniotis 2011; Versnel 2011: 444–456; Litwa 2012: 71–74; Holton 2014; Strootman 2014: 241–243; Landucci Gattinoni 2016: 52. Whitmarsh 2015: 151–152 misleadingly concludes that the text of the hymn reveals "earnest belief in the reality of a ruler-god is . . . obviously bartered in exchange for belief in the Olympian gods." Petridou 2015: 260–263, 277–279, 302-309 discusses the prominence of Demeter, Kore, and Dionysus in festivals and monuments celebrating divine presence among and care for human beings. On Duris of Samos' attitude toward Athens and Demetrius, see Baron 2011: 104; Pownall 2013; Landucci Gattinoni 2016.

45 Athenaeus 6.253C, F. For one particularly dismissive modern comment, see Olson 2008: 163, n. 262: the hymn "is indeed a singularly embarrassing incident in Athenian history." On the issue of flattery, see Erskine 2014.

46 Its earliest attestations include Empedocles fr. 40 (Hermann Diels and Walther Kranz, eds. 1951. *Die Fragmente der Vorsokratiker: Griechisch und Deutsch.* 6th ed. Berlin: Weidmann = *FVS*) referring to Selene in connection with Helios Oxybeles; Critias, *Elegies* fr. 6 (*FVS*), describing the "cheerful hope" that drinking brings; Aristophanes, *Frogs* 456, describing the chorus, on whom in the same line the sun is said

to shine; Xenophon, *Mem.* 2.7.12, describing women at work in the household of Aristarchus; *Apol.* 33, in adverbial form describing how Socrates met his death; *Symp.* 8.3, Socrates describing Hermogenes' *êthos* as cheerful as part of his self-conscious *kalokagathia; Agesilaus* 8.2, describing Agesilaus' unfailing cheerfulness, and 11.2, describing how he would look cheerful while in a state of fear and gentle while having good fortune; in later comedy, Alexis (Theodor Kock, ed. *Comicorum Atticorum Fragmenta.* Leipzig: Teubner = *CAF*) fr. 278 on wine's effect; Antiphanes fr. 80 (*CAF*); Apollodorus fr. 5 (*CAF*); Ephippus fr. 6 (*CAF*); Philemon fr. 96 (*CAF*). See also Duris (Athenaeus 12.542D) on how Demetrius of Phaleron used to dye his hair, put on perfume, and apply heavy makeup to seem *hilaros* and *hêdus* to other people in his appearance.

47 Moreover, he is like the sun, recalling a similar image applied to Athens in Hyperides, *Funeral Oration* 5; his friends are like the stars. Cf. O'Sullivan 2008a.

48 On "hearing gods," see Colless 1970: 131 (Mesopotamian deities said to have an "open ear," signifying they were "keenly responsive to the prayers" of worshippers); Versnel 1981: 26–37; Brenk 2007; Teeter 2011: 79-84 (ears signifying Egyptian gods who "hear petitions").

49 Holton 2014: 372-376.

50 On this period, see Habicht 1997: 67–97; Parker 1996: 258–264; Dreyer 2000; Kralli 2000: 117-118, 122-123; Bosworth 2002: 246–269; Lape 2004: 52–67; Waterfield 2004: 233–244; Thonemann 2005; Kuhn 2006; Grainger 2007: 127–156; O'Sullivan 2008b; Anson 2014: 165–188; Harding 2015: 63–66.

51 Plutarch, *Demetrius* 28–31 is the only surviving source to preserve any details, as is also true for the remaining events of Demetrius' career. But as Plutarch insists, he is writing biography, not history, and as a result we are very poorly informed about those years. Dunn 2016 provides a comprehensive analysis of this period of Demetrius' career.

52 On ruler cult, see, among many discussions, Bremmer 1977; Klauck 2000: 250-288; Erskine 2010, 2014; Anagnostou-Laoutides 2013; Whitmarsh 2015: 148-155. For a recent collection of studies concerning the history of Epicureanism, including theology, see Fish and Sanders 2011.

53 In Mikalson's subsequent discussion (2006: 218-219) of what he calls "'changes', a neutral term" in Hellenistic religion, he is referring neither to "a deterioration from the religion of the Classical period" nor to "positive changes leading in the direction of Christian conceptions." (For the argument that the divinization of Alexander in fact provided an influential precedent for Jesus as a god, see Amitay 2010.) Of the changes identified by Mikalson, "[f]oremost among these was ruler cult," a development that of course has direct relevance to the case of Demetrius Poliorcetes.

54 See Sørensen 2011 for the argument that, despite traditional social science models being almost exclusively ahistorical and the rejection of the relevance of psychology by historians and social scientists, "both social science and psychology are needed in order to construct and explain historical events" by investigating the tripartite relation between history, socio-cultural systems, and psychology. See also the call by Robert Hood, Jr. 2011: 34-36 for fieldwork rather than laboratory experiments to advance understanding of how people respond to failed prophecy in particular.

References

Abelson, Robert P. 1959. "Modes of Resolution of Belief Dilemmas." *The Journal of Conflict Resolution* 3: 343–352.

Adcock, Amy. 2012. "Cognitive Dissonance in the Learning Processes." In Norbert E. Seel, ed. *Encyclopedia of the Sciences of Learning.* New York: Springer Science and Business Media, pp. 588–590.

Amitay, Ory. 2010. *From Alexander to Jesus.* Berkeley, CA: University of California Press.

Anagnostou-Laoutides, Eva. 2013. "Destined to Rule: The Near Eastern Origins of Hellenistic Ruler Cult." In Richard Alston, Onno M. van Nijf, and Christina G. Williamson, eds. *Cults, Creeds, and Identities in the Greek City After the Classical Age.* Leuven: Peeters, pp. 49–84.

Anderson, Ralph. 2015. "New Gods." In Esther Eidinow and Julia Kindt, eds. *The Oxford Handbook of Ancient Greek Religion.* Oxford: Oxford University Press, pp. 309–323.

Anson, Edward. 2013. *Alexander the Great: Themes and Issues.* London: Bloomsbury Academic.

—. 2014. *Alexander's Heirs: The Age of the Successors.* Malden, MA: Wiley.

Ashton, N. G. 1984. "The Lamian War—*stat magni nominis umbra.*" *Journal of Hellenic Studies* 104: 152–157.

Atran, S. and A. Norenzayan. 2004. "Religion's Evolutionary Landscape: Counterintuition, commitment, compassion, communion." *Behavioral and Brain Sciences* 27: 713–770.

Attridge, Harold W. 2009. "Pollution, Sin, Atonement, Salvation." In Sarah Iles Johnston, ed. *Ancient Religions: Beliefs and Rituals Across the Mediterranean World.* Cambridge, MA: Harvard University Press, pp. 71–83.
Auffarth, Christoph. 1995. "Aufnahme und Zurückweisung 'Neuer Götter' im spätklassischen Athen: Religion gegen die Krise, Religion in der Krise?" In W. Eder, ed. *The Meaning of Stoicism.* Cambridge, MA: Published for Oberlin College by Harvard University Press, pp. 337–365.
Badian, Ernst. 1981. "The Deification of Alexander the Great." In H. J. Dell, ed. *Studies in Honor of Charles F. Edson.* Thessaloniki: Institute for Balkan Studies.
Bae, Bosco B. 2016. "Believing Selves and Cognitive Dissonance: Connecting Individual and Society via 'Belief'." *Religions* 7, 86; doi:10.3390/rel7070086.
Balsdon, J. P. V. D. 1950. "The 'Divinity' of Alexander." *Historia* 1: 363–388.
Banerjee, Konika and Paul Bloom. 2013. "Would Tarzan Believe in God? Conditions for the Emergence of Religious Belief." *Trends in Cognitive Sciences* 17.1: 7–8.
Barber, E. J. W. 1991. *Prehistoric Textiles: The Development of Cloth in the Neolithic and Bronze Ages with Special Reference to the Aegean.* Princeton, NJ: Princeton University Press.
—. 1992. "The Peplos of Athena." In Jennifer Neils, ed. *Goddess and Polis: The Panathenaic Festival in Ancient Athens.* Princeton, NJ: Princeton University Press, pp. 103–118.
Baron, Christopher A. 2011. "The Delimitation of Fragments in Jacoby's *FGrH*: Some Examples from Duris of Samos." *Greek, Roman, and Byzantine Studies* 51: 86–110.
Barrett, Justin L. 1999. "Theological Correctness: Cognitive Constraint and the Study of Religion." *Method and Theory in the Study of Religion* 37: 608–619.
—. 2000. "Exploring the Natural Foundations of Religion." *Trends in Cognitive Sciences* 4: 29–34 (reprinted D. J. Slone, ed. 2006. *Religion and Cognition: A Reader.* London: Equinox, pp. 86–98).
—. 2007. "Cognitive Science of Religion: What Is It and Why Is It?" *Religion Compass* 1: 768–786.
—. 2008. "Why Santa Claus Is Not a God." *Journal of Cognition and Culture* 8: 149–161.
—. 2011. "Cognitive Science of Religion: Looking Back, Looking Forward'" *Journal for the Scientific Study of Religion* 50: 229–239.
Barrett, Justin L. and Frank C. Keil. 1996. "Conceptualizing a Nonnatural Entity: Anthropomorphism in God Concepts." *Cognitive Psychology* 31: 219–247 (reprinted D. J. Slone, ed. 2006. *Religion and Cognition: A Reader.* London: Equinox, pp. 116–148).
Barrett, Justin L. and Melanie A. Nyhof. 2001. "Spreading Non-Natural Concepts: The Role of Intuitive Conceptual Structures in Memory and Transmission of Culture." *Journal of Cognition and Culture* 1: 69–100 (reprinted D. J. Slone, ed. 2006. *Religion and Cognition: A Reader.* London: Equinox, pp. 149–177).
Batson, C. 1975. "Rational Processing or Rationalization? The Effect of Disconfirming Information on a Stated Religious Belief." *Journal of Personality and Social Psychology* 32: 176–184.
Baumer, Nicolas and Pascal Boyer. 2013. "Explaining Moral Religions." *Trends in Cognitive Sciences* 17.6: 272–280.
Bayliss, Andrew J. 2011. *After Demosthenes: The Politics of Early Hellenistic Athens.* London: Continuum.
Baynham, Elizabeth. 2003. "Antipater and Athens." In O. Palagia and S. Tracy, eds. *The Macedonians in Athens 322-229 B.C.* Proceedings of an International Conference held at the University of Athens, May 24-26, 2001. Oxford: Oxbow Books, pp. 23–29.
Bearzot, Cinzia. 1993. *Vite parallele. Plutarco. Focione. Introduzione, traduzione e note.* Milan: Biblioteca Universale Rizzoli.
Beck, Roger. 2006. *The Religion of the Mithras Cult in the Roman Empire.* Oxford: Oxford University Press.
Beit-Hallahmi, Benjamin and Michael Argyle. 1997. *The Psychology of Religious Behaviour, Belief, and Experience.* London: Routledge.
Beringer, Richard E., Herman Hattaway, Archer Jones, and William N. Still, Jr. 1986. *Why the South Lost the Civil War.* Athens, GA: University of Georgia Press.
Berner, Ulrich. 2011. "Religion Explained? Lucian of Samosata and the Cognitive Science of Religion." In L. H. Martin and J. Sørensen, eds. *Past Minds: Studies in Cognitive Historiography.* London: Equinox, pp. 131–140.
Bevan, Edwyn R. 1913. *Stoics and Sceptics.* Four lectures delivered in Oxford during Hilary term 1913 for the Common University Fund. Oxford: Clarendon Press.
Billows, Richard A. 1990. *Antigonos the One-Eyed and the Creation of the Hellenistic State.* Berkeley, CA: University of California Press.

Bleeker, C. J. 1963. "Isis as Saviour Goddess." In S. G. F. Brandon, ed. *The Saviour God: Comparative Studies in the Concept of Salvation Presented to Edwin James Oliver.* Manchester: Manchester University Press.

Blok, Josine. 2014. "A 'Covenant' Between Gods and Men: *Hiera kai hosia* and the Greek Polis." In Claudia Rapp and H. A. Drake, eds. *The City in the Classical and Post-Classical World: Changing Contexts of Power and Identity.* Cambridge: Cambridge University Press, pp. 14–37.

Boedeker, Deborah. 2007. "The View from Eleusis: Demeter in the Persian Wars." In Emma Bridges, Edith Hall, and P. J. Rhodes, eds. *Cultural Responses to the Persian Wars: Antiquity to the Third Millennium.* Oxford: Oxford University Press, pp. 65–82.

Borza, Eugene. 1999. *Before Alexander: Constructing Early Macedonia.* Claremont, CA: Regina Books.

Bosworth, A. Brian. 1988. *Conquest and Empire: The Reign of Alexander the Great.* Cambridge: Cambridge University Press.

—. 2002. *The Legacy of Alexander: Politics, Warfare, and Propaganda under the Successors.* Oxford: Oxford University Press.

—. 2003. "Why Did Athens Lose the Lamian War?" In O. Palagia and S. Tracy, eds. *The Macedonians in Athens 322-229 B.C.* Proceedings of an International Conference held at the University of Athens, May 24-26, 2001. Oxford: Oxbow Books, pp. 14–22.

Bowden, Hugh. 2015. "Impiety." In Esther Eidinow and Julia Kindt, eds. *The Oxford Handbook of Ancient Greek Religion.* Oxford: Oxford University Press, pp. 325–338.

Boyer, Pascal. 1994. *The Naturalness of Religious Ideas: A Cognitive Theory.* Berkeley, CA: University of California Press.

—. 1996. "What Makes Anthropomorphism Natural: Intuitive Ontology and Cultural Representations." *Journal of the Royal Anthropological Institute* 2 (New Series): 83–97.

—. 2000. "Functional Origins of Religious Concepts: Ontological and Strategic Selection in Evolved Minds." *The Journal of the Royal Anthropological Institute* 6: 195–214.

—. 2001. *Religion Explained: The Evolutionary Origins of Religious Thought.* New York: Basic Books.

—. 2003. "Religious Thought and Behaviour as By-Products of Brain Function." *Trends in Cognitive Sciences* 7: 119–124.

Boyer, Pascal and Charles Ramble. 2001. "Cognitive Templates for Religious Concepts: Cross-Cultural Evidence for Recall of Counter-Intuitive Representations." *Cognitive Science* 25: 535–564 (reprinted D. J. Slone, ed. 2006. *Religion and Cognition: A Reader.* London: Equinox, pp. 178–214).

Boyer, Pascal and Brian Bergstrom. 2008. "Evolutionary Perspectives on Religion." *Annual Review of Anthropology* 37: 111–130.

Bremmer, Jan. 1977. "*ES KYNOSARGES.*" *Mnemosyne* 30: 369–374.

Bremmer, Jan N. and Andrew Erskine, eds. 2010. *The Gods of Ancient Greece: Identities and Transformations.* Edinburgh: Edinburgh University Press.

Brenk, Frederick E. 2007. "Zeus' Missing Ears." *Kernos* 20: 213–215.

Brock, Roger. 2013. *Greek Political Imagery from Homer to Aristotle.* London: Bloomsbury.

Bulloch, Anthony W. 1984. "The Future of a Hellenistic Illusion: Some Observations on Callimachus and Religion." *Museum Helveticum* 41: 209–230.

Burke, Edmund M. 1985. "Lycurgan Finances." *Greek, Roman, and Byzantine Studies* 26: 251–264.

Burris, Christopher T., Eddie Harmon-Jones, and W. Ryan Tarpley. 1997. "'By Faith Alone': Religious Agitation and Cognitive Dissonance." *Basic and Applied Social Psychology* 19: 17–31.

Carroll, Robert P. 1979. *When Prophecy Failed: Cognitive Dissonance in the Prophetic Traditions of the Old Testament.* New York: Seabury Press.

Cartledge, Paul. 2004. *Alexander the Great: The Hunt for a New Past.* Woodstock, NY: Overlook Press.

—. 2016. *Democracy: A Life.* Oxford: Oxford University Press.

Chaniotis, Angelos. 2003. "The Divinity of Hellenistic Rulers." In Andrew Erskine, ed. *A Companion to the Hellenistic World.* Malden, MA: Blackwell, pp. 431–445.

—. 2011. "The Ithyphallic Hymn for Demetrios Poliorketes and Hellenistic Religious Mentality." In Panagiotis P. Iossif, Andrzej S. Chankowski, and Catharine S. Lorber, eds. *More Than Men, Less Than Gods: Proceedings of the International Colloquium Organized by the Belgian School at Athens (November 1-2, 2007).* Leuven: Peeters, pp. 157–195.

Clinton, Kevin. 2010. "The Mysteries of Demeter and Kore." In D. Ogden, ed. *A Companion to Greek Religion.* Malden, MA: Wiley-Blackwell, pp. 342–356.

Colless, Brian E. 1970. "Divine Education." *Numen* 17: 118–142.

Cooper, Joel. 2007. *Cognitive Dissonance: Fifty Years of a Classic Theory.* Los Angeles, CA: Sage Publications.

Darity, Jr., William A., ed. 2008. *International Encyclopedia of the Social Sciences*. 2nd ed. Detroit, MI: Gale, Cengage Learning.

Dawson, Lorne L. 1999. "When Prophecy Fails and Faith Persists: A Theoretical Overview." *Religio* 3: 62–82.

—. 2011. "Clearing the Underbrush: Moving beyond Festinger to a New Paradigm for the Study of Failed Prophecy." In D. Tumminia and W. Swatos, Jr., eds. *How Prophecy Lives*. Leiden: Brill, pp. 69–98.

Deacy, Susan. 2010. "'Famous Athens, Divine Polis': The Religious System at Athens." In D. Ogden, ed. *A Companion to Greek Religion*. Malden, MA: Wiley-Blackwell, pp. 221–235.

Dein, Simon. 2001. "What Really Happens When Prophecy Fails: The Case of Lubavitch." *Sociology of Religion* 62: 383–401.

Dmitriev, Sviatoslav. 2011. *The Greek Slogan of Freedom and Early Roman Politics in Greece*. Oxford: Oxford University Press.

Dreyer, Boris. 2000. "Athen und Demetrios Poliorketes nach der Schlacht von Ipsos (301 v.Chr.): Bermerkungen zum Marmor Parium, FGrHist 239 B 27 und zur Offensive des Demetrios im Jahre 299/8 v. Chr." *Historia* 49: 54–66.

—. 2009. "Hero, Cults, and Divinity." In Waldemar Heckel and Lawrence A. Tritle, eds. *Alexander the Great: A New History*. Malden, MA: Wiley-Blackwell, pp. 218–234.

Dunand, Françoise. 2010. "The Religious System at Alexandria." In D. Ogden, ed. *A Companion to Greek Religion*. Malden, MA: Wiley-Blackwell, pp. 253–263.

Dunn, Charlotte Marie Rose. 2016. *Conquest, Kingship, Calamity: Demetrius Poliorcetes After Ipsus*. PhD Thesis, University of Otago.

Edelstein, Ludwig. 1966. *The Meaning of Stoicism*. Cambridge, MA: Published for Oberlin College by Harvard University Press.

Eder, Walter, ed. 1995. *Die athenische Demokratie im 4. Jahrhundert v. Chr. Vollendung oder Verfall einer Verfassungsform? Akten eines Symposiums 3.-7.* August 1992 Bellagio. Stuttgart: Franz Steiner Verlag.

Egan, Louis C., Laurie R. Santos, and Paul Bloom. 2007. "The Origins of Cognitive Dissonance: Evidence from Children and Monkeys." *Psychological Science* 18: 978–983.

Ehrenberg, Victor. 1935. "Athenische Hymnus auf Demetrios Poliorketes." *Die Antike* 7: 279–295.

—. 1946. *Aspects of the Ancient World*. New York: William Salloch.

Eidinow, Esther and Julia Kindt, eds. 2015. *The Oxford Handbook of Ancient Greek Religion*. Oxford: Oxford University Press.

Engels, Johannes. 2008. *Lykurg: Rede gegen Leokrates*. Darmstadt: Wissenschaftliche Buchgesellschaft.

Erskine, Andrew. 2010. "Epilogue." In J. Bremmer and A. Erskine, eds. *The Gods of Ancient Greece: Identities and Transformations*. Edinburgh: Edinburgh University Press, pp. 505–510.

—. 2014. "Ruler Cult and the Early Hellenistic City." In Hans Hauben and Alexander Meeus, eds. *The Age of the Successors and the Creation of the Hellenistic Kingdoms (323-276 BC)*. Leuven: Peeters, pp. 579–598.

Exline, Julie Juola. 2002. "Stumbling Blocks on the Religious Road: Fractured Relationships, Nagging Vices, and the Inner Struggle to Believe." *Psychological Inquiry* 13: 182–189.

Exline, Julie Juola and Ephraim Rose. 2005. "Religious and Spiritual Struggles." In Raymond F. Paloutzian and Crystal L. Park, eds. *Handbook of the Psychology of Religion and Spirituality*. New York: Guilford Press, pp. 315–330.

Fessler, Daniel M. T. and Edouard Machery. 2012. "Culture and Cognition." In E. Margolis and R. Samuels, eds. *The Oxford Handbook of the Philosophy of Cognitive Science*. Oxford: Oxford University Press, pp. 503–527.

Festinger, Leon. 1957. *A Theory of Cognitive Dissonance*. Evanston, IL: Row, Peterson, and Co.

—. 1964. *Conflict, Decision, and Dissonance*. Stanford, CA: Stanford University Press.

—. 1999. "Appendix B: Reflections on Cognitive Dissonance: 30 Years Later." In E. Harmon-Jones and J. Mills, eds. *Cognitive Dissonance: Progress on a Pivotal Theory in Social Psychology*. Washington, DC: American Psychological Association, pp. 381–385.

Festinger, Leon, Henry W. Riecken, and Stanley Schachter. 1956. *When Prophecy Fails: A Social and Psychological Study of a Modern Group that Predicted the Destruction of the World*. Minneapolis, MN: University of Minnesota Press (reprint 2011, Wilder Publications).

Fish, Jeffrey, and Kirk R. Sanders. 2011. *Epicurus and the Epicurean Tradition*. Cambridge: Cambridge University Press.

Flacelière, Robert and Émile Chambry, trans. 1976. *Plutarche. Vies. Tome X. Phocion—Caton Le Jeune*. Paris: Société d'Édition "Les Belles Lettres."

Flower, Michael. 1988. "Agesilaus of Sparta and the Origins of the Ruler Cult." *Classical Quarterly* 38: 123–134.

Frankfurter, David T. M. 2015. Review of B. Nongbri 2013. *Journal of Early Christian Studies* 23: 632–634.

Fredricksmeyer, Ernst. 2003. "Alexander's Religion and Divinity." In Joseph Roisman, ed. *Brill's Companion to Alexander the Great*. Leiden: Brill, pp. 253–278.

Fredriksen, Paula. 1988. *From Jesus to Christ: The Origins of the New Testament Images of Jesus*. New Haven, CT: Yale University Press.

Frost, Frank J. 1980. *Plutarch's Themistocles: A Historical Commentary*. Princeton, NJ: Princeton University Press.

Gager, John G. 1975. *Kingdom and Community: The Social World of Early Christianity*. Englewood Cliffs, NJ: Prentice-Hall.

Garland, Robert. 1992. *Introducing New Gods: The Politics of Athenian Religion*. Ithaca, NY: Cornell University Press.

Gawronski, Bertram. 2012. "Back to the Future of Dissonance Theory: Cognitive Consistency as a Core Motive." *Social Cognition* 30: 652–668.

Geertz, A. W. 2010. "Brain, Body and Culture: A Biocultural Theory of Religion." *Method and Theory in the Study of Religion* 224: 304–321.

Gervais, Will M. and Joseph Henrich. 2010. "The Zeus Problem: Why Representational Content Biases Cannot Explain Faith in Gods." *Journal of Cognition and Culture* 10: 383–389.

Goethals, George R. 1987. "Theories of Group Behavior: Commentary." In B. Mullen and G. Goethals, eds. *Theories of Group Behavior*. New York: Springer-Verlag, pp. 209–229.

Goldman, Alvin I. 2012. "Theory of Mind." In E. Margolis and R. Samuels, eds. *The Oxford Handbook of the Philosophy of Cognitive Science*. Oxford: Oxford University Press, pp. 402–424.

Grafton, A. T. and N. M. Swerdlow. 1988. "Calendar Dates and Ominous Days in Ancient Historiography." *Journal of the Warburg and Courtauld Institutes* 51: 14–42.

Grainger, John D. 2007. *Alexander the Great Failure: The Collapse of the Macedonian Empire*. London: Hambledon Continuum.

Green, Peter. 1990. *Alexander to Actium: The Evolution of the Hellenistic Age*. Berkeley, CA: University of California Press.

Grubbs, Joshua B., Julie J. Exline, and W. Keith Campbell. 2013. "I Deserve Better and God Knows It! Psychological Entitlement as a Robust Predictor of Anger at God." *Psychology of Religion and Spirituality* 5: 192–200.

Guthrie, Stewart E. 1980. "A Cognitive Theory of Religion." *Current Anthropology* 21: 181–203.

—. 1993. *Faces in the Clouds: A New Theory of Religion*. New York: Oxford University Press.

Habicht, Christian. 1979. *Gottmenschentum und griechische Städte*. 2nd ed. Munich: Beck.

—. 1997. *Athens from Alexander to Antony*. Trans. Deborah Lucas Schneider. Cambridge, MA: Harvard University Press.

Håland, Evy Johanne. 2004. "Athena's Peplos: Weaving as a Core Female Activity in Ancient and Modern Greece." *Cosmos* 20: 155–182.

Harding, Philip. 2015. *Athens Transformed: 404-262 B.C. From Popular Sovereignty to the Dominion of Wealth*. New York: Routledge.

Hardyck, Jane Allyn and Marcia Braden. 1962. "Prophecy Fails Again: A Report of a Failure to Replicate." *Journal of Abnormal and Social Psychology* 65: 136–141 (reprinted J. Stone, ed. 2000. *Expecting Armageddon: Essential Readings in Failed Prophecy*. New York: Routledge, pp. 55–63).

Harmon-Jones, Eddie. 1999. "Toward an Understanding of the Motivation Underlying Dissonance Effects: Is the Production of Aversive Consequences Necessary to Cause Dissonance?" In E. Harmon-Jones and J. Mills, eds. *Cognitive Dissonance: Progress on a Pivotal Theory in Social Psychology*. Washington, DC: American Psychological Association, pp. 71–99.

—. 2000. "An Update on Dissonance Theory, with a Focus on the Self." In A. Tesser, R. Felson, and J. Suls, eds. *Psychological Perspectives on Self and Identity*. Washington, DC: American Psychological Association, pp. 119–144.

Harmon-Jones, Eddie and Cindy Harmon-Jones. 2015. "Cognitive Dissonance Theory After 50 Years of Development." *Zeitschrift für Sozialpsychologie* 38: 7–16.

Harmon-Jones, Eddie and Judson Mills, eds. 1999. *Cognitive Dissonance: Progress on a Pivotal Theory in Social Psychology*. Washington, DC: American Psychological Association.

Harris, William V. 2001. *Restraining Rage: The Ideology of Anger Control in Classical Antiquity*. Cambridge, MA: Harvard University Press.

Harrison, Thomas. 2015. "Belief vs. Practice." In Esther Eidinow and Julia Kindt, eds. *The Oxford Handbook of Ancient Greek Religion*. Oxford: Oxford University Press, pp. 21–28.

Henrichs, Albert. 1999. "Demythologizing the Past, Mythologizing the Present: Myth, History, and the Supernatural at the Dawn of the Hellenistic Period." In Richard Buxton, ed. *From Myth to Reason? Studies in the Development of Greek Thought*. Oxford: Oxford University Press.

—. 2010. "What Is a Greek God?" In J. Bremmer and A. Erskine, eds. *The Gods of Ancient Greece: Identities and Transformations*. Edinburgh: Edinburgh University Press, pp. 19–39.

Herrman, Judson. 2009. *Hyperides: Funeral Oration*. New York: Oxford University Press.

Hess, Hans. 1938. *Textkritische und erklärende Beiträge zum Epitaphios des Hypereides*. Leipzig: Otto Harrassowitz.

Holmberg, Bengt. 1990. *Sociology and the New Testament: An Appraisal*. Minneapolis, MN: Fortress Press.

Holt, Frank L. 2012. *Lost World of the Golden King: In Search of Ancient Afghanistan*. Berkeley, CA: University of California Press.

Holton, John Russell. 2014. "Demetrios Poliorketes, Son of Poseidon and Aphrodite: Cosmic and Memorial Significance in the Athenian Ithyphallic Hymn." *Mnemosyne* 67: 370–390.

Hood, Jr., Robert. 2011. "Where Prophecy Lives: Psychological and Sociological Studies of Cognitive Dissonance." In D. Tumminia and W. Swatos, Jr., eds. *How Prophecy Lives*. Leiden: Brill, pp. 21–40.

Horrell, David G. ed. 1999. *Social-Scientific Approaches to New Testament Interpretation*. Edinburgh: T&T Clark.

—. 2002. "Social Sciences Studying Formative Christian Phenomena: A Creative Movement." In Anthony J. Blasi, Jean Duhaime, and Paul-André Turcotte, eds. *Handbook of Early Christianity: Social Science Approaches*. Walnut Creek, CA: Altamira Press, pp. 3–28.

Hoshino-Brown, Etsuko, Adam S. Zanna, Steven J. Spencer, Mark P. Zanna, Shinobu Kitayama, and Sandra Lackenbauer. 2005. "On the Cultural Guises of Cognitive Dissonance: The Case of Easterners and Westerners." *Journal of Personality and Social Psychology* 89: 294–310.

Jameson, Michael. 1998. "Religion in the Athenian Democracy." In Ian Morris and Kurt Rauflaub, eds. *Democracy 2500? Questions and Challenges*. Dubuque, IA: Archaeological Institute of America, pp. 171–195.

—. 2014. *Cults and Rites in Ancient Greece: Essays on Religion and Society*. Cambridge: Cambridge University Press.

Jaschinski, Siegfried. 1981. *Alexander und Griechenland under dem Eindruck der Flucht des Harpalos*. Bonn: Rudolf Habelt Verlag.

Johnston, Sara Iles. 2004. "Theology, Theodicy, Philosophy." In Johnston, ed. *Religions of the Ancient World: A Guide*. Cambridge, MA: Belknap Press of Harvard University Press, pp. 531–547.

—, ed. 2004. *Religions of the Ancient World: A Guide*. Cambridge, MA: Belknap Press of Harvard University Press.

Kindt, Julia. 2012. *Rethinking Greek Religion*. Cambridge: Cambridge University Press.

Kitayama, S., K. Ishii, T. Imada, K. Takemura, and J. Ramaswamy. 2006. "Voluntary Settlement and the Spirit of Independence: Evidence from Japan's 'Northern' Frontier." *Journal of Personality and Social Psychology* 91: 369–384.

Klauck, Hans-Josef. 2000. *The Religious Context of Early Christianity: A Guide to Greco-Roman Religions*. Trans. Brian McNeil. Edinburgh: T&T Clark.

Knight, George F. 1993. *Millennial Fever and the End of the World: A Study of Millerite Adventism*. Boise, IA: Pacific Press.

Kolde, Antje. 2003. *Politique et religion chez Isyllos d'Épidaure*. Basel: Schwabe & Co.

Koortbojian, Michael. 2013. *The Divinization of Caesar and Augustus: Precedents, Consequences, Implications*. Cambridge: Cambridge University Press.

Kralli, Ioanna. 2000. "Athens and the Hellenistic Kings (338-261 B.C.): The Language of the Decrees." *Classical Quarterly* 50: 113–132.

Kuhn, Annika B. 2006. "Ritual Change During the Reign of Demetrius Poliorcetes." In Eftychia Stavrianopoulou, ed. *Ritual and Communication in the Graeco-Roman World*. Liège: Centre International d'Étude de la Religion Grecque Antique.

Landucci Gattinoni, Franca. 1981. "La divinizzazione di Demetrio e la coscienza ateniese." In Marta Sordi, ed. *Religione e politica nel mondo antico*. Milan: Vita e pensiero, pp. 115–123.

—. 2016. "Duride, Samo e i Diadochi: uno storiografo nella storia." In Valérie Naas and Mathilde Simon, eds. *De Samos à Rome: personnalité et influence de Douris*. Paris: Presses universitaires de Paris Ouest, pp. 37–55.

Lape, Susan. 2004. *Reproducing Athens: Menander's Comedy, Democratic Culture, and the Hellenistic City*. Princeton, NJ: Princeton University Press.

Larson, Jennifer. 2007. *Ancient Greek Cults: A Guide*. New York: Routledge.

—. 2016. *Understanding Greek Religion: A Cognitive Approach*. New York: Routledge.

Lawson, E. Thomas. 2006. "Foreword." In T. Tremlin, *Minds and Gods: The Cognitive Foundations of Religion*. Oxford: Oxford University Press, pp. xi–xvii.

Lawson, E. Thomas and Robert N. McCauley. 1990. "Explanation and Interpretation: Problems and Promise in the Study of Religion." *Rethinking Religion. Connecting Cognition and Culture*. Cambridge: Cambridge University Press, pp. 12–31 (reprinted D. J. Slone, ed. 2006. *Religion and Cognition: A Reader*. London: Equinox, pp. 12–35).

Leach, Edmund. 1966. "Virgin Birth." *Proceedings of the Royal Anthropological Institute of Great Britain and Ireland* 1966: 39–49.

Lefkowitz, Mary. 2016. *Euripides and the Gods*. Oxford: Oxford University Press.

Lehmann, G. A. 1988. "Der 'Lamische Krieg' und die 'Freiheit der Hellenen': Uberlegungen zur Hieronymianischen Tradition." *Zeitschrift für Papyrologie und Epigraphik* 73: 121–149.

Litwa, M. David. 2012. *We Are Being Transformed: Deification in Paul's Soteriology*. Berlin: De Gruyter.

Long, A. A. 1986. *Hellenistic Philosophy: Stoics, Epicureans, Sceptics*. 2nd ed. Berkeley, CA: University of California Press.

Long, Charlotte R. 1987. *The Twelve Gods of Greece and Rome*. Leiden: Brill.

Lübke, Christian. 2007. *Alexander der Grosse—Der Lamische Krieg 323/22 v. Chr. Der vorerst letzte grosse Aufstand der Athener mit ihren Verbündeten gegen die makedonische Vorherrschaft*. Munich. GRIN Verlag.

Luhrmann, T. M. 2011. "Toward an Anthropological Theory of Mind: Overview." *Suomen Antropologi: Journal of the Finnish Archaeological Society* 36: 5–13.

—. 2012. *When God Talks Back: Understanding the American Evangelical Relationship with God*. New York: Alfred A. Knopf.

Luraghi, Nino. 2016. "Stratokles of Diomeia and Party Politics in Early Hellenistic Athens." *Classica et Mediaevalia* 65: 191–226.

Ma-Kellams, Christine. 2015. "When Perceiving the Supernatural Changes the Natural: Religion and Agency Detection." *Journal of Cognition and Culture* 15: 337–343.

Manni, Eugenio, ed. 1953. *Plutarchi Vita Demetrii Poliorcetis*. Florence: La nuova Italia Editrice.

Marcovich, Miroslav. 1988. *Studies in Graeco-Roman Religions and Gnosticism*. Leiden: Brill.

Marcus, H. R. and S. Kitayama. 1991. "Culture and the Self: Implications for Cognition, Emotion, and Motivation." *Psychological Review* 98: 224–253.

Margolis, Eric and Richard Samuels, eds. 2012. *The Oxford Handbook of the Philosophy of Cognitive Science*. Oxford: Oxford University Press.

Martin, L. H. and Sørensen, J. eds. 2011. *Past Minds: Studies in Cognitive Historiography*. London: Equinox.

Martin, Thomas R. and Sun, Ivy S. 2017. "'The Gods Were Supervising the Hardest-to-Handle Sufferings of Greece': The Meaning of *episkopein* in Plutarch, *Phocion* 28." *Rationes Rerum* 9: 93–112.

Mazlish, Bruce. 1990. *The Leader, the Led, and the Psyche*. Lebanon, NH: University Press of New England.

McKimmie, Blake M. 2015. "Cognitive Dissonance in Groups." *Social and Personality Psychology Compass* 9: 202–212.

Mikalson, Jon D. 1998. *Religion in Hellenistic Athens*. Berkeley, CA: University of California Press.

—. 2003. *Herodotus and Religion in the Persian Wars*. Chapel Hill, NC: University of North Carolina Press.

—. 2006. "Greek Religion: Continuity and Change in the Hellenistic Period." In Glenn R. Bugh, ed. *The Cambridge Companion to the Hellenistic World*. Cambridge: Cambridge University Press, pp. 206–222.

—. 2016. *New Aspects of Religion in Ancient Athens: Honors, Authorities, Esthetics, and Society*. Leiden: Brill.

Mitchell, Lynette. 2013. "Alexander the Great: Divinity and the Rule of Law." In Lynette Mitchell and Charles Melville, eds. *Every Inch A King: Comparative Studies on Kings and Kingship in the Ancient and Medieval Worlds*. Boston, MA: Brill, pp. 91–107.

Mitchell, Thomas N. 2015. *Democracy's Beginning: The Athenian Story*. New Haven, CT: Yale University Press.

Mossé, Claude. 1973. *Athens in Decline 404-86 B.C.* Trans. Jean Stewart. London: Routledge & Kegan Paul.

Mullen, Brian and George R. Goethals, eds. 1987. *Theories of Group Behavior*. New York: Springer-Verlag.

Needham, Rodney. 1972. *Belief, Language, and Experience*. Chicago, IL: University of Chicago Press.

Nongbri, Brent. 2013. *Before Religion: A History of a Modern Concept*. New Haven, CT: Yale University Press.

Norenzayan, Ara and Will M. Gervais. 2013. "The Origins of Religious Disbelief." *Trends in Cognitive Sciences* 17(1): 20–25.

Ogden, Daniel, ed. 2010. *A Companion to Greek Religion*. Malden, MA: Wiley-Blackwell.

Oliver, Graham J. 2003. "Oligarchy at Athens After the Lamian War: Epigraphic Evidence for the *Boule* and the *Ekklesia*." In O. Palagia and S. Tracy, eds. *The Macedonians in Athens 322-229 B.C.* Proceedings of an International Conference held at the University of Athens, May 24-26, 2001. Oxford: Oxbow Books, pp. 40–51.

—. 2007. *War, Food, and Politics in Early Hellenistic Athens*. New York: Oxford University Press.

Olson, S. Douglas, ed. and trans. 2008. *Athenaeus. The Learned Banqueters*. Books VI-VII. Cambridge, MA: Harvard University Press.

O'Sullivan, Lara. 2008a. "*Le Roi Soleil*: Demetrius Poliorcetes and the Dawn of the Sun-King." *Antichthon* 42: 78–99.

—. 2008b. "A Note on Clement *Protrepticus* 4.54." *Classical Journal* 103: 295–300.

—. 2009. *The Regime of Demetrius of Phalerum in Athens, 317-307 B.C.: A Philosopher in Politics*. Leiden: Brill.

Ozorak, Elizabeth Weiss. 2005. "Cognitive Approaches to Religion." In Raymond F. Paloutzian and Crystal L. Park, eds. *Handbook of the Psychology of Religion and Spirituality*. New York: Guilford Press, pp. 216–234.

Palagia, Olga and Stephen V. Tracy, eds. 2003. *The Macedonians in Athens 322-229 B.C.* Proceedings of an International Conference held at the University of Athens, May 24-26, 2001. Oxford: Oxbow Books.

Paloutzian, Raymond F. and Crystal L. Park, eds. 2005. *Handbook of the Psychology of Religion and Spirituality*. New York: Guilford Press.

Pargament, Kenneth I. 1997. *The Psychology of Religion and Coping: Theory, Research, Practice*. New York: Guilford Press.

Pargament, Kenneth I., Gene G. Ano, and Amy B. Wachholtz. 2005. "The Religious Dimension of Coping: Advances in Theory, Research, and Practice." In Raymond F. Paloutzian and Crystal L. Park, eds. *Handbook of the Psychology of Religion and Spirituality*. New York: Guilford Press, pp. 479–495.

Parker, Robert. 1996. *Athenian Religion: A History*. Oxford: Clarendon Press.

—. 2005. *Polytheism and Society at Athens*. Oxford: Oxford University Press.

—. 2011. *On Greek Religion*. Ithaca, NY: Cornell University Press.

—. 2014. "Commentary on Journal of Cognitive Historiography, Issue 1." *Journal of Cognitive Historiography* 1: 186–192.

Perrin, Bernadotte, trans. 1919. *Plutarch's Lives*. Vol. VIII. *Sertorius and Eumenes, Phocion and Cato the Younger*. Cambridge, MA: Harvard University Press.

Petridou, Georgia. 2015. *Divine Epiphany in Greek Literature and Culture*. Oxford: Oxford University Press.

Petrovic, Ivana. 2015. "Deification—Gods or Men?" In Esther Eidinow and Julia Kindt, eds. *The Oxford Handbook of Ancient Greek Religion*. Oxford: Oxford University Press, pp. 429–446.

Phillips III, Russell E., Kenneth I. Pargament, Quinten K. Lynn, and Craig D. Crossley. 2004. "Self-Directing Religious Coping: A Deistic God, Abandoning God, or No God at All?" *Journal for the Scientific Study of Religion* 9: 409–418.

Pownall, Frances. 2013. "Duris of Samos and the Diadochi." In Victor Alonso Troncoso and Edward M. Anson, eds. *After Alexander: The Time of the Diadochi (323-281 BC)*. Oxford: Oxbow Books, pp. 43–56.

—. 2016. "Folly and Violence in Athens Under the Successors." In Tim Howe and Sabine Müller, eds. *Folly and Violence in the Court of Alexander the Great and His Successors?—Greco-Roman Perspectives*. Bochum/Freiburg: projektverlag.

Prus, Robert C. 1976. "Religious Recruitment and the Management of Dissonance: A Sociological Perspective." *Sociological Inquiry* 46: 127–134.

Pyysiäinen, Ilkka. 2002. "Religion and the Counter-Intuitive." In I. Pyysiäinen and V. Anttonen, eds. *Current Approaches in the Cognitive Science of Religion*. London: Continuum, pp. 111–133.

—. 2009. *Supernatural Agents: Why We Believe in Souls, Gods, and Buddhas*. Oxford: Oxford University Press.

—. 2012. "Cognitive Science of Religion: State-of-the-Art." *Journal for the Cognitive Science of Religion* 1: 5–28.

Renfrew, Colin. 1994. "Towards a Cognitive Archaeology" and "The Archaeology of Religion." In Colin Renfrew and Ezra B. W. Zubrow, eds. *The Ancient Mind: Elements of Cognitive Archaeology*. Cambridge: Cambridge University Press, pp. 3–12 and 47–54.

—. 2007. *Prehistory: The Making of the Human Mind*. London: Weidenfeld & Nicolson.

Rhodes, P. J. 2012. "The Alleged Failure of Athens in the Fourth Century." *Electrum* 19: 111–129.
Robertson, Noel D. 1998. "The Two Processions to Eleusis and the Program of the Mysteries." *American Journal of Philology* 119: 547–575.
Rodd, Cyril S. 1981. "On Applying a Sociological Theory to Biblical Studies." *Journal for the Study of the Old Testament* 19: 95–106 (cited from David J. Chalcraft, ed. 1997. *Social-Scientific Old Testament Criticism*. Sheffield: Sheffield Academic Press, pp. 22–33).
Roisman, Joseph and Ian Worthington, eds. 2015. *Lives of the Attic Orators: Texts from Pseudo-Plutarch, Photius, and the Suda*. Trans. Robin Waterfield. Oxford: Oxford University Press.
Romm, James. 2011. *Ghost on the Throne: The Death of Alexander the Great and the Bloody Fight for his Empire*. New York: Alfred A. Knopf.
Sande, Gerald N. and Mark P. Zanna. 1987. "Cognitive Dissonance Theory: Collective Actions and Individual Reactions." In B. Mullen and G. Goethals, eds. *Theories of Group Behavior*. New York: Springer-Verlag, pp. 49–69.
Santi Amantini, Luigi, Carlo Carena, and Mario Manfredini, eds. 1995. *Plutarco. Le Vite di Demetrio and di Antonio*. Milan: Fondazione Lorenzo Valla A. Mondadori.
Scheid, John. 2016. *The Gods, the State, and the Individual: Reflections on Civic Religion in Rome*. Trans. Clifford Ando. Philadelphia, PA: University of Pennsylvania Press.
Schmitt, Oliver. 1992. *Der Lamische Krieg*. Bonn: Habelt.
Scott, Kenneth. 1928. "The Deification of Demetrius Poliorcetes: Part I" and "Part II." *American Journal of Philology* 49: 137–166, 217-239.
Scott-Kilvert, Ian and Timothy Duff, trans. 2011. *The Age of Alexander: Ten Greek Lives by Plutarch*. Rev. ed. London: Penguin Books.
Sealey, Raphael. 1993. *Demosthenes and His Time: A Study in Defeat*. Oxford: Oxford University Press.
Segal, Alan F. 1990. *Paul the Convert: The Apostolate and Apostasy of Saul the Pharisee*. New Haven, CT: Yale University Press.
Shipley, Graham. 2000. *The Greek World After Alexander 323-30 BC*. London: Routledge.
Silver, James W. 1957. *Confederate Morale and Church Propaganda*. Tuscaloosa, AL: Confederate Publishing Co.
Slone, D. Jason. 2006. "Religion and Cognition: An Introduction." In D. J. Slone, ed. *Religion and Cognition: A Reader*. London: Equinox, pp. 1–10.
—, ed. 2006. *Religion and Cognition: A Reader*. London: Equinox.
Smith, Christian and Melinda Lundquist Denton. 2005. *Soul Searching: The Religious and Social Lives of American Teenagers*. Oxford: Oxford University Press.
Sommerstein, Alan H. and Andrew J. Bayliss. 2013. *Oath and State in Ancient Greece*. Berlin: De Gruyter.
Sørensen, Jesper. 2011. "Past Minds: Present Historiography and Cognitive Science." In L. H. Martin and J. Sørensen, eds. *Past Minds: Studies in Cognitive Historiography*. London: Equinox, pp. 179–196.
Sperber, Dan. 1996. "The Epidemiology of Beliefs." In D. Sperber, ed. *Explaining Culture: A Naturalistic Approach*. Oxford: Blackwell, pp. 77–97 (reprinted D. J. Slone, ed. 2006. *Religion and Cognition: A Reader*. London: Equinox, pp. 36–53).
Sperber, Dan and Deidre Wilson. 1995. *Relevance: Communication and Cognition*. 2nd ed. Cambridge, MA: Blackwell.
Stone, Jon R. 2000. *Expecting Armageddon: Essential Readings in Failed Prophecy*. New York: Routledge.
—. 2011. "The Festinger Theory on Failed Prophecy and Dissonance: A Survey and Critique." In D. Tumminia and W. Swatos, Jr., eds. *How Prophecy Lives*. Leiden: Brill, pp. 41–68.
Strootman, Rolf. 2014. *Courts and Elites in the Hellenistic Empires: The Near East After the Achaemenids, c. 330 to 30 BCE*. Edinburgh: Edinburgh University Press.
Taylor, Nicholas H. 1998. "Cognitive Dissonance and Early Christianity: A Theory and Its Application Reconsidered." *Religion & Theology* 5: 138–153.
—. 2002. "Conflicting Bases of Identity in Early Christianity: The Example of Paul." In Anthony J. Blasi, Jean Duhaime, and Paul-André Turcotte, eds. *Handbook of Early Christianity: Social Science Approaches*. Walnut Creek, CA: Altamira Press, pp. 577–597.
Teeter, Emily. 2011. *Religion and Ritual in Ancient Egypt*. Cambridge: Cambridge University Press.
Thonemann, Peter. 2005. "The Tragic King: Demetrios Poliorketes and the City of Athens." In Olivier Hekster and Richard Fowler, eds. *Imaginary Kings: Royal Images in the Ancient Near East, Greece and Rome*. Munich: Franz Steiner Verlag.
Tracy, Stephen V. 1995. *Athenian Democracy in Transition: Attic Letter-Cutters of 340 to 290 B.C.* Berkeley, CA: University of California Press.

—. 2000. "Athenian Politicians and Inscriptions of the Years 307 to 302." *Hesperia* 69: 227–233.
Tremlin, Todd. 2006. *Minds and Gods: The Cognitive Foundations of Religion*. Oxford: Oxford University Press.
Tritle, Lawrence A. 1988. *Phocion the Good*. London: Croom Helm.
Tumminia, Diana G. 2005. *When Prophecy Never Fails: Myth and Reality in a Flying-Saucer Group*. Oxford: Oxford University Press.
—. 2011. "Introduction: How Failure Succeeds." In D. Tumminia and W. Swatos, Jr., eds. *How Prophecy Lives*. Leiden: Brill, pp. 1–8.
Tumminia, Diana G. and William H. Swatos, Jr., eds. 2011. *How Prophecy Lives*. Leiden: Brill.
Versnel, H. S. 1981. "Religious Mentality in Ancient Prayer." In H. S. Versnel, ed. *Faith, Hope and Worship: Aspects of Religious Mentality in the Ancient World*. Leiden: Brill, pp. 1–64.
—. 1990. *Ter Unus: Isis, Dionysos, Hermes. Three Studies in Henotheism*. Leiden: Brill.
—. 1995. "Religion and Democracy." In W. Eder, ed. Die athenische Demokratie im 4. Jahrhundert v. Chr. Vollendung oder Verfall einer Verfassungsform? Akten eines Symposiums 3.-7. August 1992 Bellagio. Stuttgart: Franz Steiner Verlag, pp. 367–387.
—. 2011. *Coping with the Gods: Wayward Readings in Greek Theology*. Leiden: Brill.
Waterfield, Robin. 2004. *Athens: A History. From Ancient Ideal to Modern City*. New York: Basic Books.
—. 2011. *Dividing the Spoils: The War for Alexander the Great's Empire*. Oxford: Oxford University Press.
—, trans. 2016. *Plutarch: Hellenistic Lives*. Oxford: Oxford University Press.
Weinreich, Otto. 1926. "Antikes Gottmenschentum." *Neue Jahrbücher für Wissenschaft und Jugendbildung* 2: 633–651.
Wendland, Paul. 1904. "ΣΩΤΗΡ." *Zeitschrift für die neutestamentliche Wissenschaft und die Kunde des Urchristentums* 5: 335–353.
Wernik, Uri. 1975. "Frustrated Beliefs and Early Christianity: A Psychological Enquiry into the Gospels of the New Testament." *Numen* 22: 96–130.
Westlake, H. D. 1949. "The Aftermath of the Lamian War." *Classical Review* 63: 87–90.
Whitehead, David. 2006. "Absentee Athenians: Lysias Against Philon and Lycurgus Against Leocrates." *Museum Helveticum* 63: 132–151.
—. 2009. "Spiteful Heaven: Residual Belief in Divine *Phthonos* in Post-Fifth Century Greece." *Acta Antiqua Academiae Scientiarum Hungaricae* 49: 327–333.
Whitmarsh, Tim. 2015. *Battling the Gods: Atheism in the Ancient World*. New York: Knopf.
Wicklund, Robert A. and Jack W. Brehm. 1976. *Perspectives on Cognitive Dissonance*. Hillsdale, NJ: Lawrence Erlbaum Associates.
Will, Wolfgang. 1983. *Athen und Alexander: Untersuchungen zur Geschichte der Stadt von 338 bis 322 v. Chr.* Munich: C. H. Beck.
Williams, James Maddox. 1982. *Athens Without Democracy: The Oligarchy of Phocion and the Tyranny of Demetrius of Phalerum, 322-307 B.C.* Yale University PhD.
Wilson, Charles Reagan. 1980 (new ed. 2009). *Baptized in Blood: The Religion of the Lost Cause, 1865-1920*. Athens, GA: University of Georgia Press.
Wirth, Gerhard. 1999. *Hypereides, Lykurg and die autonomia der Athener. Ein Versuch zum Verständnis eineger Redner der Alexanderzeit*. Vienna: Verlag der Österreichischen Akademie der Wissenschaften.
—. 2004. "Der Epitaphios des Hypereides und das Ende einer Illusion." In Rüdiger Kinsky, ed. *DIORTHOSEIS: Beiträge zur Geschichte des Hellenismus und zum Nachleben Alexanders des Grossen*. Munich: K. G. Saur.
Worthington, Ian. 2001. "Hyperides 5.32 and Alexander the Great's Statue." *Hermes* 129: 129–131.
—. 2004. *Alexander the Great: Man and God*. Harlow and New York: Pearson Education.
—. 2013. *Demosthenes of Athens and the Fall of Classical Greece*. New York: Oxford University Press.
Ziegler, Konrat. 1932. "Plutarchstudien." *Rheinisches Museum* 81: 51–87.
Zygmunt, Joseph S. 1972. "When Prophecies Fail: A Theoretical Perspective on the Comparative Evidence." *American Behavioral Scientist* 16: 245–267.

14

Irony in theory and practice

The test case of Cicero's *Philippics*

Luca Grillo

What is irony? Perhaps we all know it; and yet, both classical and modern attempts to respond to this question have produced quite different results. The aim of this contribution is to consider three models recently put forth by cognitive scientists and psychologists: I will argue that they offer frameworks, terminology and categories which can advance our appreciation of Latin literature, and, in particular, that they can provide a better explanation of irony than the ancient theories themselves.

To test these models and the contention that they can cast new light on classical texts, I will use Cicero's *Philippics*. Cicero constitutes a privileged author for looking at irony: being particularly witty (Plut. *Cic.* 25-6), he used it profusely; and being interested in rhetoric, he engaged with the Greco-Roman theoretical debate concerning its definition and nature. Irony pervades many of his works,[1] but in the present chapter I will focus on the *Philippics*, which, I believe, provide an exceptional case study: they form an extensive and yet self-contained corpus (fourteen speeches, composed between September 2, 44 and April 21, 43);[2] they were written with exquisite care (Plut. *Cic.* 24.6), so that they have an enduring eloquence; and we happen to possess a wealth of contextual information, thanks especially to Cicero's letters.

Permutatio ex contrario

Cicero employs various catchwords and terms of abuse to belittle his main opponent, Marc Antony:[3] he is a public and a personal enemy (*hostis* 3.6 and *inimicus* 5.3), a tyrant and a pirate (*tyrannus* and *archipirata* 13.18), a beast and gladiator (*belua* 3.28 and *gladiator* 3.18), a pest and a parricide (*pestis* 3.5 and *parricida* 4.5), and the list goes on. As for his qualities, he is out of his mind (*furens, demens* and *amens* 3.2), daring, effeminate and rotten (*audax* 3.2, *effeminatus* 3.12 and *impurus* 3.12, *impudicus* 3.12), a drunk and a criminal (*numquam . . . sobrius* 3.12 and *sceleratus* 3.9), etc. In the framework of this ongoing abuse, Cicero scatters a few compliments. For example, he praises Antony's eloquence (*homo disertus* 2.8, cf. 2.18 and 2.86), wisdom (*sapiens* 2.11, *acutus* 2.28), blameless law-abiding (5.12), outstanding authority (*uno verum optimo auctore* 1.24) and self-restraint (e.g. 3.37). Such compliments, of course, are all ironic, and the context leaves no doubt about Cicero's intentions: he means the opposite of what he says. This type of irony functions according to a classical definition which was first formulated in the *Rhetorica ad Alexandrum* (1434b, 21),[4] and later rephrased in the *Rhetorica ad Herennium*:

> *Permutatio est oratio aliud verbis aliud sententia demonstrans . . . [permutatio] ex contrario ducitur sic, ut si quis hominem prodigum et luxuriosum inludens parcum et diligentem appellet.*
>
> Allegory is a type of speech expressing one thing by the letter and one other by the meaning . . . it is drawn from contrast when one mockingly calls a spendthrift and voluptuary cheap and restrained.
>
> (Rhet. Her. *4.46)*

Noticeably, Cicero's irony *ex contrario* equally targets Antony's family members: his wife, Fulvia, who was previously married to Clodius and then to Curio, two archenemies of Cicero, is a *bona femina* (3.16); his brother, Lucius, is saluted with the honorific titles of "commander" and "victorious general" (*dux* and *imperator* 3.31); and Cicero commends the speed of his other brother, Gaius, while in fact suggesting that greed delayed his march through Macedonia (10.11). Such criticism of relatives may strike a modern audience as distasteful, but it was not uncommon in Rome; in fact, characterization of one's associates and family was just another locus of praise or blame.[5] At any rate, whether Cicero targets Antony directly or through members of his family, irony and invective join forces as two faces of the same coin. In particular, this type of irony can be seen as a variation on the theme of denigration: it adds some charm to the list of abuses and while reinforcing the desired characterization, it draws its force from its ability to engage and entertain the audience.[6]

Some modern theories help us understand what is special about irony *ex contrario* and how it functions. One can safely take Paul Grice' *Logic and Conversation* as a point of departure into the growing writings about irony by psychologists and cognitive scientists. In a lecture delivered at Harvard University in 1967 (and published in 1978), Grice set out to pinpoint the rational basis for successful communication, "a rough general principle which participants will be expected (ceteris paribus) to observe" (45). Grice named this general principle the "Cooperative Principle," which consists of four categories or maxims:

1. Quantity. "Make your contribution as informative as is required (for the current purpose of the exchange)"; "do not make your contribution more informative than is required." (45)
2. Quality. "Do not say what you believe to be false"; and "do not say that for which you lack adequate evidence." (46)
3. Relation. "Be relevant." (46)
4. Manner. "Avoid obscurity of expression; avoid ambiguity; be brief; be orderly." (46)

According to Grice, the second maxim, quality (or do not lie), is the most important, and here is where irony comes into play: irony is a deliberate and manifest way to "flout" the maxim of quality, without actually violating the Cooperative Principle. The intentional and apparent incongruity enables the addressee, without believing a lie, to appreciate the utterance as non-literal.

This understanding has many points of contact with some classical definitions, including the one quoted above, "a type of speech expressing one thing by the letter and another by the meaning" (*oratio aliud verbis aliud sententia demonstrans*). Moreover, both the *Rhetorica ad Herennium* and Grice group irony with other figures of rhetoric, like metaphor. The anonymous author of the *Rhetorica ad Herennium* explains that both irony and metaphor are types of *permutatio*, a transfer from literal to non-literal: irony is a transfer based on opposition, *ex contrario*, while metaphor is based on commonality, *per similitudinem* (4.45-6).[7] Similarly, for Grice metaphor is another way to violate the maxim of quality: metaphor involves "categorical falsity" without contradicting the Cooperative Principle, since the audience can manage to appreciate the implied meaning.[8] Lastly, both Quintilian and Grice believe that irony is meant to be understood as such, thanks especially to context and to the background, which speaker and addressee share: Quintilian identifies "delivery, character of the speaker and situation" as key factors to ensure proper comprehension (*aut pronuntiatione intellegitur aut persona aut rei natura*, 8.6.54), and Grice stresses the importance of "context, linguistic or otherwise" (1978, 50).

Highlighting the points of contact between some ancient and modern theories, however, should not blind one from appreciating the specific contribution of Grice and his pragmatics approach. Gricean pragmatics advance our appreciation of irony in at least two ways. First, the formulation of the Cooperative Principle and its four maxims aims at mapping the necessary and sufficient conditions for communication: in so doing, Grice positions irony in a very specific location – the first proposition of maxim two. Furthermore, and perhaps more importantly, Grice breaks more ground in explaining *how* the transfer from a literal to a non-literal meaning takes place. When what is implied is more than the literal, we have "implicatures"; and since there are two ways to imply more than we say, there are two types of implicatures, "conventional" and "conversational." If the words used, and nothing else, convey more than the literal, we have a conventional implicature;[9] but if the non-literal meaning is inferred from the context (e.g. the specific situation of a conversation, or a tone of voice, or a gesture, etc.), the implicature is conversational, as in the case of irony.

For example, in the second *Philippic*, Cicero works at rejecting the accusation that he instigated the murder of Caesar. One of his arguments runs as follows, "what is the difference between instigating and approving an action? And how is desiring that something happen different from rejoicing that it did happen?" (*Quid enim interest inter suasorem facti et probatorem? Aut quid refert utrum voluerim fieri an gaudeam factum?* 2.29). He continues with another rhetorical question: "therefore, is there one, except for those who enjoyed his tyranny, who either did not want this to happen or disliked the deed? (*Ecquis est igitur exceptis eis qui illum regnare gaudebant qui illud aut fieri noluerit aut factum improbarit?* 2.29). Cicero implies much more than he says. He implies that there were three groups of people in Rome: some were happy to be enslaved under Caesar (*qui illum regnare gaudebant*), and choice of *regnare* makes this a non-viable option for an audience of senators; as for the rest, some wanted Caesar killed and some approved of his murder. What Cicero actually says, however, is more palatable, thanks especially to various figures of rhetoric: a rhetorical question with a litotes (*ecquis est . . . qui . . . noluerit*), a double euphemistic periphrasis, what "happened" or the "deed" for "murder" (*illud fieri* and *factum*, which is repeated three times in two lines) and a dilemma established by *aut . . . aut* (implying that in either case the senators have to relinquish their accusation and agree with Cicero) come together to temper the "conventional implicature."

At this point, however, there is a twist: Cicero, for a change, replies to his own questions, concluding that "therefore we are all guilty," *ergo omnes in culpa sumus* (2.29). *Ergo* presents the conclusion as another "conventional implicature," since the language (*ergo*), and nothing else, implies that "being guilty" follows from both having desired and having appreciated "the deed." Equally, "we are all guilty" implies more than Cicero says. But in this case, the context, rather than the language *per se*, leaves no doubt that Cicero means the opposite of what he says, something like "you, senators who accuse me, you are all as guilty as I am; in fact, no one can be guilty for disliking slavery." This is therefore a "conversational implicature," which, of course, amounts to an ironical statement.

To sum up: how, then, does irony function? Ancient manuals of rhetoric provide a lasting definition of the phenomenon, but do not really explain how the transfer from the letter (*verba*) to the actual meaning (*sententia* or implicatum) takes place. For Grice, irony is a way to say something untrue without lying; the transfer from the literal (and untrue) wording to the ironic implicatum is triggered by a clash between an utterance and its specific, conversational context; if an utterance clashes with its context, the audience can safely assume that the speaker abides by the Cooperative Principle and therefore is not lying (or violating the maxim of quality); hence a literal statement is discarded in favor of a different or opposite meaning, and proper communication ensues, since the maxim of quality is flouted but not violated.

The *Philippics*, however, also offer many examples of ironical expressions which escape the classical-Gricean explanation; as I am going to argue in the next section, these examples require the theoretical insight of the cognitive sciences, and specifically of relevance theory.

Bella εἰρωνεία: a second type of irony

Cicero attacks various followers of Antony, including Censorinus, "who used to say he wanted to be urban praetor in words, but in practice he certainly did not want" (*qui se verbo praetorem esse urbanum cupere dicebat, re certe noluit, Phil.* 11.11). Cicero ironically juxtaposes Censorinus' alleged and real desire to be in Rome (*verbo* versus *re*) and alludes to the absurdity of a praetor *urbanus* leaving the *urbs* (to join Antony). This type of irony seems to fall outside of the definitions considered so far: praetor Censorinus *did* leave Rome, hence there is no clash between the letter and the meaning, or between *verba* and *sententia*, to put it in terms of the *Rhetorica ad Herennium*, and Grice's second maxim is not "flouted." And yet, Cicero is being ironic. This irony springs from an allusion to the cheated expectation that someone interested in becoming urban praetor show disinterest for the *urbs*.[10]

Before considering a different model, which can better account for this type of irony, it is worth realizing that instances of this sort abound in the *Philippics*. For example, the day after Antony boasted in the senate he would send his armed servants against Cicero's house, Cicero exclaims, "I have wished that Antony were present, just without his consultants" (*vellem adesset M. Antonius, modo sine advocatis*, 1.16); *advocati*, "consultants," ironically alludes to Antony's armed guards. The irony lies precisely in this allusion, while the statement *per se* remains true: Cicero would never wish to see Antony's "consultants" in the senate. Similarly, to excuse his approval of the murder of Caesar, Cicero accuses Antony (most certainly falsely) of having conceived the same thought himself (*hoc consilium . . . cepisse notissimum est* 2.34); then Cicero ironically comments "the fact that you had a good idea, for once, I praise . . . the fact that you did not act upon it, I forgive. That deed called for a true man" (*quod bene cogitasti aliquando, laudo . . . quod non fecisti, ignosco. Virum res illa quaerebat* 2.34). In the logic of the second *Philippic*, each statement rings true, what is meant does not clash against what is said, and yet the pervasive irony is hard to miss. Equally, the characterization of Lucius Antonius as general sounds literally true and ironic at once: "what destructions has he caused wherever he set foot! He massacres herds of cattle and of other flock, whatever he meets; his soldiers party on . . . the fields are devastated, the farms pillaged" (*quas effecit strages, ubicumque posuit vestigium! caedit greges armentorum reliquique pecoris quodcumque nactus est; epulantur milites . . . vastantur agri, diripiuntur villae* 3.31). In short, these few and very selective examples demonstrate that some irony escapes the standard classical definition.

Between 1981 and 2012, Julia Jorgensen, George Miller, Dan Sperber and Deirdre Wilson rejected Grice's "traditional" explanation of irony and put forth a new model, the so-called mention or echoic theory.[11] According to this theory, a speaker saying "What lovely weather!"

in the middle of pouring rain speaks ironically; but there are two ways to account for it. As seen above, the traditional view, formulated by the *Rhetorica ad Herennium* and explained by Grice, takes "the speaker to be using a figurative meaning opposite to the literal meaning of the utterance" (Jorgensen et al. 1984, 115). The mention theory, instead, takes "the speaker to be mentioning the literal meaning of the utterance" (1984, 115), "[H]owever, this literal meaning is not *used* by speakers to convey their own thoughts. Rather, it is *mentioned* as an object of contempt, ridicule or disapproval" (1984, 112-13, emphasis by authors). Thus, irony allows us to "identify the echoed material mentioned and the speaker's attitude toward it."[12] In other words, by saying "what lovely weather" in a storm, the speakers are literally talking about lovely weather, the lovely weather they do not have, and by mentioning it they also voice disappointment at their desire for such weather being frustrated. The nature of irony, then, lies in an allusion to a cheated expectation.

This different model of irony stems out of relevance theory, an understanding of human communication, which partly modifies and partly rejects Grice pragmatics. According to relevance theory, there is no need to postulate a Cooperative Principle or its maxims, since "the expectations of relevance raised by an utterance are precise enough, and predictable enough, to guide the hearer towards the speaker's meaning" (250).[13] In other words, hearers expect to receive information which is relevant to the conversation, not because they tacitly abide by a principle or some maxims, but because "the search for relevance is a basic feature of human cognition" (251). It follows that irony cannot be explained as an overt violation of the maxim of truthfulness. It must have a different nature: it echoes "a tacitly attributed thought or utterance with a tacitly dissociative attitude" (274). It also follows that irony is decoded differently. Grice envisions a two-stroke process, according to which first a hearer takes an ironic utterance literally, but then, realizing that it clashes with the context and assuming that the speaker is not lying, dismisses it in favor of the non-literal implicatum. For relevance theory, however, "verbal irony involves no special machinery or procedures not already needed to account for a basic use of language, interpretative use, and a specific form of interpretative use, echoic use" (272, emphasis by authors). Thus, pragmatics and relevance theory agree on the importance of context for producing irony,[14] but they disagree on its nature and decoding process.

An advantage of relevance theory is that it accounts both for irony *ex contrario* and for the second type, where the implicatum of an utterance does not contradict or clash with its literal meaning – indeed, praetor Censorinus did leave Rome, Cicero did wish to see Antony in the senate *without* his *advocati*, etc.;[15] however, as seen, the first definition of *ex contrario* fails to explain the second type of irony. This also means that irony *ex contrario* can be explained in at least two ways. For example, one can take Cicero complimenting Antony's intelligence, eloquence and wisdom according to the classical definition and conclude that Cicero means that Antony is stupid, unable to speak and unwise; or one can take it to mean that Cicero vents his frustration at all the intelligence, eloquence and wisdom Antony does not have, in particular voicing his expectation that a man of Antony's stature and power would possess more, and therefore expressing his disappointment. Of course, in many cases one does not need to decide for one explanation and discard the other, since both can help unpack the different layers of meaning produced by irony.

According to the mention or echoic theory, a speaker saying "what lovely weather" in a storm can ironically echo a specific utterance, say, the weather forecast promising sunshine, or, merely, "an expectation or hope that they [speaker and hearer] had shared that the weather would be good" (1984, 115). This distinction carries important implications, as two examples from Cicero can illustrate.[16] The year 43 opened in a state of unrest: some senators wanted to deal with Antony through diplomacy, hoping to avoid war, but some others, including Cicero, wanted to declare war at once, because they deemed diplomacy a waste of time, or even a delaying strategy by Antony. In January, the line of diplomacy prevailed, but Antony refused to comply with the instructions of the senate, and, as a result, on February 2, after a hot debate, the senate declared a state of emergency (*tumultus*), but not proper war (*bellum*). The following day, on February 3, Cicero pronounced the eighth *Philippic*, to voice his disappointment and criticize both the senate's failure to declare war and some senators' call for peace:

> *Sed quid plura? D. Brutus oppugnatur: non est bellum. Mutina obsidetur: ne hoc quidem bellum est. Gallia vastatur: quae pax potest esse certior? Illud vero quis potest bellum esse dicere quo consulem, fortissimum virum, cum exercitu misimus?*
>
> But why waste words? Brutus is being attacked: it's not a war. Mutina is being besieged: not even that is a war. Gaul is being devastated: which peace can be more certain? Indeed, who can claim that it is a war for which we have dispatched our most courageous consul with an army?
>
> *(8.5)*

Cicero's rhetoric gains force from the paratactic juxtaposition of "bare facts" (*Brutus oppugnatur, Mutina obsidetur* and *Gallia vastatur*) and judgments, expressed as simple statements (*non est bellum* and *ne hoc quidem bellum est*) or as a rhetorical question (*quae pax potest esse certior?*). In particular, the clash between facts and judgments creates irony, and irony targets some specific motions, crystallized as "quotations" from other senators. For example, on February 2, the day before Cicero pronounced *Philippic* 8, L. Caesar had convinced the senate not to use the word *bellum* (8.1),[17] and on the 3rd Fufius Calenus spoke before Cicero and tried to make a case for peace (8.11-12). In other words, the "literal meaning [e.g. *non est bellum*] is not *used* by speakers to convey their own thoughts. Rather, it is *mentioned* as an object of contempt, ridicule or disapproval."

Relevance theory and its understanding of irony as an allusion to a cheated expectation also helps to explain other passages in the *Philippics*. For example, having commended Brutus for opposing Gaius Antony, another brother of Marc, Cicero equally praises Gaius, but of course ironically:

> how amazing that speed, that good care and that courage of Brutus! And not even Gaius should be despised for speed, if only some inheritances falling on his path had not delayed him, one would say that he flew rather than traveled.
>
> (*quae celeritas illa Bruti, quae cura, quae virtus! Etsi ne C. quidem Antoni celeritas contemnenda est, quam nisi in via caducae hereditates retardassent, volasse eum, non iter fecisse diceres.*)
>
> (Phil. *10.11)*

On a literal level the expectation of Gaius' speed, a quintessential virtue or Roman generals, is cheated because of greed; but another allusion adds yet one more layer of irony. The image and the vocabulary recall the mythical Atalanta, who could run as fast as the wind (*celeritas, volasse*), but fatally lost a race because she slowed down (*retardassent*) to pick up some golden apples, which fell on her way (*caducae*).[18]

Historical characters can equally provide the material, which Cicero aptly "mentions," to produce irony. Accordingly, Cicero lampoons some followers of Antony, like Bestia and Decius. Bestia allegedly advanced his candidature for consulship (against Brutus, whom Cicero supported), even if he was never praetor, and Cicero comments: "this second Caesar Vopiscus, this man of the highest intellect and influence, who seeks consulship from aedilship, let him be unbound by laws" (*alter Caesar Vopiscus ille summo ingenio, summa potentia, qui ex aedilitate consulatum petit, solvatur legibus, Phil.* 11.11). Of course, by calling a follower of Antony a "man of the highest intellect and influence," who should become consul regardless of any restriction, Cicero means the opposite of what he states, and Grice pragmatics explain the transfer from literal to implicatum; but the reference to Caesar Vopiscus adds another level of irony, for which we need echoic theory. In 88, Vopiscus stood for consulship without having been praetor. He failed and never became consul (or praetor) afterwards, hence calling Bestia "a second Vopiscus" Cicero produces irony, through the allusion to the overambitious and defeated candidate.

Similarly, Cicero employs an historical reference to belittle Decius: "I have seen the auction of Publius Decius, a most distinguished man, who, following the examples of his ancestors, devoted himself . . . to debt" (*vidi etiam P. Deci auctionem, clarisismi viri, qui maiorum exempla persequens, pro alieno se aere devovit, Phil.* 11.13). Once again, Grice can explain only the first ironic statement (Decius was "a most distinguished man"); but the second (Decius was indebted) requires the echoic model. Cicero alludes to the famous *devotio* by Publius Decius Mus and his son.[19] By mentioning Decius' ancestors (*maiorum exempla persequens*) and by using *devovit*, the technical verb indicating a *devotio*, Cicero pokes fun at Publius, juxtaposing his despicable *devotio* (to debt) to the glorious sacrifice of his ancestors and implicitly passing a heavy judgment. In each of these three examples (Gaius Antony, Bestia and Decius), Cicero uses irony but also means what he writes (Gaius was delayed by greed, Bestia is a second Vopiscus and Decius is "devoted" to debt). Relevance and mention theory offer a satisfactory explanation, which involves multiple layers of irony.

According to this theory, the material which is ironically echoed does not need to be a specific utterance. It can also be a generic expectation, as some examples above illustrate. As seen, Cicero pits the Roman understanding of a good general against the military performances of Antony's bothers, Gaius and Lucius. Pointing at Gaius' delayed speed is ironic in that it alludes to a (cheated) universal hope that a good general will have speed but no greed. Similarly, the *Philippics* consistently portray Lucius as a threat to people's property (e.g. 5.21 and 5.25); hence, mentioning the *dux*'s performance of slaughtering herds of cattle and devastating farms, Cicero surely alludes to the good general he is not. The revelation about his "conquests" disappoints the expectation elicited by words like *dux* and *imperator* (3.31), while expressing contempt and disapproval at his appropriation of undeserved titles.

Remarkably, references made to specific utterances or to generic expectations mirror the dynamics of intertextuality: as seen, Cicero can activate an allusion to some specific words by other senators, thus using them as "modello testuale," to borrow Conte's terminology, or he can call to mind broader motifs or expectations, like generosity ensuing from wealth or glory from military conquests. Such motifs then function as "modello codice," a commonly accepted *topos* that can be continuously revisited, reshaped and deployed.[20] Thus, the contemporary discourse about intertextuality, which remains very lively among classicists, mirrors the terms of the echoic model proposed by some cognitive scientists. In either case, whether an author alludes to a specific passage or to a *topos*, all allusions are potentially ironic, as so many examples from Roman comedy or from Ovid and Martial or from Roman novels document.[21]

One may object that we should refrain from forcing modern models on classical texts, especially when they do not square with ancient definitions, lest we impose our views and miss what *they* meant. But in fact, in *De Oratore*, Cicero intuitively comes close to the echoic model, when he groups *dissimulatio*, a term which translates εἰρωνεία, with cheating one's expectation (*expectationibus decipiendis* 2.289). Similarly, and perhaps more importantly, Cicero himself seems to endorse the mention theory of irony, *ante litteram*. *Philippic* 2 is a model of invective *ad hominem*, whereby Cicero tries to discredit and ruin Antony; among various abuses, Cicero also alleged that Antony refused to serve in Spain under Caesar (2.75–8). To salt the wound, Cicero praised Dolabella, who was consul with Antony in 44, perhaps hoping to place a wedge between the two:[22]

> *Ter depugnavit Caesar cum civibus, in Thessalia, Africa, Hispania. Omnibus adfuit his pugnis Dolabella; in Hispaniensi etiam vulnus accepit.*
>
> Three times Caesar fought against his countrymen, in Thessaly, Africa, and Spain. Dolabella took part in all these engagements; in the Spanish war he was even wounded.
>
> *(2.75)*

Thanks to a letter to Atticus, we happen to know that originally Cicero had added that Dolabella fought "three times against his citizens in pitched battles" (*ter contra cives in acie, Att.* 16.11.2). In other words, in an earlier draft Cicero's praise of Dolabella was actually more explicitly tongue-in-cheek, lauding the consul for his great military achievements in three civil wars. From Cicero's letter, we gather that Atticus criticized this "praise," finding it excessive. Cicero took his advice (*corrigam quae a te animadversa sunt,* 16.11.2) and scrapped the line (which is not found in the manuscripts), but felt compelled to explain that:

> *At tamen est isto loco bella, ut mihi videtur, εἰρωνεία, quod eum ter contra civis in acie.*
>
> And yet, it seems to me that in this passage there is some nice irony, when I wrote that "he [fought] three times against his citizens in pitched battles."
>
> (Att. *16.11.2)*

It is remarkable that Cicero explains his first phrasing saying he was being ironic (*bella* εἰρωνεία):[23] indeed, since Dolabella did "take part in all these engagements," there is no gap between the literal meaning and the implicatum, so that this use of irony does not fit the ancient definitions (including Cicero's). The mention theory, however, accounts for it: Cicero echoes the typically Roman expectation that war buys one glory, but the expectation is cheated by the confession that, in all three cases, they were in fact *civil* wars – something worthy of shame more than of glory.

In short, the mention theory has the advantage of being capable of accounting both for instances of irony *ex contrario* and for echoic irony. The understanding that both a specific line or a general expectation can provide echoed material shares much with the dynamics and categories of intertextuality. Lastly, this model is somewhat loosely connected with ancient definitions, but in the *Philippics* Cicero validates it both by using echoic irony and by calling one instance εἰρωνεία.

Urbana dissimulatio: a third type of irony

In *Philippic* 3, Cicero tried to delegitimize Antony by annulling his allotment of the provinces,[24] arguing that the procedure had been manipulated to pilot the results. In his words:

> *Praeclara tamen senatus consulta illo ipso die vespertina, provinciarum religiosa sortitio, divina vero opportunitas ut, quae cuique apta esset, ea cuique obveniret . . . Qui sunt igitur reliqui quos sors divina delectet? T. Annius, M. Antonius. O felicem utrumque! Nihil enim maluerunt. C. Antonius Macedoniam. Hunc quoque felicem! Hanc enim habebat semper in ore provinciam. C. Calvisius Africam. Nihil felicius! Modo enim ex Africa decesserat et quasi divinans se rediturum duos legatos Uticae reliquerat.*

> Nevertheless, on that same evening the senate makes some wonderful deliberations; a scrupulous draw of the provinces, indeed by divine chance everyone received exactly what suited him best . . . Who are then the other ones blessed by divine providence? T. Annio, M. Gallius. Oh, both lucky, for there was nothing they preferred! G. Antonius got Macedonia. He too was lucky! For he always had this very province in mind. G. Calvisius received Africa. Nothing is luckier! He had just left Africa, and as if prophesizing he was going to return he left two legates in Utica.
>
> *(3.24-6)*

No doubt Cicero is being ironic, and his surprise at the amazing series of coincidences more and more explicitly suggests that these were no coincidences at all. Antony rigged the procedure. And yet, neither theory considered so far can give a satisfactory explanation for this type of irony. Substituting some words for their opposites, according to the *ex contrario*/Gricean model, would amount to replacing practically half of the words (e.g. *praeclara, religiosa sortitio, divina opportunitas*, etc.); and arguing that irony depends on an allusion to the cheated expectation of fair allotments is equally unsatisfactory.

Cicero's own theory, however, partially accounts for this type of irony. In *De Oratore*, he distinguishes between irony *ex contrario* (which he calls *inversio* 2.261-2) and *dissimulatio*, or feigned ignorance (2.275): *dissimulatio* differs from *ex contrario* in that it does not necessarily depend on a single word taking on a different/opposite meaning, but can apply to an ongoing mock-serious passage:

> *Urbana etiam dissimulatio est, cum alia dicuntur ac sentias, non illo genere, de quo ante dixi, cum contraria dicas . . . sed cum toto genere orationis severe ludas, cum aliter sentias ac loquare.*
>
> Elegant is that type of irony consisting in saying one thing and meaning something different, but not like the one mentioned above, when one says the opposite . . . rather like when one engages in a serious play with the whole tone of a speech, when one's words and thoughts are at odds.
>
> *(2.269)*

Perhaps the boundary between traditional irony (*ex contrario*) and feigned ignorance (*dissimulatio*) is not impermeable (cf. 3.202); and yet, thanks to this differentiation, Cicero provides a lasting definition of *dissimulatio*. Indeed, his understanding anticipates another theory of irony developed by psychologists.

In 1984, Herbert Clark and Richard Gerrig put forth the pretense theory of irony,[25] in an attempt to salvage the core of Gricean theory and to challenge the mention or echoic model. According to pretense theory, the speaker using irony is "pretending to be an injudicious person speaking to an uninitiated audience; the speaker intends the addressee of the irony to discover the pretense and thereby see his or her attitude." The audience "can take delight in being in on the pretense, in being a member of the inner circle."[26] In addition to helping to explain specific passages, this theory offers the advantage of contemplating different levels of audience[27] and of centering on "pretense," which is one of the original meanings of εἰρωνεία. In particular, the notion that by using irony at someone else's expense an orator can select the members of "the inner circle" and invite his audience to take sides resembles the dynamics of Roman invective.[28] Unsurprisingly, then, in the *Philippics* irony and invective join forces toward the same goal – to cast Antony as *hostis*, thus excluding him from the Roman community, and to invite the audience, starting from the senators, to step into Cicero's circle.

The *Philippics* include other instances of pretense irony, as two examples document. Fufius Calenus' calls for peace unnerved Cicero. The two had crossed swords in the past,[29] but Cicero realized that, in this case, irony provided a better weapon than invective.[30] Accordingly, he used it to rebuke his proposals and undermine his authority:

> *Uno in homine, Q. Fufi, fateor te vidisse plus quam me: ego P. Clodium arbitrabar perniciosum civem, sceleratum, libidinosum, impium, audacem, facinerosum; tu contra sanctum, temperantem, innocentem, modestum, retinendum civem et optandum. In hoc uno te plurimum vidisse, me multum errasse concedo.*
>
> In respect to one man, Q. Fufius, I confess that you did see further than I: I used to consider Clodius a dangerous citizen, lawless, licentious, treacherous, impudent, criminal; on the contrary, you considered him faultless, restrained, innocent and humble, a citizen to be kept and desired. In this one case, I acknowledge you did see much further, and I misjudged badly.
>
> *(8.16; my translation)*

The first sentence, *uno in homine . . . quam me*, does not reveal Cicero's pretense (though, knowing Cicero, it does not bode well); but as Cicero continues, the irony becomes evident and invites the senators to drift apart from Clodius, and, by association, from his "backer," Calenus.

Cicero employs the same device in a speech to the people, *Philippic* 6:

> *Quamquam, Quirites, non est illa legatio, sed denuntiatio belli, nisi paruerit . . . Facile vero huic denuntiationi parebit, ut in patrum conscriptorum atque in vestra potestate sit qui in sua numquam fuerit! Quid enim ille umquam arbitrio suo fecit? Semper eo tractus est quo libido rapuit, quo levitas, quo furor, quo vinolentia.*
>
> And yet, men of Rome, this is no embassy; it is a declaration of war if he [Antony] does not obey . . . doubtless he will find it an easy matter to obey such a message and to be at the senators' disposition and yours, a man who was never at his own! For what did he ever do of his own choice? He has always been dragged in the wake of lust, frivolity, madness, inebriation.
>
> *(6.4)*

The pretense irony of *facile vero . . . parebit* becomes clear with *qui in sua numquam fuerit*, perhaps eliciting laugher and quickly yielding to invective. In all these examples Cicero begins with the pretense of being awestruck, whether by the lucky allotment of the provinces, Calenus once having better judgment, or Antony obeying the senate – and once the pretense becomes evident, the audience "can take delight in being in on the pretense." Cicero seems unwilling to leave any ambiguity, perhaps because of the specific nature of the *Philippics*, which were "designed to influence public opinion" and "must have been copied and distributed soon after delivery, while their subject matter was still relevant to the ongoing war of words between Cicero and Antony" (J. Ramsey 2003, 18).

A specific type of pretense is the *praeteritio*, a figure by which a speaker declares to leave something unmentioned precisely while mentioning it. Cicero uses *praeteritio* when he pretends to omit Antony's private misbehavior (*sed omitto ea peccata* 2.70), his misconduct as consul (*ut omittam innumerabilia scelera urbani consulatus* 7.15), and his enmity to Rome (*omitto hostem patriae* 12.19), just to mention a few examples. Remarkably, the *Rhetorica ad Alexandrum*, allegedly the oldest manual of rhetoric preserved from the Greco-Roman world, connects this type of pretense and εἰρωνεία: "εἰρωνεία means saying something while pretending not to say it (εἰρωνεία δέ ἐστι λέγειν τι μὴ λέγειν προσποιούμενον) . . . like 'I believe it is not necessary to tell you that . . .'" (1434a = 21). As such εἰρωνεία/*praeteritio* constitutes a convenient tool, especially for recapitulating in the epilogue, when one wants to recall something without needing to explain it, and for this reason εἰρωνεία is treated among various means of recapitulation.

To return to our initial question, then, what is irony? Modern research in psychology and cognitive science has produced three main models to decode it. In my view, none of these succeeds in providing a single holistic account for irony, but each successfully advances a line of inquiry first opened by ancient manuals of rhetoric. By locating irony at maxim two of the Cooperative Principle, Grice shows how the transfer from a literal (and untrue) meaning to the implicatum takes place and how conversational implicatures allow its proper decoding; the mention theory explains how irony involves a sense of cheated expectation and accounts for some instances of irony (including one which Cicero deemed ironic) better than Cicero's own definition; the pretense theory demonstrates how irony can produce multiple levels of audience, affecting its response and splitting it between those who are in on the pretense and its victims. Each theory has much to offer to classicists, since it improves our appreciation of irony, in theory and in practice. At times, one model fits better than another, and at times more than one model work together to bring insights into different implications and levels of irony.

Notes

1 Classical treatments of irony focus more on Greek than on Latin literature, cf. Ribbeck 1876, Büchner 1941, Bergson 1971, Nünlist 2000 and Opsomer 1998 and 2000. For irony in Cicero, see Haury 1955 and Grillo forthcoming, on irony in the *post reditum* speeches; and Orlandini 2002 on *Philippics* I–II.

2 As is known, Cicero himself chose this title (Cic. *Ad Brut.* 2.3.4 and 2.4.2; cf. Plut. *Cic.* 48.6).

3 On invective, see Süss 1910, 245–63, Nisbet 1961, 192-7, Corbeill 2002, 197-217 and Craig 2004, 187–213; on invective in the *Philippics*, see Ramsey 2003, 159–60 and Manuwald 2007, 105-9, with updated bibliography.

4 Cf. Cic. *de Or.* 2.269 and 3.202 and Quint. 9.2.44.

5 For a convenient summary of the main categories listed by Süss, see Corbeill 2002, 197–217 and Craig 2004, 187-213.

6 Cicero writes that irony has a special way of creeping into people's minds (*de Or.* 3.203).

7 The *Rhetorica ad Herennium* uses either *permutatio per similitudinem* or *translatio* for metaphor, but the two terms differ in number, not in substance: when more than one metaphor (*translatio*) from the same realm occurs in the same sentence we have a *permutatio per similitudinem*. To use his example: "to what guard should we entrust our sheep, when dogs behave like wolves?" (4.46).

8 Grice 1975, 53 uses the following examples: irony is when A says "X is a fine friend," with both A and the audience knowing that X has just disappointed A; metaphor is saying "you are the cream in my coffee," and, like irony, involves two stages of interpretation. For Grice, other figures of rhetoric equally flout the second maxim: meiosis is saying "he was a little intoxicated" of a man known to have broken all the furniture; and hyperbole is saying "every nice girl loves a sailor."

9 To use Grice's own example of conventional implicature: the utterance "he is an Englishman; he is, therefore, brave" implies that being brave follows from being English, and yet, strictly, this is not what is said (Grice 1975, 44–5). The implicature "being brave follows from being English" is conventional, because it can be inferred without any background information about the "conversational situation."

10 Noticeably, to voice his disappointment and produce this type of irony, Cicero goes as far as forcing the evidence: in fact, Censorinus was a praetor, but not the *praetor urbanus*; there is evidence that in 43 the urban praetor was M. Caecilius Cornutus (Broughton *MRR* 2.338-9; cf. Shackleton Bailey 1986, 227, n. 6).

11 Sperber and Wilson 1981, 295–318; Jorgensen, Miller and Sperber 1984, 112–20; Wilson and Sperber 2012, 123-45.

12 Jorgensen, Miller and Sperber 1984, 115.
13 References in this paragraph are to Wilson and Sperber 2004, which provides a convenient summary of pragmatics and framework for echoic irony; for a full exposition of relevance theory, see Sperber and Wilson 1995, which treats echoic irony at pp. 237–43.
14 As seen above, for Grice irony derives from a conversational implicature triggering a non-literal meaning; for relevance theory, however, there is nothing specific about irony, since *any verbal communication* begins with "the recovery of a linguistically encoded sentence meaning, which must be contextually enriched in a variety of ways to yield a full-fledged speaker's meaning" (259).
15 Jorgensen, Miller and Sperber explicitly reject the proposition that there are two types of irony, standard, which I called *ex contrario*, and echoic. Their goal is to find one model capable of explaining irony *tout court* (1984, 114).
16 For examples from other speeches by Cicero, cf. Grillo 2015, 146 and forthcoming.
17 Probably L. Caesar advanced his motion in a previous senatorial meeting and he was not present when Cicero pronounced the 8th *Philippic*, perhaps due to illness (*Fam.* 12.2.3; cf. Manuwald 2007, 920-1).
18 For this reference, cf. Shackleton Bailey (1986, 257, n. 9).
19 *Devotio* indicated a rare and highly commended sacrifice by Roman generals, who in a critical situation would offer their life and die in battle to spare their army from defeat.
20 Conte 1986, 31; cf. Hinds 1998, 41.
21 I find this line of inquiry particularly stimulating and intend to explore the interaction between mention theory and intertextuality further in a monograph on irony in Latin literature.
22 Ramsey 2003, 266–7.
23 Cf. Orlandini 2002, 220.
24 The provinces were allotted on November 28, 44 (3.38), and *Philippics* 3 was delivered to the senate on December 20 (Marinone 2004).
25 1984, 121-6.
26 1984, 121. As one can see, what drives these scholars is the quest for *the* model to explain irony *tout court*; what interests me, however, is whether any given model can help to explain classical texts; therefore, multiple models can coexist, and in certain cases more than one model simultaneously applies even to the same passage, as I argue at the end of this chapter.
27 For the notion of double audience, Clark and Gerrig cite Fowler: "irony is a form of utterance which postulates a double audience, consisting of one party that hearing shall hear and shall not understand, and another party that, when more is meant than meets the ear, is aware both of that more and of the outsiders' incomprehension" (1965, 305-6).
28 E.g. Corbeill 2002.
29 As tribune of the plebs in 61, Calenus supported Clodius, he appeared in court against two clients of Cicero, Caelius (Cic. *Cael.* 19) and Milo (Ascon. 44-5), and then he followed Caesar in Gaul (*B Gall* 8.39.4) and in Spain (*MRR* 2.267).
30 Calenus was the father-in-law of Vibius Pansa, who was consul in 43, and, as a sign of respect, Pansa had chosen him to be the first to be called on in the senate throughout 43.

References

Bergson, L. (1971). "Eiron und Eironeia," *Hermes*, 99: 409–22.
Büchner, W. (1941). "Über den Begriff der Eironeia," *Hermes*, 4: 339–58.
Clark, H. and Gerrig, R. (1984) "On the pretense theory of irony," *Journal of Experimental Psychology*, 113.1: 121–6.
Conte, G.B. (1986). *The Rhetoric of Imitation: Genre and Poetic Memory in Virgil and Other Latin Poets*. Ithaca, NY.
Corbeill, A. (2002). "Ciceronian invective," in J. May (ed.), *Brill's Companion to Cicero: Oratory and Rhetoric*. Leiden: 197–217.
Craig, C. (2004). "Audience, expectations, invective, and proof," in J. Powell and J. Patterson (eds.), *Cicero the Advocate*. Oxford: 187–213.
Fowler, H. (1965). *A Dictionary of Modern English Usage*. Oxford.
Grice, H. P. (1975). "Logic and conversation," in P. Cole and J. Morgan (eds.), *Syntax and Semantics*. Vol. 3. New York: 41–58.
Grice, H. P. (1978). "Further notes on logic and conversation," in P. Cole (ed.), *Syntax and Semantics*. Vol. 9. New York: 113–28.

Grillo, L. (2015). *Cicero's* De Provinciis Consularibus Oratio: *Introduction and Commentary*. Oxford.
Grillo, L. (Forthcoming). "Irony in Cicero's *post reditum* speeches," in Opsomer (ed.), *Psychology and the Classics*. Leiden.
Haury, A. (1955). L'ironie et l'humour chez Cicéron. Leiden.
Hinds, S. (1998). *Allusion and Intertext: Dynamics of Appropriation in Roman Poetry*. Cambridge.
Jorgensen, J., Miller, G., Sperber, D. and Wilson, D. (1984). "Test of the mention theory of irony," *Journal of Experimental Psychology: General* 113.1: 112–20.
Manuwald, G. (2007). *Cicero, Philippics 3–9*. Leiden.
Marinone, N. (2004). *Cronologia ciceroniana*. Bologna.
Nisbet, R. (1961). *M. Tulli Ciceronis in Calpurnium Pisonem Oratio*. Oxford.
Nünlist, R. (2000). "Rhetorische Ironie – Dramatische Ironie: Definitions und Interpretationsprobleme," in J. P. Schwindt (ed.), *Zwischen Tradition und Innovation: poetische Verfahren im Spannungsfeld klassischer und neuerer Literatur und Literaturwissenschaft*. Munich: 67–87.
Opsomer, J. (1998). "The rhetoric and pragmatics of irony/εἰρωνεία," *Orbis* 40.1: 1–34.
Opsomer, J. (2000). "εἰρωνεία and the corpus Plutarcheum (with an appendix on Plutarch's irony)," in L. Van der Stockt (ed.), *Rhetorical Theory and Practice in Plutarch*. Leuven: 309–29.
Orlandini, A. (2002). "Pour une approche pragmatique de l'ironie (Cicéron, *Philippiques*, livres 1–2)," *Pallas* 59: 209–24.
Ramsey, J. (2003). *Cicero: Philippics I–II*. Cambridge.
Ribbeck, O. (1876). "Ueber den Begriff der εἴρων," *Rheinsches Museum*, 31: 381–400.
Shackleton Bailey, D. R. (1986). *Cicero Philippics*. Chapel Hill, NC.
Spengel, L. (1953). *Rhetores Graeci. Vol. 3*. Lipsiae.
Sperber, D. and Wilson, D. (1981). "Irony and the use-mention distinction," in P. Cole (ed.), *Radical Pragmatics*. New York: 295–318.
Sperber, D. and Wilson, D. (1995, 2nd ed.). *Relevance, Communication and Cognition*. Oxford.
Wilson, D. and Sperber, D. (2004). "Relevance theory," in L. Horn and G. Ward (eds.), *A Handbook of Pragmatics*. Oxford: 249–90.
Wilson, D. and Sperber, D. (2012). *Meaning and Relevance*. Cambridge.

15

Roman ritual orthopraxy and overimitation

Jacob L. Mackey

Introduction

The orthopraxy of Roman ritual has frequently been remarked upon, not only by modern scholars,[1] but also by the Romans themselves,[2] as well as by their Greek visitors.[3] This chapter explores the cognitive roots of this fastidiousness about ritual. I appeal to "overimitation," the cross-culturally well-attested tendency of children and adults to imitate with extraordinarily high fidelity. Overimitation is widely seen as crucial to the transmission of cultural practices, which often consist of action sequences that are teleologically and causally opaque:[4] that is, directed toward no transparent *end* and having no transparent causal efficacy as *means*. No "rational" imitator, motivated by an instrumental means-ends calculus, would imitate such actions. Without overimitation, teleologically and causally opaque cultural practices, lacking transparent instrumental value, would presumably not achieve wide distribution.

Bringing the research on overimitation into contact with our evidence for Roman ritual allows me to offer proposals relevant to the fields of both classics and psychology. As to classics, I submit that the orthopraxy of Roman cult, whose teleological and causal opacity seems indisputable, is ideally suited to explanation by way of overimitation. I shall argue that in addition to explaining Roman orthopraxy, the findings on overimitation help make sense of the Roman penchant for multiple aetiologies. For the opaque yet manifestly intentional actions of Roman ritual lacked not only intuitive rationale but also authoritative explanation. In these conditions, multiple aetiologies represented a cognitively autonomous search for explanations. Finally, counterintuitive though it may seem in an orthoprax system, overimitation helps explain Roman ritual change. For overimitation is precisely what preserved opaque ritual technologies for addition, amendment, and innovation in the face of new religious exigencies.

As to my proposal for psychology, bringing the cultural dataset that is our evidence for Roman ritual into contact with cognitive theory will allow me to offer a solution to a problem in overimitation studies. For the underlying motivation that drives overimitation is disputed.[5] On the "causal" account, imitators encode opaque actions as non-obvious but nonetheless instrumentally indispensable *causal* components in an overall action sequence, so they imitate such actions with precision.[6] A competing theory, the "normative" account, holds that imitators encode opaque actions as *normatively* indispensable, as mandated by the conventions of ritual performance, and thus imitate with high fidelity.[7] The related "affiliative" account proposes that imitators copy with high fidelity in order to connect or *affiliate* with others or with a social group.[8] I attempt to reconcile these competing theories by suggesting that *affiliation* with others, which almost always includes adhering to and endorsing a group's *norms*, is indeed often *causal*, but the causality is of the psychological and social rather than physical sort.

Roman imitation

Let us begin by briefly surveying thinking, ancient and modern, about imitation and overimitation. The Romans gave imitation a place of primacy among learning mechanisms.[9] For example, Quintilian held that a robust faculty of *imitatio* was the surest sign, after a quick and exact memory, that a child possessed a "teachable nature."[10] *Imitatio* served Quintilian for a general theory of cultural transmission.[11] There can be no doubt "but that a great part of *ars* consists in imitation," given that "it is useful to follow what has been well-invented."[12] Not only do aspiring orators, musicians, and painters imitate in order to learn, but even peasants follow the *exempla* of experience-tested agricultural practices.[13] For Quintilian, *imitatio* transmitted more than just instrumental knowledge and skills. He drew causal connections between imitation and the youthful *mens*[14] as well as between imitation and *mores*. Indeed, as he writes, *frequens imitatio transit in mores*, "regular imitation turns into character."[15] In this, he was surely looking back to Cicero, who supposed that we imitate, *imitamur*, others' inclinations and practices, *studia* and *instituta*, as well as their habits and character, *consuetudo* and *mos*.[16] These citations testify to a Roman tradition in which imitation is not mere behavioral mimicry but rather a source of our *mores*, that is, our attitudes, dispositions, and habits.

Imitation and overimitation

In the modern social sciences, imitation has at times fared less well. Durkheim, for example, dismissed imitation as nothing but "mechanical monkey-business that makes us copy the movements that we witness" and thus as insufficient to account for anything as collective and social as a community's "moeurs."[17] Naturally, Durkheim opposed his competitor and chief proponent of imitation, Gabriel Tarde, who had formulated a sociology based on causal interactions among individual agents, notably through imitation.[18] As Tarde put it, "where there is any social connection between two living beings, there is imitation." For him, imitation amounted to "an action at a distance from one spirit upon another," which could be "willed or not, passive or active." This "inter-spirituelle" relation among agents transmitted from one agent to another not only external forms of behavior but also "une certaine dose de *croyance* et de *désir*."[19] Tarde's insight was that imitation alone was a powerful enough mechanism to account for the observable patterns of thought and behavior in a given community. Tarde lost his fight with Durkheim but recent research suggests that he was on to something, and that the quality of our conception of imitation determines the quantity of work imitation can do in explaining observable facts about collective practices and the transmission of culture.

Today, imitation is recognized as a highly flexible cognitive and behavioral capacity, possessing both epistemic and social functions. It begins very early, perhaps within minutes of birth.[20] Imitation depends upon a rich social-cognitive endowment, including a capacity for bidirectional bodily and psychological mapping between self and other, that is, the awareness that others are in salient physical and mental respects "like me."[21] This self-other equivalence includes an at least implicit awareness that others act intentionally, in pursuit of goals. This "intentional stance"[22] is revealed in imitation. For example, children as young as eighteen months of age distinguish between intended and accidental actions, and decline to imitate the latter.[23] Again, at fourteen months, children can imitate "rationally"; that is, they can omit causally irrelevant actions when copying an adult.[24] The implication is that they perceive the adult's goal in acting and do not copy those actions that appear unnecessary as means. Moreover, children under two years of age intuit the goal toward which an adult is striving even when the adult's actions do not succeed. They can imitate so as to achieve the goal that had eluded the adult.[25]

So far, I have only mentioned examples of children modulating their imitation in light of what they take to be their model's intentions. And yet children may also imitate without modulation, that is, with great fidelity. That is, they do not select from among or alter their model's actions but instead imitate their model precisely. This faithfulness in copying has been dubbed "overimitation."[26] An experiment was devised to probe its causes.[27] Three- to five-year-old children were first shown a task (opening a jar to get a toy) featuring causally irrelevant actions (tapping the jar with a feather before removing the lid). When the children had succeeded in verbally identifying the action unnecessary to retrieving the toy, they were shown a new toy-retrieval task, which also featured some causally irrelevant actions. Despite their competence in discriminating causally necessary from unnecessary actions, when asked to retrieve the toy themselves, the children scrupulously imitated even the needless actions. Indeed, they did so even when explicitly told not to copy such actions. The conclusion was that these findings reveal a distortion in children's causal learning. They unconsciously encode all of an adult's actions, even the seemingly causally irrelevant, as efficacious and thus imitate accordingly, despite their discursive ability to discriminate between relevant and irrelevant actions.

This "causal" or "instrumental" account of overimitation has not found universal favor. Many researchers prefer one or a blend of two *social* accounts, the "normative" and the "affiliative." On the normative account, between the ages of two and three years of age, children begin to adopt a "normative stance"[28] or "ritual stance"[29] toward the intentional but not transparently instrumental actions that they see adults perform. That is, they implicitly interpret purposeful but opaque actions as the conventional or ritualized behaviors of their social group, rather than as mechanically instrumental, and they imitate accordingly.

To test this thesis, an experiment was devised with two conditions, each of which featured causally opaque actions. In one condition, the model's actions were introduced as goal-oriented and instrumental ("here's how to ring this bell") and in the other, the model's actions were introduced as method-oriented and conventional ("here's how we 'dax'," where "daxing" amounts to ringing a bell).[30] Children overimitated in both cases, but did so more in the method-oriented condition. In addition, especially in the method-oriented condition, they spontaneously protested when observing a third party omit the causally opaque action. Moreover, when asked, children explicitly described the third party's omissions as "wrong," especially in the method-oriented test. The conclusion is that under appropriate conditions, children tend to focus on the *how* or method rather than the *goal* or instrumental rationale of behavior. They encode causally opaque intentional actions as normatively indispensable parts of a conventionalized or ritualized activity.

Another social theory of overimitation is the "affiliative" account. In one study purporting to provide evidence for this thesis, children observed as two adults retrieved a toy from a wooden box. One adult opened the box after performing causally irrelevant actions upon it, such as tapping the side three times, whereas the other adult opened the box without performing any irrelevant actions. The children could now choose between two action sequences as models for imitation when it was their turn to retrieve the toy from the box. When the "efficient" adult, who had performed no irrelevant actions, stayed in the room and the "irrelevant" adult, who had engaged in causally opaque behavior, such as tapping the side of the box, departed, the children tended not to imitate the causally opaque actions. But when the "efficient" adult departed and the "irrelevant" adult remained in the room, children tended to include the causally opaque actions in their imitation, and they did so at roughly the same rate as children in another test condition who had seen both adults perform the causally irrelevant actions and who therefore had no "efficient" action sequence available for imitation. The conclusion is that this shows that imitation can be "a means of promoting shared experience" and "aligning oneself with one's cultural in-group."[31]

These summaries of selected studies that I have provided for the sake of illustration have all focused on contexts in which children observed adults who demonstrated actions in more or less "pedagogical" ways. However, it turns out that children will also overimitate actions that they have witnessed as third-party observers, that is, in contexts not featuring child-directed pedagogy.[32] Moreover, I have discussed only experiments with children, but researchers have found, surprisingly, that overimitation increases with age into adulthood.[33] Finally, overimitation occurs across cultures, with some curious variations currently under investigation.[34]

Regardless of which account they favor, researchers broadly agree that overimitation is important to the cultural transmission of knowledge and skills unlikely to be discovered or invented through processes of individual rather than social learning, that is, learning through observation of and collaboration with others. Such knowledge and skills are also unlikely to be reliably passed on through "emulation",[35] that is, low-fidelity or "lossy" imitation.[36] In contrast, relatively lossless overimitation facilitates mastery of specialized, causally opaque action repertoires that must be mastered by culture-learners quickly and accurately, including such normative activities as language use,[37] social conventions, rituals, and so forth, as well as such instrumental activities as are found in material technologies, tool use, agricultural practices, and the like.

I have mentioned that overimitation is triggered by actions that appear purposeful but possess no transparent means-end rationale. Let us close this section by considering this fact in greater detail. Both children and adults adopt the so-called "ritual stance" in the face of "teleologically opaque" or "causally opaque" but nonetheless apparently intentional actions. That is, our social-cognitive faculties inexorably lead us to interpret the behavior of other agents as purposely selected by them as means that causally conduce to their goals. However, the behaviors involved in many cultural activities, including rituals and other conventionalized activities,[38] may be opaque with respect to both means and goal. That is, some actions are *teleologically opaque* in that they have no obvious goal or end. Likewise, some actions are *causally opaque* in that their *goal* is relatively clear but the actions' causal contribution as *means* is not.[39] Both teleological and causal opacity may trigger the ritual stance in culture-learners. Taking the ritual stance leads behaviorally to overimitation or high-fidelity copying of actions with obscure ends or obscure causal relevance.

In the next section, we explore how the teleological and causal opacity of Roman cult action, along with children's and adults' propensity to overimitate such action, contributed to Roman religion's orthopraxy.

Overimitation and Roman orthopraxy

Consider the following scene of cult in honor of the god of boundaries, Terminus. Here we find children practicing alongside their elders in what would surely have been for them a powerful context of cultural learning (Ov. *F.* 2.645–652 and 658):[40]

> *ara fit: huc ignem curto fert rustica testo*
> *sumptum de tepidis ipsa colona focis.*
> *ligna senex minuit concisaque construit arte,*

et solida ramos figere pugnat humo;
tum sicco primas inritat cortice flammas;
stat puer et manibus lata canistra tenet.
inde ubi ter fruges medios immisit in ignes,
porrigit incisos filia parva favos.
. . .
et cantant laudes Termine sancte tuas.

An altar is made. Here the country wife herself brings / in a potsherd fire taken from the warm hearth. / The old man chops wood and arranges the pieces with skill / and struggles to fix branches in the hard ground. / Then he encourages the first flames with dry bark. / His son stands and holds a wide basket in his hands, / then, when he has tossed grain thrice into the flames, / his little daughter offers sliced honeycombs. // And they (*sc.* the family and their neighbors) sing your praises, holy Terminus.

Ovid depicts all these cult actions as purposeful. Surely they were: there are no sneezes or unintended behavioral tics in the tableau as he presents it. And yet we may ask, for example, to what end does Ovid's *camillus* throw grain into the fire? And why three times? This action has no transparent teleology, no clear end or goal, no "pragmatic basis," in the words of Jörg Rüpke.[41] Or consider the hymn to Terminus. Ovid tells us that its goal is praise of the god, *laudes*. Even if its teleology was clear enough to participants and observers, we nonetheless know from Pliny the Elder that the Romans were accustomed—and perhaps Ovid's family should be understood—to raise their right hands to their lips and to spin themselves around when praying.[42] Why? The causal contribution of such gestures to the end of praising or entreating a god is obscure.

We should not dismiss the hypothesis that these and countless other teleologically and causally opaque gestures of Roman cult were preserved simply because they appeared to be *effective*. As Clifford Ando explains it, "in light of the terrifying superiority of the gods, and knowing what had worked before, one had an overwhelming obligation scrupulously to recreate precisely that earlier performance."[43] The critical phrase here is "what had worked." The Romans' was an *efficacious* system of cult. Pliny, for example, could connect orthoprax prayer with desired ritual outcomes and, conversely, defective prayer with ritual failure.[44] He held that even *mutae religiones*, ritual gestures without prayer, could work their effects.[45] So perhaps something like the causal account of overimitation best explains Roman orthopraxy.

Indeed, ritual efficacy deriving from orthoprax ritual performance may easily be seen in many cases of magic,[46] where we find words and gestures that are undoubtedly efficacious *ex opere operato*, that is, as long as they are properly performed, no matter how and perhaps in some inscrutable way because causally opaque. Take an example from Cato (*Agr.* 160):[47]

Luxum siquod est, hac cantione sanum fiet. Harundinem prende tibi viridem P. IIII aut quinque longam, mediam diffinde, et duo homines teneant ad coxendices. Incipe cantare: "motas vaeta daries dardares astataries dissunapiter," usque dum coeant.

If you suffer any dislocation, let it be healed by this incantation. Take a green reed 4 or 5 feet long, split it in the middle, and let two people hold it to your hips. Begin to chant: "*motas vaeta daries dardares astataries dissunapiter*," until they come together.

The causal opacity of these vocalizations and gestures is almost perfect. What could the pressing of a split green reed to the hips, or the chanting of *motas vaeta daries dardares astataries dissunapiter*, possibly contribute to the *physical* effect of fixing a dislocated limb? Because the causal relationships among the ritual, the incantation, and the healing of the dislocation are so difficult to perceive or infer, the entire sequence must be overimitated in the causal account's sense of overimitation. That is, culture-learners must encode all actions and vocalizations as instrumentally necessary, in some non-transparent way, to obtain their putative effect in the physical world. I such cases, orthopraxy flows from the practitioner's desire to get what she wants in combination with her inability to penetrate the causal relations among words, deeds, and desired outcome.

A second and perhaps similar example comes from Pliny the Elder. He relates a story from L. Calpurnius Piso Frugi to the effect that Numa used to call down lightning "by means of certain rites and prayers."[48] Numa's successor, Tullus Hostilius, attempted the same ritual but failed *to imitate* the ritual *accurately* (*imitatum parum rite*) and so was struck by lightning.[49] Pliny implies that Tullus' failure to imitate precisely certain causally opaque actions resulted in his failure to produce precisely the effects he desired in the physical world. However, if we compare Pliny's version of the story to the version found in Livy, who also paraphrases Frugi, we find that Livy explicitly preserves the reason for the rite's failure. The "improper ritual" excited Jupiter's anger—*ira Iovis sollicitati prava religione*—and the god communicated his wrath by striking Tullus with lightning.[50]

Livy's version of the story removes the failed ritual from the domain of strictly physical causality and supports the *normative* account of overimitation. In this version, it is not some causal mechanism of nature but rather Jupiter that protests Tullus' breach of orthoprax ritual norms. So, Tullus' botched ritual does not reduce to physical causation of an unintended physical effect. Instead, his failed performance violated ritual norms shared socially among gods and mortals and he was punished for this norm-violation.

These results do not suggest that we need to discard the causal account of overimitation. Instead, Livy shows that we may unite the normative and causal accounts. For it is precisely by adhering to certain norms, in this case norms of ritual practice, that one *causes* certain effects. However, these effects are not in the first place *physical* but rather *psychological*. A normatively proper ritual would not have produced its lightning through some obscure causal mechanism of the strictly physical domain; rather, a normatively proper ritual would in the first place have affected Jupiter's *mind* in just the right way, eliciting in the second place his intervention in the physical world. So the normative is causal, but the effects of proper norm-following are in the first place psychological and only secondarily physical. That is, the physical effects follow upon and from the psychological. Normative actions, even when quite opaque, have real effects, but these effects are realized in the minds of others, in this case gods, who recognize and share the same norms.

This hypothesis conjoining the normative and causal accounts finds support in the distinctively Roman institution of *instauratio*.[51] As Livy's Camillus puts it, *instauratio* is the repetition from scratch of a ritual or ceremony "because something from the ancestral rite (*ex patrio ritu*) has been omitted (*praetermissum*) due to negligence (*neglegentia*) or chance."[52] Cicero dilates upon *instauratio* in *De haruspicum responsis*, a speech he delivered before the *collegium pontificum*.[53] Certain portents were interpreted by Etruscan diviners, *haruspices*, as evidence that "games (*ludi*) have been performed without due attention and have been desecrated." It was up to the *septemviri epulonum*, a board of priests, to determine if anything had in fact been "omitted (*praetermissum*) or done wrong," and then it was for the *pontifices* to decide whether "those same *sacra* are to be celebrated anew and from the beginning (*instaurata*)."[54] Cicero goes on to list in detail some of the deceptively minor infractions that could cause games or ceremonies to require *instauratio*: "if a dancer has stopped, or if a flute-player suddenly falls silent, or if a boy . . . has let go of the rein of a chariot." In such cases the games have not been correctly performed (*non rite facti*).[55]

The rationale behind *instauratio* Cicero gives as follows: repetition from scratch "expiates ritual errors and the minds of the gods are appeased" (*mentes deorum immortalium . . . placantur*).[56] Surely *instauratio* represents a very extreme cultural capture and deployment of the urge to overimitate. For *instauratio* presupposes that orthopraxy is not enough. It is not enough just to get the rites *right*. Instead, one must *start all over again* in the event of the slightest error. Here we see the practical antidote to Piso Frugi's cautionary tale about Tullus Hostilius. Incorrect ritual, *prava religio*, excites divine anger. Fortunately, *instauratio*, correct re-performance from scratch, can turn aside that anger and appease the *mentes deorum*.

Here we must pause to consider the nature of the divine minds so appeased. If the rituals through which these minds are affected are causally opaque, and hence must be imitated with high fidelity, then it seems likely that the divine minds to which the rituals are addressed are relatively *psychologically opaque*. I say "relatively" because the transparency of the human mind was a commonplace in Roman antiquity. The human face, the *vultus*, speaks the "silent speech of the mind," the *sermo tacitus mentis*,[57] disclosing the contents of one mind to another directly and clearly. Take the following two texts by Cicero on the intersubjective transparency of agents' psychological states, the first from *De legibus* and the second from *De Oratore*:

> *Leg.* 1.26–27: *speciem ita* (sc. *natura*) *formavit oris, ut in ea penitus reconditos mores effingeret.* [27] *nam et oculi nimis argute quem ad modum animo affecti simus loquuntur et is qui appellatur vultus, qui nullo in animante esse praeter hominem potest, indicat mores.*
>
> Nature has so shaped the appearance of the face that it has portrayed on it the character hidden deep inside. [27] For the eyes tell all too clearly how we have been affected in our mind, and that which is called the expression, which can exist in no living thing except the human being, reveals our character.

> *De Or.* 3.223: *isdem enim omnium animi motibus concitantur et eos isdem notis et in aliis agnoscunt et in se ipsi indicant.*
>
> The minds of all people are excited by the same emotions and people recognize these emotions by the same signs in others as they reveal them in themselves.

On this view, even without sharing a language, human beings are united by and made intelligible to one another by deep cognitive, affective, expressive, and bodily commonalities.[58] We simply intuit in the eyes, expression, and comportment of others what is happening inside of them and the nature of their *mores*, or character.[59]

In contrast to this intersubjective transparency of one human being to another, the divine mind is psychologically opaque. We have no unmediated access to a divine *vultus* upon which to glimpse the disposition of the divine mind. What we have instead is ritual, which is like language if in no other respect then at least insofar as it is governed by norms. And as Cicero pointed out, language is opaque to the minds of those who do not share linguistic norms, without which neither party can move or affect the mind of the other: "words move only those who are joined together by partnership in language."[60] Indeed, even young children are cognizant of the psychological, as opposed to behavioral, impact upon others of their communicative attempts.[61] Where the gods are concerned, what we have for communication is the language of ritual, a system of norms for producing effects in, and for reciprocally reading the state of, the divine mind.

We indicate our will in prayer and cult action, while the gods indicate their will through the entrails of sacrificial victims or by means of *auspicia* and prodigies. This language of entrails, *auspicia*, and prodigies is itself a language to be mastered through high-fidelity imitation.[62] But even given mastery of this language, we must ensure that our communications are free of solecisms and ambiguities in order to guarantee that we cause the desired effect in minds so opaque. Hence the insistence on orthopraxy in ritual action. Hence the cautionary tale of Tullus Hostilius, inadvertently inciting the divine mind to anger through a breach of ritual norms. Hence, finally, *instauratio*, which allows for "conversational repair," the mending of ritual norm breaches, the restarting of communications with the gods, in order to produce in their minds propitious effects.

Allow me to summarize baldly my argument thus far: in order reliably to exert causality in the domain of divine psychology, Romans were constrained to overimitate in their ritual performances. Only adherence to action sequences that were normatively determined, that were mutually endorsed by mortals and gods, could ensure achievement of the desired effects in the divine mind. Only overimitation of causally opaque ritual actions could guarantee the desired psychological effect, because the psychology to be affected, the divine mind, was not directly but only mediately available, through ritual.

So, we should view overimitation as at once normative and causal. In the social world, especially in a social world in which gods and mortals are deeply affiliated as if in community,[63] the culturally *normative* aspects of our actions are often instrumental to their *causality*. And that causality is not first of all *physical* but rather *psychological*. As Roman cult shows, adhering to ritual norms conduces to divine placation, while breaching those norms creates divine disapproval. And this causality, psychological though it be, is every bit as real and consequential as the push-pull causality of the physical world.

This understanding of overimitation as driven by motivations at once normative and causal positions us to explain Roman orthopraxy itself. If Roman ritual was a system of causally opaque, normative actions, then its very *opacity* will have triggered an overimitation response in Roman culture-learners, children and adults alike. Overimitation in response to the causal opacity of Roman ritual will have ensured high-fidelity copying of the ritual and hence faithful cultural transmission. Roman ritual, with its cautionary tales about and its "instaurative" remedy for under-imitation, is a paradigmatic and perhaps even extreme case of causally opaque cultural knowledge transmitted by overimitation. So, to state the thesis simply, the causal opacity of Roman ritual led culture-learners to adopt the ritual stance, which then led, as behavioral consequence, to overimitation. This overimitation resulted in high-fidelity ritual copying, that is to say, the orthopraxy that has long been known to characterize Roman religion.

There are two additional fallouts from my overimitation account of orthopraxy, at which we can only glance here. First, we know that Romans declined to insist upon authoritative explanations of their rituals. I would submit that this explanatory vacuum in the face of causal opacity encouraged the proliferation of multiple aetiologies, that is, just the sort of relatively unconstrained search for explanations one might expect from a relatively cognitively autonomous religious system. Second, as paradoxical as it may sound, overimitation accounts for ritual change. For overimitation preserved instrumentally opaque ritual technologies in exquisite detail for later, context-sensitive innovation and refinement. Let me dilate briefly upon each of these points in turn.

First, as I said, the causal opacity of Roman ritual helps to explain the Roman penchant for multiple aetiologies. Without authoritative explanations for how opaque rituals *work*, for how their actions and accoutrements caused their effects, Roman inference about ritual causality could remain open-ended. Such inference took the form of origin-stories, moralizing explanations, and naturalizing explanations, all of which may be observed at Plutarch, *Quaestiones Romanae* 93, which I present analytically and in abbreviated form here (Q indicates a question and A_n indicates an answer):

Q: Διὰ τί γυψὶ χρῶνται μάλιστα πρὸς τοὺς οἰωνισμούς;

A_1: πότερον ὅτι καὶ Ῥωμύλῳ δώδεκα γῦπες ἐφάνησαν ἐπὶ τῇ κτίσει τῆς Ῥώμης;

A_2: ἢ ὅτι τῶν ὀρνίθων ἥκιστα συνεχὴς καὶ συνήθης οὗτος; . . . διὸ καὶ σημειώδης ἡ ὄψις αὐτῶν ἐστιν.

A_3: ἢ καὶ τοῦτο παρ᾽ Ἡρακλέους ἔμαθον; . . . ὅτι πάντων μάλιστα γυψὶν ἐπὶ πράξεως ἀρχῇ φανεῖσιν ἔχαιρεν Ἡρακλῆς, ἡγούμενος δικαιότατον εἶναι τὸν γῦπα τῶν σαρκοφάγων πάντων· πρῶτον μὲν γὰρ οὐδενὸς ἅπτεται ζῶντος οὐδ᾽ ἀποκτίννυσιν ἔμψυχον οὐδὲν . . .

A_4: ἀνθρώποις δὲ . . . ἀβλαβέστατός ἐστιν, οὔτε καρπὸν ἀφανίζων οὔτε φυτὸν οὔτε ζῷον ἥμερον κακουργῶν.

A_5: εἰ δ’ ὡς Αἰγύπτιοι μυθολογοῦσι, θῆλυ πᾶν τὸ γένος ἐστὶ καὶ κυΐσκονται δεχόμενοι καταπνέοντα τὸν ἀπηλιώτην . . . καὶ παντάπασιν ἀπλανῆ τὰ σημεῖα καὶ βέβαια γίνεσθαι πιθανόν ἐστιν ἀπ᾽ αὐτῶν.

Q: Why do they (*sc.* Romans) especially use vultures in auspication?

A_1: Is it because twelve vultures appeared to Romulus at the founding of Rome?

A_2: Or because this is the rarest and least everyday of birds? . . . Therefore the sighting of them is significant?

A_3: Or because they learned this from Hercules? . . . Because Hercules rejoiced most of all when vultures appeared at the beginning of an undertaking, supposing the vulture to be the most just of all flesh-eaters; for, first, it touches no living thing and does not kill anything that's alive . . .

A_4: [Or because] it is the most harmless to human beings, neither stealing grains or herbs nor hurting tame animals.

A_5: But if, as in Egyptian myth, their kind is entirely female and they conceive when they receive the blowing east wind . . . it is wholly plausible that the significations that come from them are steady and sure (*sc.* because not made erratic by courtship and sex).

Here, as so often in this text, and as indeed in Ovid, Propertius, and elsewhere, we see in florid form the Roman penchant for multiple explanations. Plutarch reports two origin-stories for the Romans' use of vultures in taking the auspices. First, he reports that they adapted vulture-auspication from Romulus' city-founding augury and, second, that they learned it from Hercules. This latter explanation segues into a moralizing rationalization according to which the vulture was an especially just bird. On a related theory, the vulture is harmless to human beings and therefore, presumably, uniquely suited to aid them in their communications with the gods. Naturalistic accounts, such as that the vulture is very rare and therefore significant to see, blend with the admittedly fanciful, such as the "Egyptian" explanation that vultures, all of them being female, are not made flighty and erratic by courtship and sex and are therefore reliable birds.

Such proliferation of explanations appears to be quintessentially Roman. It is surely significant in this connection that Plutarch's *Quaestiones Graecae* almost entirely lack multiple aetiologies, while his *Quaestiones Romanae* feature very few questions that do not receive multiple answers, among which, as here, Plutarch declines to adjudicate any one as authoritative. Indeed, I would suggest that the Roman habit of multiple explanation has its own explanation, which stands out for its lack of competition. I have in mind the story of the discovery and subsequent burning of Numa's seven books on Roman cult, a story that captured the imagination enough to be told repeatedly.[64] In Livy's version of the tale,[65] a chest was excavated below the Janiculum that contained seven books by Numa *de iure pontificum*, on the pontifical law, alongside seven books of philosophy. When the books were read it was felt that their contents, if widely circulated, would "destroy (*dissolvere*) religious practices." Thus, the books were burned publicly by the *victimarii*, official assistants at sacrifice. The upshot is that Numa's own authoritative explanation of the cult practices, norms, and regulations that he himself had introduced were lost, and purposely so. The story amounts, I submit, to an aetiology, necessarily *post hoc*, designed to explain the Roman ritual system's very *lack* of authoritative explanations.

In this Roman explanation for Roman cult's lack of explanations, we see a chief difference between orthoprax and orthodox religions. It is not that orthodox religions must lack orthoprax performances of causally opaque rituals; rather, what makes orthodox religions orthodox is that they endorse and prescribe authoritative explanations for such rituals. In contrast, orthoprax traditions such as the Romans' decline to do so. Orthopraxy permits (but does not necessitate) polydoxy, and polydoxy expressed itself at Rome under the aspect of multiple aetiology.[66] The mind begs for interpretation of purposive actions that lack transparent instrumental rationale. The opacity of Roman cult, largely composed of just such actions, was a spur to such interpretation, and thus to multiple aetiologies.

Let us turn now to my proposal that overimitation helps to explain ritual change. Its orthopraxy notwithstanding, Roman religion was constantly evolving.[67] In older accounts Roman religion changed due to decadence, decay, the disruptive force of Greek rationalism, ritualism's failure to meet some innate human spiritual need, and so forth. I prefer a perspective from overimitation. In an orthoprax system, the relatively stable norms of current practice open up what we might call a "zone of latent ritual solutions", that is, a *terra ignota* of possible modifications to the ritual technology that are, given the prevailing norms of the ritual system, within reach, just beyond the *terra nota* of established practice. Overimitation, and the orthopraxy it makes possible, by dint of their very conservativeness, *preserve* ritual technologies to be built upon. Faithfully transmitted ritual technologies, however much they are supposed or assumed to remain unchanged, will inevitably suggest possibilities for their own repurposing and refinement, albeit still in accord with the norms of the orthoprax system, in order to address new religious problems. The result is a "ratchet effect" in which new ritual technologies are made possible by, because they are built upon, older ones. The result is a "cumulative" ritual culture of innovations constructed upon existing technologies.[68]

Now, exactly how did Romans modify ritual? Their tolerance for multiple understandings of opaque ritual allowed them a relative freedom to explore within the zone of latent ritual possibilities, simply by dint of reinterpretation. Take, for example, Macrobius on what he terms *permutatio sacrificii*, change of sacrifice, or *emendatio sacrificiorum*, correction of sacrifices.[69] We learn that it had been Apollo's will that the *Compitalia* be celebrated with "heads on behalf of heads," *pro capitibus capita*. Like all Apolline oracles, this recommendation is hermeneutically opaque. The god's intentions, his psychology, hide behind the darkness of his words. King Tarquin took Apollo to mean that human heads should be offered on behalf of human heads, that is, behead some to preserve the heads of many.

Into this scene of ritual norms and ritual practice Brutus stepped with an alternative interpretation. He proposed that "heads" of garlic and poppy be offered in place of (*pro*) human heads. Absent a consensus about the spirit of the ritual prescription, Brutus' reinterpretation followed its letter scrupulously. Orthopraxy was maintained but "head" was now seen in a new perspective, this perspective came to guide practical attitudes, and these attitudes then guided actual practice. In an orthodox tradition, such reinterpretation might well have occasioned a schism. But their orthoprax tradition, which permitted speculation as to the rationale of causally opaque ritual action, allowed Romans to strike out into the zone of latent cognitive and practical possibilities afforded by existing cult norms and thus to modify practice.

We may now sum up these arguments about Roman cult. As we have seen, overimitation contributes to an understanding of the sources of Roman ritual orthopraxy and in so doing, offers new perspectives on the Roman habit of multiple aetiologies as well as Roman ritual change. We considered reasons for supposing that overimitative orthopraxy was driven by *causal* factors. But we also found that cult was primarily *psychologically* causal and only secondarily *physically* causal. That is, physical effects followed upon the psychological effects worked by ritual in the divine mind. Most importantly, we saw that cult was psychologically causal because it was *normative*. Producing effects in the divine mind required that opaque actions stipulated by mutually endorsed norms be imitated with high fidelity. As a system of causally opaque behaviors, Roman cult engendered in cognition a ritual stance whose behavioral fallout was a faithful overimitation that reciprocally worked to sustain cult. In Rome, these overimitated ritual behaviors—orthopraxy itself—sponsored open-ended inference about causes, or multiple aetiologies. Finally, orthopraxy allowed for ritual change both because it preserved ritual technology faithfully for modification and innovation and because it afforded the ritual system a flexibility to respond to the inference and speculation that it itself engendered and refused to curtail.

At the beginning of this chapter, I promised a proposal for psychologists, so here it is. Instead of asking whether overimitation results from the encoding of opaque actions as causally relevant in the *physical* domain, or whether overimitation results from the encoding of opaque actions as *normatively* necessary or, finally, whether overimitation serves an affiliative function, forging connections between individuals and identifying individuals with groups, we ought instead to view overimitation as a response to opaque actions that is at once causal, normative, and affiliative.

Many actions are causal not mechanically but only insofar as they are appropriately normative. Moreover, they can only be recognized as failing or succeeding to meet a normative standard among affiliated individuals, affiliated especially, perhaps, in their sharing of certain norms. For example, a thanking of the gods, a *gratulatio*, exerts a social and psychological force on the gods: they *get* thanked and the assumption is that they appreciate it. But the gods only succeed in getting thanked if mortals follow prescribed practices of *gratulatio* and, moreover, if those getting thanked and doing the thanking are affiliated in their access to, endorsement of, and tendency to recognize the force of certain norms and norm-governed performances.

So, I would suggest that in overimitation we see an inextricable convergence of causal, normative, and affiliative factors, and hence of causal, normative, and affiliative motives on the part of overimitators. That is, in the social world, even or perhaps especially in a social world that embraces gods along with humans, the *normative* aspects of actions performed by *affiliated*, in-group members can be instrumental to their *causality*. The causality in question may not be *physical* but *psychological*. As Roman ritual shows, getting the *norms* of ritual right causes divine pleasure, while getting the norms wrong causes divine displeasure. This causality, psychological though it be, is very real and every bit as consequential as the physical causality involved in opening boxes or making tools.

In summary, then, the Roman evidence allows us to preserve the intuitions that motivate all three modern accounts of overimitation: Roman ritual orthopraxy was *causal* because psychologically (and therefore sometimes physically) causal; it was *normative* because it was only psychologically or physically causal provided that arbitrary, opaque, but conventional actions were imitated with high fidelity; and it was *affiliative* because ritual orthopraxy marked a fellow insider, a person affiliated with the Roman society of gods and mortals by her observation of the *patrius ritus*.

Notes

1 It would be perverse to attempt a comprehensive list of such statements, but see, recently, Dowden 1992: 8; Beard, North, & Price 1998: 1.42; Turcan 2000: 2; Warrior 2006: xv; Rüpke 2007: 86; Ando 2008: 14.

2 E.g., pretending to legislate orthopraxy, Cic. *Leg.* 2.19: *ritus familiae patrumque seruanto*. For ritual repetition due to imperfect performance, Liv. 5.52.9: *recordamini . . . quotiens sacra instaurentur, quia aliquid ex*

patrio ritu neglegentia casuve praetermissum est; cf. Cic. *Har. Resp.* 21. For orthopraxy in prayer, see Plin. *N.H.* 28.3.11 ... *videmusque certis precationibus obsecrasse summos magistratus et, ne quod verborum praetereatur aut praeposterum dicatur, de scripto praeire aliquem*, etc.

3 D. H. *Ant. Rom.* 7.70.3–4: οὐθὲν ἀξιοῖ καινοτομεῖν εἰς αὐτὰ ...: "They (*sc.* Romans) deem no innovations to them (*sc.* rites) should be made ...".

4 Gergely & Csibra 2005 is crucial to the literature on the *opacity* of many socially and culturally transmitted behaviors. See, too, Boyer 2001: 232–233; Whitehouse 2012; Herrmann, Legare, Harris, & Whitehouse 2013; Legare, Wen, Herrmann, & Whitehouse 2015. I draw the distinction between teleological and causal opacity from Csibra & Gergely 2011.

5 See Over & Carpenter 2012 for an overview and assessment of the findings and literature.

6 Lyons, Young, & Keil 2007; Lyons & Keil 2013.

7 E.g., Nielsen 2012; Keupp, Behne, & Rakoczy 2013; Nielsen, Kapitány, & Elkins 2015.

8 E.g., Over & Carpenter 2009; Nielsen & Blank 2011; Watson-Jones *et al.* 2014; Legare *et al.* 2015.

9 Hence the Roman emphasis on *exempla*: see Roller 2004. For a wide-ranging study of Roman imitation, see Germany 2016.

10 Quint. *Inst.* 1.3.1: *ingenii signum in parvis praecipuum memoria est ... proximum imitatio: nam id quoque est docilis naturae.* See Morgan 1998: 240–270.

11 Cf. Fantham 1995: 131 on Quintilian's "recognition of imitation as the method by which writing and all physical crafts are developed—the method which also shapes the first steps in every intellectual discipline."

12 Quint. *Inst.* 10.2.1: *neque enim dubitari potest quin artis pars magna contineatur imitatione. nam ut invenire primum fuit estque praecipuum, sic ea quae bene inventa sunt utile sequi.*

13 Quint. *Inst.* 10.2.2: *rustici probatam experimento culturam in exemplum intuentur.*

14 For the danger that *initatio* of bad models can "infect" (*inficiunt*) the young *mens*, see Quint. *Inst.* 1.11.2: *nec vitia ebrietatis effingat neque servili vernilitate imbuatur nec amoris, avaritiae, metus discat adfectum; quae ... mentem, praecipue in aetate prima teneram adhuc et rudem, inficiunt.*

15 Quint. *Inst.* 1.11.3.

16 Cic. *Off.* 1.118.

17 Durkheim 1897: 113: "la singerie machinale qui nous fait reproduire les mouvements dont nous sommes les témoins" (my translation). Cf. *ibid.* 110: "C'est la singerie pour elle-même." The book from which these quotations come, *Le suicide*, features an entire chapter ("L'imitation," pp. 107–138) devoted to discrediting imitation.

18 On the recent revival of interest in Tarde, and for the debate between Durkheim and Tarde, see Candea 2010. I closely follow the interpretation of Tarde offered by Schmid 2009: 197–214.

19 Tarde 1895: viii: "où il y a un rapport social quelconque entre deux êtres vivants, il y a imitation"; p. 157: "une action à distance d'un esprit sur un autre"; "voulue ou non, passive ou active."

20 So Meltzoff & Moore 1983. But see Heyes 2016.

21 Meltzoff 2013.

22 Term borrowed from Dennett 1989.

23 Carpenter, Akhtar, & Tomasello 1998.

24 Gergely, Bekkering, & Király 2002, responding to Meltzoff 1988. But see Beisert *et al.* 2012.

25 Meltzoff 1995.

26 A coinage of Lyons, Young, & Keil 2007, where the assumption that informs the value-laden "over-" is that fidelity in imitating causally irrelevant actions results from a *distortion* in causal learning processes.

27 Lyons, Young, & Keil 2007. Cf. Lyons *et al.* 2011; Lyons & Keil 2013, and cf. especially Nielsen & Tomaselli 2010, recording experiments with two- to thirteen-year-old Bushmen children.

28 Rakoczy & Schmidt 2013.

29 Herrmann, Legare, Harris, & Whitehouse 2013.

30 Keupp, Behne, & Rakoczy 2013. Cf. Keupp *et al.* 2015.

31 Nielsen & Blank 2011.

32 Nielsen, Moore, & Mohamedally 2012.

33 McGuigan, Makinson, & Whiten 2011; Flynn & Smith 2012; Whiten *et al.* 2016. Relevant, too, is Gergely & Csibra 2005.

34 Nielsen & Tomaselli 2010; Nielsen *et al.* 2014; Berl & Hewlett 2015; Nielsen *et al.* 2016; Klinger, Mayor, & Bannard 2016; Corriveau *et al.* 2017.

35 Tomasello 1996.

36 See, e.g., Whiten *et al.* 2009.

37 For the role of overimitation in language and communication, see Subiaul *et al.* 2016; Klinger, Mayor, & Bannard 2016; Corriveau *et al.* 2017.
38 See, with the discussion of Gergely & Csibra 2005, Boyer 2001: 232–233; Whitehouse 2012; Herrmann, Legare, Harris, & Whitehouse 2013; Legare, Wen, Herrmann, & Whitehouse 2015.
39 See Csibra & Gergely 2011 for the distinction between teleological and causal opacity.
40 I discuss this scene as a context of cultural learning at some length in Mackey (forthcoming).
41 Rüpke 2007: 87–90.
42 Plin. *H.N.* 28.5.25: *in adorando dextram ad osculum referimus totumque corpus circumagimus.* On this passage, see Corbeill 2004: 28–29.
43 Ando 2008: 14.
44 Plin. *N.H.* 28.3.11.
45 Plin. *N.H.* 28.5.25: *etiam mutas religiones pollere manifestum est.*
46 See Versnel 1991, on the utility of continuing to employ, if advisedly, the venerable religion-magic distinction.
47 On this ritual, see Versnel 2002, with full references.
48 Plin. *N.H.* 2.140: *sacris quibusdam et precationibus vel cogi fulmina vel impetrari.*
49 Plin. *N.H.* 2.140: *tradit L. Piso . . . quod imitatum parum rite Tullum Hostilium ictum fulmine.*
50 Liv. 1.31.8: *sed non rite initum aut curatum id sacrum esse . . . ira Iovis sollicitati prava religione fulmine ictum cum domo conflagrasse.*
51 See Cohee 1994.
52 Liv. 5.52.9: *recordamini . . . quotiens sacra instaurentur, quia aliquid ex patrio ritu neglegentia casuve praetermissum est.*
53 On the background of this speech before the *collegium pontificum*, and its participation in religious, political, and other discourses, see recently Beard 2012.
54 Cic. *Har. Resp.* 21.
55 Cic. *Har. Resp.* 23.
56 Cic. *Har. Resp.* 23: *errata expiantur et mentes deorum immortalium ludorum instauratione placantur.*
57 Cic. *Pis.* 1.1.
58 Cf. Cic. *De Or.* 3.221: *imago animi vultus, indices oculi: nam haec est una pars corporis, quae, quot animi motus sunt, tot significationes et commutationes possit efficere.* Cf., e.g., *De Or.* 3.222.
59 Cf. Fantham 2004: 296. See Fögen 2009b on the universal "language" of gesture, vultus, and non-verbal vocalization in Roman thought.
60 Cic. *De Or.* 3.223: *verba enim neminem movent nisi eum, qui eiusdem linguae societate coniunctus est.*
61 Shwe & Markman 1997.
62 On the language of *auspicia* in particular, see Linderski 1986 and 2006.
63 Cicero speaks of the *deorum et hominum communitas et societas inter ipsos* (*Off.* 1.153).
64 Varr. *Curio de cultu deorum* fr. 3 Cardauns (= Aug. *Civ.* 7.34–35); Val. Max. 1.1.12; Plin. *N.H.* 13.84–87; Plu. *Num.* 22; Lact. *Div. Inst.* 1.22; Aur. Vict. *Vir. ill.* 3.2.
65 Liv. 40.29.3–14.
66 I borrow and adapt this term from Reines 1987 and Keller & Schneider 2011.
67 For an account of this process, see, most recently and thoroughly, Rüpke 2012 and, most succinctly and pungently, Ando 2015: 72–81.
68 For the ratchet effect and cumulative culture, see Tennie, Call, & Tomasello 2009.
69 Macr. *Sat.* 1.7.34–36, discussed with respect to its significance for ritual change by Ando 2015: 74–75.

References

Psychology

Beisert, M., N. Zmyj, R. Liepelt, F. Jung, W. Prinz, & M. M. Daum (2012) "Rethinking 'Rational Imitation' in 14-Month-Old Infants: A Perceptual Distraction Approach," *PLoS ONE*, 7.3: e32563.

Berl, R. E. W. & B. S. Hewlett (2015) "Cultural Variation in the Use of Overimitation by the Aka and Ngandu of the Congo Basin," *PLoS ONE*, *10*.3: e0120180.

Boyd, R., P. J. Richerson, & J. Henrich (2011) "The Cultural Niche: Why Social Learning Is Essential for Human Adaptation," *Proceedings of the National Academy of Sciences of the United States of America*, 108, 10918–10925.

Buchsbaum, D., A. Gopnik, T. L. Griffiths, & P. Shafto (2011) "Children's Imitation of Causal Action Sequences Is Influenced by Statistical and Pedagogical Evidence," *Cognition*, 120, 331–340.

Callaghan, T., *et al.* (2011) *Early Social Cognition in Three Cultural Contexts*. Boston, MA: Wiley-Blackwell.

Carpenter, M. (2006) "Instrumental, Social, and Shared Goals and Intentions in Imitation," in S. J. Rogers & J. H. G. Williams (eds.) *Imitation and the Social Mind: Autism and Typical Development*. New York: The Guilford Press: 48–70.

Carpenter, M., J. Call, & M. Tomasello (2005) "Twelve- and 18-Month-Olds Copy Actions in Terms of Goals," *Developmental Science*, 8: F13–F20.

Carpenter, M., M. Tomasello, & T. Striano (2005) "Role Reversal Imitation and Language in Typically-Developing Infants and Children with Autism," *Infancy*, 8: 253–278.

Clay, Z., H. Over, & C. Tennie (2018). "What Drives Young Children to Over-Imitate? Investigating the Effects of Age, Context, Action Type, and Transitivity," *Journal of Experimental Child Psychology, 166*: 520–534.

Corriveau, K. H., C. J. DiYanni, J. M. Clegg, G. Min, J. Chin, & J. Nasrini (2017) "Cultural Differences in the Imitation and Transmission of Inefficient Actions," *Journal of Experimental Child Psychology, 161*: 1–18.

Dautenhahn, K. & C. Nehaniv (2002) "The Agent-Based Perspective on Imitation," in K. Dautenhahn & C. Nehaniv (eds.) *Imitation in Animals and Artifacts*. Cambridge, MA: MIT Press: 1–40.

Flynn, E. & K. Smith (2012) "Investigating the Mechanisms of Cultural Acquisition: How Pervasive is Overimitation in Adults?," *Social Psychology, 43*.4: 185–195.

Gergely, G. (2006) "Sylvia's Recipe: The Role of Imitation and Pedagogy in the Transmission of Cultural Knowledge," in N. J. Enfield & S. C. Levenson (eds.) *Roots of Human Sociality: Culture, Cognition, and Human Interaction*. Oxford: Berg: 229–255.

Gergely, G. & G. Csibra (2005) "The Social Construction of the Cultural Mind: Imitative Learning as a Mechanism of Human Pedagogy," *Interaction Studies, 6*.3: 463–448.

Gergely, G., H. Bekkering, & I. Király (2002) "Rational Imitation in Preverbal Infants," *Nature, 415*: 755.

Graf, F., *et al.* (2014) "Imitative Learning of Nso and German Infants at 6 and 9 Months of Age: Evidence for a Cross-Cultural Learning Tool," *Journal of Cross-Cultural Psychology, 45*.1: 47–61.

Henrich, J., S. J. Heine, & A. Norenzayan (2010) "The Weirdest People in the World?," *Behavioral and Brain Sciences, 33*: 61–135.

Herrmann, P. A., C. H. Legare, P. L. Harris, & H. Whitehouse (2013) "Stick to the Script: The Effect of Witnessing Multiple Actors on Children's Imitation," *Cognition, 129*: 536–543.

Heyes, C. (2016). "Imitation: Not in Our Genes," *Current Biology, 26*.10: R412–R414.

Hoehl, S., M. Zettersten, H. Schleihauf, S. Gratz, & S. Pauen (2014) "The Role of Social Interaction and Pedagogical Cues for Eliciting and Reducing Overimitation in Preschoolers," *Journal of Experimental Child Psychology, 122*: 122–133.

Kenward, B. (2012) "Over-Imitating Preschoolers Believe Unnecessary Actions Are Normative and Enforce Their Performance by a Third Party," *Journal of Experimental Child Psychology, 112*: 195–207.

Kenward, B., M. Karlsson, & J. Persson (2011) "Over-imitation Is Better Explained by Norm Learning Than by Distorted Causal Learning," *Proceedings of the Royal Society B: Biological Sciences, 278*: 1239–1246.

Keupp, S., T. Behne, & H. Rakoczy (2013) "Why Do Children Overimitate? Normativity Is Crucial," *Journal of Experimental Child Psychology, 116*: 392–406.

Keupp, S., T. Behne, J. Zachow, A. Kasbohm, & H. Rakoczy (2015) "Over-Imitation Is Not Automatic: Context Sensitivity in Children's Overimitation and Action Interpretation of Causally Irrelevant Actions," *Journal of Experimental Child Psychology, 130*: 163–175.

Klinger, J., J. Mayor, & C. Bannard (2016). "Children's Faithfulness in Imitating Language Use Varies Cross-Culturally, Contingent on Prior Experience," *Child Development, 87*.3: 820–833.

Legare, C. & P. L. Harris (2016) "The Ontogeny of Cultural Learning," *Child Development, 87*: 633–642.

Legare, C., N. J. Wen, P. A. Herrmann, & H. Whitehouse (2015). "Imitative Flexibility and the Development of Cultural Learning," *Cognition, 142*: 351–361.

Lyons, D. E. & F. C. Keil (2013) "Overimitation and the Development of Causal Understanding," in M. R. Banaji & S. A. Gelman (eds.) *Navigating the Social World*. Oxford: Berg: 145–149.

Lyons, D. E., A. G. Young, & F. C. Keil (2007) "The Hidden Structure of Overimitation," *Proceedings of the National Academy of the Sciences, 104*.50: 19751–19756.

Lyons, D. E., D. H. Damrosch, J. K. Lin, D. M. Macris, & F. C. Keil (2011) "The Scope and Limits of Overimitation in the Transmission of Artefact Culture," *Philosophical Transactions of the Royal Society B: Biological Sciences, 366*: 11580–1167.

Matheson, H., C. Moore, & N. Akhtar (2013) "The Development of Social Learning in Interactive and Observational Contexts," *Journal of Experimental Child Psychology, 114*: 161–172.

McGuigan, N. & A. Whiten (2009) "Emulation and "Overemulation" in the Social Learning of Causally Opaque Versus Causally Transparent Tool Use by 23- and 30-Month-Olds," *Journal of Experimental Child Psychology, 104*: 367–381.

McGuigan, N., J. Makinson, & A. Whiten (2011) "From Over-Imitation to Super-Copying: Adults Imitate Causally Irrelevant Aspects of Tool Use with Higher Fidelity Than Young Children," *British Journal of Psychology, 102*: 1–18.

McGuigan, N., A. Whiten, E. Flynn, & V. Horner (2007) "Imitation of Causally Opaque Versus Causally Transparent Tool Use by 3- and 5-Year-Old Children," *Cognitive Development, 22*: 353–364.

Meltzoff, A. N. (1988) "Infant Imitation After a 1-week Delay: Long-Term Memory for Novel Acts and Multiple Stimuli," *Developmental Psychology, 24*: 470–476.

Meltzoff, A. N. (1995) "Understanding the Intentions of Others: Re-Enactment of Intended Acts by 18-Month-Old Children," *Developmental Psychology, 31*: 838–50.

Meltzoff, A. N. (2013) "Origins of Social Cognition: Bidirectional Self-Other Mapping and the 'Like-Me' Hypothesis." In M. R. Banaji & S. A. Gelman (eds.) *Navigating the Social World: What Infants, Children, and Other Species Can Teach Us*. Oxford: Oxford University Press: 139–144.

Meltzoff, A. N. & M. K. Moore (1983) "Newborn Infants Imitate Adult Facial Gestures," *Child Development, 54*: 702–709.

Moraru, C., Gomez, J., & McGuigan, N. (2016) "Developmental Changes in the Influence of Conventional and Instrumental Cues on Over-Imitation in 3-to 6-Year-Old Children," *Journal of Experimental Child Psychology, 145*: 34–47.

Nielsen, M. (2012) "Imitation, Pretend Play, and Childhood: Essential Elements in the Evolution of Human Culture?," *Journal of Comparative Psychology, 126*: 170–181.

Nielsen, M. & V. Slaughter (2007) "Multiple Motivations for Imitation in Infancy," in K. Dautenhahn & C. L. Nehaniv (eds.) *Imitation and Social Learning in Robots, Humans and Animals: Behavioural, Social and Communicative Dimensions*. Cambridge, UK: Cambridge University Press: 343–360.

Nielsen, M. & E. W. E. Susianto (2010) "Failure to Find Over-Imitation in Captive Orangutans (Pongo Pygmaeus): Implications for Our Understanding of Cross-Generation Information Transfer," in Johan Håkansson (ed.) *Developmental Psychology*. New York: Nova Science Publishers: 153–167.

Nielsen, M. & K. Tomaselli (2010) "Over-Imitation in Kalahari Bushman Children and the Origins of Human Cultural Cognition," *Psychological Science, 21*: 729–736.

Nielsen, M. & C. Blank (2011) "Imitation in Young Children: When Who Gets Copied Is More Important Than What Gets Copied," *Developmental Psychology, 47*.4: 1050–1053.

Nielsen, M., R. Kapitány, & R. Elkins (2015) "The Perpetuation of Ritualistic Actions as Revealed by Young Children's Transmission of Normative Behavior," *Evolution and Human Behavior, 36*.3: 191–198.

Nielsen, M., C. Moore, & J. Mohamedally (2012) "Young Children Overimitate in Third-Party Contexts," *Journal of Experimental Child Psychology, 112*: 73–83.

Nielsen, M., I. Mushin, K. Tomaselli, & A. Whiten (2014) "Where Culture Takes Hold: Overimitation and Its Flexible Deployment in Western, Aboriginal, and Bushmen Children," *Child Development, 85*.6: 2169–2184.

Nielsen, M., I. Mushin, K. Tomaselli, & A. Whiten (2016) "Imitation, Collaboration, and Their Interaction Among Western and Indigenous Australian Preschool Children," *Child Development, 87*: 795–806.

Nielsen, M., F. Subiaul, A. Whiten, B. Galef, & T. Zentall (2012) "Social Learning in Humans and Non-Human Animals: Theoretical and Empirical Dissections," *Journal of Comparative Psychology, 126*: 109–113.

Over, H., & M. Carpenter (2009) "Priming Third-Party Ostracism Increases Affiliative Imitation in Children," *Developmental Science, 12*.3: F1–F8.

Over, H., & M. Carpenter (2012) "Putting the Social into Social Learning: Explaining Both Selectivity and Fidelity in Children's Copying Behavior," *Journal of Comparative Psychology, 126*: 182–192.

Rakoczy, H. & M. F. H. Schmidt (2013) "The Early Ontogeny of Social Norms," *Child Development Perspectives*, 7.1: 17–21.

Rakoczy, H., F. Warneken, & M. Tomasello (2008) "The Sources of Normativity: Young Children's Awareness of the Normative Structure of Games," *Developmental Psychology, 44*: 875–881.

Rogoff, B., R. Paradise, R. Mejía Arauz, M. Correa-Chávez, & C. Angelillo (2003) "Firsthand Learning through Intent Participation," *Annual Review of Psychology, 54*: 175–203.

Shwe, H. I. & E. M. Markman (1997) "Young Children's Appreciation of the Mental Impact of Their Communicative Signals," *Developmental Psychology, 33.4*: 630–636.
Sommerville, J. A. & A. J. Hammond (2007) "Treating Another's Actions as One's Own: Children's Memory of and Learning From Joint Activity," *Developmental Psychology, 43.4*: 1003–1018.
Subiaul, F., K. Winters, K. Krumpak, & C. Core (2016) "Vocal Overimitation in Preschool-Age Children," *Journal of Experimental Child Psychology, 141*: 145–160.
Tennie, C., J. Call & M. Tomasello (2009) "Ratcheting Up the Ratchet: On the Evolution of Cumulative Culture," *Philosophical Transactions of the Royal Society B: Biological Sciences, 364*: 2405–2415.
Tomasello, M. (1999) *The Cultural Origins of Human Cognition*. Cambridge, MA: Harvard University Press.
Tomasello, M. (2008) *The Origins of Communication*. Cambridge, MA: MIT Press.
Tomasello, M. (2009) *Why We Cooperate*. Cambridge, MA: MIT Press.
Tomasello, M. & M. Carpenter (2005) "Intention Reading and Imitative Learning," in S. Hurley and N. Chater (eds.) *Perspectives on Imitation: From Neuroscience to Social Science, Vol. 2: Imitation, Human Development, and Culture*. Cambridge, MA: MIT Press: 133–148.
Tomasello, M., M. Carpenter, J. Call, T. Behne, & H. Moll (2005) "Understanding and Sharing Intentions: The Origins of Cultural Cognition," *Behavioral and Brain Sciences, 28*: 675–735.
Vivanti, G., D. R. Hocking, P. Fanning, & C. Dissanayake (2017). "The Social Nature of Overimitation: Insights from Autism and Williams Syndrome," *Cognition, 161*: 10–18.
Watson-Jones, R. E., C. H. Legare, H. Whitehouse, & J. M. Clegg (2014) "Task-Specific Effects of Ostracism on Imitative Fidelity in Early Childhood," *Evolution & Human Behavior, 35*: 204–210.
Whitehouse, H. (2012) "Ritual, Cognition, and Evolution," in R. Sun (ed.) *Grounding Social Sciences in Cognitive Sciences*. Cambridge, MA: MIT Press: 265–284.
Whiten, A., N. McGuigan, S. Marshall-Pescini, & L. M. Hopper (2009) "Emulation, Imitation, Over-Imitation and the Scope of Culture for Child and Chimpanzee," *Philosophical Transactions of the Royal Society B: Biological Sciences, 364*: 2417–2428.
Whiten, A., Allan, G., Devlin, S., Kseib, N., Raw, N., & McGuigan, N. (2016) "Social Learning in the Real-World: 'Over-Imitation' Occurs in Both Children and Adults Unaware of Participation in an Experiment and Independently of Social Interaction," *PLoS One, 11.7*: e0159920.

Roman studies

Ando, C. (2008) *The Matter of the Gods: Religion and the Roman Empire*. Berkeley, CA: University of California Press.
Ando, C. (2015) *Roman Social Imaginaries: Language and Thought in Contexts of Empire*. Toronto: University of Toronto Press.
Beard, M. (2012) "Cicero's 'Response of the Haruspices' and the Voice of the Gods," *JRS, 102*: 20–39.
Bloomer, W. M. (2011) "Quintilian on the Child as a Learning Subject," *CW, 105.1*: 109–137.
Bremmer, J. N. (1995) "The family and Other Centres of Religious Learning in Antiquity," in J. W. Drijvers & A. A. MacDonald (eds.) *Centres of Learning: Learning and Location in Pre-Modern Europe and the Near East*. Leiden: Brill: 29–38.
Cohee, P. (1994) "Instauratio Sacrorum," *Hermes, 122.4*: 451–468.
Corbeill, A. (2004) *Nature Embodied*. Princeton, NJ: Princeton University Press.
Fantham, E. (1995) "The Concept of Nature and Human Nature in Quintilian's Psychology and Theory of Instruction," *Rhetorica: A Journal of the History of Rhetoric, 13.2*: 125–136.
Germany, R. (2016) *Mimetic Contagion*. Oxford: Oxford University Press.
Katajala-Peltomaa, S. & V. Vuolanto (2011) "Children and Agency: Religion as Socialisation in Late Antiquity and the Late Medieval West," *Childhood in the Past, 4.1*: 79–99.
Liebeschuetz, J. H. W. G. (1979) *Continuity and Change in Roman Religion*. Oxford: Oxford University Press.
Linderski, J. (1986) "The Augural Law," *ANRW II, 16.3*: 2146–2312.
Linderski, J. (2006) "Founding the City," in S. B. Faris & L. E. Lundeen (eds.) *Ten Years of the Agnes Kirsopp Michels Lectures at Bryn Mawr College*. Bryn Mawr: Bryn Mawr College: 88–107.
Mackey, J. L. (2016) "Children as Religious Agents," in C. Laes & V. Vuolanto (eds.) *Children and Everyday Life in the Roman and Late Antique World*. London: Routledge.
Mackey, J. L. (in proofs) "Developmental Psychologies in the Roman World: Change and Continutity," *History of Psychology*. Berlin: Walter De Gruyter.

Mackey, J. L. (forthcoming) *Belief & Cult: From Intuitions to Institutions in Roman Religion*. Princeton, NJ: Princeton University Press.
Mantle, I. C. (2002) "The Roles of Children in Roman Religion," *G&R*, *49*.1: 85–106.
McWilliam, J. (2014) "The Socialization of Roman Children," in J. E. Grubbs, T. Parkin, & R. Bell (eds.) *The Oxford Handbook of Childhood and Education in the Classical World*. Oxford: Oxford University Press: 264–285.
Morgan, T. (1998) *Literate Education in the Hellenistic and Roman Worlds*. Cambridge: Cambridge University Press.
North, J. A. (1976) "Conservatism and Change in Roman Religion," *PBSR*, *44*: 1–12.
Osgood, J. (2011) "Making Romans in the Family," in M. Peachin (ed.) *The Oxford Handbook of Social Relations in the Roman World*. Oxford: Oxford University Press: 69–83.
Prescendi, F. (2007) *Décrire et comprendre le sacrifice: Les réflexions des Romains sur leur proper religion à partir de la littérature antiquaire*. Stuttgart: Steiner.
Prescendi, F. (2010) "Children and the Transmission of Religious Knowledge," in V. Dasen & T. Spath (eds.) *Children, Memory, and Family Identity in Roman Culture*. Oxford: Oxford University Press: 73–93.
Rawson, B. (2003) *Children and Childhood in Roman Italy*. Oxford: Oxford University Press.
Roller, M. B. (2004) "Exemplarity in Roman Culture: The Cases of Horatius Cocles and Cloelia," *CP*, *99*.1: 1–56.
Rüpke, J. (2007) *Religion of the Romans* (trans. and ed. R. Gordon). Cambridge: Polity Press.
Rüpke, J. (2012) *Religion in Republican Rome: Rationalization and Ritual Change*. Philadelphia, PA: University of Pennsylvania Press.
Scheid, J. (1987–1989) "La parole des dieux: L'originalité du dialogue des Romains avec leurs dieux," *Opus*, 6–8: 125–136.
Scheid, J. (2007) "Les Sens des Rites. L'Exemple Romain," in J. Scheid (ed.) *Rites et Croyances dans les Religions du Monde Romain*. Genève: Fondation Hardt: 39–63.
Tulloch, J. (2012) "Visual Representations of Children and Ritual in the Early Roman Empire," *Studies in Religion / Sciences Religieuses*, *41*.3: 408–438.
Versnel, H. S. (1991) "Some Reflections on the Relationship Magic-Religion," *Numen*, *38*.2: 177–197.
Versnel, H. S. (2002) "The Poetics of the Magical Charm: An Essay in the Power of Words," in P. Mirecki & M. Meyer (eds.) *Magic and Ritual in the Ancient World*. Leiden: Brill: 105–158.

Other

Keller, C. & L. C. Schneider (eds.) (2011) *Polydoxy: Theology of Multiplicity and Relation*. London: Routledge.
Reines, A. J. (1987) *Polydoxy: Explorations in a Philosophy of Religion*. Amherst, NY: Prometheus Books.

16

Theory of mind from Athens to Augustine

Divine omniscience and the fear of God

Paul C. Dilley

Theory of mind, a key concept in contemporary cognitive science, hinges on how we infer the mental activity of others, human or divine, and reflexively, how they infer ours. In this chapter, I briefly discuss an important aspect of theory of mind and its behavioral implications: the knowledge of divine agents, and sometimes gifted human ones, about the private actions and thoughts of others; and the fear of divine punishment based on this knowledge. Although ancient theories of mind were diverse, several trends are discernable across Mediterranean societies, including from Greek and Roman sources: first, that exceptional knowledge is often attributed to divine agents, who in some contexts, such as oracular delivery, are even credited with omniscience; but only in the later Christian context was divine knowledge of private thoughts emphasized, through cultivation of the fear of God. This was in turn related to an increased moral concern over mental purity, in which the entertaining of evil thoughts, even without acting upon them, led to divine retribution.[1]

Theories of mind

Within developmental psychology, much of the research over the past several decades is based on Wimmer and Perner's "false belief" test, in which children are interviewed to determine whether they understand that other people can possess opinions that differ from their own, for example inaccurate knowledge about the location of chocolate which has been moved.[2] These experiments concluded that, around the age of four, children begin to recognize that others have particular mental states to which they and others have no direct access, but which can be inferred. More generally, philosopher Alvin Goldman defines theory of mind as "the cognitive capacity to attribute mental states to self and others," in which "mental states" include "perceptions, bodily feelings, emotional states, and propositional attitudes (beliefs, desires, hopes, and intentions)."[3]

What can theory of mind tell us about the ancient world? While cognitive science proceeds from the assumption that humans have the universal capacity to infer mental states, cognitive anthropologists have observed that these mental states are conceived in culturally diverse ways. For example, drawing on a substantial body of ethnographic work, anthropologist Tanya Luhrmann has proposed six different theories of mind which can be identified across cultures:

the Euro-American secular theory of mind; the Euro-American modern supernaturalist theory of mind; the opacity theory of mind; the transparency of language theory; the mind-control theory; and perspectivism.[4]

According to the Euro-American secular view,

> Entities in the world, supernatural or otherwise, do not enter the mind, and thoughts do not leave the mind to act upon the world . . . At the same time, what is held in the interior of the mind is causally important. Intentions and emotions are powerful and can even make someone ill.

This strongly bounded view of the mind is modified in the Euro-American supernaturalist theory as follows:

> The mind-world boundary becomes permeable for God, or for the dead person, or for specific "energies" that are treated as having causal power and, usually, their own energy. The individual learns to identify these supernatural presences, often through implicit or even explicit training.[5]

Luhrmann's distinction between the "secular" and "supernaturalist" types in the modern West implies that various theories of mind can co-exist within a single culture or political sphere, a situation which likely attained in the ancient Mediterranean and Graeco-Roman world; and that a developmentally acquired theory of mind can be modified through training, for which there is ample evidence in at least one ancient Mediterranean group, namely early Christian monasteries.

Indeed, Luhrmann's typology can be extended to describe other theories of mind, whether in ancient or modern cultures, by considering alternative permutations of what she calls the "dimensions of mind": whether the mind is perceived as "bounded" or "porous"; the significance attached to "interiority" – that is, "are emotions and thoughts understood to be causally powerful and significant?"; the "epistemic stance" – namely, the status of imagination as either a privileged path to reality or simply a transient figment; the "sensorial weighting," for instance the particular emphasis given to sight in the modern West; and "relational access" to the thoughts of others, including "relational responsibility" to act upon this knowledge.[6]

It is impossible to reconstruct a full theory of mind with all these dimensions in any single culture of the ancient Mediterranean, Greek, Roman, or otherwise.[7] Evidence is limited; what is available stems from widely differing regions and times, and takes varied forms. In this chapter, I focus on a single dimension of theory of mind, "relational access," which is perhaps the most broadly attested dimension in ancient sources, and track several developmental trends through Late Antiquity. I am thus providing a diachronic study of a single aspect of theory of mind, rather than a full synchronic thick description of a single culture or group. In particular, I am interested in exploring divine omniscience, and to a lesser extent the enhanced knowledge of special humans, as an important arena for cultural diversity within the universal human capacity for theory of mind; and how divine omniscience was related to punishment for breaking ethical norms or social contracts, which may be unknown to humans, but not to the gods.

Toward divine omniscience and punishment

Divine omniscience is rarely invoked explicitly in Greek literature. Even Zeus is said to have been tricked by Prometheus in a text so well known as the etiology of animal sacrifice in

Hesiod's *Works and Days* 548–558.[8] This less-than-perfect divine knowledge is explored by anthropologist Pascal Boyer in his cognitive theory of religion, in which he argues that the attribution of "situational knowledge" – that is, information related to the current situation of the worshipper – is far more common than omniscience. Situational knowledge is superhuman, yet it emphasizes information relevant to the particular situation of the ritual agent.[9] Jennifer Larson draws on Boyer's concept of "situational knowledge" in her analysis of oracular deities in ancient Greece, noting that "there was variation among individual petitioners in the amount of information each deems it necessary to provide as context."[10] The most explicit assertion of divine omniscience is Croesus of Lydia's question to multiple oracles concerning his current activities, which only Apollo answers suitably, prefacing his response with "I know the number of the sands and the dimensions of the sea."[11]

The potential of divine omniscience for enforcing societal codes is explicitly asserted in Greek literature in a surviving fragment of the satyr-play *Sisyphos*, by Kritias, which suggested that humans first invented laws to eliminate violence and crime; when people violated these laws in secret, the idea of an all-seeing God who punishes hidden infractions was invented.[12] On the other hand, Zeus and other gods are frequently invoked to enforce oaths, so that their knowledge of infractions and willingness to punish them are assumed. As Raffaelle Pettazzoni argues in a classic if largely overlooked study, *The All-Knowing God*, the concept of divine omniscience is widely attested in the ancient Mediterranean through the all-seeing eye of sky gods, which he attributes to anthropomorphism; these gods are usually associated with punishment, and often named in oaths.[13]

Within the Cognitive Science of Religion, several universal human mental features have been invoked to explain this widespread preoccupation with the divine judge. Justin Barrett has proposed that humans are innately equipped with a "hyperactive agent detection device" (HADD), a cognitive module leading them to assume that gods are watching them and serving as witnesses to moral or immoral behavior.[14] More specifically, concern about *post-mortem* judgment has also been explained by an appeal to adapted human cognitive architecture: Istvan Czachesz argues that people are predisposed to believing in the continuance of cognition and emotion after death.[15] Whatever the accuracy of these universalizing claims, they simply point to innate tendencies, while drawing on the controversial "just-so" stories of evolutionary psychology, which often appeal to diachronic and reductionist explanations.[16] They fail to address how beliefs in divine omniscience and post-mortem cognition are acquired and maintained in specific cultural environments, whether ancient or modern.[17]

The fear of gods as agents of divine judgment is evident in the so-called "confession inscriptions" of Asia Minor, which Angelos Chaniotis identifies as one of several "epigraphic memorials of divine justice," noting that

> all these texts were inscribed in sanctuaries as manifestations of the belief in punishment by vengeful gods for human misdemeanors. Their emotive effect on those who read them – or listened to the texts being read aloud – must have been awe and fear.[18]

One particular form of divine omniscience and punishment seems to have been difficult for some to accept: namely that God, and even some especially pious individuals, could see private thoughts; furthermore, evil thoughts that were not properly engaged subjected the wayward thinker to divine punishment. The novelty of this concept is nowhere better illustrated than in a comment made by Constantine to an assembly of bishops: "For previously there were in me things which seemed lacking in justice, nor did I think that a higher power saw anything which I carried inside, secreted away in my heart."[19] Thus the idea of God monitoring and assessing his thoughts struck the emperor as surprising. Not that this idea was completely unprecedented: oracular petitioners might expect the gods to read their thoughts;[20] and a number of passages in the Hebrew Bible suggest that God was aware of inner thoughts and dispositions.[21] At the same time, other biblical verses suggest a degree of skepticism about this very claim.[22]

The fear of God: private thoughts and divine punishment

Constantine was one of many in Late Antiquity whose embrace of Christianity – at whatever pace it occurred – included the training and incorporation of new ideas, including alternative theories of mind. There is substantial evidence for this process in the literature of early Christian monasticism, especially communal monasticism (cenobitism), where new disciples were subject to extensive training in the fear of God.[23] Large-scale monastic groups rose to prominence in fourth-century Egypt, under the leadership of figures such as Pachomius, who led the Koinonia, a federation of monasteries in Upper (southern) Egypt. Pachomius and his successors Theodore and Horsiesius wrote in Coptic, but their writings were later translated into Greek and Latin; many elements of Egyptian monasticism are reflected in the works of Jerome and Augustine.

The fear of God is perhaps best described as a practice of imagining, especially through visualization and audition, the shame and guilt felt by sinners before the divine tribunal, with all of their hidden deeds, including private sinful thoughts, exposed to others. Through constant meditation on this last judgment, as well as undergoing related forms of discipline, such as public shaming and corporal punishment, monks acquired the "fear of God" as an enduring disposition, to be drawn upon especially in moments of cognitive distress. In particular, tempting thoughts and affective urges could be tempered and ultimately rejected by harnessing the fear of God in the active imagination of post-mortem condemnation.

In some cases the fear of God might be described as a feeling of utter helplessness in the presence of the terrible divine majesty. But it is a mistake to reduce this fear to a particularly "religious" emotion, such as Rudolf Otto's famous concept of *mysterium tremendum*.[24] Christian, and especially monastic, authors describe the fear of God not as an involuntary and unprovoked reaction, but as a personal attitude that must be acquired and maintained. Once acquired, it motivates disciples to obey the commandments, as noted by Horsiesius, the third leader of Pachomius's community: "Fear God and guard his commandments, for this is [incumbent on] everyone, because God will bring every deed to judgement."[25] Thus, even if monastic directors overlooked breaches in discipline, God would reveal and punish all offenses at the last judgment.

Evocative descriptions of the post-mortem punishment of sinners were a central feature of the monastic care of souls. When Theodore, a prominent disciple of Pachomius,

> saw someone unwilling to take care over his own life and showing disregard, he warned him with great patience about the fearful judgments of God. For it is terrifying to fall into the hands of the living God (Heb. 10:31).[26]

Lengthy *ekphrases* of this scene are found in Pachomian homilies, such as the following:

> Guard your childhood so that you will be able to protect your old age, lest you experience shame and regret in the valley of Josaphat, while God's entire creation watches you and reproaches you, saying: "We thought that you were a sheep, we have found you in this place a wolf. Walk, now, into the gulf of Hell; cast yourself now into the heart of the earth!" Oh, the enormity! You walked in the world, being praised as an elect; in the hour you arrived at the valley of Josaphat, the place of judgement, you were found naked,

> while everyone gazes upon your sins and your unseemliness, which is revealed to God and humans. Woe to you in that moment! Where will you turn your face? Will you open your mouth? What will you say? Your sins are etched in your soul, which is as dark as a hairshirt. What will you do in that moment? Are you weeping? Your tears will not be accepted. Are you praying? Your prayers will not be accepted, because those to whom you were given have no mercy . . .

Disciples who were unaccustomed to ideas of divine omniscience, or for whom this idea was not salient, became accustomed to the fear of God through constantly listening to similar images, discussing them, and thinking about them.

In descriptions of the divine judgment scene, monks are encouraged to imagine themselves under God's scrutinizing gaze, as well as the judgmental eyes of monastic colleagues or family members. As in classical models, shame derives from being seen by others.[27] The fear of God, however, extends the notion of what is shameful from sins committed in public to secret acts and even to private thoughts, despite their invisibility to others. This shift in perception is accomplished through a simple analogy: monks are told to imagine the divine eye watching their every move and piercing through to their inner thoughts and feelings, especially in situations of temptation. God's privileged view will be extended to others at the day of judgment, when every sin will be revealed to all.

In an extended discussion on disciplining the lustful gaze in his monastic *Rule*, Augustine reminds his disciples that they are being constantly observed by God: "But look, even if he goes undetected and is not seen by any person, what will he do about that inspector from above, from whom nothing can be concealed?"[28] The monastic voyeur should thus consider that his own gaze is under God's surveillance, trumped by the divine panopticon. In a sermon, Augustine further encouraged his audience to imagine God's all-seeing gaze piercing through the intimate confines of houses and bedrooms:

> He must be feared in public, he must be feared in private. You go out, you are seen; you come in, you are seen. The lamp burns, he sees you; the lamp is out, he sees you. You enter your bedroom, he sees you; you turn things over in your heart, he sees you. Fear him, him whose concern is that he see you; and at least through fearing, be chaste.[29]

For Augustine, then, sin can be avoided by meditating upon the ubiquitous gaze of the divine judge.

Ascetic teachers proposed various thought experiments for conceptualizing how unseen sins, and especially mental activity, might be visible to God and others at the last judgment. Origen describes how everyone will have to give an account of their actions, words, and thoughts, even those which are hidden: while the books of the heart are "rolled up and covered, containing writing which we carry, worked over with certain remarks of the conscience, yet not known to anyone but God alone," at the last judgment, "inscribed tablets containing the letters of our deeds and thoughts will be read, as we have said, by every rational creature."[30] Augustine asserts that the resurrection body will be transparent, with the pure thoughts of the pious laid bare for all to see:

> Someone will answer me, saying: "If they are covered, how will it not be possible to lay hidden? Will not our hearts and our inner organs lay hidden?" Everyone will see the thoughts, my siblings, the thoughts that no one sees now except God, in that society of saints. There no one wants what he thinks to be hidden, because no one thinks evil.[31]

Such cognitive exercises facilitated expansion of the sense of shame by imagining how the deceased's life history will be open and visible/legible to all.

In addition to considering these demonstrations of divine omniscience, various forms of institutional discipline encouraged monks to develop this crucial aspect of the fear of God. Larger cenobitic monasteries included extensive surveillance networks, through which informants reported sins to the abbots, or other members of the monastic hierarchy. Ideally, surveillance was based on the direct and continuous observation of disciples. For instance, the *Rule of the Master* stipulates that the dean was to sleep surrounded by his ten disciples, such that he could monitor the behavior of them all.[32] Occasional inspection of cells is attested in Pachomian monasteries, presumably to ensure that monks were not hoarding food or other illicit belongings. The system of surveillance also depended on the obligation of all monks to report the infractions of other disciples to their superiors. If this was not done voluntarily, leaders might resort to interrogation, especially if violations of the rules were suspected.[33]

In contrast to rules, hagiography usually eschews direct mention of the surveillance apparatus, instead focusing on the miraculous charisms of monastic leaders such as Pachomius, especially the controversial gift of clairvoyance, which features frequently in the *Lives*. Simply stated, the faculty of clairvoyance (*to dioratikon*) is an enhanced perception of sin, whether remote or hidden actions; cardiognosticism, "knowledge of hearts," an enhanced perception of evil thoughts, is a special case of clairvoyance, though effectively it can be difficult to distinguish between the two.[34] Thus, when a monk confesses his sins after Theodore remarks that one of them has become distracted by a cooking vessel, whether he is referring to the young monk Patlole's temptation or act, both of which are mentioned, is uncertain.[35]

Pachomius frequently detects and exposes hidden sin. In a typical anecdote, he visits a monastery in which a disciple, hungered by fasting, has stolen five figs and hidden them in a jar. He then states before the community: "I was sent here today because of the health of a soul, and I found what I came here for in an earthenware vessel," prompting the brother to reveal his sin. According to the *Life*, the other disciples "marvel at the Spirit of God which was in our father Pachomius and at his perfect gaze."[36] Cardiognosticism was directed at disciples undergoing temptation, especially when they refused to confess their thoughts. Thus, a brother consistently hides his troubled heart from Pachomius, who declares that the demon who was attacking his soul has now fully possessed him. When asked how to expel it, Pachomius asserts that the disciple must simply follow his teachings.[37] In both instances, Pachomius is performing divine omniscience, training the sinful monk and other onlookers in the fear of God.

The controversy surrounding Pachomius's clairvoyance is another salutary reminder of the refusal of some people, including powerful Christians, to accept the idea of divine omniscience, at least as exercised through holy men. Pachomius was eventually forced to defend his practice of clairvoyance as a gift of the Spirit before skeptical bishops at the Council of Latopolis.[38] Theodore used a similar argument in his defense of clairvoyance recorded in the *Life of Pachomius*.[39] And in the *Letter of Ammon*, the bishop describes how, as a young man, he was particularly drawn to Theodore's power of cardiognosticism, despite his doubts about it: "I asked Ausonius to convince me, from the scriptures, whether it is at all possible for someone to see what is hidden in the hearts of others."[40] Ausonius duly provides a series of biblical passages in which cardiognosticism is displayed, including Samuel's first encounter with Saul (1 Kings 9.19–20), and Peter's assessment of Simon in Acts 8:23.[41] And many of the subsequent episodes recorded in Ammon's *Letter* emphasize Theodore's knowledge of hidden sin and evil thoughts in order to justify his disciplinary practices, including expulsion.[42]

In addition to teaching and discipline, monks themselves were expected to incorporate a sense of divine omniscience into their actions and thoughts by practicing various forms of self-scrutiny. Thus, Pachomius is said to have urged, "If we place before ourselves at each moment our weaknesses and our evil thoughts, so that we regret them in this age, we will escape eternal shame, the fire, and unceasing rebuke." Monks were thus urged to imagine themselves in the place of God, judging their secret sins and thoughts, and to adopt a repentant attitude while alive so that they would not be punished after death.

Isaiah of Scetis, in his *Discourse* 15, presents a multi-step cognitive exercise for the self-scrutiny of private thoughts. Although Isaiah admits that one's thoughts are hidden from other disciples, he recommends that the monk call to mind God's all-seeing eye: "Consider that God pays attention to your every action, that he sees your thought." He then asks the members of his audience to imagine their shame if the secret thought were to be acted out in public: "Whatever you are ashamed to do in front of people, it is also shameful to think in secret, for 'from its fruit the tree is known'" (Mt 12:33). This is the same feeling of shame a monk should have before God regarding the thought. Finally, the monk should admonish his or her own soul: "So say to your soul, 'If you are ashamed that sinners like yourself see you sin, how much more should you be ashamed of God who pays attention to the secrets of your heart'?" According to Isaiah, this exercise ensures that "the fear of God is revealed in your soul."[43]

Conclusion

Theory of mind, "the cognitive capacity to attribute mental states to self and others," is a universal human capacity. On the other hand, various "dimensions" of theory of mind can vary widely across cultures; even if some concepts are shared, they might be given vastly different emphasis from one culture to another. Thus, while the idea of divine omniscience and punishment is implicit in Zeus's role as oath-maker, and explicitly rationalized by the Athenian dramatist Kritias, Christianity encouraged constant meditation on God's ability to see hidden sins (clairvoyance), including private thoughts (cardiognosticism). Moreover, different groups in the same culture can have different theories of mind: new Christians such as Constantine were surprised to learn of God's cardiognosticism, and a group of bishops challenged the monastic leader Pachomius's claim to the same divinely mediated ability. For novice monks, the fear of God, including the accessibility of their private thoughts and their accountability for them, had to be learned and developed: by listening to and contemplating the detailed ekphrases of the divine panopticon and last judgment in monastic homilies; by subjecting themselves to the surveillance network, as well as their leader's appeal to the charisms of clairvoyance and cardiognosticism; and finally, by the cognitive exercise of self-scrutiny, in which disciples imagined their deeds and thoughts under constant observation by God and other monks, evaluating whether they conformed to monastic rule.

Notes

1 For a detailed discussion of theory of mind in Late Antique Christianity, and especially early monasticism, see Dilley 2017.
2 Wimmer and Perner 1983, 103–128. In this experiment, children observe a puppet show in which one of the characters, Maxi, is outside playing while someone moves a piece of chocolate; when Maxi comes back inside, the children are asked where the puppet will look for the chocolate. While three to four-year-olds point to its actual location, four to five-year-olds point to its former location, because they infer that Maxi will incorrectly assume that it has remained in that spot.
3 Goldman 2012, 402.
4 Luhrmann 2011, 7.
5 Luhrmann 2011, 6–7.

6 Luhrmann 2011, 7–8.
7 Of course, various ancient models of the mind are recoverable, even if they do not address all components of modern theory of mind, especially the mental states of others (human or divine). See, e.g., Habinek 2011 and Short 2012.
8 See discussion in Larson 2016, 95.
9 Boyer 2001, 152 and 158.
10 Larson 2016, 98.
11 Herodotus 1.47.3, 6.125.2.
12 Kritias, *Frag.* 25 (Diels), quoted in Sextus Empiricus, *Against the Mathematicians* 9.54. See discussion in Chaniotis 2012, 207.
13 Pettazzoni 1956, at 22–23. This anticipates the arguments of Guthrie 1980, 181–203.
14 Barrett 2004, 40–42, 47–49.
15 Czachesz 2012, 42–45.
16 This term is applied to evolutionary psychology in Gould 1997. For a Darwinian approach to divine omniscience that avoids evolutionary psychology, see now Johnson 2016.
17 Cf. Luhrmann 2012, xii: "Evolutionary psychology does not explain how God *remains* real for modern doubters. This takes faith, which is often the outcome of great intellectual struggle." For a similar critique of the Cognitive Science of Religion, see Laidlaw 2007, 211–246.
18 Chaniotis 2012, 216.
19 Constantine, *Epistula ad episcopos catholicos* (CSEL 26: 208). Despite its rarity in the pre-Christian Mediterranean, there is cross cultural evidence, e.g. the attribution of "mind-reading" to ancestors in the ritual of the contemporary Pomio Kivung religious movement in Papua New Guinea, on which see Whitehouse 2007, 247–280.
20 Larson 2016, 95–102, at 96.
21 The evidence is collected in Pettazzoni 1956, 97–114, with a preponderance of texts from the Psalms and wisdom literature. God is also described as observing the heart in the prophets, especially Jeremiah, and in some early Babylonian religious texts.
22 See, in particular, Psalms 73:11, 94:7, 10:11, Job 24:15, Sirach 23:18–20, Sirach 16:16–17 and 20–21 (as a temptation for the pious).
23 The fear of God is a biblical phrase, occurring most famously in Proverbs 9:10: "the fear of God is the beginning of wisdom."
24 Otto 1958, 12–24. For a critique of Otto's concept of *sui generis* emotions from the perspective of cognitive psychology and religious studies, see Taves 2009, 16–55.
25 Horsiesius, *Testament* 56 (Boon 1932: 147).
26 *First Greek Life of Pachomius* 132 (Halkin 1932: 83). For Pachomius's personal meditation on the fear of God, see, e.g., *First Greek Life of Pachomius* 18.
27 For treatments of shame in classical Greek literature, see e.g. Williams 1993.
28 Augustine, *Rule* 4:5 (Lawless 1987, 88).
29 Augustine, *Sermon* 132:2 (PL 38: 736).
30 Origen, *Commentary on the Epistle to the Romans* 9:6 (PG 14: 1242B-C): note that Origen imagines the book of the heart alternately as a roll and a tablet, but not a codex.
31 Augustine, *Sermon* 243:5 (PL 38: 1145).
32 *Rule of the Master* 29.
33 Theodore often did so for Pachomius: see *Great Coptic Life of Pachomius* 64, 74, 77 (following Lefort 1925 for text, Veilleux 1980 for reconstruction).
34 The revelation of hidden deeds is described in *Great Coptic Life of Pachomius* 59, 64, 77, *First Greek Life of Pachomius* 74, 89 (Pachomius) and *Great Coptic Life of Pachomius* 148, 185 (Theodore); hidden thoughts/demons in *Great Coptic Life of Pachomius* 106, 107, 111, 122 (Pachomius) and *Great Coptic Life of Pachomius* 195 (Theodore). Clairvoyance (*dioratikon*) is itself defined as the ability "to see the thoughts (*enthumēmata*) of souls" in *First Greek Life of Pachomius* 48 (Halkin 1932: 31), but for the purposes of this chapter I use cardiognosticism to denote the perception of hidden thoughts in particular.
35 *Great Coptic Life of Pachomius* 87; cf. *Great Coptic Life of Pachomius* 108.
36 *Great Coptic Life of Pachomius* 72 (CSCO 107: 74–75).
37 *Great Coptic Life of Pachomius* 102.
38 Cf. the account in *First Greek Life of Pachomius* 112.
39 In *First Greek Life of Pachomius* 135, Theodore emphasizes the Spirit as the source of revelation and the importance of humility.

40 *Letter of Ammon* 16 (Goehring 1986: 135).
41 The other cited passages are 1 Kings 16:6–12; 4 Kings 4:27; 4 Kings 5:25–27; Proverbs 27:23; Acts 14:8–10.
42 *Letter of Ammon* 17, 19–20, 22–24, 26.
43 Isaiah of Scetis, *Discourse* 15:2 (Augoustinos: 83).

References

Barrett, Justin, *Why Would Anyone Believe in God?* (Alta Mira Press, 2004).
Boon, Amand, ed., *Pachomiana Latina* (Bureaux de la Revue, 1932).
Boyer, Pascal, *Religion Explained: The Evolutionary Origins of Religious Thought* (Basic Books, 2001).
Chaniotis, Angelos, "Constructing the Fear of Gods: Epigraphic Evidence from the Sanctuaries of Greece and Asia Minor," in idem, ed., *Unveiling Emotions: Sources and Methods for the Study of Emotions in the Greek World* (Franz Steiner Verlag, 2012), 205–234.
Czachesz, Istvan, *The Grotesque Body in Early Christian Discourse: Hell, Scatology, and Metamorphosis* (Routledge, 2012).
Dilley, Paul, *Monasteries and the Care of Souls in Late Antique Christianity: Cognition and Discipline* (Cambridge University Press, 2017).
Goehring, James, *The Letter of Ammon and Pachomian Monasticism* (Walter de Gruyter, 1986).
Goldman, Alvin, "Theory of Mind," in E. Margolis, R. Samuels, and S. Stich, eds., *The Oxford Handbook of Philosophy of Cognitive Science* (Oxford University Press, 2012), 402–424.
Gould, Stephen J., "Evolution: The Pleasures of Pluralism," *The New York Review of Books*, June 26, 1997: www.nybooks.com/articles/archives/1997/jun/26/evolution-the-pleasures-of-pluralism.
Habinek, Thomas, "Tentacular Mind: Stoicism, Neuroscience, and the Configurations of Physical Reality," in B. Stafford, ed., *Bridging the Gap Between Neuroscience and the Humanities: A Field Guide to a New Meta-Field* (University of Chicago Press, 2011), 64–83.
Halkin, François, ed., *Sancti Pachomii vitae graecae* (Société des Bollandistes, 1932).
Johnson, Dominic, *God Is Watching You: How the Fear of God Makes Us Human* (Oxford University Press, 2016).
Laidlaw, James, "A Well-Disposed Social Anthropologist's Problems with the 'Cognitive Science of Religion,'" in H. Whitehouse and J. Laidlaw, eds., *Religion, Anthropology and Cognitive Science* (Carolina Academic Press, 2007), 211–246.
Larson, Jennifer, *Understanding Greek Religion* (Routledge, 2016).
Lawless, George, *Augustine of Hippo and his Monastic Rule* (Clarendon Press, 1987).
Lefort, L. Théophile, ed., *Sancti Pachomii vita bohairice scripta*, CSCO 89 (L. Durbecq, 1925).
Luhrmann, Tanya, "Towards an Anthropological Theory of Mind," *Anthropological Association* 36.4 (2011): 5–69.
Luhrmann, Tanya, *When God Talks Back: Understanding the American Evangelical Relationship with God* (Verso, 2012).
Monachos, Augoustinos, ed., *Tou hosiou patros hemon abba Esaiaou logoi 29* (Jerusalem, 1911).
Otto, Rudolf, *The Idea of the Holy*, trans. John Worley (Oxford University Press, 1958).
Pettazzoni, Raffaele, *The All-Knowing God: Researches into Early Religion and Culture* (Methuen, 1956).
Short, William Michael, "A Roman Folk Model of Mind," *Arethusa* 45 (2012): 109–147.
Taves, Ann, *Religious Experience Reconsidered: A Building-Block Approach to the Study of Religion and Other Special Things* (Princeton University Press, 2009).
Veilleux, Armand, *Pachomian Koinonia, vol. 1* (Cistercian Publications, 1980).
Vogüé, Adalbert de, ed., *La Règle du Maître*, SC 105–107 (Éditions du Cerf, 1964–1965).
Whitehouse, Harvey, "Towards the Integration of Ethnography, History, and the Cognitive Science of Religion," in H. Whitehouse and J. Laidlaw, eds., *Religion, Anthropology and Cognitive Science* (Carolina Academic Press, 2007), 247–280.
Whitehouse, Harvey and James Laidlaw, eds., *Religion, Anthropology and Cognitive Science* (Carolina Academic Press, 2007).
Williams, Bernard, *Shame and Necessity* (University of California Press, 1993).
Wimmer, H. and J. Perner, "Beliefs about Beliefs: Representation and Constraining Function of Wrong Beliefs in Young Children's Understanding of Deception," *Cognition* 13 (1983): 103–128.

Part IV
Performance and cognition

17

Sappho's kinesthetic turn

Agency and embodiment in archaic Greek poetry

Sarah Olsen

At the outset of her monograph on agency, kinesthesia, and culture, Carrie Noland describes how personal experience motivated and shaped her inquiry:

> It was while watching a graffiti writer that I first began to perceive how agency might work. As I observed the writer, his gestures revealed themselves to be simultaneously a repetitive routine and an improvisational dance; a script was obviously at the root of the performance and a script was its ultimate, durable product, but in between, as I could plainly see, a body was afforded a chance to feel itself moving through space.[1]

She goes on to argue that "gestures, learned techniques of the body, are the means by which cultural conditioning is simultaneously embodied and put to the test," thus exploring how movement both enacts and subverts the acculturation of the human body.[2] Cognitive theory as such does not play a prominent role in Noland's book, yet some of its insights resonate deeply with her observations and anecdotes. By situating her claims within recent work on cognition and embodiment, I seek here to extend the historical reach of her analysis. This process of contextualization will enable me to put Noland in dialogue with Sappho and Alcman – that is, to explore how Greek song represents movement and gesture as capable of both reinforcing and exceeding the bounds of cultural conditioning and somatic routine. Through a series of close readings, I will demonstrate that this modern paradigm of agency, acculturation, and embodied awareness sheds light on strategies of self-reference, description, and audience engagement in archaic Greek poetry.

Agency, embodiment, dance, and cognition

Noland's anecdote about watching the graffiti artist at work highlights some crucial elements of her argument and methodology. First, it offers a striking example of kinesthetic empathy: the process whereby a spectator feels an intimate, embodied connection with the movement of another. I watch someone dance, and my muscles tighten or my feet tap as though I am dancing along with them. I see an athlete reach up to catch a ball, and my own spine lengthens in response. The precise nature of such embodied response to movement has been widely discussed and debated, with the intriguing possibilities of neuroimaging research bringing a fresh perspective to the longstanding interests of phenomenologists.[3] Deidre Sklar, for example, defines kinesthetic empathy as "the process of translating from visual to kinesthetic modes," which in turn generates "the capacity to participate with another's movement or another's sensory experience of movement."[4] For the purposes of my analysis here, I take Sklar's formulation as a working definition, acknowledging the challenges of defining empathy itself and locating its role in human cognition.[5] While Noland does not make the empathetic element of her own experience explicit, her description of her response to watching the graffiti artist at work projects her into a shared kinesthetic space. She imagines what it would feel like to move as the artist moves, and this imaginative leap enables her to think about embodied agency in a new way.

Kinesthetic empathy, as a cognitive phenomenon and theoretical concept, has the potential to exemplify the first half of Noland's thesis: that cultural conditioning influences individual embodied experience. Ongoing research in cognitive science investigates the neurological mechanisms that underlie such experiences of kinesthetic identification with the movement of others, and suggests that training, socialization, and lived experience inform individual cognition. In one well-known study, scientists used MRI technology to study the brain activity of individuals viewing performances of ballet and capoeira. Their subjects included practitioners of both ballet and capoeira, as well as inexpert viewers. Their research revealed that the brain activity of the subjects differed when they viewed a movement practice in which they had been trained as compared with one of which they had no personal experience. Based on the regions of the brain activated, the investigators concluded that the human brain processes and understands movement by integrating the observed actions of others with an individual's own prior motor experience.[6] There is apparently a neurologically significant difference between viewing an action that one has previously *seen* and viewing an action that one has previously *performed*.

Engaging with this research, Susan Foster argues that dance performances cultivate empathy in their audiences by exploiting the assumptions about body, mind, and subjectivity prevalent in their historical moment.[7] She notes that while modern scientific approaches offer promising insight into the interaction of somatic training and visual stimuli on a neurological level, the concept of kinesthetic empathy itself predates neuroimaging: the 20th-century dance critic John Martin ascribes the power of dance to a process of "inner mimicry" in the observer; the ancient Greek philosopher Plato suggests that elderly spectators enjoy dance because it stimulates a kind of embodied memory.[8] On the one hand, the fundamental elements of such embodied and cognitive experience must be trans-historical, at least insofar as the structure and functioning of the human brain remain constant across time and culture. At the same time, Foster's work demonstrates that these experiences are described and configured in ways thoroughly embedded in specific cultural contexts.[9] I have argued elsewhere that in ancient Greek poetry and philosophy, the concept of kinesthetic empathy is deployed to explain how choral performance creates group cohesion.[10] I will explore some further intimations of this effect below.

Noland's imaginative identification with the graffiti artist, however, seems to pull in the opposite direction. For her, the experience of kinesthetic empathy supports the development of the second half of her thesis: the claim that gesture and embodiment can "put cultural conditioning to the test."[11] She goes on to suggest that even as we move through the world in ways shaped and scripted by broader cultural forces, we possess the ability to "turn inward, toward the kinesthetic sensations" of our bodies, allowing us to "'unbraid' movement practices from ideological ends and open up the possibility of no longer perpetuating 'social structures at the level of the body.'"[12] In developing this argument, Noland builds on Sklar's analysis of somatic routine, visual imagination, and embodied awareness.

Sklar, critiquing Pierre Bourdieu, argues that "the hold of the habitus is not absolute, and we do sometimes transcend its automatic and efficient grip."[13] She offers examples from both everyday life and dance, observing that:

> Pressing the brakes for the tenth time in the middle of a traffic jam, we may question the reason we own cars, calculating the cost and effort of maintaining them, envisioning the natural resources mined to make and run them, seeing the socioeconomic system that requires getting places quickly, and bringing to mind the millions of people in nonindustrial circumstances who don't require them. Performing a plié in the studio, perhaps dancers, too, have lucid moments of seeing themselves, as if from a distance, lined up among the others, holding onto a wooden pole in order to "gracefully" drop and rise over and over again, all agreeing to the perceptual, ideological, and aesthetic conventions of a sociocultural system that values "ballet." Perhaps the lucid moments occur in the opposite direction, consciousness diving inward and immersing in the minute sensations of toes gripping, quads clenching, spine extending, wrist softening, breath suspending. In the first kind of lucidity, one calls on visual imagination to project across distances to "see" the larger system, one's own body bobbing up and down at the barre to keep the system going; in the second, one calls on proprioception, turning awareness inward to "feel" one's body as a continuum of kinetic sensations. In either case, the hold of the habitus is broken, inviting opening beyond routine.[14]

Sklar sees a dynamic process of acquiescence and resistance at work in both ordinary action and choreographed movement, akin to Noland's experience of watching the graffiti artist and reflecting upon the possibility of embodied agency. Noland further suggests that "such a critical sensitivity to our acts . . . demands isolation, a willed disconnection from the purposive, instrumental, or communicative contexts into which we, as cultural beings, are almost always being thrust."[15] Sklar's ballerina at the barre offers one example of the ways in which disconnection and dissonance may occur even in the process of performing socially conditioned action. Noland's book is punctuated by the image of the graffiti writer: she posits that the "sensuous choreography of graffiti" brings to life the way in which "culture as gesture" is "always producing more and other than it intends."[16]

As Noland notes, cognitive theorists do not agree on the mechanisms and parameters of this kind of kinesthetic awareness.[17] In locating human agency here, Noland and Sklar push past the boundaries of current research on cognition, embodiment, and subjectivity. At the same time, I believe that certain elements of that research can allow us to more confidently extend their claims across time. Cognitive theory is compelling in part because it encourages us to explore, deconstruct, and reimagine the possible relationships between self and society, individual experience and cultural resonance.[18] For example, Edward Warburton combines cognitive science with phenomenology in order to develop a new account of "dance marking" (a technique whereby dancers commit choreography to memory through a compressed series of gestures and movements).[19] Building on an enactive model of cognition, he suggests:

> Consider, for example, a dancer "marking" a movement during the process of learning a new combination. As any dancer can attest, this activity often demands fast and continuously evolving responses to rapidly changing conditions. At the perceptuo-motor level, movement coordination requires continuous reciprocal influence between perceptual flow and motor commands; the dancer is undeniably situated in relationship to self (and instructor) and thinking "in real time." At the same time, the dancer is manipulating

> her environment, for instance using movement reductions such as a small hand gesture to indicate a turning movement on count eight, to exploit predictability in the task situation and automatize what was a formerly effortful process of skill acquisition. In short, she can use her implicit memory to prime in correct sequence a "turning" motor program by taking it off-line. The fact that the success of this activity depends crucially on each individual's own cognitive mapping of somatic, kinesthetic, and mimetic knowledge and skills underscores the ways dancers actively generate and maintain their identities, and thereby enact talk from the body in a unique cognitive activity such as dance marking.[20]

Warburton's account of marking echoes Sklar's observations about the interplay of automatic and intentional action and the assertion of individual identity and agency within the context of somatic training or routine, but he grounds his claims in a specific understanding of how the human brain processes sensory and motor experience. His engagement with cognitive research lends support to the more experiential arguments of Sklar and Noland.

Let us begin, then, from the premise that Noland's theory of embodied agency and acculturation is both accurate and trans-historical, even if it must be calibrated for different historical and cultural contexts. In the remainder of this chapter, I will explore how her framework illuminates strategies of self-reference and description in select songs of Alcman and Sappho, and thereby deepens our phenomenological understanding of archaic song.

Managing maidens: Alcman 1 and 3 *PMG*

Ancient Greek *choreia* was a potent and complex socializing institution. On the one hand, the process of singing and dancing in a chorus worked to acculturate individual bodies to their roles within systems and hierarchies of age, gender, and social status. At the same time, we should take care not to overstate the coercive force of the chorus, as communal *mousikē* was also a dynamic and adaptable form, capable of enacting group cohesion, negotiating among different group identities (e.g., local and panhellenic), and still allowing for individual experience and expression.[21] Choral performance thus provides a good test-case for the ancient relevance of modern notions about kinesthetic agency.

When an ancient Greek choral dancer learned to sing and dance in a group, she embedded herself in a particular social, ritual, and civic context and often sang lyrics that reinforced her role within that context, whether through explicit choral self-reference, mythic exempla, or allusions to contemporary figures and events. As Noland's model would suggest, her movement and gesture reinforce a specific kind of social conditioning. But the very nature of multimedia performance also tests the limits of such conditioning. Intonation, music, and dance are flexible media: the poet (composer, choreographer) can certainly instruct his performers to pronounce, play, and move in very specific ways, but bodies, voices, and even instruments are infinitely variable and capable of making their own choices – ranging from a barely noticeable flat note to a dramatic change in position – in the course of live performance. Speaking again of the graffiti artist, Noland remarks that:

> the body we observe in the act of writing may indeed be communicating a message or completing a task, but it is simultaneously measuring space, monitoring pressure and friction, accommodating shifts of weight. These kinesthetic experiences that exceed communicative or instrumental projects affect the gestures that are made and the meanings they convey.[22]

Greek choreuts likewise complete a scripted task: singing and dancing within the parameters of their choreographed performance. But what kind of agency might have manifested itself in the course of that performance? If Noland and Sklar are correct, then it would have been possible for participants to experience the embodied acculturation of *choreia* while also engaging in a kind of kinesthetic awareness that surpassed, complicated, or resisted it. By exploring how the language of choral song describes and engages with the embodied experiences of performers and spectators, we can begin to imagine some of the ways in which this experience could have worked on the ground. In this section, I will consider how Alcman's *partheneia* strive to reinforce the cultural conditioning of the body through words and performance. I will further suggest that, even within this choreographic regime, we can identify places where these songs leave space for the kind of kinesthetic awareness theorized by Noland and Sklar.

Alcman's *partheneia* were probably performed in archaic Sparta by choruses of young women as part of a ritual related to preparation for marriage, with the poet acting as composer, choreographer, and chorus trainer.[23] These songs have long been understood as paradigmatic examples of how *choreia* works to inculcate its participants into their social and cultural roles.[24] Here, I am specifically interested in how that process of socialization draws upon the ability of the language of the choral song to frame and affect the embodied experience, presence, and perception of the performers and spectators.

Following a lengthy and fragmentary mythic narrative, Alcman's first *partheneion* turns to a self-referential description of choral actors and action (1.39–43 *PMG*):[25]

> [...] ἐγὼν δ' ἀείδω
> Ἀγιδῶς τὸ φῶς· ὁρῶ
> F' ὥτ' ἅλιον, ὅνπερ ἇμιν
> Ἀγιδὼ μαρτύρεται
> φαίνην· [...]
>
> [...] And I sing
> of Agido's light. I see her
> like the sun, which
> Agido calls to shine as a witness
> for us [...]

This formulation foregrounds the authority of song over the expression and perception of the performing body. The chorus sings Agido into being, defining her as "light" (40), just "like the sun" (41), before describing her as an agent in her own right (42-43). While Agido is presumably present, the choral song adds a layer of descriptive commentary to the movement of her body, encouraging the audience to interpret Agido through the filter of imagination.[26]

The chorus then turns to its other leader, Hagesichora, singing (Alc. 1.45-59 *PMG*, trans. Campbell, modified):

> [...] δοκεῖ γὰρ ἤμεν αὔτα
> ἐκπρεπὴς τὼς ὥπερ αἴτις
> ἐν βοτοῖς στάσειεν ἵππον
> παγὸν ἀεθλοφόρον καναχάποδα
> τῶν ὑποπετριδίων ὀνείρων ·
> ἦ οὐχ ὁρῇς; ὁ μὲν κέλης
> Ἐνετικός· ἁ δὲ χαίτα

τᾶς ἐμᾶς ἀνεψιᾶς
Ἀγησιχόρας ἐπανθεῖ
χρυσὸς [ὠ]ς ἀκήρατος·
τό τ' ἀργύριον πρόσωπον,
διαφάδαν τί τοι λέγω;
Ἀγησιχόρα μὲν αὕτα·
ἁ δὲ δευτέρα πεδ' Ἀγιδὼ τὸ Fεῖδος
ἵππος Ἰβηνῶι Κολαξαῖος δραμήται.

[. . .] for she [Hagesichora] herself seems
pre-eminent, just as if
one were to put a horse
among grazing herds,
a sturdy, thunderous-hoofed prize-winner,
one of those seen in rock-sheltered dreams.
Why, don't you see? The racehorse
is Venetic, but the hair
of my cousin Hagesichora
has the bloom
of undefiled gold,
and her silver face –
why do I tell you openly?
This is Hagesichora here;
and the second in beauty after Agido
will run like a Colaxaean horse against an Ibenian.

Here, the chorus calls upon the audience to look at Hagesichora herself ("this is Hagesichora here," 57) – but only after offering a rapid series of descriptive references, wherein they first liken Hagesichora to a racehorse, then call attention to the way she "blooms" (53) like "undefiled gold" (54) and displays a "silver face" (55). These descriptions do not add up to a coherent image of a specific female body. They praise the *chorēgos* through images of motion and light, again priming the audience to perceive the performer's body in a specific way. Peponi observes that in these lines, Hagesichora becomes "a construction, a spectrum of constant visual metamorphoses."[27] In her reading, these images and similes engage with the audiences' powers of imagination, stimulating a process of comparison and "imaginative visualization."[28] But while such prompting may enrich an audience's experience of the dance, it also exercises a controlling force, encouraging certain comparisons and generating specific images in a way that privileges those comparisons and images over other, unexpressed possibilities.[29]

Alcman's descriptive strategies interfere with the cognitive action described by Sklar, wherein the individual ballerina envisions her own motion at the barre and consequently reflects upon the social conditioning and cultural norms being inscribed upon her body.[30] When Alcman gives his chorus words for describing and constructing both their own performance and the actions of their leaders, he anticipates the kind of "visual imagination" described by Sklar and turns it to his own ends, reinforcing the "hold of habitus" with language.[31] By choreographing motion *and* scripting words, the *chorodidaskalos* exerts a powerful form of authority over the expression and experience of his choreuts. This does not mean that an individual Spartan choreut could not have experienced the kind of kinesthetic resistance proposed by Sklar and further theorized by Noland, but rather that in order to do so, she would have had to distance herself not only from the motions being performed by her body, but also from the words about those motions coming out of her mouth.[32]

We might recall, at this point, that "Hagischora" ("chorus-leader") and "Agido" ("leader") were surely not historical women, but conventional roles adopted by preeminent maidens in successive ritual re-performances of these songs.[33] The dazzling and diffuse images of the performing body in this song reinforce the exchangeability of the individual girl. By aestheticizing the performer in this particular way, Alcman diverts the focus from the unique and specific body of the dancer herself, and toward abstract images and repeatable roles. For the performers and audience members alike, it may have been challenging to reject the soft-focus allure of these radiant images and really "see," in the mode proposed by Noland and Sklar, the gestures and kinesthetic experiences of the individual body.[34]

Peponi further argues that the phrase "as if someone set up a horse among the grazing beasts" (ὤπερ αἴτις ἐν βοτοῖς στάσειεν ἵππον, 47) constitutes an oblique reference to the *chorodidaskalos*' creative vision and the process of choral production: the subject of *staseien* is Alcman, who "sets up" the dancer like a horse among the herds.[35] I find this analysis convincing. In addition, I think there is more to be gleaned from the poet's claim to "set up," or choreograph, this particular horse "among the grazing beasts" (ἐν βοτοῖς, 47). This image positions Hagesichora as an outstanding figure among a group of undifferentiated others, like a *chorēgos* among her choreuts.[36] This element of the image is hardly surprising in this context. But grazing animals are also located outside – specifically, on the margins of settlements and societies. By likening Hagesichora to a horse among the herds, the poet places her in the traditional outdoor haunt of playful, dancing virgins.

Such spaces, however, have sinister overtones. While it is unlikely that a 7th-century Spartan poet-choreographer would have known the Homeric poems and the *Homeric Hymns* in exactly the form we read them today, those texts do suggest that the "maiden in the meadow" (or at the seaside) was a kind of type-scene, an image active in the Greek cultural imagination from a fairly early period.[37] Figures like Nausicaa and Persephone represent the precarious position of the girl at play in the wild, whose preeminent position among her companions reflects her social status and readiness for marriage, yet also singles her out as a likely candidate for rape.[38] When Alcman sets Hagesichora up in a field, "among the grazing beasts," listener and performer alike may experience a moment of fearful instability. Again, the descriptive imagery of the song generates a script for understanding the position and status of the performer, and situates her dancing within a specific social context. If the choreography of the dance emphasized the singularity of the soloist in some way, it would further intensify the sense that the performer herself embodies the literary trope. At the same time, I wonder whether the hint of instability embedded within the image of the outstanding maiden dancer might also generate the kind of dissonance or disconnection that Noland sees as the root of embodied resistance to cultural norms.

Alcman's second *partheneion*, though even more fragmentary than the first, offers a tantalizing reference to gesture and movement that resonates with the models that I have been discussing. At one point, the chorus speaks longingly of its *chorēgos*, Astymeloisa, singing "if only, coming close, she might take hold of my soft hand" (ἇσ]σο̣ν [ἰο]ῖσ' ἀπαλᾶς χηρὸς λάβοι, Alc. 3.80 *PMG*). As in the first *partheneion*, the chorus praises the outstanding beauty of its leader while effacing itself.[39] This song also vividly dramatizes the potential rupture between the chorus and the *chorēgos*, describing Astymeloisa as a soloist poised between her fellow-performers and the city as a whole.[40] Peponi posits that

> the explicit reference to the holding of the hand [at line 80], signaling the desired return of the *chorēgos* to the female choral ensemble, functions as an implicit female counter-act to the anticipation of the same gesture by a male, marking in a different ritual context the final stage of the young woman's transition from the female chorus to the male *oikos*.[41]

Peponi's analysis highlights the fact that hand-holding is a common element of female *choreia* in both literary paradigms and iconography, and at the same time a gesture with significant nuptial associations.[42]

Whether Alcman's choreuts actually join hands in performance or merely imagine doing so, the song evokes a gesture that already exceeds a singular cultural meaning. Astymeloisa's imagined hand extends in two directions – back toward her fellow choreuts, and out toward her future husband, who is potentially contained within the civic audience. On one level, the imagery of the gesture facilitates and controls her social transition, as she is figuratively "handed off" from maidenhood to marriage. But I would further suggest that the chorus' longing for Astymeloisa's hand, especially if accompanied by related choreography, could be an example of Noland's central claim: that gesture can both reinforce and exceed the forces of social conditioning. Hand-holding may be codified choreography: an action serving to support the coordination of the dance, express the coherence of the group, and demarcate the social position of the preeminent *chorēgos*. But the individual dancer would also have been able to wrap her fingers a bit tighter or direct her attention to the warm contact of skin with skin – thereby bringing individual intimacy or meaning to the experience of performing with the group. Perhaps that visceral experience of corporeal intimacy even offered an opportunity for a kinesthetic experience of female bonds both beyond men and beyond words. Alcman's descriptive strategies thus reinforce the notion that choral dance accomplishes the cultural conditioning of the body, while also creating imaginative openings for individual experience and agency.

Alcman, Sappho, and kinesthetic empathy

Alcman and Sappho display some striking affinities, despite the meaningful differences in their corpora and poetic personae.[43] Alcman fr. 26 and Sappho fr. 58 offer an important example, especially since some scholars have recently argued that they were both solo songs sung to the accompaniment of the kithara, functioning as preludes to a choral performance.[44] If so, we might consider how they encourage their audiences to engage with the coming *choreia*.

In Alcman fr. 26, the speaker addresses a chorus of young women (παρσενικαὶ, 1), lamenting that (Alc. 26 *PMG*, trans. Campbell):

> οὔ μ' ἔτι, παρσενικαὶ μελιγάρυες ἱαρόφωνοι,
> γυῖα φέρην δύναται· βάλε δὴ βάλε κηρύλος εἴην,
> ὅς τ' ἐπὶ κύματος ἄνθος ἅμ' ἀλκυόνεσσι ποτήται
> νηδεὲς ἦτορ ἔχων, ἁλιπόρφυρος ἱαρὸς ὄρνις.
>
> No longer, honey-toned, strong-voiced girls,
> can my limbs carry me. If only, if only I were a cerylus,
> who flies along with the halcyons over the flower of the wave
> with resolute heart, strong, sea-blue bird.[45]

Sappho fr. 58 is likewise addressed to a female chorus, and the speaker similarly expresses regret for the limitations of her own body (Sappho fr. 58.1-8, ed. and trans. West 2005):[46]

> Υ̓́μμες πεδὰ Μοίσαν ἰ]οκ[ό]λπων κάλα δῶρα, παῖδες,
> σπουδάσδετε καὶ τὰ]ν φιλάοιδον λιγύραν χελύνναν·
> ἔμοι δ' ἄπαλον πρίν] ποτ' [ἔ]οντα χρόα γῆρας ἤδη
> ἐπέλλαβε, λεῦκαι δ' ἐγ]ένοντο τρίχες ἐκ μελαίναν·
> βάρυς δέ μ' ὀ [θ]ῦμος πεπόηται, γόνα δ' [ο]ὐ φέροισι,

> τὰ δή ποτα λαίψηρ' ἔον ὄρχησθ' ἴσα νεβρίοισι.
> τὰ <μὲν> στεναχίσδω θαμέως· ἀλλὰ τί κεν ποείην;
> ἀγήραον ἄνθρωπον ἔοντ' οὐ δύνατον γένεσθαι.
>
> [You for] the fragrant-bosomed [Muses'] lovely gifts
> [be zealous,] girls, [and the] clear melodious lyre:
> [but my once tender] body old age now [has seized;]
> my hair's turned [white] instead of dark;
> my heart's grown heavy, my knees will not support me,
> that once on a time were fleet for the dance as fawns.
> This state I oft bewail; but what's to do?
> Not to grow old, being human, there's no way.

The speaker of fr. 26 gestures briefly to the past at the beginning of the fragment, when he says that his limbs "no longer" (οὔ . . . ἔτι, 1) support him – implying that they once did. Fr. 58 offers a similar but more expanded image, as the speaker remarks: "my knees will not support me, / that once on a time were fleet for the dance as fawns" (5-6). She thus engages in a form of kinesthetic empathy with the performers being addressed, recalling her own past experiences of dance. She momentarily and poignantly imagines herself as a fellow-dancer, separated from the performers by age alone.[47] If this song was performed as a prelude, the singer offers a model of kinesthetic empathy and imaginative spectatorship to the audience.

Alcman fr. 26 and Sappho fr. 58 both feature speakers who long to join the chorus, although Sappho's version more clearly emphasizes the affinity between the singer and the performers. But perhaps Alcman's song included similar reflections in verses that have not survived. Even if it did not, the two songs are probably more alike than not. They lament the limitations of age, especially as viscerally and acutely felt when imagining, viewing, or singing about *choreia*. They direct attention to embodied experience and sensation – current weakness and weight in the body (γυῖα φέρην δύναται, 26.2; βάρυς δέ μ' ὁ [θ]ῦμος πεπόηται, γόνα δ' [ο]ὐ φέροισι, 58.5), past agility in dance (τὰ δή ποτα λαίψηρ' ἔον ὄρχησθ' ἴσα νεβρίοισι, 58.6), and the potential to imaginatively "fly along" in performance (ἅμ' ἀλκυόνεσσι ποτήται, 26.3). These two songs do not seem to thrust their singers and listeners into a specific cultural context in the way that Alcman's *partheneia* do, although perhaps this impression is largely due to our limited knowledge of their performance context and the fragmentary state of the songs themselves.

By so vividly describing the singer's personal somatic response to choral performance, these songs offer cues to their listeners, subtly prompting them to think about their own limbs and memories. Sappho fr. 58 and Alcman fr. 26 model, and thus perhaps encourage, a kind of mindful spectatorship – one which attends to the subjective position of the viewer and her kinesthetic experience. This is the kind of experiential viewing described by both Noland and Sklar when they speak of the graffiti artist in action, or a line of ballet dancers at the barre, or the individual driver pressing on the brakes.[48]

Anactoria's vanishing act: Sappho fr. 16

I will now turn to a song that shifts from acknowledging kinesthetic awareness to subtly testing the limits of embodied acculturation: Sappho fr. 16. Scholars have debated whether this song was originally monodic or choral, and if the latter, some have proposed a wedding celebration as one possible performance context – a choral scenario not identical to that of Alcman's *partheneia*, but similar in some important respects.[49] As we will see, fr. 16 exhibits thematic and stylistic affinities with maiden song. But there are other ways of accounting for this affinity. Timothy Power develops the concept of "parachoral monody" to describe Sappho's songs, suggesting that many of them may be solo songs that deliberately and strategically evoke chorality.[50] I share Power's sense that the complexity of Sappho's relationship to performance context may be as much poetic strategy as a symptom of modern ignorance. If archaic Greek poetry almost inevitably "thrusts" its listeners "into purposive, instrumental, or communicative contexts,"[51] I want to explore how fr. 16 endeavors to push and pull its internal characters and external audience in and out of context, invoking familiar tropes surrounding the representation of the female dancer while simultaneously complicating and subverting them.

In an influential analysis, John Winkler explores traces of "double consciousness" in Sappho's songs: the acknowledgment of masculine norms and expectations alongside allusions to woman-centered spaces and sexual subjectivity.[52] He highlights the multiplicity of meaning in Sappho's lyrics, whereby she is able to restate dominant cultural messages while also inserting her own interpretive perspective. Fr. 16 is a crucial example of his argument, as he explores how Sappho displays her fluency in the masculine discourse of war and battle, then dramatically reframes that discourse with the revelation that "all valuation is an act of desire."[53]

While Winkler's analysis of fr. 16 focuses exclusively on lines 1-12, the song's double consciousness takes a kinesthetic turn in the subsequent stanzas, our knowledge of which has been expanded by recent papyrological work. The fullest version of the song we possess runs thus:

ο]ἰ μὲν ἰππήων στρότον, οἰ δὲ πέσδων
οἰ δὲ νάων φαῖσ' ἐπ[ὶ] γᾶν μέλαι[ν]αν
ἔ]μμεναι κάλλιστον, ἔγω δὲ κῆν' ὄτ-
τω τις ἔραται·
πά]γχυ δ' εὔμαρες σύνετον πόησαι 5
π]άντι τ[ο]ῦτ', ἀ γὰρ πόλυ περσκέθοισα
κάλλος [ἀνθ]ρώπων Ἐλένα [τ]ὸν ἄνδρα
τὸν [. . .] ἄρ]ιστον
καλλ[ίποι]ς' ἔβα 'ς Τροίαν πλέοισα·
κωὐδ[ὲ πα]ῖδος οὐδὲ φίλων τοκήων 10
πά[μπαν] ἐμνάσθη, ἀλλὰ παράγαγ' αὔταν
]σαν
γν]αμπτον γὰρ [. . .] νόημμα
[. . .] κούφως τ [. . .] νοήσηι
]με νῦν Ἀνακτορί[ας] ὀνέμναι- 15
σ' οὐ] παρεοίσας,
τᾶ]ς κε βολλοίμαν ἔρατόν τε βᾶμα
κἀμάρυχμα λάμπρον ἴδην προσώπω
ἢ τὰ Λύδων ἄρματα κἀν ὄπλοισι
πεσδομ]άχεντας. 20

Some say a host of cavalry, others of infantry,
and others of ships, is the most beautiful thing
on the black earth, but I say it is whatever
one loves.
It is perfectly easy to make this understood
by everyone: for she who far surpassed humankind
in beauty, Helen, left
her most noble husband
and went sailing off to Troy
with no thought at all for her child or dear parents,

> but (love? Kypris?) led her astray . . .
> for [she (sc. Kypris?) with un]bending mind
> [. . .] lightly [. . .] thoughts
> (and she?) has put me in mind now of Anactoria
> who is not here;
> I would rather see her lovely walk
> and the bright sparkle of her face
> than the Lydians' chariots and
> armed infantry.[54]

There are a few reasons to believe that Sappho's speaker is imagining Anactoria specifically as a dancer. Anne Pippin Burnett observes that "the word βᾶμα might refer to Anactoria's manner of dancing," but that is not the only hint of dance here.[55] The suggestive adverb "lightly" (κούφως, 14) is unfortunately without context, but it is used elsewhere in Greek song of both military and choral motion.[56] It may, therefore, be a striking example of what Winkler terms Sappho's "bilingualism": her talent for choosing language that resonates in different ways for different groups.[57] But even if she does not apply this adverb directly to Anactoria's motion, Sappho's speaker gives us a subsequent somatic description evocative of the maiden choral leader. She calls attention to the allure of Anactoria's step (17) and face (18), a mode of praise standard for the performing *parthenos*.[58] She also recalls Anactoria as an individual, contrasting her singular beauty with plural military objects (19-20). Anactoria thus stands out from the mass of chariots and infantry like a maiden *chorēgos* among her companions.

As we have seen, Alcman's first *partheneion* similarly highlights the beautiful face of the outstanding maiden dancer (ἀργύριον πρόσωπον, Alc. 1.55 *PMG*), and positions Hagesichora as a racehorse standing out from the grazing herd (Alc. 1.45-59 *PMG*). Yet if Alcman's image projects the maiden *chorēgos* into a familiar type-scene, associating her with animals and the natural landscape, Sappho's version generates a striking dissonance. Chariots and infantry are not typical choral companions: Anactoria is singular, isolated.

Anactoria's motion gestures in another direction as well. Within the larger structure of the poem, there is a parallel between Anactoria's "lovely step" (ἔρατόν . . . βᾶμα, 17) and Helen's travel to Troy ("she went," ἔβα, 9), established by the common use of *bainō* and *basis*. But Helen is a complicated, even transgressive, model. While she is often represented and celebrated as an archetypal *parthenos*, she is also a woman who fails to remain stable within her marriage. In fr. 16, Sappho represents her as an active agent, motivated by desire to "step" toward Troy.[59] Without explicit surviving references to Anactoria's future, the song leaves us wondering where her "lovely step" will bring her, with Helen's extra-nuptial travels in the opening lines complicating the allusions to maidenly corporeality and dance in lines 17-20.

Sappho's representation of Anactoria plays with the imagery of the maiden *chorēgos*. She offers her audience images of Anactoria that are at once familiar and strange. I am not the first to notice this dimension of Sappho's poetry, but I want to connect it specifically with the concept of kinesthetic agency.[60] As I have mentioned before, Noland argues that the "critical sensitivity to our acts" required for the experience of embodied and kinesthetic agency "demand[s] isolation, a willed disconnection from the purposive, instrumental, or communicative contexts into which we, as cultural beings, are almost always being thrust."[61] Sklar's description of the ballet dancer catching sight of her movement in the mirror suggests that such disconnection can arise in moments of sudden dissonance and self-awareness, the kind of kinesthetic turn I have traced thus far in both Alcman and Sappho. But I would suggest that fr. 16 goes a step further by making Anactoria herself "absent" (οὐ] παρεοίσας, 16).

The negative οὐ in line 16 is a supplement. But assuming that it is correct, it offers a contrast to Alcman 1 *PMG*, wherein the chorus insistently calls attention to the presence of Hagesichora.[62] I have argued that the strategies of self-reference and presence in Alcman's *partheneia* often endeavor to complete the work of cultural conditioning upon the body, thrusting both performers and audience members into a specific, if also repeatable, social context. Sappho, by removing Anactoria from the here-and-now of her song, reverses the process. She brings the work of imaginative prompting to the fore and thereby further destabilizes the ability of a song to fix the meaning of a body. Anactoria's "lovely step" (ἔρατον . . . βᾶμα, 17) is not a movement occurring before our eyes, but one recalled by the speaker.

When Sklar describes the ballet dancer watching her own movement at the barre, she depicts the practice of dance as both the embodiment of cultural values (specifically the aesthetics of ballet) *and* a way for the individual to identify the grip of those values – in her framework, the first step toward resistance. In fr. 16, Sappho makes a similar move, situating Anactoria within the somatic value-system of maiden chorality, then removing her again, using absence and isolation to create space for reflection. The audience is reminded of the potential gaps between description and action, script and performance, cultural conditioning and individual kinesthetic experience. Perhaps Sappho's song impacted some members of its ancient audience in ways akin to the effect of the graffiti writer upon Noland – a prompt to reconsider and reimagine the intersection of agency, embodiment, and acculturation.

Conclusion

Noland's model acknowledges the power of embodied cultural conditioning and prompts us to consider how social and historical forces inform individual experiences of movement and action. But she also insists on the possibility of seeing and feeling gesture as "available for *but not equivalent to* social meanings."[63] She demonstrates that experiences of kinesthetic agency, just like experiences of embodied acculturation, are culturally specific in that they rely upon an individual subject's sense of her own movement and positionality in relation to specific contexts and frameworks. She thus complicates and even dismantles the straightforward separation of the individual body from the society in which it resides while yet preserving the possibility of meaningful individual agency. As I noted above, this resonates with ongoing research in cognitive science and enables us to analyze how trans-historical physiological phenomena function within specific social, cultural, and historical contexts.

My goal for this chapter, therefore, has been to explore the intersection of phenomenology, cognitive science, and archaic Greek poetry and examine the potential impact of Alcman's and Sappho's descriptive strategies in performance. In the process, I have also reflected on the various ways in which scholars of the humanities and of the past can contextualize, expand, and nuance the insights of cognitive research by bringing them into dialogue with particular theoretical, historical, and literary problems. Our fragments of Alcman and Sappho do not directly tell us how ancient Greeks felt when they watched or participated in *choreia*. But through their representation of embodied action, affect, and response, they point to some of the ways in which culture, embodiment, and agency were imagined, configured, and perhaps even experienced in archaic Greece.[64]

Notes

1 Noland 2009: 1.
2 Noland 2009: 2.
3 See, e.g., Foster 2008 and 2011, Hagendoorn 2008, Warburton 2011: 70–75, Reason and Reynolds 2012, and Meineck 2017: 120–153.

4 Sklar 2001: 199 n. 3.
5 See Meineck 2017: 18–22, with further bibliography.
6 Calvo-Merino et al. 2005. See also Brown et al. 2005, Calvo-Merino et al. 2006, and Cross et al. 2006. On the incorporation of this research into the study of performance, culture, and representation (ancient and modern), see, e.g., Meineck 2011, 2012, and 2017: 120-153, Reason and Reynolds 2012, and De Preester 2013.
7 Foster 2011.
8 On Martin, see Foster 2011: 155–162. On Plato (*Leg.* 657d), see Jackson 2016 and Olsen 2017: 165–167.
9 Foster 2008 and 2011.
10 Olsen 2017.
11 Noland 2009: 2.
12 Noland 2009: 210.
13 Sklar 2008: 91. For Bourdieu's model of embodied experience and the construction of culture, see Bourdieu 1977: 93–94.
14 Sklar 2008: 91.
15 Noland 2009: 210.
16 Noland 2009: 17.
17 Noland 2009: 3-4 and n. 3, with further bibliography.
18 Cf., e.g., Anderson 2015 and Garratt 2016.
19 Warburton 2011: 69.
20 Warburton 2011: 70, building primarily on Thompson 2005.
21 For this understanding of *choreia*, see, e.g., Calame 1997, Stehle 1997, Kowalzig 2007, Kurke 2007, and Olsen 2015.
22 Noland 2009: 2.
23 Calame 1997: 184–185, 227–228.
24 Cf., e.g., Calame 1977 [1997], Clark 1996, Peponi 2007, and Ferrari 2008: 105–126.
25 Greek translations are my own where not otherwise indicated.
26 As Peponi observes, this description "dematerializes" Agido (2004: 299). Peponi demonstrates that this dematerialization has a complex and powerful aesthetic effect (2004: 299-303), but I would add that it also emphasizes the power of language to shape the audience's reception of dance.
27 Peponi 2004: 302.
28 Peponi 2004: 301. On this process see also Peponi 2015.
29 Paxton remarks that language "can certainly influence our point of view [of, e.g., a dance] and may even suggest what *can* be thought about – that is, limit our perception or experience to the form encompassed by language. It does seem to me that if we spend much time communicating with others via language about a painting, music, or dance, we accustom our minds to the language version of the experience" (2001: 422, emphasis in original).
30 Sklar 2008: 91.
31 E.g., by composing words that reinforce the specific social position of *parthenoi*, a position which would have also been displayed by their ritual performance of dance (e.g., Calame 1997 [1977], Stehle 1997: 30–39, 74–88).
32 On a similar note, dancers themselves could experience reciprocal and mutually reinforcing forms of kinesthetic empathy; cf., e.g., Peponi 2007 on the interplay of performers, spectators, and performers-as-spectators in Alcman 3 *PMG*.
33 Nagy 1990: 347.
34 I do not mean that auditory cues simply override visual perception, but rather that the descriptive language of the song encourages one particular way of understanding the multisensory experience of performance. On the complexities of processing multisensory information, see further Calvert, Spence, and Stein 2004; on the interaction of imagination and sensation in the viewing of (primarily dramatic) *choreia*, see Weiss 2018: 11–18.
35 Peponi 2004: 313-316.
36 See Calame 1977 II: 68, Lonsdale 1993: 200–201, Clark 1996: 156, Hutchinson 2001: 87, and Ferrari 2008: 75.
37 On this trope, see Lonsdale 1993: 222–233, 2004, Rosenmeyer 2004, Murnaghan 2013: 156–158, and Reitzammer 2016: 42.
38 Cf. *Il.* 16.179–192, *Ody.* 6.99–109, *Hom. Hymn. Ven.* 117–120, and *Hom. Hymn. Dem.* 20.
39 Stehle 1997: 88–93.

40 Peponi 2007: 357–362.
41 Peponi 2007: 361–362.
42 Peponi 2007: 359. See, e.g., *Hom. Hymn. Ap.* 194-196, Aesch. *Eum.* 307, and Arist. *Thesm.* 954–956, as well as Lonsdale 1993: 214–218 for examples from visual art. For hand-holding as a nuptial gesture (specifically the placing of the hand upon the wrist, or *xeir epi karpōi*, though note that the same phrase is used of female choreuts with no marital connotations in *Hom. Hymn. Ap.* 194–196), see *Hom. Hymn. Ven.* 117–121, Lonsdale 1993: 213–217, Oakley and Sinos 1993: 32, 45, and 137 n. 71, and Peponi 2007: 361.
43 Cf. Hallett 1996: 140–142 and Stehle 1996: 147–149. I am not arguing here for a precise intertextual relationship (that is, one poet's awareness of another), but rather considering how they engage with similar themes in distinct but overlapping ways.
44 Power 2010: 202–203, Kousoulini 2013: 436–437.
45 I follow most editors (including Page 1962 and Campbell 1988) in reading ποτήται in line 3 (the cerylus "flies along" with the halcyons), rather than φορεῖται (the cerylus is "carried"). But for a defense of the latter reading, with earlier bibliography on both sides of the question, see Vestrheim 2004.
46 For other possible reconstructions of this song, see Di Benedetto 2004 and Lidov 2009. On the similarities between these poems see Calame 1983: 474, Lardinois 2009: 51–53, and Kousoulini 2013: 439–440.
47 See further Olsen 2017: 163–165.
48 Sklar 2008: 91 and Noland 2009: 1. We might also consider whether a musical prelude is formally akin to the kind of choreographic marking discussed by Warburton 2011, and if so, whether it would similarly allow for the expression and experience of individual identity with the constraints of training and routine.
49 Hallett 1996: 140–142, Lardinois 1996: 166–167 and 2001: 83–85.
50 Power forthcoming.
51 Noland 2009: 210.
52 Winkler 1990: 162–187.
53 Winkler 1990: 177.
54 The text I print here is that of Voigt 1971, modified to accommodate the additions and changes suggested by Burris, Fish, and Obbink 2014 and Obbink 2016: 17–18. I have adapted the translation of Campbell 1988 accordingly, with reference to Obbink 2016: 28–29. For further discussion of the supplements, readings, and state of the papyrus, see Burrish, Fish, and Obbink 2014 and Obbink 2016.
55 Burnett 1983: 280 n. 5.
56 Cf. *Il.* 13.158, [Hes.], *Scut.* 323, Pindar *Olympian* 14.17 and *Nemean* 8.19, Eur, *Alc.* 584, Ar., *Thesmo.* 954, and Bierl 2011: 430 n. 51.
57 Winkler 1990: 174–175.
58 Head and face: *Ody.* 6.107, Alc. 1.51–54 *PMG*; feet: Alcman 2.70 *PMG*, Pindar *Paean* 6.17–18. On this point, see Segal 1998: 73.
59 On Helen as *parthenos*, see Calame 1997: 191–202, Swift 2010: 218-238, and Murnaghan 2013: 164–165. For discussions of Sappho's Helen in relation to her epic and masculine-sympotic counterparts, see Winkler 1990: 176–178, duBois 1996: 86-87, Stehle 1996: 221–223, Williamson 1996: 261–262, and Segal 1998: 64–72.
60 Cf., e.g., DuBois 1996 and Segal 1998: 74-78.
61 Noland 2009: 210.
62 Cf. Alc. 1.57 *PMG*: "she is Hagesichora" (Ἁγησιχόρα μὲν αὔτα) and Alc. 1.78-79 *PMG*: "isn't fair-ankled Hagesichora here?" (οὐ γὰρ ἁ κ[α]λλίσφυρος / Ἁγησιχ[ό]ρ[α] πάρ' αὐτεῖ).
63 Noland 2009: 54, emphasis in original.
64 Many thanks to Jennifer Deveraux, Peter Meineck, and the participants of the 2016 Ranieri Colloquium on Classics and Cognitive Theory for their insightful feedback on this project. I am also grateful to Seth Estrin, Erin Lam, and Naomi Weiss for their comments on the final version of the chapter.

References

Anderson, Miranda. 2015. *The Renaissance Extended Mind.* London.
Bourdieu, Pierre. 1977. *Outline of a Theory of Practice.* Trans. Richard Nice. Cambridge.
Burris, Simon, Jeffrey Fish, and Dirk Obbink. 2014. "New fragments of Book 1 of Sappho." *ZPE* 189: 1–28.

Calame, Claude. 1977. *Les choeurs de jeunes filles en Grèce archaïque (Vols. I and II)*. Rome.
—. 1983. *Alcman: Introduction, texte critique, témoignages, traduction et commentaire*. Rome.
—. 1997. *Choruses of Young Women in Ancient Greece: Their Morphology, Religious Role, and Social Functions*. Trans. Derek Collins. Lanham, MD.
Calvert, Gemma, Charles Spence, and Barry Stein, eds. 2004. *The Handbook of Multisensory Processes*. Cambridge, MA.
Calvo-Merino, B., D.E. Glaser, J. Grèzes, R.E. Passingham, and P. Haggard. 2005. "Action observation and acquired motor skills: An FMRI study with expert dancers." *Cerebral Cortex* 15: 1243–1249.
Calvo-Merino, B., J. Grèzes, D.E. Glaser, R.E. Passingham, and P. Haggard. 2006. "Seeing or doing? Influence of visual and motor familiarity in action observation." *Current Biology* 16: 1905–1910.
Campbell, David. 1988. *Greek Lyric II*. Cambridge, MA.
Clark, Christina. 1996. "The gendering of the body in Alcman's *Partheneion* 1: Narrative, sex, and social order in archaic Sparta." *Helios* 23: 143–172.
De Preester, Helena, ed. 2013. *Moving Imagination: Explorations of Gesture and Inner Movement*. Amsterdam.
Di Benedetto, Vincenzo. 2004. "Osservazioni sul nuovo papiro di Saffo." *ZPE* 149: 5–6.
duBois, Page. 1996. "Sappho and Helen." In *Reading Sappho: Contemporary Approaches*. E. Greene, ed. Berkeley, CA. 79–88.
Ferrari, Gloria. 2008. *Alcman and the Cosmos of Sparta*. Chicago, IL.
Foster, Susan. 2008. "Movement's contagion: The kinesthetic impact of performance." In *The Cambridge Companion to Performance Studies*. Tracy Davis, ed. Cambridge. 46–59.
—. 2011. *Choreographing Empathy: Kinesthesia in Performance*. New York.
Garratt, Peter, ed. 2016. *The Cognitive Humanities: Embodied Mind in Literature and Culture*. London.
Greene, Ellen, ed. 1996. *Reading Sappho: Contemporary Approaches*. Berkeley, CA.
Greene, Ellen, and Marilyn Skinner, eds. 2001. *The New Sappho on Old Age: Textual and Philo sophical Issues*. Cambridge, MA.
Hagendoorn, Ivar. 2008. "Dance, choreography, and the brain." In *Art and the Senses*. Francesca Bacci and David Melcher, eds. Oxford. 499–514.
Hallett, Judith. 1996. "Sappho and her social context: Sense and sensuality." In *Reading Sappho: Contemporary Approaches*. E. Greene, ed. Berkeley, CA. 125–142.
Hutchinson, G.O. 2001. *Greek Lyric Poetry: A Commentary on Selected Larger Pieces*. Oxford.
Jackson, Lucy. 2016. "Greater than *logos*? Kinaesthetic empathy and mass persuasion in the choruses of Plato's *Laws*." In *Emotion and Persuasion in Classical Antiquity*. Ed Sanders and Matthew Johncock, eds. Stuttgart. 147–164.
Kousoulini, Vasiliki. 2013. "Alcmanic hexameters and early hexametric poetry: Alcman's poetry in its oral context." *GRBS* 53: 420–440.
Kowalzig, Barbara. 2007. *Singing for the Gods: Performance of Myth and Ritual in Archaic and Classical Greece*. Oxford.
—. 2013. "Broken rhythms in Plato's *Laws*: Materialising social time in the *Khoros*." In *Performance and Culture in Plato's Laws*. A.-E. Peponi, ed. Cambridge. 171–211.
Kurke, Leslie. 2007. "Visualizing the choral: Epichoric poetry, ritual, and elite negotiation in fifth century Thebes." In *Visualizing the Tragic: Drama, Myth, and Ritual in Greek Art and Literature*. Christina Kraus, Simon Goldhill, Helene P. Foley, and Jás Elsner, eds. Oxford. 63–101.
Lardinois, André. 1996. "Who sang Sappho's songs?" In *Reading Sappho: Contemporary Approaches*. E. Greene, ed. Berkeley, CA. 150–174.
—. 2001. "Keening Sappho: Female speech genres in Sappho's poetry." In *Making Silence Speak: Women's Voices in Greek Literature and Society*. André Lardinois and Laura McClure, eds. Princeton. 75–92.
—. 2009. "The new Sappho poem (P. Köln 21351 and 21376): Key to the old fragments." In *The New Sappho on Old Age: Textual and Philosophical Issues*. E. Greene and M. Skinner, eds. Cambridge, MA. 41–57.
Lidov, Joel. 2009. "Acceptance or assertion? Sappho's new poem in its books." In *The New Sappho on Old Age: Textual and Philosophical Issues*. E. Greene and M. Skinner, eds. Cambridge, MA. 84–102.
Lonsdale, Steven. 1993. *Dance and Ritual Play in Greek Religion*. Baltimore, MD.
Mauss, Marcel. 1979. "Body techniques." In *Sociology and Psychology, Essays*. Ben Brewster, trans. London. 97–123.
Meineck, Peter. 2011. "The neuroscience of the tragic mask." *Arion* 19.1: 113–158.
—. 2012. "The embodied space: Performance and visual cognition at the fifth century Athenian theater." *New England Classical Journal* 39.1: 3–46.

—. 2017. *Theatrocracy: Greek Drama, Cognition, and the Imperative for Theatre*. London.
Murnaghan, Sheila. 2013. "The choral plot of Euripides' *Helen*." In *Choral Mediations in Greek Tragedy*. Renaud Gagné and Marianne Hopman, eds. Cambridge. 155–177.
Nagy, Gregory. 1990. *Pindar's Homer: The Lyric Possession of an Epic Past*. Baltimore, MD.
Noland, Carrie. 2009. *Agency and Embodiment: Performing Gestures/Producing Culture*. Cambridge, MA.
Oakley, John and Sinos, Rebecca. 1993. *The Wedding in Ancient Athens*. Madison, WI.
Obbink, Dirk. 2016. "The newest Sappho: Text, apparatus criticus, and translation." In *The Newest Sappho: P. Sapph. Obbink and P. GC inv. 105, frs. 1-4*. Anton Bierl and André Lardinois, eds. Leiden. 13–33.
Olsen, Sarah. 2015. "Conceptualizing *Choreia* on the François Vase: Theseus and the Athenian youths." *Mètis: Anthropologie des mondes grecs anciens* N.S.13: 107–121.
—. 2017. "Kinesthetic choreia: Empathy, memory, and dance in ancient Greece." *CP* 112: 153–174.
Page, D.L. 1962. *Poetae Melici Graecae*. Oxford.
Paxton, Steve. 2001. "Improvisation is a word for something that can't keep a name." In *Moving History/Dancing Cultures: A Dance History Reader*. Ann Dils and Ann Cooper Albright, eds. 421–426. Middletown, CT.
Peponi, Anastasia-Erasmia. 2004. "Initiating the viewer: Deixis and visual perception in Alcman's lyric drama." *Arethusa* 37: 295–316.
—. 2007. "Sparta's prima ballerina: *Choreia* in Alcman's second *Partheneion* (3 PMGF)." *CQ* 57: 351–362.
—. 2015. "Dance and aesthetic perception." In *A Companion to Ancient Aesthetics*. Pierre Destrée and Penelope Murray, eds. Malden, MA. 204–217.
Power, Timothy. 2010. *The Culture of* Kitharōidia. Cambridge, MA.
—. "Sappho's parachoral monody." In *The Genres of Archaic and Classical Greek Poetry: Theories and Models*. Margaret Foster, Leslie Kurke, and Naomi Weiss, eds. Leiden.
Reason, Dee and Matthew Reynolds, eds. 2012. *Kinesthetic Empathy in Creative and Cultural Practices*. Chicago, IL.
Reitzammer, Laurialan. 2016. *The Athenian Adonia in Context*. Madison, WI.
Rosenmeyer, Patricia. 2004. "Girls at play in early Greek poetry." *AJP* 125: 163–178.
Segal, Charles. 1998. *Aglaia: The Poetry of Alcman, Sappho, Pindar, Bacchylides, and Corinna*. Lanham, MD.
Sklar, Deidre. 2001. Dancing with the Virgin: Body and Faith in the Fiesta of Tortugas, New Mexico. Berkeley, CA.
—. 2008. "Remembering kinesthesia: An inquiry into embodied cultural knowledge." In *Migrations of Gesture*. Carrie Noland and Sally Ann Ness, eds. Minneapolis, MN. 85–111.
Stehle, Eva. 1996. "Romantic sensuality: A response to Hallett on Sappho." In *Reading Sappho: Contemporary Approaches*. Ellen Greene, ed. Berkeley, CA. 143–149.
—. 1997. *Performance and Gender in Ancient Greece*. Princeton, NJ.
Swift, Laura. 2010. *The Hidden Chorus: Echoes of Genre in Tragic Lyric*. Oxford.
Thompson, Evan. 2005. "Sensorimotor subjectivity and the enactive approach to experience." *Phenomenology and the Cognitive Sciences* 4: 407–427.
Vestrheim, Gjert. 2004. "Alcman fr. 26: A wish for fame." *GRBS* 44: 5–18.
Voigt, Eva-Maria. 1971. *Sappho et Alcaeus. Fragmenta*. Amsterdam.
Warburton, Edward. 2011. "Of meanings and movements: Re-languaging embodiment in dance phenomenology and cognition." *Dance Research Journal* 43: 605–623.
Weiss, Naomi. 2018. *The Music of Tragedy: Performance and Imagination in Euripidean Theater*. Berkeley, CA.
West, M.L. 2005. "The new Sappho." *ZPE* 151: 1–9.
Williamson, Margaret. 1996. "Sappho and the other woman." In *Reading Sappho: Contemporary Approaches*. Ellen Greene, ed. Berkeley, CA. 248–264.
Winkler, John. 1990. *The Constraints of Desire: The Anthropology of Sex and Gender in Ancient Greece*. New York.

18

What do we actually see on stage?

A cognitive approach to the interactions between visual and aural effects in the performance of Greek tragedy[1]

Anne-Sophie Noel

In this chapter, I propose to *look at* and *hear about* a few theatrical objects in Aeschylus' *Oresteia*, for which visual effects and verbal cues most likely did not match in the course of the performance. I offer close readings of these passages, grounded in a neuroscientific framework and in philosophical theories on the integration of multisensory inputs. In doing so, I aim to shed new light on the multisensory treatment of objects in Aeschylus' dramas and on its reception by its original audience.

Owing to the limitations of the scientific knowledge acquired in this field up to today, this investigation inevitably entails some speculative thinking. However, I hope to demonstrate its heuristic value: gaining a better understanding of the multisensory interactions occurring in ancient theatrical shows can be determining to recast a new definition of *performance* in antiquity. Adopting this perspective seems particularly relevant in the case of Aeschylus' extant dramas, which contain, I will argue, an implied poetics of fluid interactions between aural and visual effects. At a broader level, my goal is to challenge the overemphasis that has been placed upon visual effects, as opposed to verbal effects, in the most current scholarship about the performance of ancient theater.[2] This focus on actual visual effects produced on stage has been historically important: Performance Studies applied to ancient theater have consisted mainly (and fairly) in re-habilitating the spectacular dimension (ὄψις) against more than two thousand years of scholarly neglect largely deriving from Aristotle's *Poetics*.[3] However, time has come for a more comprehensive investigation of spectacle that may benefit from recent research on the brain activity involved in the integration of multimodal sensations.[4] I will specifically focus on the experience of incongruence between the aural and visual sensations, since this seems to be a specific feature of Aeschylus' dramaturgy.

Multisensory integration in neuroscience and brain theory

Research on embodied cognition and processing of emotions has been applied to performance studies for about a decade now. The main orientation, as traced by the ground-breaking works of Bruce McConachie and Rhonda Blair,[5] is the exploration of the roles played by the 'mirror neuron system', empathy and embodiment in shaping the relationship between actors and spectators. How the knowledge brought by the neuroscience can impact actor's pedagogy, acting techniques and applied theater is also under investigation in the most recent volumes published in this field.[6] However, neither multisensory interactions nor the interplay between the senses and imagination have come into focus in this still burgeoning domain.

On the side of functional MRI and multisensory studies, however, a lot of recent research has been done to grasp a better understanding of our daily-life, real-world actions that often stimulate several senses simultaneously.[7] The spatial organization of multisensory brain regions has been under scrutiny: for instance, the bimodal superior temporal cortex (bSTC) was identified as a portion of the brain that reacts to both visual and auditory inputs and whose responses are enhanced when the inputs are multisensory (auditory-visual).[8] Researchers in this area admit to still facing important methodological issues: some acknowledge that most experiences, based on the processing of abstract stimulus material (like button press, dot and tone),[9] fail to reach the complexity of the integration of audiovisual inputs in a real-life context (and *a fortiori*, in an aesthetic work of art). The methodology for reaching more micro-scale results in brain mapping is also still under discussion.[10] In spite of these limitations (that point toward further research), a convergent body of experiences has shown the flexibility of the multisensory system.[11] Multisensory integration, defined as the set of processes by which information arriving from individual sensory modalities (e.g. vision, audition, touch) interacts with and influences processing in other sensory modalities, is now well established and has crucial implications for contemporary sensory neuroscience.[12] The idea that sensory inputs were processed separately in specific unisensory areas before being unified has been abandoned in favor of fluid relationships between interconnected, multisensory brain areas.

In the field of brain theory and philosophy of mind, the conceptualization of perception as a 'constructive process'[13] may also be an interesting theory to re-think the spectator's experience. It argues that far from being a passive 'recording' of sensory inputs, perception is an active process in which the brain 'optimizes the mutual information between its sensory signals and some parsimonious neuronal representations'.[14] The underlying idea is that of the Bayesian brain,[15] which postulates that the brain works as an 'inference machine', possessing an internal representation of the world that is not only shaped by sensory inputs, but also shapes them in return to make them cohere with this representation. This has also been termed 'Prediction Processing' by Andy Clark:[16] in our perceptual apprehension of the exterior world, we are constantly making predictions about sensory inputs, according to the represented model that we have interiorized. If I hear a banging noise, I will spontaneously make inferences about all the possible sources for this potentially alarming signal and then select the sensory inputs relevant to identify efficiently this signal in given circumstances. Building on this, Andy Clark phrases in a striking way the fact that perception entails imagination:

> Such perceivers are thereby imaginers too: they are creatures poised to explore and experience their worlds not just by perception and gross physical action but also by means of imagery, dreams, and (in some cases) deliberate mental simulations.[17]

This neuroscientific research and theoretical models allow us to reconsider what it means for a spectator to simultaneously experience multisensory inputs triggered by a theatrical performance, and Peter Meineck has paved the way for applying 'Prediction Processing' and the processing of multisensory inputs to ancient Greek drama in his recent ground-breaking monograph.[18] His experimental research on masks convincingly supports the idea that ancient spectators may have been more receptive to the 'multimodal and sensorimotor facets of speech processing and movement',[19] because mask-acting requests, for an efficient communication, the combination of 'highly coordinated movement, gestures, and heightened poetic language'.[20] An hypothesis that I am inclined to challenge, yet, is the emphasis placed on vision at the expense of aural effects: drawing on the McGurk effect, Meineck states that in the performance, 'the predominantly higher-order cognitive processing of the spoken word is subordinated to the lower-order processing of visual or other somatosensory data'.

The McGurk effect is a multisensory illusion triggered by the processing of a voice articulating a consonant (like [b]), dubbed with a face articulating another consonant (like [g]): the acoustic speech signal is systematically heard as another consonant ([d]), showing that the visual information alters the auditory one.[21] If this effect has been widely observed at the level of a consonant or a syllable, it has, however, hardly been proven at the level of a short sentence.[22] One may legitimately wonder whether this 'trumping'[23] of auditory input by the visual one is transferrable to such a highly complex aesthetic experience as a theatrical performance.

Having said that, even without resorting to the McGurk effect, it seems evident that vision is a specific trait of drama that is often emphasized in the Greek plays as a more reliable sense,[24] and I certainly agree with Green when he writes that 'there is a good deal of evidence, both literary and pictorial, to suggest that what people perceived as one of the most exciting things about theatre when it was first being invented was the visual spectacle'.[25] While acknowledging this dimension, I am keen to bring up excerpts from Aeschylus' plays that comment on the interactions between visual and aural effects: without downplaying the power of visual perception, they seem to postulate extremely fluid relationships and reciprocal cross-modal influences between them in the performance.

In the parodos of the *Seven Against Thebes*, the women of the chorus run onto the stage and cling to the statues of the gods, expressing their fear of the war to come. This famous passage emphasizes the 'visual construal of the sound'[26] and, reciprocally, the auditory construal of the sight.

> ἀκούετ᾽ ἢ οὐκ ἀκούετ᾽ ἀσπίδων κτύπον;
> πέπλων καὶ στεφέων <πότε> ποτ᾽ εἰ μὴ νῦν
> ἀμφὶ λιτανὰ <βαλεῖν > ἕξομεν;
> κτύπον δέδορκα· πάταγος οὐχ ἑνὸς δορός.[27]
>
> Do you hear, or do you not, the clatter of shields? When, when, if not now, shall we be able <to adorn the gods> with robes and garlands as prayer-offerings? I see the banging noise—it is the clatter of many spears![28]

κτύπον δέδορκα, 'I see the banging noise': the phrase suggests a complete merging of the aural and the visual. The women of the chorus address each other in a collective moment of panicking, but it seems rather clear that this address can also be construed as an apostrophe to the audience, prompted to process to the same effort of imagination, stimulated by a particularly powerful series of plosive alliterations miming the clash of weapons.[29] In his 2012 article,[30] Agis Marinis employs the Aristotelian concept of ἐνέργεια, 'actualization', to comment on the way this Aeschylean phrase 'impels the audience to visualize images, enabling them to participate in the persuasive process through their sensory reaction to words'.[31] In Andy Clark's terms, one could say that the women's speech prompts the spectator's brain to combine their prior knowledge and internal representation of war with the incoming sensory evidence, to generate a visualization of the arrival of Polyneikes' army at the seven gates of Thebes—a staging that was, by the way, impossible to realize concretely.[32]

This enacted thinking seems all the more important in the *Seven* because it also underlies the famous shields' *Redepaare*. The shields are material objects that have a dramatic visual impact both on the internal and external audience, yet they are entirely construed by the speeches of both herald and Eteocles.[33] On some shields, inscribed words are materialized, becoming gold or bronze letters, hence being apprehended by senses of audition, sight but also touch, inasmuch as metal may speak to some spectators in terms of texture, hardness and coolness. Along the same aesthetic principle, when the corpses of Eteocles and Polyneikes are finally brought under the eyes of the audience, the women of the chorus state that the discourse of the messenger (λόγος) has been made 'visible' (προὖπτος):[34] theatrical action consists in making the words heard before become a visual discourse on stage, but everything remains language, be it auditory or visual. Anticipating modern cognitive findings about the cross-modal influences occurring in the processing of multimodal sensory inputs, Aeschylus thus stimulates the vision of the audience with sound and other non-visual inputs (like touch and proprioception), while manipulating the auditory sensation by non-auditory inputs. Therefore, his dramaturgy, I argue, can be considered as an enacted refutation of 'visual dominance', the domination of vision over the other human senses.

Multisensory objects in the *Oresteia*

For two significant objects in the *Oresteia*, Apollo's bow and Clytemnestra's net cloth, the audience most probably faced the task to process discordant auditory and visual inputs. Its response to it must have been inevitably various. In her study combining art criticism and neuroscience, Gabrielle Starr notes that all people do not have the same capacity to visualize images in their mind precisely.[35] An ancient audience, however trained to the active listening required by regular poetic performances and contests, was probably composed of spectators more or less able to shape internal images eliciting cross-modal sensations. Auditory images, Starr goes on, also differ when they are prompted by plain speech, metrical speech or music[36]—a diversity that ancient dramatists could exploit fully in a performance that involved meter and music. For the two objects under focus, I am keen to argue that Aeschylus could draw on 'somatosensory' language[37] to affect the visual processing of the spectators watching a performance: the 'auditory imagery' prompted by this powerful affective and sensory language competed with real visual effects on stage.[38]

In the beginning of *Eumenides* by Aeschylus, the god Apollo, staged in the previous plays of the trilogy as a statue, appears in person to defend his protégé Orestes. He violently chases away the Erinyes who have followed Orestes inside his Delphic sanctuary.

> Ἔξω, κελεύω, τῶνδε δωμάτων τάχος
> χωρεῖτ', ἀπαλλάσεσθε μαντικῶν μυχῶν,
> μὴ καὶ λαβοῦσα πτηνὸν ἀργηστὴν ὄφιν,
> χρυσηλάτου θώμιγγος ἐξορμώμενον,
> ἀνῇς ὑπ´ ἄλγους μέλαν´ ἀπ´ ἀνθρώπων ἀφρόν,
> ἐμοῦσα θρόμβους οὓς ἀφείλκυσας φόνου.[39]

> Out, I tell you, get out of this house at once! Get away from my inner prophetic sanctum, in case you find yourself on the receiving end of a winged flashing snake speeding from my golden bowstring, and vomit out in agony black foam taken from human bodies, bringing up the clots of blood that you have sucked.[40]

This aggressive stage entrance might be a reminiscence of the first appearance of Apollo on Olympus, in the first *Homeric Hymn*:[41] the god of the bow, Apollo τοξοφόρος or ἑκατηβόλος, according to his traditional Homeric and lyric epithets, boldly aims his taut weapon at the frightened assembly of the gods before being gently disarmed by his mother. In the Aeschylean play, although the text does not hint explicitly at the use of the bow as a theatrical prop on stage, it seems reasonable to suppose that the actor who embodied Apollo did hold a bow and direct it against the Erinyes in a threatening way. The weapon is an emblematic and iconic attribute of the god and was very likely used on stage to make him recognizable. A few decades later, Euripides parodied the systematic use of the attribute in the prologue of his *Alkestis*.

> ΘΑ. Τί δῆτα τόξων ἔργον, εἰ δίκην ἔχεις;
> ΑΠ. Σύνηθες αἰεὶ ταῦτα βαστάζειν ἐμοί.
>
> Death: If justice, then what need for your bow and arrows?
> Apollo: It is my custom always to carry them.[42]

This may well be a playful metatheatrical comment made by Euripides to criticize an abuse of the use of props, without any larger dramatic purpose. But in the case of the *Eumenides*, the presence of the bow is dramatically meaningful. It lends to Apollo's aggressive orders to the Erinyes a much more threatening aspect—μὴ καὶ λαβοῦσα suggests that if they don't leave at once, they will be shot.[43] The silhouette of the actor wielding a bow has a salient character: Apollo's war pose could arguably have startled a predominantly hoplite audience. This offensive posture also clearly has a visual reference: this staged Apollo seems an iconic imitation of paintings or sculptures of the vengeful Apollo, directing his arrows against the Giants, or against the children of Niobe, as on the famous Niobids Crater.[44] But on stage, he is like a painting that moves and speaks, which might be attention-grabbing and, in passing, realizes the famous aphorism attributed by Plutarch to Simonides.[45] The arrow maintained under tension, while he addresses the group of Erinyes, also has the potential to stimulate proprioception and create a compelling tension in the audience itself: according to the Prediction Processing model, this visual cue was a prompt to anticipate in imagination the shooting that may happen—an effect that contemporary film directors constantly exploit by zooming in on hands grasping firing weapons. If one adds that to be heard, the masked actor would have had to face the audience in a frontal way,[46] one may legitimately raise the hypothesis that the spectators could have been afraid of being struck themselves, as is reported by Macrobius about a Roman pantomime of *Hercules Furens*.[47] The bow hence stood out from its surroundings and might well have been what neuroscientists call a 'salient object',[48] or 'a locus for foveal vision', as Meineck puts it—foveal vision being used for focusing on details and scrutinizing objects.[49]

Apollo's bow is thus one of these significant props whose presence on stage is highly meaningful.[50] What strikes me then is the discordance that could appear between the actual object, which was most probably seen on stage, and the way the god describes it. Apollo does not speak of bow and arrows but of a 'winged flashing snake speeding from [his] golden bowstring'. The verbal cues here are vigorously multisensory: vision is stimulated by the suggestion of the hybrid shape of a 'winged snake' (πτηνὸν ὄφιν).[51] Snake and arrow were symbolically associated in the collective imagination of the audience—we may think of the arrows of Heracles, anointed with the poisoned blood of the Lernaean hydra. The shapes of a snake and an arrow were compared in later treatises on animals and hunting, to the extent that P. Monbrun can speak of the snake as the 'doublet animal de la flèche'.[52]

Aeschylus nevertheless adds wings to this snake-like arrow; on one level, it may be taken as a metonymic association prompted by the flight of the arrow through the air, and by the fletching or feathers at the back of the arrow. However, the metaphorical formulation may also arouse in the audience's imagination the visualization of a winged serpent, facilitated by the existence of folkloric beliefs about such a composite creature. Herodotus describes winged serpents (πτερωτοὶ ὄφιες) as a species supposedly living in Arabia.[53]

The visual stimulation is also achieved through several mentions of color: ἀργηστὴν, 'flashing', alludes to the shimmering aspect of the snake's skin.[54] But it also means white. White is juxtaposed to the golden color, brought into focus with the compound χρυσηλάτου, that relates the bow of this theatrical Apollo to the golden or silver bow that is his attribute in epics and hymnic poetry.[55] A chromatic contrast then stands out, between the bright white and gold and the 'black foam' that the Erinyes are going to vomit up if they are hit by an arrow. The words of Apollo reactivate a visual leitmotiv that runs throughout the whole trilogy—the contrast between black (black-clad Erinyes, φαιοχίτωνες),[56] the white attire of the Olympians[57] and the ominous color of blood, masterfully staged by Aeschylus in the famous 'carpet scene'.[58] The color of blood is also hinted at in our passage through the allusion to the blood clots—θρόμβους φόνου (182).

In addition to these multiple visual cues, the kinesthetic sense, the sense of motion or 'proprioception', is also stimulated by the allusion to the hurling of the arrow, signaled by the participle ἐξορμώμενον, 'sent forth with speed'. The allusion to the bowstring was also possibly meant to stimulate auditory sensations, especially because the twang of the bowstring was commonly associated with the vibration of the strings of the lyre, the other emblematic attribute of Apollo.[59] In the Homeric *epos*, Odysseus, an expert with the bow, is described as an ἀνὴρ φορμίγγος ἐπιστάμενος[60] and his bowstring makes the shrieking sound of a swallow.[61] In *Iliad* Book I, the arrows that Apollo hurls at the Achaeans speed with a 'terrible sound or scream' (δεινὴ κλαγγή, 49).

Thus, verbal cues strategically enrich the visual spectacle of Apollo bending his bow through an amalgamation of meaningful visual, auditory and somatosensory inputs. This first encounter between Apollo and the Erinyes is shaped as a kind of supernatural and cosmic battle, in which glistening winged snakes oppose the whistling snakes that adorn the hair of the Chthonic deities. The god of the Sun threatens the daughters of Night; white and gold are opposed to black and red, *drakôn* against *drakôn*, to borrow a chapter title from Ogden's monograph.[62] It could also be construed as a creative remake of the initial battle of Apollo with the serpent Pythô.[63]

Although this brief two-verse passage was heard by the spectators fleetingly, one can see how Aeschylus stimulates the mind's eye of the audience, by building on prior cultural memories rooted in the imagination of his audience. The same discrepancy between what was actually seen on stage and what was heard by the spectators is also evident in the case of an even more famous and polysemic object in the *Oresteia*: the many-colored cloth (ποικίλα)[64] that Clytemnestra uses as a trap for Agamemnon, which certainly had the visual relief of a 'salient object' onstage. Its versatile nature and appearance are made clear through its material reconfigurations in the so-called 'carpet scene' and the two tableau scenes of the exposition of the corpses, at the end of *Choephori* and *Eumenides*.[65]

One may assume that pieces of purple-colored cloth were used to stage this object, but it is difficult to draw any further conclusion. Its actual nature eludes any monolithic depiction, insofar as through the course of the trilogy, it is named with twenty-four different denominations ranging from plain textiles spread on the ground (πετάσματα) to a variety of clothes (πέπλος, the long tunic; εἷμα, the cloak; καλύμματα, the veils), or even various types of fetter.[66] Its changing designations transform it successively into a spider web (ἀράχνης ὕφασμα),[67] a hunting or fishing net (δίκτυον, ἄρκυς, ἀμφίβληστρον, βρόχος, ἄγρευμα)[68] and a shroud (φᾶρος, δροίτης κατασκήνωμα),[69] or turn it into chains, fetters or manacles (δεσμός, πέδαι et ξυνωρίς).[70] This profusion of words, superimposed on the successive material reconfigurations of this object, opens out its nature and significance in an overwhelming way. An ἄπειρον ἀμφίβληστρον, this treacherous cloth is an 'endless net', 'endless' having both a physical and symbolic meaning here: it does not have any opening for arms or head; it is also 'endless' since its identity is never fixed in a unique shape.

This superabundance of verbal tags certainly prompted plenty of visual responses in the audience's imagination but also a variety of multisensory stimuli. Melissa Mueller rightly speaks of the 'sensory overload' of Clytemnestra's fabrics. She convincingly argues that the strong smell of the *porphyra* dye could have been an extra sensual input provided by the performance.[71] Complementarily, I suggest, the sense of touch was highly stimulated too, not only *in addition to* auditory and visual inputs but actually *through* them. The tactile qualities of the net cloth are triggered off by the verbal cues enumerating the diverse materials—the delicate weaving of the spider web, the coarser feel of the stitching of the net, or the metallic touch of the chains. Moreover, in the scene where the ποικίλα are spread on the ground, the dramatic action consists in observing which way Agamemnon will choose to enter the palace; in other terms, in viewing what his bare feet will touch or not touch[72] and what impact this haptic contact will have on the fabrics themselves, explicitly associated to his *oikos*' fortune,[73] as well as on his own life. Fearing the jealousy of the gods, Agamemnon is explicitly searching for an appropriate way to lay his feet on the cloth, delicately, unlike one would do for 'foot-wipers' (ποδοψήστρων). Moreover, if Aeschylus had actually used murex for these stage tapestries—we cannot have any certitude about this, but it seems quite plausible that spectators could themselves have been in a comparable state of doubt[74]—the scene would have aroused the added threat of the actor destroying the delicate dye as well.

Another layer of multisensory stimulation, spurring the spectator's haptic sense, is the well-known visual and symbolic association between this path of cloth and a flow of blood pouring from the cursed palace of the Atrids.[75] The polysemy of the twice-occurring term πόρος (πορφυρόστρωτος πόρος)[76] has rarely been noticed: the path is not only an earthly one, it also runs like a purple stream, a purple river or a strait, all meanings carried by the substantive πόρος.[77] The metaphorical transmutation of cloth into a fluid element is supported by this wording and induces a corollary new set of sensual experience: Agamemnon's feet do not only touch the textiles, but are also splashed by a flow of blood. The epithet πορφυρόστρωτος (a notable *hapax*) also reinforces the aqueous nature of the stream of cloth: applied to πόρος, it is reminiscent of the famous and puzzling homeric 'purple sea',[78] which is not to be understood as a sea reflecting the color of sunset, but as a mass of water animated by the 'bubbling of fermentation' and a 'shimmering surface'[79] similar to that of the *porphura.* Scholars interested in the history of the senses and synesthesia recently reconsidered the question: A. Grand-Clément notes that in archaic poetry, 'la référence chromatique se trouve parfois associée à d'autres notions, qui ne sont pas nécessairement visuelles, comme le mouvement, les matières et les textures, les odeurs, les sonorités'.[80] This is very close to what M. Bradley wrote, defining color as an 'object-centered experience',[81] 'imbued with sensory property beyond the visual domain' in ancient Greece.[82] In the Aeschylean drama, this first association of purple and stream (πορφυρόστρωτος πόρος) anticipates the later connection made by Clytemnestra between the purple color and the endless flow of the sea:[83] this verbal imagery has a great coherence and strongly heightens the sensory properties of the prop seen on stage. Thus, Aeschylus allows his audience to simultaneously witness and build in their mind a vivid image of Agamemnon's walk home that foreshadows the fatal bath in which he is going to die behind the doors of the palace. In that regard, he manipulates not only the sensory responses induced by this versatile object but also the different time dimensions, since past and future murders are united in this flowing stream of blood—giving to Heraclitus' claim about ever-changing water a gruesome scenic incarnation.[84]

Like Apollo's bow, the net cloth is therefore another striking case where powerful, sensory-loaded words overwhelm the visual spectacle represented on stage. Even in today's big-budget theatrical productions of the *Oresteia*, staging the 'carpet scene' and the net cloth, these two sides of the same versatile object, is a challenging task.[85] There is no material object that can match the versatility and hybridity of the object constructed by the auditory and visual interactions. Hence it may be worth asking the following questions: when the spectators of the *Oresteia* left the theater, what would they have remembered? The simple prop likely used to materialize the bow on stage, or the snake-like winged arrow, threateningly aimed by Apollo against the Erinyes? The material pieces of red clothes actually spread on stage, or the monstrous, hybrid and versatile object that the multisensory cues prompted in their imagination?

I am inclined toward the latter and this conviction may now be supported by neuroscience and contemporary brain theory. As stated in the beginning of this chapter, perception is now interpreted as a 'constructive process' in which prior knowledge and sensual predictions about what is being seen play a crucial role. The progress made in functional brain imaging also conduced to good evidence that the mental imagery and visual perception activate many of the same brain early processing areas:[86] imagined sensation and actual perception involve the sensory modes in a comparable manner,[87] which leads Clark to assume that 'perceiving, imagining, understanding, and acting are now bundled together, emerging as different aspects and manifestations of the same underlying prediction-driven, uncertainty-sensitive, machinery'. Of note is also the dramaturgic construction of the trilogy, drawing on mirror scenes,[88] anticipations and resurgences of visual and verbal leitmotivs: auditory images, which are organized temporally rather than spatially,[89] can benefit from the fact that they are reminded and reconfigured smoothly, gaining a growing pervasiveness throughout the course of the trilogy.[90] Therefore, if I do not want to reduce the diversity of subjective sensory experiences encountered by the spectators, I am defending an interpretation of Aeschylus' plays as works profoundly engaged with multisensory interactions that challenged visual dominance, in favor of the co-working of perception and imagination.

This investigation can have broad implications for the study of ancient performance. Visual effects are currently too restrictively understood as visible material staging. The stress should rather be placed on the interplay between the senses allowed by the combination of aural cues and visual effects, on the one hand, and the cognitive processes that go on in the imagination of the spectators of the performance. This shift of paradigm for performance studies as applied to ancient Greek tragedy may well bring new arguments against negative judgments still circulating today: Greek tragic stagecraft would produce an 'austere' theatrical art, a 'minimalist' or 'highly formal' performance with very few spectacular effects.[91] Such preconceived ideas seem to derive from an essentially visually focused definition of a performance that urgently needs to be questioned.

Notes

1 This chapter has greatly benefited from the resources of the Center for Hellenic Studies: I am grateful to CHS director Gregory Nagy, the senior fellows and all the CHS staff for making them available to me. Special thanks to Laura Slatkin, David Konstan and Peter Meineck for their reading and judicious critique, and for the audience at NYU for their encouraging reactions and thought-provoking comments.

2 For instance, in the recent reference volume edited by Harrison and Liapis (2013), performance studies are defined as being about 'the non-verbal constituents of ancient theater' (1). In her study of tragic

props, Chaston (2010) also privileges the visual impact of props as 'images' which may 'serve a cognitive function in thinking and problem solving' (1). Therefore, although focused on tragic objects and adopting a cognitive perspective, Chaston's monograph proposes neither any dramaturgical analysis of the tragic stagecraft nor a study of cognitive processing of multisensory stimuli in the performance of Greek tragedy.

3 Aristotle's comments on the value of ὄψις are more complex and nuanced than a simple dismissal. For a discussion about this, see Taplin 1977 (477–479); Frazier 1998; Billault 2001; de Marinis 2009; Sifakis 2013.
4 Telò 2013 (53-69) pursued a comparable goal in a chapter about synesthesia in ancient Greek comedy. He states that the importance of 'the more carnal senses of touch, taste, and smell' in comedy calls 'into question the epistemological centrality ascribed in ancient (as well as modern) times to sight and to foster an alternative aesthetic regime that enhances the inherently visual quality of theatrical performance'. Although I focus mainly on audio-visual interactions in this chapter, I believe other senses were also involved in the experience of tragic spectators (like touch and proprioception), as will be seen below.
5 McConachie and Hart 2006; McConachie 2008, 2013, 2015; Blair 2008, Blair and Cook 2016.
6 Falletti *et al.* 2016.
7 Straube 2017 (2); Clark 2015 (64).
8 Gentile *et al.* 2017 (10104).
9 Straube *et al.* 2017 (20).
10 Gentile *et al.* 2017.
11 Gentile *et al.* 2017 (10111).
12 Clark 2015 (86).
13 Friston 2010 (3).
14 Friston 2010 (6).
15 Doja *et al.* 2007.
16 Clark 2015.
17 Clark 2015 (85).
18 Meineck 2017.
19 Meineck 2017 (188).
20 Meineck 2017 (187).
21 McGurk and McDonald 1976; Meineck 2017 (181, 'the audience's visual processing system trumps the audience's auditory processing system').
22 Van Engen *et al.* 2017. They show that further research on audiovisual integration should involve different types of text material.
23 Meineck 2017 (181).
24 Cf. Chaston 2010 (27).
25 Green 1994 (16-17).
26 Marinis 2012 (26).
27 Aeschylus, *Seven*, 100-103.
28 Trans. Sommerstein, Loeb, 2008 modified.
29 These are proper 'somatosensory words', to put it in Meineck's terms (2017 (191-195)). Music and metric also probably impacted greatly auditory images, see Starr 2013 (78); for a metrical analysis of this choral ode, see Muñoz 2012.
30 Marinis 2012; see also Trieschnigg 2016.
31 Marinis 2012 (40).
32 Cf. Clark 2015 (85).
33 Cf. Chaston 2011 (67); Pallidini 2016 (113-114) *contra* Wiles 1990.
34 Aeschylus, *Seven*, v. 847.
35 Starr 2013 (77).
36 Starr 2013 (78).
37 Meineck 2017 (191-195).
38 Taplin 1977 (325).
39 (*Eum.*, 179-185)
40 Trans. Sommerstein (Loeb 2008).
41 *HH Ap.*, 1-9.
42 Trans. Kovacs (Loeb 1994).
43 Apollo has promised to Orestes that 'he will not be soft to [his] enemies' (*Eum.*, 65-66).

44 Attic red-figure Kalyx Krater attributed to the Niobid painter, ca. 460–450 BC, Paris, Musée du Louvre, inv. G 341; see Noel 2014.
45 *Plutarch, De gloria Atheniensium*, 3.347a.
46 My grateful thanks go to Peter Meineck for drawing my attention to mask frontality here; see Meineck 2011 (139–142).
47 Macrobius, *Saturnalia* 2.7.16–18: 'When Pylades acted the part of the insane Hercules and some thought he wasn't maintaining the gait appropriate to an actor, he took his mask off and scolded the people who were laughing by saying "Idiots! I'm dancing a madman." In the same play, he shot an arrow into the audience. And when he was acting Hercules at Augustus' behest in his dining hall, he aimed his bow and shot some arrows—and Augustus thought it only fair to find himself in the same position, vis-à-vis Pylades, as the Roman people has been'.
48 Talsma et al. 2011 (401): 'Visual objects are said to be highly salient when they have a particularly distinctive feature with respect to the neighboring items and the background, or if they occur suddenly'.
49 Meineck 2012 (122).
50 Taplin 1977 (36): 'It is not difficult to think of stage properties which are so imbued with dramatic significance that it would do positive harm to the play in performance if they were not straightforwardly represented; the purple cloth in A. *Ag*, the sword in S. *Aj*, the bow in S. *Phil*, or E. *Her*, the image of Artemis in *IT* and many others'. See also Taplin 1978 (18–19).
51 In his 1936 essay, Stanford rightly labeled this type of metaphor as '*synaesthetic* or *intersensal* metaphor' (48); this synesthetic quality must nevertheless be considered as triggering a real multisensory experience and not only be seen as a rhetorical ornamentation. On synesthesia in Sophocles, see also Segal 1977.
52 See *Nicander's Theriaca* and *Oppian's Cynegetics*, mentioned in Monbrun 2007 (234–243).
53 Herodotus II, 75, 76.
54 There might be wordplay here too, because the rare substantive ἀργῆς could designate a kind of serpent, in a literal or metaphorical way (Achae. 1, Trag. Adesp. 199; ὄφις ἀργῆς, Hippocrates, *Epidemiae* 5, 86).
55 *Iliad*, II, 764; V, 509; XXIV, 605-606; Callimachus, *Hymn to Apollo*, 32–35.
56 *Cho.*, 1049.
57 *Eum.*, 351, πάνλευκοι πέπλοι.
58 On this scene, see further on and the discussion in Lebeck 1971 (74–79; 80–91); Taplin 1977 (314–316); Jenkins 1985; Halm-Tisserand 2002; Noel 2013; Mueller 2016 (48–57).
59 Euripides, *Heracles*, 1064 (τοξήρει ψαλμῷ); *Bacch.*, 783–784; Aeschylus, fr. 57, l. 7 *TrGF*3 Radt; Telestes, fr. 6; Diogène trag., fr. 1.
60 *Od.*, XXI, 406.
61 *Od.*, XXI, 411–412.
62 Ogden 2013 (215).
63 Ogden 2013 (192–195).
64 *Ag.*, 923, 935, 946, 957.
65 That the tapestry and the 'net cloth' are different avatars of the same object is already suggested by Lebeck 1971 (81, 85) and Taplin 1977 (314); more recently, see Noel 2013 (164-168); Mueller 2016 (60–62).
66 For a complete list, see Noel 2013 (166-167).
67 *Ag.* 1492.
68 Δίκτυον (*Ag.*, 1115, *Cho.*, 999); ἄρκυς (*Ag.*, 1116, *Cho.*, 1000); ἄπειρον ἀμφίβληστρον (*Ag.*, 1382); ἀμφίβληστρον (*Cho.*, 492); βρόχος (*Cho.*, 557); ἄγρευμα θηρός (*Cho.*, 998). See the typology of Halm-Tisserand 2002.
69 *Cho.*, 999, 1011; *Eum.*, 634.
70 *Cho.*, 493, πέδαι δ'ἀχάλκευτοι; *Cho.*, 981, τὸ μηχάνημα, δεσμὸν ἀθλίῳ πατρί; *Cho.*, 983, πέδαις τε χειροῖν καὶ ποδοῖν ξυνωρίς; *Cho.*, 1000, ποδιστῆραι πέδαι. See *Od.*, VIII, 266sq. on Hephaistos' net. On affinities between cloth and metal in Greek thought, see Détienne and Vernant 1974 (279); Frontisi-Ducroux 1975; Jenkins 1985 (121); Noel 2013.
71 Mueller 2016 (56).
72 Agamemnon does take off his *arbulai* (944-945) and the act of walking on the pieces of cloth is evoked verbally several times before being effectively performed by Agamemnon (906-907, 924–925, 936, 948, 957). Clytemnestra prevents him from touching the ground with his feet (μὴ χαμαὶ τιθείς τὸν σὸν πόδ', 907).
73 *Ag.*, 948–950; 958–965.

74 The purple colour could be obtained with cheaper material than the very expensive murex (for instance with woad and madder; see Cleland et al. 2007 (155)), but the difference was not necessarily visible in a one-time performance.
75 Lebeck 1971; Taplin 1977.
76 Ag., 910, 921.
77 Cf. *Persians*, 453, 875, *Cho.*, 366, *Eum.*, 452, *PE*, 532.
78 *Iliad*, XVI, 91; also 'purple wave', *Iliad*, I, 482; XIV, 16; XXI, 326; *Od*, II, 428; XI, 243.
79 Bradley 2013 (133).
80 Grand-Clément 2013 (129).
81 Grand-Clément 2013 (132).
82 Bradley 2013 (132-135).
83 *Ag.*, v. 958-960. Building on Ferrini (2000), Grand-Clément (2013) affirms the long-term association between πορφύρεος and the sea throughout archaic and early classical poetry (134).
84 Heraclitus, DK 22b12.
85 A history of this prop and its material reconfigurations on the modern stage is still needed. Modern directors sometimes suppress it (as did Ariane Mnouchkine in her historical production of the *Oresteia* at the Théâtre du Soleil, Vincennes, 1990-1992); or they privilege only one aspect of this proteiform object (the net for Peter Hall, London, Olivier Theatre, 1981 and Epidaurus, 1982; a long red piece of cloth for Olivier Py, *L'Orestie d'Eschyle*, Théâtre de l'Odéon, Paris, 2008). Others reinvent the meaning of the path of cloth in a personal way: Peter Stein (1994, Moscow) staged a mound of marron shirts symbolizing the victims of war; Katie Mitchell (London, 1999) replaced the purple path with 'a long blasphemous patchwork of the little dresses that had once belonged to Iphigenia' (Paul Taylor, *Monday Review*, December 6, 1999, 10). These are powerful reinterpretations of the original object, yet they also (inevitably) reduce its extraordinary versatile identity. In a TV miniseries entitled *Helen of Troy* (John Kent Harrison, 2003, Universal Home Entertainment), Aeschylus' net cloth makes an unexpected appearance: the film director had her Clytemnestra travel to Troy to take her revenge against Agamemnon. She throws a net on him while he is bathing in a pool of the palace of Priam.
86 Clark 2015 (95), who refers to Kosslyn et al. 1995; Ganis et al. 2004.
87 Starr 2013 (75) draws on this to develop crucial aspects of her aesthetic theory.
88 Taplin 1978 (122-139).
89 Starr 2013 (75).
90 As when Orestes recalls the 'embroidered net' of Clytemnestra when put on trial at the Aeropagus, *Eum.* 460.
91 See for example Kovacz, in Roisman 2014 (952).

References

Billault A., 2001, 'Le spectacle tragique dans la *Poétique* d'Aristote', in A. Billault and C. Mauduit, *Lectures antiques de la tragédie grecque*. Lyon. 43–59.

Blair R., 2008, *The Actor, Image, and Action: Acting and Cognitive Neuroscience*. London, New York.

Blair R. and Cook A., 2016, *Theatre, Performance and Cognition: Languages, Bodies and Ecologies*. London.

Bradley M., 2013, 'Colour as a synaesthetic experience in antiquity', in S. Butler and A. Purves, *Synaesthesia and the Ancient Senses*. Durham, NC. 127–140.

Bradley M., 2015, *Smell and the Ancient Senses*. Abingdon, New York.

Brosch M., Selezneva E., Scheich H., 2005, 'Nonauditory events of a behavioral procedure activate auditory cortex of highly trained monkeys'. *Journal of Neuroscience*, 25, 6797.

Butler S. and Purves A., 2013, *Synaesthesia and the Ancient Senses*. Durham, NC.

Calame C., 2006, *Pratiques poétiques de la mémoire: représentations de l'espace-temps en Grèce ancienne*. Paris.

Chaston C., 2010, *Tragic Props and Cognitive Function: Aspects of the Function of Images in Thinking*, Mnemosyne Supplements 317. Leiden/Boston, MA.

Clark A., 2015. *Surfing Uncertainty: Prediction, Action, and the Embodied Mind*. New York.

Cleland L., 2007, *Greek and Roman Dress from A to Z*. London, New York.

De Marinis M., 2009, 'Aristotele teorico dello spettacolo nella *Poetica*', online publication on 22/10/2009, http://drammaturgia.fupress.net/saggi/saggio.php?id=4275.

Detienne M. and Vernant J.P., 1974, *Les Ruses de l'intelligence, la mètis des Grecs*. Paris.

Falletti C., Gabriele S. and Jacono V., 2016, *Theatre and Cognitive Neuroscience*. London.

Ferrini M.F., 2000, 'La porpora e il mare', *AIV*, 158 (1), 47–94.
Frazier F., 1998, 'Public et spectacle dans la *Poétique* d'Aristote', *Cahiers du GITA*, 11, 123–144.
Frontisi-Ducroux F., 1975, *Dédale: Mythologie de l'artisan en Grèce ancienne*. Paris.
Ganis G., Thompson W.L., Mast F. and Kosslyn S.M., 2004, 'The brain's mind's images: The cognitive neuroscience of mental imagery'. In M.S. Gazzaniga (ed.), *The Cognitive Neurosciences*. Cambridge, MA: MIT Press. 931–941.
Gentile F., van Atteveldt N., De Martino F. and Goebel, R., 2017. 'Approaching the ground truth: Revealing the functional organization of human multisensory STC Using ultra-high field fMRI'. *Journal of Neuroscience*, 37(42), 10104–10113.
Goheen Robert F., 1955, 'Aspects of dramatic symbolism: Three studies in the *Oresteia*', *The American Journal of Philology*, 76.2, 113–137.
Grand-Clément A., 2011, *La Fabrique des couleurs: histoire du paysage sensible des Grecs anciens: VIIIe-début du Ve s. av. n. è.* Paris.
Grand-Clément A., 2013, 'La mer pourpre: façons grecques de voir en couleurs. Représentations littéraires du chromatisme marin à l'époque archaïque'. *Pallas*, January, 92, 143–161.
Green J.R., 1994, *Theatre in Ancient Greek Society*. London, New York.
Halm-Tisserand M., 2002, 'Le filet: un "leitmotiv" dans l'*Orestie* d'Eschyle'. *Ktèma*, 27, 293–305.
Harrison G.W.M. and Liapis V., 2013, *Performance in Greek and Roman Theatre*. Leiden, Boston, MA.
Jenkins I.D., 1985, 'The ambiguity of Greek textiles', *Arethusa*, 18, 109–132.
Kosslyn S.M., Behrmann M. and Jeannerod, M., 1995. 'The cognitive neuroscience of mental imagery'. *Neuropsychologia*, 33(11), 1335–1344.
Lebeck A., 1971, *The Oresteia: A Study in Language and Structure*. Cambridge, MA, Harvard University Press.
Lee M., 2004, '"An evil wealth of garments", deadly πέπλοι in Greek tragedy'. *CJ*, 99.3, 253–279.
Mackay A., 2008, *Orality, Literacy, Memory in the Ancient Greek and Roman World*. Leiden, Boston, MA.
Mackie C.J., 2004, *Oral Performance and Its Context*. Leiden, Boston, MA.
Malafouris L., 2013. *How Things Shape the Mind: A Theory of Material Engagement*. Cambridge, MA.
Marinis A., 2012, 'Seeing sounds: Synaesthesia in the parodos of *Seven Against Thebes*'. *Logeion, A Journal of Ancient Theatre*, 2, 26–59.
Mate J., Allen R.J. and Baqués J., 2012, 'What you say matters: Exploring visual–verbal interactions in visual working memory', *The Quarterly Journal of Experimental Psychology*, 65, 3.
McConachie B., 2008, *Engaging Audiences: A Cognitive Approach to Spectating in the Theatre*. New York.
McConachie B., 2013, *Theatre and Mind*. Basingstoke, New York.
McConachie B., 2015, *Evolution, Cognition, and Performance*. Cambridge.
McConachie B. and Hart E., 2006, *Performance and Cognition, Theatre Studies and the Cognitive Turn*. London, New York.
McGurk H. and McDonald J., 1976, 'Hearing lips and seeing voices', *Nature*, 264, Dec., 23–30.
Meineck P., 2011, 'The neuroscience of the tragic mask', *Arion*, 19, 1, 113–158.
Meineck P., 2012, 'The embodied space: Performance and visual cognition at the fifth century theatre'. *New England Classical Journal*, 39.1, 3–46.
Meineck P., 2015, 'The affective sciences and Greek drama', SCS talk, www.academia.edu/20484133/The_Affective_Sciences_and_Greek_Drama.
Meineck P., 2017, *Theatrocracy: Greek Drama, Cognition, and the Imperative for Theatre*. Abingdon, New York.
Minchin E., 2001, *Homer and the Resources of Memory: Some Applications of Cognitive Theory to the Iliad and the Odyssey*. Oxford.
Monbrun P., 2007, *Les Voix d'Apollon, L'arc, la lyre et les oracles*. Rennes.
Mueller M., 2016, *Objects as Actors: Props and the Poetics of Performance in Greek Tragedy*. Chicago, IL.
Muñoz, A.I., 2012. 'Métrique et tropos dans deux tragédies D'Eschyle: les sept contre thèbes et les Perses'. *Classica - Revista Brasileira de Estudos Clássicos*, 25, 1, 175–218.
Noel A.-S., 2013, 'Le vêtement-piège et les Atrides: métamorphoses d'un objet protéen', in B. Le Guen and S. Milanezi (eds.), *L'appareil scénique dans les spectacles de l'Antiquité*. Paris. 161–182.
Noel A.-S., 2014, 'L'arc, la lyre et le laurier d'Apollon: de l'attribut emblématique à l'objet théâtral', *Gaia, revue interdisciplinaire sur la Grèce archaïque*, 'Objets de la mythologie (2), 17, 105–128.
Ogden D., 2013, *Drakôn: Dragon Myth and Serpent Cult in the Greek and Roman Worlds*. Oxford.
Pallidini L.P., 2016, *A Cloud of Dust: Mimesis and Mystification in Aeschylus' Seven Against Thebes*. Oxford.
Revermann M., 2006, 'The competence of theatre audiences in fifth- and fourth-century Athens', *JHS*, 126, 99–124.

Roisman H.M. (ed.), 2014, *The Encyclopedia of Greek Tragedy*, 2 vols. Chichester.
Schmid C., Büchel C. and Rose M., 2011, 'The neural basis of visual dominance in the context of audio-visual object processing'. *Neuroimage*, 55, 304–311.
Segal C., 1977, 'Synaesthesia in Sophocles'. *Illinois Classical Studies*, II, 88–96.
Shaughnessy N. (ed.), 2013, *Affective Performance and Cognitive Science: Body, Brain, and Being*. London.
Sifakis G.M., 2013, 'The misunderstanding of *opsis* in Aristotle's *Poetics*', in G. Harrison and V. Liapis, *Performance in Greek and Roman Theatre*. Leiden, Boston, MA. 45–61.
Small J.P., 1997, *Wax Tablets of the Mind: Cognitive Studies of Memory and Literacy in Classical Antiquity*. London, New York.
Squire M., 2016, *Sight and the Ancient Senses*. London, New York.
Stanford W.B., 1936, *Greek Metaphor: Studies in Theory and Practice*. Oxford.
Starr G.G., 2013, *Feeling Beauty: The Neuroscience of Aesthetic Experience*. Cambridge, MA, London.
Straube B., van Kemenade B.M., Arikan B.E., Fiehler K., Leube D.T., Harris L.R. and Kircher T., 2017, 'Predicting the multisensory consequences of one's own action: BOLD suppression in auditory and visual cortices'. *PloS One*, 12, 1, e0169131.
Talsma D., Senkowski D., Soto-Faraco D. and Woldorff M.G., 2011, 'The multi-faceted interplay between attention and multisensory integration'. *Trends in Cognitive Sciences* 14, 9, 400–410.
Taplin O., 1977, *The Stagecraft of Aeschylus: The Dramatic Use of Exits and Entrances in Greek Tragedy*. Oxford, London.
Taplin O., 1978, *Greek Tragedy in Action*. Berkeley, CA.
Telò M., 2013, 'Aristophanes, Cratinus and the smell of comedy'. In S. Butler and A. Purves, *Synaesthesia and the Ancient Senses*. Durham, NC. 53–69.
Trieschnigg C., 2016, 'Turning sound into sight in the chorus' entrance song of Aeschylus', in V. Cazzato and A. Lardinois (eds), *The Look of Lyric, Greek Song and the Visual*. Leiden, Boston, MA, 217–237.
Van Engen K., Xie, Z. and Chandrasekaran B., 2017, 'Audiovisual sentence recognition not predicted by susceptibility to the McGurk effect'. *Attention, Perception, & Psychophysics*, 79, 2, 396–403.
Watson J., 2001, *Speaking Volumes: Orality and Literacy in the Greek and Roman World*. Leiden, Boston, MA.
Wiles D., 1990, 'Le décor des *Sept contre Thèbes* d'Eschyle. Approche structuraliste', trans. P. Ghiron-Bistagne, *Cahiers du GITA* 6 (1990/1), 145–160.

19

Mirth and creative cognition in the spectating of Aristophanic comedy

Angeliki Varakis-Martin

What can positive psychology tell us about the ancient audience's experience of Aristophanic comedy? Was the comedy a vehicle for the expression of the poet's own political views, or did the festive experience of the comic performance broaden the mind of its spectators, enhancing their ability to process the words and action of the play in a creative and individual way?

Scholars have long debated Aristophanes' political intentions and the influences of his plays on Athenian life. On one side there are those who see Aristophanes' comedy as playful and festive with no serious political intentions (Heath, 1987; Reckford, 1987; Halliwell, 2008), and on the other side sit those who approach Aristophanes as a serious dramatist who exerted political influence (MacDowell, 1995; Henderson, 1998; Vickers, 1997, 2015). In looking to situate current scholarship within longstanding historical debates, Philip Walsh has observed that more recent studies have tried to 'navigate between the traditional binaries', seeing the meaning of Aristophanes as something elusive; a constant negotiation between individual and collective thinking that is dependent on each audience member's engagement with the comic. As Walsh observes, this middle approach 'cautions against dogmatism and champions diversity' (Walsh, 2009: 69). Informed by the criticism of Bakhtin,[1] some of these approaches have sought to understand Aristophanic comedy's political orientation in its nexus of conflicting forces and unstable interactions that promote the active engagement of the spectator in the meaning making of the play. These interactions are partly expressed through the ambivalence of meaning produced by the intertextual dimensions of the text. For example, in his seminal study *Aristophanes and the Definition of Comedy*, Michael Silk examines in detail the expansive and discontinuous quality of Aristophanic comedy with reference to its language, style and characterisation, describing the poet's comic vision as open and 'pregnant with possibility', a world 'where incompatible opposites are found freely co-existing or freely exchanged' (Silk, 2000: 93), destabilising the contemporary world. Echoing Silk's observations regarding the openness and expansive quality of Aristophanic comedy, Charles Platter in his book *Aristophanes and the Carnival of Genres* argues that the Aristophanic text is a 'carnivalesque medley of genres' that continually forces audiences to readjust their perspectives.[2] In his view, the constant interaction of literary genres (e.g. tragedy, iambic poetry) prevents the emergence of a single dominant interpretive construct, denying the comedy the existence of a fixed interpretive point of view, making it open to multiple readings and meanings. According to Platter, this style of reading gives access to the 'joyful relativity' of Aristophanic comedy whilst at the same time making it appealing to a broad and diverse audience (Platter, 2007: 41).

The 'openness'[3] of Aristophanic comedy has also been considered with reference to its staging, with Joe Park Poe arguing that the frequent instances of unmotivated appearances and fusion of performance styles, especially in the poet's earlier comedies, far from being intrusive, add to the effect of inconsequential stage action which Aristophanes strives to achieve (Poe, 2000).

Although the aforementioned studies raise some very interesting points about the diverse and unlimited interpretive possibilities of Aristophanic comedy, they do not consider the play as an embodied performance event, excluding from their analysis an important dimension of theatre; the experiential dimension. By experiential dimension, I mean the aspect that looks at theatre, as an event, that is participatory and emotive. From a cognitive point of view, approaching comic performance as an embodied experience and not solely a place of representation is most important in order to get a fuller picture of audience engagement.

The absence of an analysis of the experiential dimension of theatre in most theoretical studies of ancient drama is not without justification. As Felix Budelmann correctly puts it:

> People's experience – in writing, rehearsing, performing, watching, or reading ancient drama – is something that cultural history finds hard to analyse with the tools at its disposal. We have learned that studying the subjective experience of persons past or present is a hazardous endeavour, and more often than not we wisely avoid the topic.
>
> *(Budelmann, 2010: 109)*

Without claiming that I can offer definitive answers regarding the subjective experience of comic audiences, it is my belief that recent advances in the realm of neuroscience can offer scholars new perspectives. These fresh perspectives, which are grounded on hard data about the brain's neural activity, offer a way into this elusive but undeniably significant subject by offering a more rounded account of theatre spectatorship, past and present.

In the present chapter, I will approach Aristophanic comedy as an emotive environment that was predominantly mirthful and uplifting. Drawing from theories in cognitive science[4] and positive psychology that suggest that positive emotion[5] broadens cognition, I shall argue that mirth and feelings of joy were not merely responses that measured the comedy's success, but embodied emotions with a fundamental role to play in shaping the audience's mode of thinking, as well as in enhancing the latter's level of creative engagement, turning spectators into co-authors in the meaning making of the event.[6]

Aristophanic performance as a positive environment and mirthful performance

Comic performance is normally associated with and used to trigger spontaneous laughter, which is considered the best indicator of positive emotion. As Silk observes:

> Although comedy as a mode of art is not recognisable by a cluster of fixed textual references, yet it remains recognisable. It has its repertoire of 'family characteristics'. The first one is that comedy tends to amuse. Any version of comedy may at some point be serious in some sense of that word but it still tends to amuse. This characteristic effect may be associated with feelings of joy.
>
> *(Silk, 2000: 93)*

In ancient Athens, Aristophanic theatre was particularly buoyant due to the wider festive framework and scale of the event. Audiences were not only theatre audiences who were witnessing staged celebrations, but also festival participants who were licenced to laugh in a free-spirited manner.[7] Stephen Halliwell describes this collective laughter as playful. In his study *Greek Laughter: A Study of Cultural Psychology from Homer to Early Christianity*, he repeatedly stresses the importance of the festive framework in the appreciation of the comedy's humour, particularly its obscenities and characters' excessive behaviours. As he states:

> within the fully demarcated festive framework the pressures of shame are temporarily lifted. By cultural conventions the theatre audience is not just permitted but encouraged to laugh at everything . . . institutionalising and ritualising a potentially negative force (of shame) into communal celebrations of joy.
>
> *(Halliwell, 2008: 247)*

Halliwell's analysis sees the broader festival ethos as instrumental in determining the way in which ancient spectators perceived Aristophanic comedy and how they may have responded to its humour. In Halliwell's festive analysis, Aristophanic laughter is playful, as he sees it as a unifying force with the power to relax audience's attitudes and inhibitory controls.

Although one cannot dismiss the presence of other emotions throughout the duration of the comic performance, it is, however, broadly agreed that a mirthful response was the key objective of Aristophanic comedy, which is in contrast to the later Menandrian comedy, aimed to generate laughter from the very beginning.[8] As the comic poet himself tells us in *Assembly Women*, 'Those who like a hearty laugh should vote for me – I give them laughs' (Halliwell, 2016, trans. 1159: 196).

Niall Slater in his article 'Making the Aristophanic audience' makes a convincing case when he argues that Aristophanes, from the very start of the performance, wanted to warm up his audiences. As part of the comic poet's strategy to bring spectators into a positive and uplifting mood, he mentions warm-up jokes, often a series of shorter jokes, none of which is absolutely essential to understanding the rest of the play (Slater, 1999: 354), or the use of an individual on stage as a model for audience behaviour, as happens in the beginning of *Frogs*, with the presentation of a laughing Heracles (Slater, 1999: 364). In this example laughter is used as a cue and link with the audience. One can only imagine the potential impact of this opening scene, especially when considering the contagious nature of laughing.[9] In large open-air theatres, emotional contagion, including contagious laughter, takes place with some ease, as audiences are in full view of each other, unifying in this way diverse groups of people. The oldest theatre is that of Dionysus Eleuthereus that was built on the south slope of the Akropolis. The good acoustics of the space might also contribute to this effect. In a study on laughter Scott *et al.* (2014) see laughter not only as a product of humour but as a social emotion associated with bonding, affection and emotional regulation. The study emphasises the distinction between reactive laughter that is driven by outside events, and deliberate laughter that is described as an intentional communicative act that is more controlled (Scott *et al.*, 2014: 618). Ian Ruffel in his essay 'Audience and emotion' highlights the social and collective dimension to the experience of laughter in Aristophanic theatre, drawing attention to these distinctions. He sees the social dimension of laughter as twofold. On one hand, there is the passive side of the transaction where laughter is contagious and comes as a result of being in the presence of laughing spectators. On the other hand, there is the active side of the transaction, where laughter is used by the performer as a cue and link with the audience, creating an 'act of complicity', or even solidarity, between both sides. Ruffell's observation is tied to his broader argument regarding the distinction between tragedy and comedy, arguing that although comedy and tragedy in Ancient Greece were collective experiences, 'the potential for a coherent emotive response was more plausible in comedy owing to the social and public dimension of laughter' (Ruffell, 2008: 49).

On a visual level, the distorted appearance and exaggerated physicality of the stage characters which aimed to make people look 'ridiculous' would have created a humorous atmosphere. In Greek performance, the visual appearance of the stage figures and their exaggerated physicality would have captured the audience's attention, generating a mirthful response, even before listening to what the characters had to say or do. We know, for example, on the basis of visual evidence, that Aristophanic performers, in comparison to tragic performers, were not only distorted and funny in appearance 'τὸ γελοῖον πρόσωπον αἰσχρόν τι καὶ διεστραμμένον ἄνευ ὀδύνης', 'the laughable mask is something ugly and twisted, but not painfully' (Aristot. *Poet.* 1449a35, LCL 199: 44), but were probably much more energetic on stage, with what Richard Green describes as 'rapid and even violent movement' evoking a heightened sense of excitement (Green, 2002: 111).[10] Every form of communication where gesture, tone of voice and posture are perceptible can transmit affect and this is important to keep in mind when discussing the experience of comic performance in which the visual and aural, somatosensory elements are key components of the performance experience. In the realm of psychology and neuroscience, there is substantial evidence that perception, cognition, emotion and judgement all involve corresponding activation of bodily states, be it during actual encounters with stimuli or during offline processing. For example, in a study conducted to discover movements that could help regulate emotion, researchers identified a set of motor characteristics that predict the elicitation or enhancement of each of the emotions: anger, fear, happiness and sadness. More specifically, a feeling of happiness was predicted by *jumping* and *rhythmic movements*. Happiness was also predicted to be enhanced by *free flow*. Additional motor elements that predicted feeling happy 'were enlarging the shape of the body in the horizontal (*spread*) and vertical (*rise*) direction as well as *upward* movements in space' (Shafir, Tsachor and Welch, 2015).

Although comic audiences were not themselves moving rapidly or jumping on stage, they were exposed to stage figures who were physically active and fast-paced. We must assume that this kind of exposure would have affected the spectators' levels of energy and affective state through empathetic engagement. This type of engagement in theatre has been scientifically explained through the Mirror Neuron System.[11] As Vittorio Gallese explains with reference to this system,

> Whenever we are looking at someone performing an action, besides the activation of various visual areas [of the brain], there is concurrent activation of the motor circuits that are recruited when we ourselves perform the action. Although we do not overtly reproduce the observed action, nevertheless our motor system becomes active as if we were executing that very same action that we are observing.[12]

Lending support to such observations, according to which action perception involves covert motor activity, it has also been proposed that actions belonging to the motor repertoire of the observer are mapped on the observer's motor system. More specifically, Calvo-Merino *et al.* (2005) compared the brain activity of dancers who watched their own dance style versus other styles, revealing the influence of motor expertise on action observation. The researchers found greater bilateral activations when expert dancers viewed movements that they had been trained to perform compared to movements they had not. The results show 'that the human mirror system might be sensitive to the degree of correspondence between the observed action and the motor capability of the observer' (Calvo-Merino *et al.*, 2005).

These findings are particularly interesting to consider with reference to Aristophanic performance, where we know that comic choruses comprised Athenian citizens. It could be argued that particular popular dances that featured in the comedy may have figured in the motor repertoire of those who had previously trained to be part of a comic chorus or who were familiar with these dances from other festive contexts (e.g. wedding celebrations), making Athenian audiences further engaged in the comic action.

Aside from exaggerated and lively movements, song, dance and musicalised speech were also essential elements of Aristophanic comedy. Music-making typically involves rhythm and always involves tempo, and these both affect emotional experience. These elements would have certainly added to the overall uplifting and energetic atmosphere of the ancient performance. Bruce McConachie in his insightful volume *Engaging Audiences: A Cognitive Approach to Spectating in the Theatre* (2008) explains how, similarly to speech, music and tempo are linked to gesture and intention and musicians can communicate general emotions to listeners regardless of the auditors' musical training. So, for example, in the presence of celebratory music audiences will recognise and embody the emotion of joy. He continues by observing that 'the more spectators join together in one emotion, the more empathy shapes the emotional response of the rest' (McConachie, 2008: 97). Juslin's BRECVEMA model of musical cognition (named after the first letters of the eight mechanisms included: brain stem reflexes, rhythmic entrainment, evaluative conditioning, emotional contagion, visual imagery, episodic memory, musical expectancy, and aesthetic judgement) is the most comprehensive attempt to date to explain how music arouses emotions, adding scientific ground to McConachie's observations.[13] The model theorises that there are several underlying induction-mechanisms, which range from simple reflexes to complex judgements that mediate between a musical event and an aroused emotion. One of these mechanisms is called *evaluative conditioning*. As outlined by John Sloboda, 'Evaluative conditioning refers to a process whereby an emotion is induced by a piece of music because the stimulus has been paired repeatedly with other positive or negative stimuli' (Sloboda, 2014: 622). Another equally significant mechanism is *emotional contagion* which refers to a process whereby an emotion is induced because the listener perceives the emotional expression of the music and then 'mimics' this expression internally in a similar manner to what happens when one listens to an emotional speech (Sloboda, 2014: 622). Both mechanisms may explain why audiences who are exposed to familiar celebratory tunes experience emotions of joy.

There is evidence from the Aristophanic texts that Aristophanic performance featured popular songs that were already familiar to audiences from different social and ritual settings. The most common musical accompaniment to these songs was the aulos, an instrument which 'probably sounded something like an oboe' (Robson, 2009: 141). Such songs included victory songs, ritual hymns like the phallic hymn in *Acharnians* (263–279) and other celebratory songs, such as the wedding songs in *Birds* (1731) and *Peace* (1319). In the final wedding scene of *Peace*, wedding songs merge with rustic festivities associated with the vintage, showing connections with the popular tradition of grape-harvesting songs (Karanica, 2014: 213).

In her book, *The Songs of Aristophanes*, Laeticia Parker notes:

> A proportion of Aristophanes' song is of virtually no poetic significance. It is in lyric metre, because that is how choruses express themselves, because lyric metre and song confer of themselves a certain impetus and heightening of excitement.
>
> *(Parker, 1997: 123)*

Although some may disagree with Parker's view, she does make an important point about the ability of music to create 'impetus and heightening of excitement' which was hugely important in the broader festive environment of Aristophanic theatre. For example, in *Frogs*, the songs of both comic choruses are described as *thaumatos* in the play (185, 398), hinting at their emotive impact. The arrival-song of the initiates in particular is described as extremely lively, resembling the ecstatic dances of the Eleusinian mysteries. The words of the chorus suggest a Bacchic experience with the initiates organized in *thiasoi* that engage in wild dance (345–348). A few lines later (350–352) the chorus leader exhorts the chorus to call with its songs to Iacchus, 'its partner in the dance' (396). In the first stanza Iacchus is celebrated as the inventor of this 'marvelous festive song' (398–400).

When discussing the creative possibilities of conveying Aristophanic comedy's immediacy and emotive impact on the modern stage, Silk mentions the mode of music as something that could best capture the comedy's impact, observing a structural resemblance of Aristophanic comedy to many of the popular American musicals of the modern era (Silk, 2007: 292).

In a predominantly upbeat performance environment, where audiences from the very beginning were exposed to energetic stage figures of a humorous nature, popular tunes and vibrant dances, as well as an overall fast tempo, the experience of positive emotion, on a conscious (the feeling component) and subconscious level, would have influenced reasoning.[14] As I shall show in the following sections it was not only the knowledge of cultural conventions and 'festival ethos' that relaxed the audience's attitudes towards the comic action (Halliwell, 2008: 247), but the embodied experience itself which facilitated a more open and flexible way of thinking and of processing the words and actions of the comedy.

Positive emotion and cognitive broadening: the key theories

Decades of research have suggested that emotive experiences, such as emotional feelings and moods, influence not only the content but also the structure of thought, shaping how individuals think as opposed to merely what they think. The most influential figure in the revival of scientific interest in emotion has been the neuroscientist Antonio Damasio, who explored the close relationship of emotion to cognition and how emotion influences reasoning, challenging the distinction between body and mind (Damasio, 1994). Scientific research is constantly presenting us with new evidence for various types of collaboration between affective and cognitive processes.[15] Social psychologist Joseph Forgas explains how contemporary cognitive theories are able to scientifically explain the precise mechanisms for the infusion of affect into thinking:

> The idea that affect influences cognition has been around for a long time. How and why does this 'affect infusion' occur, and what are the psychological mechanisms that produce it? Unlike earlier conditioning and psychoanalytic explanations, contemporary cognitive theories postulate precise mechanisms responsible for the infusion of affect into thinking and judgments. Affect congruence posits that affect can influence the content of cognition through two complementary mechanisms: inferential processes and memory processes. In addition, affect can also influence how the information is processed.
>
> *(Forgas, 2008: 96)*

With reference to the latter point, a large amount of research in the realm of positive psychology has shown convincingly that positive emotions, such as mirth and joy, modulate one's mode of thinking, enhancing performance on many cognitive tasks, especially creative problem tasks that are open-ended.[16] If we are to approach theatre as a complex cognitive process that is carried out within a highly emotive environment (as seen in the previous section), these findings are important to consider. As Evelyn Tribble correctly states in her introduction to her book *Cognition in the Globe*,

> Watching a performance of an unknown play is a complex cognitive process, demanding that audiences attend to identifying characters, following and predicting action, interpreting spatial cues, and listening to complex patterned language.
>
> *(Tribble, 2011: 9)*

In regards to Aristophanic comedy, with its many allusions and interpretive possibilities, watching could be described as a cognitive process that requires one to constantly move beyond the information provided from the text and stage action through associative thinking, making it comparable to a creative cognitive task. Kidd, in his study *Nonsense and Meaning*, uses the metaphor of centre and periphery to explain the distinctiveness of one's experience when engaging with Aristophanic comedy:

> The deeper one delves into the richness of comedy's meaning, the more one unearths a complex set of allusions or allegories, the further one seems to drift from the comedy itself. It is as if in the pursuit of meaning, one becomes distanced from the comedy's centre, not closer to it.
>
> *(Kidd, 2014: 187)*

Here, Kidd suggests that the ability to make broad and remote associations is an essential part of one's cognitive process when engaging with the Aristophanic world. If this is true, how far does our emotive state facilitate this type of associative thinking?

Before addressing this question in more detail, it is important to review some of the key cognitive theories which have informed my analysis. These theories propose that cognitive broadening is facilitated and promoted when one is experiencing feelings of joy and mirth. Although these theories have never been tested in the context of theatre, they do show how positive emotion (that is, normally evoked from viewing comic clips or listening to happy tunes) influences one's performance in creative cognitive tasks and open-ended problems. The latter could be described as comparable to some of the playful and loosely connected Aristophanic stage actions and language.

Alice Isen's extensive work on positive emotion, which extended over two decades, provides the strongest evidence that positive emotion broadens cognition and facilitates creative thinking (Isen, Daubman and Nowicki, 1987). Isen hypothesised that the reason positive affect increases creativity is because it primes and broadens the associative process that increases the odds of remote associations that are original.[17] Her conclusions fit with Barbara Friedricskon's 'broaden and build theory' which has proposed that positive affect, including mirth and joy, broadens one's scope of attention and thought-action repertoire (Fredrickson, 2001), and Rowe and colleagues' work that contends that positive emotion, such as feeling amused, results in relaxation across the semantic and visuospatial cognitive domains, increasing the breadth of information available for cognitive processing (Rowe, Hirsh and Anderson, 2006).

But what exactly is meant by cognitive broadening and how might this enhance creative thinking?

Rowe proposed that in the visual spatial domain, positive affect results in a 'leaky' attentional filter, allowing peripheral, irrelevant information to be more fully processed. In the context of an open-air theatre this could mean that one is more able to process visual information that extends beyond the frame of the stage and into the real world (world of audience and broader landscape). In the conceptual domain, positive affect is associated with an increased capacity to generate remote associates for familiar words, indicating that access to internal semantic information is facilitated by positive affect in such a way that more distant and thus more novel semantic associations become closer and more easily accessible. For example, it has been shown that when participants in psychology experiments are asked to come up with semantic associations for the word 'pen', those in a positive affective state generate a number of unusual and remote associations, such as 'barn' and 'pig', stemming from their unusual interpretation of pen as a fenced enclosure, as compared to those in a neutral state, who come up with the same high-frequency associates, such as 'paper'.[18] It is more likely therefore that those in a positive emotive state will come up with a greater number of diverse and novel responses. As I shall show in the following section, this is particularly interesting to consider when thinking of Aristophanes' playful use of language and stage actions.[19]

Ashby and Isen were the first to connect the observed cognitive broadening ('leaky' visual attention and ability to make remote semantic associations) to increased levels of dopamine, which is a neurotransmitter involved in controlling movement and posture and modulating mood, and which plays a central role in positive reinforcement and dependency. Their study provided for the first time a description of the neuropsychological mechanisms that underlie the influence of positive affect, including mirth, on cognition, bridging the literatures on positive affect and cognitive broadening and novel thinking. Ashby and Isen's theory is known as the 'dopaminergic theory of positive affect'[20] and assumes that creative problem solving is increased and creativity is enhanced partially 'because increased dopamine release in the anterior cingulate improves cognitive flexibility and facilitates the selection of cognitive perspective' (Ashby *et al.*, 2002: 249). Although in the last few years there have been a number of critical reviews[21] of Ashby's popular hypothesis and scientists have sought to refine the dopaminergic theory of positive affect (Rowe *et al.*, 2006), it is still broadly agreed that there is not only a strong correspondence between an increase in dopamine levels and flexible thinking (Zabelina *et al.*, 2016), but also a link between positive affect, including mirth, and dopamine release.

Building on from Ashby and colleagues' dopaminergic theory, it has also been proposed that induced smiles increase dopaminergic activity in various brain regions, among them the midbrain dopamine system and the prefrontal cortex, suggesting that the expressive component of an emotion could both induce positive emotion and broaden cognition (Wiswede *et al.*, 2009).[22] Thus, a facial movement, such as smiling, is not only understood as a response to environmental stimuli but can also alter the way in which we perceive that stimuli. This is particularly important to consider when trying to analyse the experience of comic performance which has as its prime objective a mirthful response.

Finally, it has been suggested by cognitive scientists that flexible thinking is not only a by-product of mirth and joy but that it also induces positive emotion, increasing the probability that those who have adopted a flexible mode of thinking will sustain their positivity over prolonged periods of time (Chermahini and Hommel, 2011). In the context of Aristophanic performance, it also points towards the possibility that, once comic audiences were brought into a joyful emotive state (e.g. through warm-up jokes), they were more likely to sustain it, not only because of the positive atmosphere and funny appearance and actions of the comic characters, but also because of their flexible and creative way of processing those actions. In discussing the open-air theatre performance experience, Peter Meineck argues that the expansive views of Greek theatre may have also facilitated abstract thinking and memory retrieval 'that transcends both spatial and temporal frames of reference' through the release of dopamine, which is 'the main neurotransmitter used for controlling the gaze when scanning distant space' (Meineck, 2017: 72). Thus, the Greek audience may have been able to sustain a flexible mode of thinking, not only because of their emotive state, but also because of the 'dopamine-inducing' open-air environment in which they were situated.

Taken together, all the aforementioned studies propose that emotions and performance environments can have an effect on how the brain processes visual and conceptual information.[23] These findings are important when looking at comic performance, as they connect positive emotive states and the expressive component of positive emotion (smiling) to associative and flexible thinking. This kind of thinking is highly desirable when viewing a comedy which is allusive, inter-theatrical and playful. Although it is important to acknowledge that theatre is not a laboratory and words presented as autonomous items on a piece of paper (as happens in most creative cognitive tasks that are carried out as part of psychology experiments) are perceived differently to those that are found within a particular dramatic and performance context – where the prosodic (intonation) and paralinguistic (volume and gestures) signals affect understanding – it is undeniable that words and actions carry a particular set of meanings depending on the diverse abilities and experiences of the audience. As I shall show in the following section, in Aristophanic comedy, where the narrative structure is relatively loose and does not always follow a consequential logic, the physical actions and allusive language of the characters can more easily be approached as separate and autonomous items open to free association and multiple interpretations. Thus, a broadening of attention, as a result of feeling mirthful, can be seen to facilitate a more creative engagement with the play, allowing for a fuller appreciation of its interpretive potentialities.

Aristophanes' 'multiplicity' of meaning and creative cognition

In Aristophanic comedy, especially in the poet's earlier comedies (e.g. *Acharnians*), the narrative structure is relatively loose and does not always follow a consequential logic. Silk's analysis, for example, talks of Aristophanic comedy's use of discontinuity and accumulation as part of the poet's broader comic vision that 'sees and conveys the world in sequences as much as consequences' (Silk, 2000: 157). As Silk observes, the sequences of autonomous items that are observed in many of Aristophanes' passages are 'hard for many modern sensibilities to grasp unmediated by conceptualised and abstracted causality' (Silk, 2000: 157). These passages follow a simple parataxis and a reluctance to subordinate words, phrases and clauses.[24] Such accumulation of words often requires one to find alternative and novel ways to connect various and seemingly unrelated items. In comparing Aristophanic comedy to Classical tragedy and Menandrian comedy, Silk observes,

> Classical tragedy and Menandrian comedy conform to laws of decorum, laws of homogeneity and laws of processive coherence, from which departure is difficult and probably destructive. Aristophanic comedy, with its own coherence, yet has at least some freedom to conform – or not. If comedy means freedom, this is comedy at its most comic.
>
> *(Silk, 2000: 159)*

This seemingly non-causal accumulation of items in the text is often reflected in Aristophanic staging. In his article 'Multiplicity, discontinuity, and visual meaning in Aristophanic comedy', Poe talks of Aristophanic comedy's distinctive feature of 'multiplicity' which means that Aristophanes gives his public a great deal to see: people and things doing things simultaneously on stage. These actions are not always mentioned in the text (Poe, 2000: 258). This, in Poe's view, is an expression of the comedy's openness and absurd nature. For example, frequently actors run into the *skene* in the middle of a dialogue, normally to fetch something, like Dicaeopolis' basket of charcoal in *Acharnians*, returning soon, if not immediately, while spoken dialogue continues. In Poe's view, in tragedy gestures and physical actions are pointed out in the text to emphasise their significance, something that does not happen in comedy. Tragedy's tendency to hierarchise and focus on what is most meaningful encourages it to ration action for the sake of emphasis, encouraging concentrated attention (Poe, 2000: 259). Contrary to this, in Aristophanic comedy the action seems more various and undirected.[25] With reference to comedy's multiple entrances and exits, Poe observes,

> What is humorous about a succession of exits and entrances to fetch an altar, a basket, a goat, a lustral water? What is funny about the spectacle of the comings and goings of neutral figures like mute attendants? I think that there are two complementary answers to these questions. One is simply that part of Old Comedy's humour seems to lie in its violation of the dramatic norms of tragedy. The second, more basic answer, however, is that comic openness can extend beyond absurdity to meaninglessness.
>
> *(Poe, 2000: 280)*

Poe's observations highlight the openness of the comedy in terms of interpretation. What Poe describes as meaningless I would argue is the exact opposite, especially if one accepts that audiences adopt a cognitively flexible mode of thinking and a broadening of attention as a result of feeling mirthful. The comic action of these exits and entrances, I would argue, is not meaningless but open to unlimited readings, allowing the spectator to engage creatively with the performance and to bind diverse and seemingly unrelated items in novel ways. The suggestion that positive emotion enhances a more holistic processing of external visual space, which involves attending to more visual stimuli simultaneously, would have made this task all the more effortless, especially in an ancient open-air theatre where audiences were much more aware of their surroundings. As already mentioned, positive mood states are associated with greater global or holistic processing (i.e., seeing the forest before the trees) vs. local processing (i.e., seeing the trees before the forest). This involves attending to more stimuli in both external visual space and internal semantic space, potentially derived from long-term memory, allowing access to more information to simultaneously influence creative solution efforts. As Rowe and colleagues conclude with reference to the examined thesis, positive affect may serve to broaden the scope of attentional filters:

> Positive moods facilitate tasks requiring a more global and encompassing style of information processing . . . but impair those calling for a narrow, focused style, such as selective visual attention. A buoyant mood may represent a fundamental shift in the breadth of information processing, the result of which would be to cultivate a more open and exploratory mode of attention to both exteroceptive and interoceptive sources of information.
>
> *(Rowe et al., 2006)*

In this respect, we can see how Aristophanes' 'multiplicity of action', which is a task requiring a global style of information processing that may have been facilitated by the broader joyful atmosphere of the performance (distorted stage figures, vibrant dances and popular songs), would have contributed to broadening the spectator's mode of thinking, which would have in turn generated original interpretive solutions.

The juxtaposition and often accumulation of many seemingly incompatible and/or unrelated items in Aristophanic comedy, especially in the poet's earlier plays (which also happen to be some of his most buoyant comedies), create a climate of radical ambivalence 'forcing audiences to choose from a broad range of interpretive possibilities' (Platter, 2007: 31). This choice of possibilities would have been maximised if the spectator's attention was broadened. When one adapts an exploratory mode of attention, as a result of feeling mirthful, the lack of an Aristotelean causality is not 'difficult to grasp' because the brain is instantly able to consider a broader array of interpretive options and discover alternative and novel ways in which words, sounds, actions, stage objects and sentences could relate to each other. If in a relatively neutral state, however, like when trying to read the comedy in a more neutral emotive state, it is most likely that this discontinuity and accumulative narration may at first cause frustration, or may require more effort, in an attempt to find alternative ways to make sense of the comedy. As a result, people may reject to interpret what at first seems incomprehensible, labelling it as 'meaningless'. Kidd, for example, argues that some of Aristophanes' passages are meaningless and scholars must resist the temptation to try and find meaning in everything they read because one of the playful features of Aristophanic comedy could be found in this very lack of 'serious' and 'contextual' meaning (Kidd, 2014: 10). Interestingly, Kidd acknowledges that these passages do not lack meaning, but lack a particular kind of 'serious' meaning that, I would add, is directly connected to a certain type of concentrated thinking, which is focused, systematic and logical, instead of broadened, creative and free flowing.

Michael Vickers in his volumes *Pericles on Stage* and *Aristophanes and Alkibiades* builds his key argument about Aristophanes' allusions to well-known political personalities around the premise that audiences were able to appreciate Aristophanic characters as operating on many different levels. He talks about 'polymorphic characterisation' in which different facets of an individual's personality or public image could be played by a different character of the play, or alternatively, a single dramatic character might embody features of different well-known historical personalities. As he argues, audiences would have been able to draw upon these features which could have acted as a common point of reference binding the play in a rather unconventional manner. In his view, the most damaging thing to a proper understanding of Aristophanes is the principle of 'one thing at a time', according to which we must not turn our attention to two different levels of humour operating simultaneously (Vickers, 2015: vxiii). A broad associative ability and flexibility of thought inhibits such 'one thing at a time' reasoning, encouraging a more expansive interpretation of the comic world and its characters. At the same time, it facilitates the imaginative integration of disparate information into creative interpretive solutions, promoting an active engagement in the meaning making of the play. The pleasure that comes from Aristophanic comedy often depends upon the freedom and non-prescriptive way in which the play encourages one to creatively engage with the comedy, catering for diverse audiences of different abilities. The person's ability to understand and gain pleasure from Aristophanes' playful and interpretatively open humour would seem to depend, at least partly, upon ideational fluency (the ability to produce many ideas to fulfil certain requirements), associative fluency and cognitive flexibility, which are creative cognitive abilities that are increased when one is experiencing emotions of joy.

Conclusion

The most influential cognitive theories to date regarding the relationship of positive affect and cognition are relatively consistent in their findings that positive emotions, such as mirth and joy, cause a broadened mode of thinking which facilitates creativity. This knowledge can add valuable insights to our understanding of the elusive topic of Aristophanic spectatorship. Studies which describe Aristophanic comedy as an 'open' rather than 'closed' world could be validated not only through the dialogic and playful discourse of the play with its many allusions, or through the aesthetic of open-air performance, but also through the global and broadened perceptual focus from which a mirthful audience would have experienced the comedy's expansive world.[26]

In discussing the mode of thinking of comic characters more generally, John Moreall highlights the fact that all comic protagonists are divergent thinkers by whom a dozen imaginative answers to a problem may be found. As he explains,

> In contrast to the rigidity of tragic thinking comic thinking is flexible. Its protagonists do not approach life with simple concepts and narrow category systems into which every experience has to fit. They have messier sets of concepts which apply here but not there, today but not yesterday . . . The flexible thinking in comedy matches the complexity, diversity and movement of life itself.
>
> *(Morreall, 1998: 340)*

If we are to rely on the existing scientific evidence, we can safely assume that flexible thinking is experienced by mirthful audiences. A mirthful response is, thus, not only a response to the humour of the comedy, but also an emotive state that shapes the way in which audiences experience and make sense of the comic world. When approaching Aristophanic theatre as a predominantly positive environment, and not simply a text or a site of representation, we can appreciate its broader social and political benefits. As an embodied emotion, mirth allows people to focus more broadly and to make more lateral connections, and these were clearly Athenian assets, related to creativity and novel thinking.

Notes

1 Bakhtin's favoured example of this type of dialogic discourse in the field of literature was the novel which he defines 'as a diversity of speech types and a diversity of individual voices, artistically organised', tracing novelistic dialogism back to folk practices of carnival. Carnival thus provided the novelist with a reservoir of dialogic practices, allowing for the 'carnivalisation' of speech (Bakhtin, 1981: 426). The utopian communicative conditions of carnival with its freedom and festive laughter provided Bakhtin with an archetype for dialogic discourse and it is on the basis of such a model that he undertakes his analysis of the novel.

2 Platter's analysis is centred around language, and more specifically the carnivalisation that happens in Aristophanes' language. He uses Bakhtin's model of carnival as a concept that assumes a metaphorical character that is pre-dominantly literary and sees 'carnivalisation' as a systemic feature of discourse (Platter, 2007).

3 In his analysis of comic stage action Poe uses the term 'open' to explain the absurdity of Aristophanic comedy. 'Comedy is open in the sense that the laws of probability are suspended . . . as comedy becomes more open, action becomes more absurd, for openness creates theoretically unlimited comic possibilities, anything goes' (Poe, 2000: 270).

4 Cognitive science helps us understand how humans perceive the world and construct their experience. In the field of cognitive science a new paradigm emerged during the 1980s: that of embodied cognition. This type of cognition holds that the mind is embodied and embedded in our environments. As the research into embodied cognition has developed over the past few decades there has been a turn towards cognitive science in the humanities. In the realm of theatre studies this 'cognitive turn' has started to expand significantly (McConachie, 2008). See also Shapiro (2014).

5 The use of the terms positive emotion and positive affect are often used indiscriminately by psychologists and cognitive scientists in the scientific literature. My own usage of the terms emotion and affect in this chapter will refer to both the conscious experience of positive emotion (the feeling component) and the responses of the autonomous nervous system that accompany affective states and which have a direct impact on cognitive processing. As D. S. Levine nicely summarises, 'There is some confusion

in the scientific literature between three commonly used and closely related terms: emotion, affect, and mood. In clinical terminology, "emotion" tends to be used for what a person is feeling at a given moment. Joy, sadness, anger, fear, disgust, and surprise are often considered the six most basic emotions, and other well-known human emotions ... "Affect" tends to be used for the outward, physical signs of emotion as in the phrase "flat affect" to describe the presentation of some types of schizophrenics. "Mood" tends to be used for a pervasive emotion over a longer period of time, anywhere from a day to several years. However, psychologists and modelers are far from consistent in their use of these terms; in particular the words "emotion" and "affect" are often interchanged. Many recent scholars have attempted to formulate a comprehensive definition of emotion. Yet by and large they have not been able to improve on Aristotle's statement (in *Rhetoric*) that "Emotion is that which leads one's condition to become so transformed that his judgment is affected, and which is accompanied by pleasure and pain' (Levine, 2007: 38). For LeDoux, 'The terms "emotion" and "feeling" are, in fact, often used interchangeably. In English we have words like fear, anger, love, sadness, jealousy, and so on, for these feeling states, and when scientists study emotions in humans they typically use these "feeling words" as guideposts to explore the terrain of emotion' (LeDoux, 2012: 653). Goschke and Bolte propose a working definition according to which 'emotion' can be conceived 'as psychophysiological response patterns which involve several components including an evaluation of the significance of an event in the light of one's needs, motives and goals; the physiological responses of the autonomous nervous system as indicated by different indicators of increased arousal; specific facial and postural expressions and finally the feeling component' (Goschke and Bolte, 2014: 404). It is important to note that it is possible various emotions were categorised differently in ancient Greece and this is something we must bear in mind when analysing the ancient experience of emotion. See Plamper (2015), Meineck (2017) and Salvo, Cairns and Fulkerson (2015).

6 Unlike most studies on Aristophanic laughter that examine mirth as an element that contributes to and is reflective of the aesthetics and narrative of the comedy, this chapter will treat mirth as an embodied emotion and affective state that facilitates creative thinking. Halliwell, for example, describes comic laughter as an emotive response that 'punctuates a performance with appropriate rhythms of involvement and approval, thereby complementing and enhancing the work of the actors' (2004: 190). Sommerstein's analysis of laughter draws upon the Aristophanic text through a careful consideration of a number of passages from the surviving plays specifically referring to different kinds of laughter, including 'shared laughter'. He describes 'shared laughter' as spontaneous and caused by a pleasurable experience, 'namely a festival'. This kind of laughter, in his view, is closely related to the pattern of Old Comedy, which includes moments of feasting and festive events. As he states, 'All those in Aristophanes, who laugh when they escape from suffering, laugh as a group. Corresponding to this, the schemes of Aristophanes' heroes are only successful when they are capable of creating shared pleasure' (Sommerstein, 2009: 112).

7 The festival context of Greek theatre has been well documented in studies such as Pickard-Cambridge (1968) and Csapo and Slater (1995).

8 Menandrian comedy, in comparison to Aristophanic comedy, became more organised in its form with an interest in realism and romantic feelings. As Kidd mentions, 'The wild cognitive storm that is Old Comedy had subsided while more reasonable sources of pleasure arose' (Kidd, 2014: 189).

9 Provine suggests that there is a possibility 'that human beings have auditory feature detector neural circuits that respond exclusively to this species-typical vocalisation ... in turn the feature detector triggers the neural circuits that generate the stereotyped action pattern of laughter' (1996: 38). On contagious laughter see also Provine (1992, 2000).

10 For an approach to Greek drama through vase-paintings that depict comic stage figures see Taplin (1993). Also Green (2002).

11 In his discussion of science and cognitive theory Mark Fortier observes how mirror neurons pinpoint the difference between reading a play and seeing a production. As he states: 'the emotional power of this effect has led some cognitive theorists to posit empathy and feeling as inherently, biologically, dominant, even unavoidable in theatre. In theatre we cannot escape empathy, and intellect and understanding are secondary' (Fortier, 2015: 190). For a survey on mirror neurons see Rizzolatti and Sinigaglia (2016); also Kilner, Friston and Frith (2016) and Binder *et al.* (2017).

12 Gallese (2001: 35).

13 See Juslin (2013).

14 In some climactic moments of the performance where the intensity of the emotion may have been extreme, such as during exhilaration of laughter, we must assume that cognition would have been deregulated and cognitive performance decreased. The Yerkes–Dodson law is an empirical relationship

between arousal and performance. The law states that performance increases with physiological or mental arousal, but only up to a point. When levels of arousal become too high, performance decreases. The process is displayed as an inverted U-curve which increases and then decreases with higher levels of arousal. See Yerkes and Dodson (1908). Stephen Kidd in his recent study on *Nonsense and Meaning in Ancient Greek Comedy* draws a connection between the exuberant context of particular comic moments and the appearance of long compound Aristophanic words. He presents as an example a 62-syllable unintelligible word at the celebratory end of *Assembly Women*, arguing that the more heightened the emotions, the greater the risk a comedian could take with his audience because of the joyous ambience in which such a word would have materialised (Kidd, 2014: 150).

15 From a neuroanatomical viewpoint it has been indicated that none of the brain's structures or regions are exclusively devoted to emotion or cognition; instead their respective systems most probably overlap. 'Affect is represented by a widely distributed functional network that includes both sub-cortical (affective) and anterior frontal regions (cognitive). Thus no brain area can be designated specifically as cognitive or affective . . . The affect-cognition distinction is phenomenological rather than ontological' (Duncan and Barrett, 2007). LeDoux proposes that emotions are in fact part of higher-order cognitive processing and not distinct: 'Emotional states of consciousness, or what are typically called emotional feelings, are traditionally viewed as being innately programmed in subcortical areas of the brain, and are often treated as different from cognitive states of consciousness, such as those related to the perception of external stimuli. We argue that conscious experiences, regardless of their content, arise from one system in the brain. In this view, what differs in emotional and nonemotional states are the kinds of inputs that are processed by a general cortical network of cognition, a network essential for conscious experiences' (LeDoux and Pine, 2016).

16 In the sphere of literary theory, Bakhtin, in his book *Rabelais and His World*, emphasises the importance of creative thinking through laughter, within the context of community celebrations. In Bakhtin's analysis, the broader environment is deeply attached to cognition and refers to the carnival's immediate context of collective participation and festive laughter, eliminating the distinction between social hierarchies, but also fixed roles, like between those who dress up and engage in revelry behaviour and those who observe the carnival festivities. What is of particular interest is the connection Bakhtin makes between positive emotion through festive laughter and people's ability to overcome fear and most importantly inhibition, not only in terms of physical behaviour, but also in terms of thinking. 'Laughter has a deep philosophical meaning . . . it is a peculiar point of view relative to the world; Through laughter the world is seen anew, no less (and perhaps more) profoundly than when seen from the serious standpoint' (Bakhtin, 1968: 94).

17 Many of the demonstrated effects of positive affect – such as flexible, creative and unusual thinking – might be better conceptualised as consequences of a broadening of focus (e.g. accessing more conceptual and visual information), rather than synonyms of it. As Zabelina *et al.* (2016: 4) argue when discussing the relationship between a creative mind set and 'leaky' attention, 'indeed for creativity leaky attention may help people notice information that is outside their focus of attention, and integrate this information into their current information processing, leading to a creative thought. This mechanism is akin to reduced latent inhibition, or reduced ability to screen or inhibit from conscious awareness stimuli that were previously experienced as irrelevant. In other words reduced latent inhibition may enhance creativity by enlarging the range of unfiltered stimuli available in conscious awareness, thereby increasing the odds of synthesizing novel and useful combinations of stimuli'.

18 The Isen *et al.* (1987) data and the creative problem solver both agree that the most common associate is not necessarily lost or unavailable to the positive affect subjects. Instead, both the data and the model suggest that other responses are also cued, so that the dominant response becomes somewhat less dominant than it is under neutral affect conditions, and less typical responses become relatively more accessible. See Ashby *et al.* (2002).

19 For more on Aristophanes' use of comic wordplay see Kidd (2014: 137–147).

20 The theory proposes that moderate increases of positive affect are increased with phasic dopamine release in midbrain areas (ventral tegmental area, VTA) and that ascending dopamine projections to the anterior cingulate cortex and basal ganglia mediate effects of positive affect on executive attention and cognitive set switching, thereby enhancing creative and flexible thinking. However, this theory does not assume that positive affect simply turns dopamine on or off. Instead, it is assumed that moderate levels of dopamine are present even under neutral affect conditions. The induction of mild positive affect is assumed only to slightly increase these normal dopamine levels. See Ashby *et al.* (2002). Ashby's and Isen's approach is inspired by insights into the neurobiology of reward, the encounter of which

has been shown to induce both positive mood and phasic increases of dopamine. The added factor of motivational intensity that differentiates between positive emotions with high motivational intensity (appetitive) and those with low motivational intensity (consummatory) is also important to consider in relation to cognitive broadening. As Subramaniam and Vinogradov (2013) note, 'We must emphasize that the relationship between positive mood, reward processing, dopamine neuromodulation, and cognitive control is likely to be highly non-linear and complex. Studies from animals and humans reveal that the prediction and receipt of rewards is associated with phasic activation of midbrain dopaminergic firing, while a positive mood is associated with overall increased prefrontal dopamine release each with distinct implications for cognition'. (See also Seth and Friston, 2016.)

21 These critical studies consider the possibility that there exists an optimal level of dopamine release depending on the nature of the cognitive task and one's individual dopamine level (Chermahini and Hommel, 2011). Taken together, these neuropsychological cognitive studies provide strong support for the general theory that dopamine mediates the influence of positive affect on human cognitive abilities and that positive affect expands and enhances human cognitive abilities. These approaches use spontaneous eye-blink rates, an indirect but well-established clinical marker of the individual dopamine level, to monitor the impact of mood manipulation.

22 The facial-feedback hypothesis claims that the facial expression itself – that is, the movement of the facial muscles producing the expression – triggers both the physiological arousal and the conscious feeling associated with the emotion. The idea that the muscular movements involved in certain facial expressions trigger the corresponding emotions was first introduced by Sylvian Tomkins. See Elkman (1995). Another study on smiling proposed that wilful and spontaneous facially expressed positive emotions (Duchene Smiles), the act of smiling, broaden cognition (Johnson *et al.*, 2010). This study is based on the broaden hypothesis, part of Fredrickson's (1998) broaden-and-build theory, and proposes that positive emotions lead to broadened cognitive states. For the relationship of mirth and laughter and insights into the neural basis of laughter, see also Caruana *et al.* (2015).

23 More recently it has also been proposed that benign situations that do not trigger conscious feelings of positive emotional arousal could also expand attentional scope, suggesting how implicit positive affective cues may instigate psychological responses analogous to those traditionally associated with conscious emotional arousal. This is important to consider in theatre environments where frequently the atmosphere of the performance may be uplifting and full of energy without audiences continuously being conscious of their feelings (Friedman and Förster, 2010).

24 Silk gives a number of examples of such Aristophanic lists with inconsequential ordering from a number of plays describing it as a key characteristic of Aristophanes' writing. See Silk (2000: 98–158).

25 'The more open comedy becomes, confronting the spectators with the unexpected, the unmotivated and the impertinent, the more amorphous the dramatic situation becomes and the more the dramatic illusion fades into the background. It sometimes therefore is very weak, so that a dramatically unmotivated appearance of someone who has no role in the plot, far from being inappropriate or intrusive, adds to the effect of undirected, inconsequential coming and going of mutes which in places Aristophanes strives to achieve' (Poe, 2000: 276). *Contra* English (2005: 2). In her article 'The evolution of Aristophanic stagecraft', Mary English argues that some visual elements, including stage objects, could have acted as links between the somewhat disjointed scenes. 'While this "distancing" effect might be active in some instances, many of the visual elements of Aristophanic comedy work to highlight the larger themes of the play and act as a link (albeit not the only one) between the somewhat disjointed scenes'.

26 As Silk observes, 'We may ponder the significance of the way that Aristophanic comic drama can break its illusion and establish a complicity between actors and spectators then return to the illusion as if nothing has happened. In a sense nothing has happened. There is no real breach, because actually or potentially the complicity is always there . . . the frame it breaks is already open – open to and open like the life outside it' (Silk, 2000: 91).

References

Ashby, F. G., Isen, A. M. & Turken, A. U. 1999. 'A neuropsychological theory of positive affect and its influence on cognition'. *Psychological Review*, 106, 3: 529–550. doi: 10.1037/0033-295x.106.3.529.

— 2002. 'The effects of positive affect and arousal on working memory and executive attention' in S. Moore & M. Oaksford (Eds.), *Emotional Cognition: From Brain to Behaviour*. Amsterdam: John Benjamins: 245–287.

Bakhtin, M. 1968. *Rabelais and His World*. Trans. Hélène Iswolsky. Cambridge: Cambridge University Press
— 1981. *The Dialogic Imagination: Four Essays*. Ed. Michael Holquist, trans. Caryl Emerson and Michael Holquist. Austin, TX: University of Texas Press.
Barajas, Mark S. 2014. 'Thinking and feeling: The influence of positive emotion on human cognition'. *The Hilltop Review*, 7, 1, Article 3.
Binder, E., Dovern, A., Hesse, M. D., Ebke, M., Karbe, H., Saliger, J., Fink, G. R. & Weiss, P. H. 2017. 'Lesion evidence for a human mirror neuron system'. *Cortex*, 90: 125–137. doi: 10.1016/j.cortex.2017.02.008.
Budelmann, Felix. 2010. 'Bringing together nature and culture: On the uses and limits of cognitive science for the study of performance reception' in E. Hall & S. Harrop (Eds.), *Theorising Performance: Greek Drama, Cultural History and Critical Practice*. London: Duckworth: 108–122.
Calvo-Merino, B., Glaser, D. E., Grèzes, J., Passingham, R. E. & Haggard, P. 2005. 'Action observation and acquired motor skills: An FMRI study with expert dancers'. *Cerebral Cortex*, 15: 1243–1249. doi: 10.1093/cercor/bhi007.
Caruana, Fausto, Avanzini, Pietro, Gozzo, Francesca, Francione, Stefano, Cardinale, Francesco & Rizzolatti, Giacommo. 2015. 'Mirth and laughter elicited by electrical stimulation of the human anterior cingulate'. *Cerebral Cortex*, 71: 323–331. doi: 10.1016/j.cortex.2015.07.024.
Chermahini, S. A. & Hommel, B. 2010. 'The (b)link between creativity and dopamine: Spontaneous eye blink rates predict and dissociate divergent and convergent thinking'. *Cognition*, 115, 3: 458–465.
— 2011. 'Creative mood swings: Divergent and convergent thinking affect mood in opposite ways'. *Psychological Research*, 76, 5: 634–640. doi: 10.1007/s00426-011-0358-z.
Colzato, L. S. 2010. 'Search dopaminergic control of attentional flexibility: Inhibition of return is associated with the dopamine transporter gene (DAT1)'. *Frontiers in Human Neuroscience*. doi: 10.3389/fnhum.2010.00053.
Csapo, E. & Slater, W. J. 1995. *Context of Ancient Drama*. Ann Arbor, MI: The University of Michigan Press.
Damasio, A. R. 1994. *Descartes' Error: Emotion, Reason, and the Human Brain*, 3rd ed. New York: Putnam Pub Group.
Duncan, S. & Barrett, L. F. 2007. 'Affect is a form of cognition: A neurobiological analysis'. *Cognition & Emotion*, 21, 6: 1184–1211. doi: 10.1080/02699930701437931.
Elkman, P. 1995. 'Sylvian Tomkins and facial expression' in V. E. Demos & S. S. Tomkins (Eds.) *Exploring Affect: The Selected Writings of Silvan S. Tomkins*. Cambridge: Cambridge University Press: 209–225.
English, M. 2005. 'The evolution of Aristophanic stagecraft'. *Leeds International Classical Studies*, 4.03: 1–16.
Forgas, J. P. 2008. 'Affect and cognition'. *Perspectives on Psychological Science*, 3, 2: 94–101. doi: 10.1111/j.1745-6916.2008.00067.x.
Fortier, M. 2015. *Theory/Theatre*. London: Routledge.
Fredrickson, B. L. 1998. 'What good are positive emotions?' *Review of General Psychology*, 2, 3: 300–319. doi: 10.1037//1089-2680.2.3.300.
— 2001. 'The role of positive emotions in positive psychology: The broaden-and-build theory of positive emotions'. *American Psychologist*, 56, 3: 218–226. doi: 10.1037/0003-066x.56.3.218.
Friedman, R. S., & Förster, J. 2010. 'Implicit affective cues and attentional tuning: An integrative review'. *Psychological Bulletin*, American Psychological Association (APA), 136, 5: 875–893.
Gallese, V. 2001. 'The "shared manifold" hypothesis: From mirror neurons to empathy'. *Journal of Consciousness Studies*, 8 (v-vii): 33–35.
Goschke, T. & Bolte, A. 2014. 'Emotional modulation of control dilemmas: The role of positive affect, reward, and dopamine in cognitive stability and flexibility'. *Neuropsychologia*, 62: 403–423. doi: 10.1016/j.neuropsychologia.2014.07.015.
Green, Richard. 2002. 'Towards a reconstruction of performance style' in Pat Easterling & Edith Hall (Eds.), *Greek and Roman Actors: Aspects of an Ancient Profession*. Cambridge: Cambridge University Press: 93–126.
Halliwell, S. 1995. *Aristotle's Poetics, LCL*. Cambridge, MA, Harvard University Press.
— 2008. *Greek Laughter: A Study of Cultural Psychology from Homer to Early Christianity*. Cambridge: Cambridge University Press.
— 2014. 'Laughter' in Martin Revermann (Ed.), *The Cambridge Companion to Greek Comedy*. Cambridge: Cambridge University Press: 189–205.
— 2016. *Frogs and Other Plays*. Oxford: Oxford University Press.

Heath, M. 1987. *Political Comedy in Aristophanes*. Hypomnemata, 87. Göttingen: Vandenhoeck & Ruprecht.
Henderson, J. 1998. 'Attic Old Comedy, frank speech and democracy' in K. Raaflaub & D. Boedeker (Eds.), *Democracy, Empire and the Arts in Fifth Century Athens*. Cambridge: Cambridge University Press: 255–273.
Isen, A. M., Daubman, K. A., & Nowicki, G. P. 1987. Positive affect facilitates creative problem solving. *Journal of Personality and Social Psychology*, 52, 6: 1122–1131. doi: 10.1037/0022-3514.52.6.1122.
Isen, A. M., Niedenthal, P. M., & Cantor, N. 1992. 'An influence of positive affect on social categorization'. *Motivation and Emotion*, 16, 1: 65–78. doi: 10.1007/bf00996487.
Johnson, K. J., Waugh, C. E. & Fredrikson, B. L. 2010. 'Smile to see the forest: Facially expressed positive emotions broaden cognition'. *Cognition and Emotion*, 24.10: 299–321.
Juslin, P. N. 2013. 'From everyday emotions to aesthetic emotions: Towards a unified theory of musical emotions'. *Physics of Life Reviews*, 10, 3: 235–266. doi: 10.1016/j.plrev.2013.05.008.
Karanica, A. 2014. *Voices at Work: Women Performance, and Labor in Ancient Greece*. Baltimore, MD: Johns Hopkins University Press.
Kidd, S. E. 2014. *Nonsense and Meaning in Ancient Greek Comedy*. Cambridge: Cambridge University Press.
Kilner, James M., Friston, Karl J. & Frith, Chris D. 2016. 'An account of the mirror neuron system'. *Discovering the Social Mind: Selected Works of Christopher D. Frith*. London: Routledge.
Konstan, D. 2015. 'Affect and emotion in Greek literature'. *Oxford Handbooks online*. doi: 10.1093/oxfordhb/9780199935390.013.41.
LeDoux, J. 2012. 'Rethinking the emotional brain'. *Neuron*, 73, 4: 653–676. doi: 10.1016/j.neuron.2012.02.004.
LeDoux, J. E. & Pine, D. S. 2016. 'Using neuroscience to help understand fear and anxiety: A two-system framework'. *American Journal of Psychiatry*, 173.11: 1083–1093.
Levine, D. 2007. 'Neural network modeling of emotion'. *Physics of Life Reviews*, 4/1: 37–63.
Limone, J. 2011. 'Time in theatre' in B. Reynolds (Ed.), *Performance Studies: Key Words, Concepts and Theories*. New York: Palgrave Macmillan: 215–225.
McConachie, B. A. 2008. *Engaging Audiences: A Cognitive Approach to Spectating in the Theatre*. New York: Palgrave Macmillan.
MacDowell, D. M. 1995. *Aristophanes and Athens: An Introduction to the Plays*. Oxford: Oxford University Press.
Meineck, P. 2017. *Theatrocracy: Greek Drama, Cognition and the Imperative for Theatre*. London and New York: Routledge.
Morreall, J. 1998. 'The comic and tragic visions of life'. *Humor: International Journal of Humor Research*, 11, 4. doi: 10.1515/humr.1998.11.4.333.
— 1999. *Comedy, Tragedy, and Religion*. Albany, NY: State University of New York Press.
— 2014. 'The comic vision of life'. *The British Journal of Aesthetics*, 54, 2: 125–140.
Parker, Laeticia. 1997. *The Songs of Aristophanes*. Oxford: Oxford University Press.
Pickard-Cambridge, W. 1968. *The Dramatic Festivals of Athens*. 2nd ed., rev. J. Gould-D. M. Lewis. Oxford: Oxford University Press.
Plamper, J. 2015. *The History of Emotions: An Introduction*. Oxford: Oxford University Press.
Platter, C. 2007. *Aristophanes and the Carnival of Genres*. Baltimore, MD: Johns Hopkins University Press.
Poe, J. P. 2000. 'Multiplicity, discontinuity and visual meaning in Aristophanic comedy'. *Rheinisches Museum Für Philologie*, 143.3, 4: 256–295.
Provine, R. R. 1992. 'Contagious laughter: Laughter is a sufficient stimulus for laughs and smiles'. *Bulletin of the Psychonomic Society*, 30, 1: 1–4. doi: 10.3758/bf03330380.
— 1996. 'Laughter'. *American Scientist*, 84: 38–47.
— 2000. *Laughter: A Scientific Investigation*. London: Penguin.
Reckford, K. J. 1987. *Old and New Comedy*. London: University of North Carolina Press.
Rizzolatti, G. & Craighero, L. 2004. 'The mirror neuron system'. *Annual Review Neuroscience*, 27: 169–192.
Rizzolatti, G. & Sinigaglia, C. 2008. *Mirrors in the Brain: How Our Minds Share Actions and Emotions*. Oxford: Oxford University Press.
—. 2016. 'The mirror mechanism: A basic principle of brain function'. *Nature Reviews Neuroscience*, 17.12: 757–765.
Robson, J. 2009. *Aristophanes: An Introduction*. London and New York: Bloomsbury.
Rowe, G., Hirsh, J. B., & Anderson, A. K. 2006. 'Positive affect increases the breadth of attentional selection'. *Proceedings of the National Academy of Sciences*, 104/1: 383–388. Proceedings of the National Academy of Sciences. doi: 10.1073/pnas.0605198104.

Ruffell, I. 2008. 'Audience and emotion' in M. Revermann & P. Wilson (Eds.), *Performance, Iconography, Reception: Studies in Honour of Oliver Taplin*. Oxford: Oxford University Press.
Runco, M. A. & Pritzker, S. R. 1999. *Encyclopedia of Creativity*. San Diego, CA: Academic Press.
Salvo, I., Cairns, D. L., & Fulkerson, L. (Eds.). 2015. *Emotions between Greece and Rome* (BICS supplements: vol. 125). London: Institute of Classical Studies, School of Advanced Study.
Scott, S., Lavan, N., Chen, S., & McGettigan, C. 2014. 'The social life of laughter'. *Trends in Cognitive Sciences*, 18, 12: 618–620. http://doi.org/10.1016/j.tics.2014.09.002.
Seth, Anil K. & Friston, Karl J. 2016. 'Active interoceptive inference and the emotional brain'. *Philosophical Transactions of the Royal Society B*, 371.1708: 20160007. doi: 10.1098/rstb.2016.0007.
Shafir, T., Tsachor, R. P., & Welch, K. B. 2015. 'Emotion regulation through movement: Unique sets of movement characteristics are associated with and enhance basic emotions'. *Frontiers in Psychology*, 6: 2030. http://doi.org/10.3389/fpsyg.2015.02030.
Shapiro, L. 2014. *The Routledge Handbook of Embodied Cognition*. London: Routledge.
Silk, M. S. 2000. *Aristophanes and the Definition of Comedy*. New York: Oxford University Press.
— 2007. 'Translating/transposing Aristophanes' in E. Hall & A. Wrigley (Eds.), *Aristophanes in Performance, 421 BC–AD 2007: Peace, Birds and Frogs*. Oxford: Legenda: 287–308.
Slater, N. W. 1999. 'Making the Aristophanic audience'. *American Journal of Philology*, 120, 3: 351–368. Baltimore, MD: Johns Hopkins University Press.
Sloboda, J. 2014. *Handbook of Music and Emotion: Theory, Research, Applications*. Oxford: Oxford University Press.
Sommerstein, A. H. 2009. *Talking about Laughter: And Other Studies in Greek Comedy*. New York: Oxford University Press.
Stets, Jan E. & Turner, Jonathan H. 2007. *Handbook of the Sociology of Emotions*. New York: Springer-Verlag New York.
Subramaniam, Karuna & Vinogradov, Sophia. 2013. 'Improving the neural mechanisms of cognition through the pursuit of happiness'. *Frontiers in Human Neuroscience*, 7. DOI: 10.3389/fnhum.2013.00452.
Taplin, O. 1993. *Comic Angels*. Oxford: Oxford University Press.
Tribble, E. 2011. *Cognition in the Globe*. London: Palgrave.
Van Holstein, M., Aarts, E., van der Schaaf, M. E., Geurts, D. E. M., Verkes, R. J., Franke, B., van Schouwenburg, M. R., *et al.* 2011. 'Human cognitive flexibility depends on dopamine D2 receptor signaling'. *Psychopharmacology*, 218, 3: 567–578.
Vickers, M. 1997. *Pericles on Stage: Political Comedy in Aristophanes' Early Plays*. Austin, TX: University of Texas Press.
— 2015. *Aristophanes and Alcibiades: Echoes of Contemporary History in Athenian Comedy*. Berlin. Walter de Gruyter GmbH.
Wadlinger, H. A. & Isaacowitz, D. M. 2006. 'Positive mood broadens visual attention to positive stimuli'. *Motivation and Emotion*, 30, 1: 87–99.
Walsh, Philip. 2009. 'Debates over Aristophanes' politics and influence'. *Classical Receptions Journal*, 1, 1: 55–72.
Wiswede, D., Münte, T. F., Krämer, U. M., & Rüsseler, J. 2009. 'Embodied emotion modulates neural signature of performance monitoring'. *PLoS ONE*, 4/6: e5754. Public Library of Science (PLoS). DOI: 10.1371/journal.pone.0005754.
Yerkes, R. M. & Dodson, J. D. 1908. 'The relation of strength of stimulus to rapidity of habit-formation'. *Journal of Comparative Neurology and Psychology*, 18: 459–482. doi: 10.1002/cne.920180503.
Zabelina, D. L., Colzato, L., Beeman, M., & Hommel, B. 2016. 'Dopamine and the creative mind: Individual differences in creativity are predicted by interactions between Dopamine genes DAT and COMT'. *PLoS ONE*, 11, 1: 1–16.

Part V
Artificial intelligence

20

The extended mind of Hephaestus

Automata and artificial intelligence in early Greek hexameter

Amy Lather

Early hexameter poetry is littered with fantastical objects, and Hephaestus' creations are foremost among these. In addition to Achilles' shield, in Homer Hephaestus is also credited with the production of numerous automata: the tripods, bellows, and maidens that serve as his assistants, and in Hesiod, he is responsible for the creation of Pandora, the most notorious automaton of all. As the presence of such artifacts in early epic attests, the concept of non-human forms of intelligence has existed since antiquity and has been a source of fascination ever since.[1] This is because the very idea of artificial intelligence begs the question as to what constitutes a human mind and what might distinguish human cognition from that of a robot. For this reason, the depictions of automata that appear in archaic hexameter offer insight into how the Greeks of this period defined human intelligence and conceptualized mental activity. This chapter will argue that the accounts of automata discussed here imagine different ways in which cognitive processes may extend out of the brain and into the world to materialize in products of technology.

One of the most basic criteria that defines not only cognition but organic life more generally is the ability to act with a degree of independence, hence the term automaton literally denotes a "self-moving" (αὐτο- + μα-) creature.[2] The description of the gates of Olympus provides a succinct account of the features that commonly define automata in early epic. The gates are explicitly characterized as *automatai* because they open themselves and do so in accordance with the will of the Hours: "And the gates of heaven groaned open of their own accord, which belong to the Horae, to whom great heaven and Olympus are entrusted in order either to open the dense cloud or shut it" (αὐτόμαται δὲ πύλαι μύκον οὐρανοῦ ἃς ἔχον Ὧραι, / τῇς ἐπιτέτραπται μέγας οὐρανὸς Οὔλυμπός τε / ἠμὲν ἀνακλῖναι πυκινὸν νέφος ἠδ' ἐπιθεῖναι, *Il.* 5.749–51=8.393–5).[3] The gates of Olympus thus present a hybrid of features. For one, they are made of cloud, but nonetheless are said to "groan" as though mechanical and composed of metal or wood. Elsewhere, the verb μυκάομαι is often used of cattle lowing (e.g. *Il.* 18.580, *Od.* 10.413), and given the association of this term with a distinctive animal sound, its appearance here underscores the animate quality of the gates by suggesting that they have a kind of voice in addition to being self-movers. This description of the gates identifies two characteristics that will resurface in the passages discussed below: 1) that automatic objects may be defined as such for their ability to act in accordance with the decisions of their users; and 2) in this way they seem to accrue lifelike characteristics by performing tasks usually allotted to humans and/or by looking or sounding like animate creatures. In brief, what the gates typify in their connection to the will of the Hours is the idea that cognition can take place not just within an individual's mind, but also in the phenomenal world.

Like the gates of Olympus, the automata on which I focus here (which also conceptualize automata as products that act out users' wills) can be interpreted anew in light of the theory of extended mind first advanced in a 1998 paper by Andy Clark and David Chalmers.[4] In summary, this view contends that humans operate in a coupled system with their environments. This is one where "the human organism is linked with an external entity in a two-way interaction . . . if we remove the external component the system's behavioral competence will drop, just as it would if we removed part of its brain" (Clark 2008: 222). The example they adduce in favor of this view is the figure of Otto, who has Alzheimer's disease and so always carries with him a notebook in which he continually writes down new information. When Otto decides one day to visit the Museum of Modern Art, he accesses the relevant page in his notebook where the address is written, and so makes his way there. Inga, by contrast, also decides to visit an exhibit at the same museum, but manages to recall the address of the museum and navigate to it without the aid of a notebook, simply by remembering that it was on 53rd street. The point of this comparison is to show that there is nothing that makes Inga's recollection substantively different from Otto. The only difference between the two processes is that one takes place in the head (Inga) and one relies on an external prop (Otto's notebook). In other words, the only differences between them are superficial, and so there is nothing to disprove the conclusion that Otto's notebook forms as much a part of his cognitive processing as Inga's memory does for hers, even though one is external, while the other is not.

Given the interactivity between users and objects that will be discussed here, the theory of extended mind can thus be used as an interpretative framework for illuminating the cognitive mechanisms at work in these examples.[5] For it is nowhere clearer than in these portrayals of automata that objects need not be passive and inert, but may constitute a crucial part of cognitive processes. Further, that Hephaestus' automata are forged from precious material like gold and crafted with consummate artistry in addition to being always at hand to aid their master aligns with Clark's characterization of the objects involved in cognition as "artifacts" in his 1997 publication *Being There*. Here he stipulates that for an object to count as a feature of extended cognition, it will exist in "a rather special kind of user/artifact relationship – one in which the artifact is reliably present, frequently used, personally 'tailored,' and deeply trusted" (1997: 217).[6]

I concentrate primarily on those automata created by Hephaestus for two reasons: 1) because these passages portray the interaction between automata and someone who is both their creator and user; and 2) because another defining characteristic of Hephaestus is his handicap: he is κυλλοποδίων, often translated as "club-footed."[7] The association between automata and a god distinctive for his lameness thus illuminates the potential for such artifacts to interact in a causal way with their users by expanding their physical and mental capacities and enabling them to pursue complex tasks. Moreover, this is where Hephaestus' crafts differ from those of Daedalus. While the latter is credited with the capability of making moving statues, Hephaestus also creates a number of automated *tools*.[8] These are forms of artificial intelligence whose automation was designed to fulfill a practical purpose and which thus are designed to interact in a specific way with their user in the sort of coupled system adumbrated by Clark and Chalmers.

Moreover, that Clark formulates the cognitive relation between user and artifact in a chapter entitled "Language: The Ultimate Artifact" (1997: 193–218) highlights the fact that Hephaestus' automata exist only as verbal images – the products of poetic language. Clark argues that language is an integral part of the human cognitive apparatus, not only because it enables us to communicate with one another, but also because it "enables us to exploit our basic cognitive capacities for pattern recognition and transformation in ways that reach out to new behavioral and intellectual horizons" (1997: 194).[9] Indeed, language is also characterized as "a kind of magic trick by which to go beyond the bounds of our animal natures" (Clark 2003: 81). This conception of language thus aligns with the presentation of Hephaestus' automata as operating by magic (since there is no indication as to any mechanism responsible for their automation), which seems perhaps to highlight how extended cognition (such as human use of language) operates with such ease and fluidity so as to seem magical rather than mechanical. As I will demonstrate here, human language (and poetic language in particular) likewise functions in these depictions of artificial intelligence as a form of technology that enables both poet and audience to "think about thinking," which is often cited as the defining characteristic of human cognition.[10]

The ergonomics of automation

Given that most of Hephaestus' automata exist in order to assist his metallurgic craft, his animated creations feature most prominently in the lengthy portrayal of his workshop that appears in *Iliad* 18. Most important for Hephaestus' metalworking are his automated bellows, which emit hot air whenever he needs it:

> τὰς δ' ἐς πῦρ ἔτρεψε κέλευσέ τε ἐργάζεσθαι.
> φῦσαι δ' ἐν χοάνοισιν ἐείκοσι πᾶσαι ἐφύσων
> παντοίην εὔπρηστον ἀϋτμὴν ἐξανιεῖσαι,
> ἄλλοτε μὲν σπεύδοντι παρέμμεναι, ἄλλοτε δ'αὖτε,
> ὅππως Ἥφαιστός τ' ἐθέλοι καὶ ἔργον ἄνοιτο.
>
> He turned these into the fire and ordered them to get to work. And all twenty got to blowing into the melting-pots, emitting every kind of strong-blowing blast, and were ready to hand at one time then another, in whatever way he desired, as he bustled about and his task was carried out.
>
> (Il. *18.469–73)*[11]

Although it is not made explicit, it appears that the bellows have some form of sensory perception since, after all, they are able to heed Hephaestus' command to get to work (κέλευσέ τε ἐργάσεςθαι, 469) and issue their blasts when he needs them (ἄλλοτε μὲν σπεύδοντι παρέμμεναι, ἄλλοτε δ' αὖτε, 472). The bellows are thus integral to Hephaestus' art because they enhance his ability by freeing up his hands to pursue the crafting of the shield proper: "he took in one hand a great hammer, and in the other he took up the tongs" (γέντο δὲ χειρὶ / ῥαιστῆρα κρατερήν, ἑτέρηφιδὲ γέντο πυράγην, 476–7). If we interpret this description in light of the theory of extended mind, then the bellows seem to count as a part of Hephaestus' mental machinery because they are so closely connected to *his* functionality. This is made clear from the explicit connection the narrator draws between the automated movements of the bellows and Hephaestus' task, his *ergon* (ὅππως Ἥφαιστός τ' ἐθέλοι καὶ ἔργον ἄνοιτο, 473). In other words, the bellows play a distinctively ergonomic role by facilitating Hephaestus' craftsmanship and enabling his products to take shape.

While Hephaestus' bellows do not have the exact same causal role attributed to Otto's notebook (which served as a physical extension of Otto's memory and so enabled Otto to access information he had acquired sometime before), they nonetheless are presented as part of a two-way interaction with their progenitor (the "coupled system" mentioned above). This is evident from the fact that they do not depend on conscious directives issued by Hephaestus:

they know what Hephaestus wants them to do even without being told.[12] And this is significant for two reasons. First, it illustrates how objects and living beings were imagined to participate in the same cognitive system, an aspect that becomes clear from the fact that these automata operate in such a fantastical and magical way (that is, without any apparent mechanism or medium of communication with Hephaestus). Second, this underscores how the bellows are also extensions of a different sort: namely, the products of the poet's imagination. And it is in book 18, where these descriptions of automata proliferate, that the Homeric poet's own form of craftsmanship is on heightened display.

The *ergon* to which the bellows contribute is the shield of Achilles, the description of which forms the centerpiece of *Iliad* 18 and is widely regarded as a metapoetic reflection on the poet's own activity.[13] One of the reasons that this ekphrasis irresistibly lends itself to a metapoetic interpretation is because it seems to depict a shield that couldn't exist in reality, but only in poetic fantasy.[14] This is due to the way the narrator describes the images contained therein, which suggests that its figures are moving and speaking.[15] This mode of narration exemplifies the role a variety of sensory domains play in producing vividness. However one interprets this ekphrasis, though, it is difficult not to read the passage as a description of the poet's activity. The language of the ekphrasis, in other words, draws attention to the fact that it is a creation of the poet's mind, the language being analogous to the movements of Hephaestus' bellows, reflecting the mental workings of Hephaestus as he moves about his work. Common to the depiction of the bellows and the shield is the apparition of motion, which contributes to the perception of *life*, not just lifelikeness.

It is possible to observe another interaction between language, cognition, and artificial intelligence in the description of Hephaestus' tripods. While the tripods, like the bellows, are also automatic in the strict sense of the word (as they are able to progress back and forth between Hephaestus' workshop and the assembly of the gods), Hephaestus is still in the process of creating them. This is significant because the automatic nature of the tripods is therefore only revealed because the narrator elucidates Hephaestus' plan to make them so:

> τὸν δ' εὗρ' ἱδρώοντα ἑλισσόμενον περὶ φύσας
> σπεύδοντα· τρίποδας γὰρ ἐείκοσι πάντας ἔτευχεν
> ἑστάμεναι περὶ τοῖχον ἐϋσταθέος μεγάροιο,
> χρύσεα δέ σφ' ὑπὸ κύκλα ἑκάστῳ πυθμένι θῆκεν, (375)
> ὄφρά οἱ αὐτόματοι θεῖον δυσαίατ' ἀγῶνα
> ἠδ' αὖτις πρὸς δῶμα νεοίατο θαῦμα ἰδέσθαι.
> οἳ δ' ἤτοι τόσσον μὲν ἔχον τέλος, οὔατα δ' οὔ πω
> δαιδάλεα προσέκειτο· τά ῥ' ἤρτυε, κόπτε δὲ δεσμούς.
> ὄφρ' ὅ γε ταῦτ' ἐπονεῖτο ἰδυίῃσι πραπίδεσσι.

> And Thetis found him sweating and bustling to and fro about his bellows in haste, for he was crafting tripods, twenty altogether, to stand around the wall of his well-built hall, and he placed golden wheels beneath the base of each, so that they might enter the assembly of the gods of their own accord and come back home again, a marvel to see. And these were completed so far but their cunningly wrought handles were not yet attached: Hephaestus was preparing these and was forging their bonds. While he was laboring over these things in his visionary mind . . .
>
> *(Il. 18.372–80)*

Here an explicit link between Hephaestus' cognitive faculties and his automata emerges through the allusion to his "visionary mind" (ἰδυίῃσι πραπίδεσσι, 380; I here cite the translation of ἰδυίῃσι conveyed in Frontisi-Ducroux's (2002) "diaphragme visionnaire"). This is a formulaic expression used to describe Hephaestus' craftsmanship, appearing later in book 18 in the ekphrasis of the shield (482) as well as at *Od.* 7.92 in the description of the magical guard-dogs Hephaestus fashioned for Alcinous' palace. In each case, the dative plural is always used, which can be construed as either an instrumental or a locative use of the case.[16] This is because πραπίδες denotes both a cognitive faculty as well as a physiological component. Although its precise referent is unclear, it most likely designates the area in and around the lungs and diaphragm. The semantics of πραπίδες in connection with Hephaestus' craft thus typify how his efforts (ταῦτ' ἐπονεῖτο) involve both mind and body simultaneously in a coupled system. The fact that this formula could be sensibly construed either as an instrumental or locative use of the dative is telling, because this testifies to how Hephaestus' craftsmanship entails the extension of his mind out into the world. His labor takes place both *in* his πραπίδες and *by means of* this faculty simultaneously manifesting in the form of the tripods.

The adjective ἰδυίῃσι encourages this reading because it intimates how Hephaestus' plans are both cognitive as well as physical, insofar as they find visible form in his creation of the automata. As Françoise Frontisi-Ducroux (2002) has argued, ἰδυίῃσι in the context of Hephaestus' πραπίδες encompasses both seeing and knowing ("son savoir, aussi sensible qu'intellectuel, est de l'ordre du visual," 479) and so emphasizes Hephaestus' ability to use his craft to bring to life the images in his mind ("une vaste boîte à images," ibid.). And the visual nature of his craft is further emphasized with the narrator's interjection of θαῦμα ἰδέσθαι, "wondrous to see," which stresses how vision is the primary mode through which the marvelous quality of the tripods will be encountered.

It is thus Hephaestus' *mental* image of self-propelling tripods that the narrator reveals to his audience by elucidating the intended purpose of Hephaestus' work of affixing wheels to the base of each tripod, "that they might enter the assembly of the gods of their own accord and come back home again" (18.376–7). This typifies the capacity of language to illuminate what would otherwise be imperceptible: the will and intentions of another's mind. This is an aspect of language and artificial intelligence that will be discussed in detail in the next section, where the fabulous exterior of Pandora serves as a disguise to mask the wickedness of her true, inner nature. In this case, however, the poet's ability to ascertain Hephaestus' intentions lends credence to his claim of being divinely inspired in his speech and so having access to information otherwise unavailable to mortals. And the visual nature of both Hephaestus' and the poet's craft is emphasized by the narrator's interjection of his own voice with the phrase θαῦμα ἰδέσθαι, "a marvel to see," which also, paradoxically, reminds the audience that these automata do not yet exist as such: they are as yet still figments of Hephaestus' (and the poet's) imagination.

Moreover, it is significant that what makes Hephaestus' tripods "wondrous to see" is not just the addition of wheels (which is in itself not extraordinary, given that such objects are attested elsewhere, cf. *Od.* 4.131–2), but rather the fact that they are normal tripods in form, just with the capacity to move to and fro. In other words, there is nothing apparently automatic about them. They are, in today's parlance, "smart tripods": seemingly everyday objects endowed with intelligent characteristics that enable them to discern their master's will and act accordingly. The process illustrated in the description of their construction, whereby the narrator combines and innovates on existing forms of technology in order to imagine a new, automatic hybrid (where the wheeled tripods aren't just movable but self-moving), nicely illustrates the scaffolding role that language can play in structuring creativity and innovation.

As Clark (2008: 58–60) has argued, language is technological to the extent that it provides an external and material set of symbols that enable us to formulate thoughts and thereby put those thoughts into a form accessible to both ourselves and others. It is for this reason that language functions as "a key cognitive tool by means of which we are able to objectify, reflect upon, and hence knowingly engage with our own thoughts, trains of reasoning, and cognitive and personal characters" (Clark 2008: 59) because it allows the transformation of thought into objects.[17] I suggest that we can see this kind of scaffolding at work in this portrayal of the tripods, where the innovative combination of the language of technology (wheels, tripods, etc.) allows the narrator to create an entirely new kind of automatic object from the combination of these features.

For not only do the Homeric portrayals of Hephaestus' automatic tools envision an extended mind at work, it is also possible to see in these descriptions the specific ways in which language functions as an analogous tool for the narrator, one that enables him to imagine a divine version of human forms of technology. In so doing, he illuminates a magical quality to language itself and his own special skill at using it, one that consists in the possibility words offer to vividly depict such imaginary creations as Hephaestus' automata.[18] In this way, these descriptions nicely prefigure the ekphrasis of Achilles' shield by demonstrating how it is only in a verbal medium that artifacts can become genuine automata.

Homo loquens: the problem of language in artificial intelligence

It is in humanoid creations, however, that the relationship between language and artificial intelligence becomes more complex, as the addition of language and voice to human-like automata makes it more difficult to determine what is artificial about their cognitive faculties and thus more difficult to discern what distinguishes robotic forms like Hephaestus' helper maidens and Pandora from real-life humans.[19]

Hephaestus is assisted in his workshop by a group of robotic golden handmaidens who exhibit features of human cognition:

> δῦ δὲ χιτῶν', ἕλε δὲ σκῆπτρον παχύ, βῆ δὲ θύραζε
> χωλεύων: ὑπὸ δ' ἀμφίπολοι ῥώοντο ἄνακτι
> χρύσειαι ζωῇσι νεήνισιν εἰοικυῖαι.
> τῇς ἐν μὲν νόος ἐςτὶ μετὰ φρεςίν, ἐν δὲ καὶ αὐδὴ
> καὶ σθένος, ἀθανάτων δὲ θεῶν ἄπο ἔργα ἴσαςιν.
> αἳ μὲν ὕπαιθα ἄνακτος ἐποίπνυον· αὐτὰρ ὃ ἔρρων.
>
> He put on a tunic, and took up a stout staff, and went haltingly to the door. And the golden servants, looking just like young living maids, hastened to support their master. In them was mind together with understanding, and in them was also voice and strength, and they have knowledge of handiwork, endowed by the immortal gods. And these were busying about around their master as he limped.
>
> (*Il. 18.416–21)*

The contrast between Hephaestus' movements and those of his automata is marked: he is twice said to be limping (χωλεύων, ἔρρων), while the maidens conduct themselves dexterously and rapidly (ῥώοντο, ἐποίπνυον). While the maidens are thus more physically capable than Hephaestus, he is also repeatedly characterized as their "master" (ἄνακτι, ἄνακτος), which emphasizes that the function of the maidens is primarily to assist him and so compensate for his physical defects, analogous to the way that Otto's notebook serves to ameliorate the effects of his Alzheimer's.

What enables the maidens to function in this way? It is not just the fact that they are automatic, but because they have distinctively human cognitive characteristics. It is clear that they, like Hephaestus' automated tools, have some form of perception, since they rush unbidden to aid Hephaestus. But they have additional cognitive abilities besides. The narrator indicates that "in" them is "mind together with understanding" (τῆς ἐν μὲν νόος ἐστὶ μετὰ φρεσίν) as well as "voice and strength" (ἐν δὲ καὶ αὐδὴ καὶ σθένος) and they have a degree of practical knowledge (ἔργα ἴσασιν) besides. The repetition of ἐν here is significant because it underscores how these faculties (mind, understanding, voice, and strength) are located *within* the bodies of the maidens. And given the emphasis on their resemblance to real-life maidens (ζωῇσι νεήνισιν εἰοικυῖαι), it seems clear that it is not only the presence of these faculties that contributes to this resemblance, but also the fact that they are located internally. The golden material of the maidens *belies* the mental and physical capacities located within them that enable them to behave just like living women. This will also be the case with Hesiod's depiction of Pandora in the *Works and Days*, whose deceptive nature is represented in part by the disjunction between her pleasing outer appearance and destructive inner, invisible cognitive characteristics.

Moreover, the repetition of ἐν also indicates that these faculties are distinct elements that have been implanted into their forms, analogous to the images that Hephaestus repeatedly adds "into" the shield of Achilles later in the same book (e.g. 483, 490, 535, 541). This depiction of human intelligence suggests, therefore, that the human mind is not a single, unified entity, but a conglomeration of different faculties (understanding, speech, and strength) that together serve to animate the body from within. This raises the question, then: what exactly makes the intelligence of Hephaestus' maidens "artificial"? Even their voices are presented as distinctively human, as is clear from the fact that they have αὐδή in particular. As Jenny Strauss Clay (1974) has shown, αὐδή in Homer is a word reserved specifically for the human voice.[20] The only feature that distinguishes the maids from living girls is that they have bodies fashioned from gold instead of flesh and blood. What this intimates, therefore, is the constitutive role of the biological body in defining intelligence as uniquely human rather than artificial.

For this reason, this depiction of the maidens' cognitive profile reveals how lifelike, animated artifacts can aid in thinking *about* thinking, and in particular about the interactive relationship between the body and cognition. This kind of self-reflective cognition is often termed "higher-order thinking" and treated as the cognitive faculty that distinguishes humans from all other primates.[21] Analogous to the way that language enables the Homeric narrator to imagine animated versions of existing forms of technology and artifacts, verbal depictions of artificial intelligence are also themselves kinds of mind tools that allow poet and audience to imagine alternative forms of cognition. Language, then, for both poet and audience, is integral to the conceptualization of cognitive properties and processes.

So far my aim has been to demonstrate how the automata that populate Hephaestus' workshop function as extensions of Hephaestus' mind and so depict how artifacts can helpfully contribute to and participate in cognitive processes in the way described by Clark. And these automata work so seamlessly in concert with Hephaestus, in fact, that the inner workings of these objects are left completely opaque. While this supports the idea that Hephaestus' automata were understood to operate by means of magic rather than mechanism, this seemingly magical quality is significant. For the very fact that Hephaestus creates automatic versions of objects that, in the human world, were products of craft and mechanics conveys the idea that superlative forms of technology (i.e. native to the divine and produced by magic and not by craft) blur the boundaries between body and mind, functioning by totally opaque means. It is this opacity that will come to the fore in the figure of Pandora, who is classified as a trap as well as "inescapable" in both the *Theogony* and *Works and Days* precisely because her evil nature is so perfectly concealed by her fabulous appearance.[22]

In fact, the two distinctive accounts of Hephaestus' creation of Pandora that appear in the *Works and Days* and *Theogony* can be read as a case study in how language can function as a tool that enables the kind of higher-order thinking that gives humans the unique ability to reflect on their own cognitive processes. It is possible to see in these depictions, in other words, how "language allows us to make our own thought processes into objects for further scrutiny" (Clark 2003: 79), specifically by providing the medium through which artifacts like Pandora can acquire human-like cognitive characteristics. While the first section demonstrated how both automata and language alike are forms of technology that both reflect and facilitate the process of extended cognition, Hesiod's descriptions of Pandora reveal two different ways of conceptualizing the relation between artificial intelligence and human cognition. In so doing, the figure of an animated, human-like artifact like Pandora emerges as a rich resource for delineating the contours and dynamics of cognition and the human body.[23]

While in the *Works and Days* Pandora is a kind of automaton akin to Hephaestus' golden maidens, the Pandora of the *Theogony* is more akin to a "dummy," as West (1978: 158, 61–2) characterizes her. This makes it all the more striking, then, that she exerts a powerful effect on viewers and so can serve as a "sheer trap" (δόλον αἰπύν, *Theog.* 590, *Op.* 83) despite having no mind or voice of her own. For this reason, Pandora typifies how even apparently inanimate objects can be deeply implicated in cognitive processes.[24] Thus, the figure of Pandora in the *Theogony* functions as a way of conceptualizing the interrelation between human minds and objects.

In the *Theogony*'s version, Pandora is presented as an artifact with a cognitive life that is defined by her paradoxical nature, one encapsulated in the phrase "beautiful evil" (καλὸν κακόν, *Theog.* 585). Moreover, she is also a "likeness" or "copy" (ἴκελον, 572) as well as a prototype, and she is both person and object. The fact that she is made to resemble a "shy maiden" (παρθένῳ αἰδοίῃ, 572), a being that does not yet exist in the *Theogony*'s universe (since Pandora is here explicitly cited as the source of all mortal women, ἐκ τῆς γὰρ γένος ἐστὶ γυναικῶν θηλυτεράων, 591), calls attention to the way that images invite interaction by prompting viewers to consider what they look *like*. And even though there is no "real-life" model for Pandora, the fact of her creation testifies to how artifice (in this case, earth) allows for imaginary beings to have a lifelike appearance.[25]

That image-making offers the opportunity to craft new, seemingly animate objects is clearly expressed in the description of the headband Hephaestus makes for Pandora (578–84). While her veil as well as headband are both "marvels to see" (θαῦμα ἰδέσθαι, 575, 581), the headband is as such specifically because of the figures Hephaestus has depicted on it. Like Odysseus' brooch (*Od.* 19.225–31) and Achilles' shield (*Il.* 18.466–7), which also evoke *thauma* because the images they portray seem to be alive and moving, the headband contains "as many terrible creatures as the earth and sea produce" (τῇ δ' ἔνι δαίδαλα πολλὰ τετεύχατο, θαῦμα ἰδέσθαι, / κνώδαλ' ὅσ' ἤπειρος δεινὰ τρέφει ἠδὲ θάλασσα, 581–2) which are "wondrous things, resembling living beings with voices" (θαυμάσια, ζωοῖσιν ἐοικότα φωνήεσσιν, 584). The animals are so lifelike, in other words, that they even seem to be making noise. It has also been observed that this piece, like Achilles' shield, is no artifact that could exist in reality because of the sheer vastness of its portrayal.[26]

This paradoxical quality to the headband is mirrored in Pandora's characterization as a καλὸν κακόν (585) and also accounts for the wonder, *thauma*, that accompanies the sight of Pandora as well as her headband and veil: "And wonder gripped the immortal gods and mortal men" (θαῦμα δ' ἔχ' ἀθανάτους τε θεοὺς θνήτους τ' ἀνθρώπους, 588). In the context of the *Theogony*'s narrative, this response takes on greater significance because it dramatizes the first encounter between men and crafted images as one in which viewers are passive to the effects of wonder as it "takes hold of" or "grips" men and gods alike. And this is what makes Pandora a "sheer trap, irresistible to men" (δόλον αἰπύν, ἀμήχανον ἀνθρώποισιν, 589) because to behold her beauty is to fall victim to its stupefying effect.[27] The cognitive effect of Pandora is also articulated in her characterization as "irresistible" or "impossible" (ἀμήχανον) for men, a term that speaks to the impossibility of Pandora falling prey to any form of human intelligence or craft. This formulation of Pandora's effects on viewers thus suggests that objects can powerfully impact those who interact with them, and it is in this way that Pandora acquires a kind of agency in spite of her motionless, voiceless nature.[28]

Unlike the depiction of her that we will see in the *Works and Days*, this Pandora is all surface: she does not have the "voice and strength" (αὐδὴ καὶ σθένος) of Hephaestus' robot maidens, for instance, and is not clearly distinguished from her accessories.[29] The absence of any cognitive or perceptual faculties in her creation, by underscoring how it is only through her appearance that she functions as a trap, thus indicates that what makes Pandora dangerous is that her form preys on the gullible nature of the human mind that emerges from its susceptibility to seeing life in inanimate objects.[30] And the vulnerability of the mortal intellect to such illusions is precisely the idea that the Muses in the proem to the *Theogony* communicate, when they tell Hesiod "we know how to speak many false things that seem to be genuine, but we also know how to proclaim the truth, when we wish to" (ἴδμεν ψεύδεα πολλὰ λέγειν ἐτύμοισιν ὁμοῖα, / ἴδμεν δ' εὖτ' ἐθέλωμεν ἀληθέα γηρύσασθαι, 27–8). Thus, Pandora's efficacy as a snare derives from the susceptibility of the human intellect to the mind-boggling effects of lifelike artifacts.

In addition, the paradoxical use of the term "likeness" to describe the first iteration of a mortal maiden reminds us that Pandora also represents in this narrative the first human image created in the mortal world. For this reason, she models the triumph of *technē* over *phusis*, because she is here explicitly characterized as the source of all human women.[31] In other words, this portrayal suggests how the production of images and likenesses raises the possibility for biological life (human women) to imitate art (Pandora) rather than the other way around.

A striking contrast emerges from the *Works and Days* portrayal of Pandora, for this account indicates that Pandora deceives because of her *resemblance* to humans. This is because the *Works and Days* emphasizes the cognitive and intellectual abilities that Hephaestus places within her form that enable her to move, talk, and think like a person. Like his robotic maidens, the Pandora of the *Works and Days* is endowed with voice and strength (αὐδὴ καὶ σθένος, 62–3), as well as mind and a "wily" character (νόον καὶ ἐπίκλοπον ἦθος, 67) and practical skills (ἔργα διδασκῆσαι, 64). But in this case, unlike with Hephaestus' helpers, these features do not just contribute to her lifelikeness, but are specifically supposed to make her a bane for mortal men, because these are coupled with her "doglike mind" (κύνεόν τε νόον) and "wily character" (ἐπίκλοπον ἦθος). Moreover, Hermes also implants in her "lies and guileful words" in addition to her wickedness (ψεύδεά θ' αἱμυλίους τε λόγους καὶ ἐπίκλοπον ἦθος, 78). And whereas in the *Theogony* the lifelike character of Pandora was marked in the *thauma* that her appearance elicits, this version focuses instead on her erotic powers. Aphrodite contributes "charm" and "longing" and "limb-gnawing cares" (καὶ χάριν ἀμφιχέαι κεφαλῇ χρυσέην Ἀφροδίτην /καὶ πόθον ἀργαλέον καὶ γυιοβόρους μελεδώνας, 65–6), which makes it clear that it is Pandora's ability to evoke sexual desire that accounts for her characterization here as a "sheer and implacable snare" (δόλον αἰπὺν ἀμήχανον, 83). Moreover, the fact that these are "poured over her head" (ἀμφιχέαι κεφαλῇ, 65) communicates how these effects are generated specifically by Pandora's beautiful appearance: her face "like that of an immortal goddess" and her body that of a lovely maiden (ἀθανάτῃς δὲ θεῇς εἰς ὦπα ἐίσκειν / παρθενικῆς καλὸν εἶδος ἐπήρατον, 62–3).

While her physical exterior thus ensnares viewers by evoking desire, everything else animating Pandora is hidden within her (ἐν θέμεν, 61; ἐν δὲ θέμεν, 67; ἐν δ' ἄρα οἱ στήθεσσι, 77; ἐν δ' ἄρα, 79). This repetition of ἐν underscores how these are physically inside Pandora so as to form a contrast with her outer appearance. And this emphasis on the hiddenness of her wicked character thus puts a different slant on why Pandora is a "sheer and implacable snare" (δόλον αἰπὺν ἀμήχανον, 83): because her beauteous exterior gives no indication as to her hateful and destructive nature. As a result, this Pandora deceives in a way markedly unlike that of the *Theogony*.

While that version indicated that Pandora beguiled because she "lies like the truth" attributed to the Muses, and seems disarmingly genuine and lifelike although she was endowed with no cognitive faculties, the Pandora of the *Works and Days* deceives because her appearance conceals the type of mind working within her. In other words, Pandora is evil not because she is a form of artificial intelligence, but because of how her adorned appearance ensures that her true nature remains completely inscrutable to others.

In this sense, Pandora models a problem endemic to humans and divinities alike: namely, that it is always possible for appearances to deceive and to conceal a person's true intentions. As a beautiful statue that is also a literal vessel for evil, just like the jar that accompanies her (*Op.* 94–105), this description of Pandora demonstrates how the figure of the animated statue functions as a fertile image for depicting the disconnect between surface and interior that is a trope of archaic portrayals of deception.[32] For to conceive of deceitfulness as a form of concealment is to posit that the human body serves as a kind of hollow shell for the collection of faculties that animate the body from within, analogous to the way that Hephaestus animates his maidens and Pandora in the *Works and Days* by implanting voice, strength, and mind within their frames.

Conclusions

By interpreting portrayals of automata in early hexameter as forms of extended mind, I hope to have drawn attention to the cognitive processes that become apparent in the interactions between beings living and artificial. The first section's focus on Hephaestus' animated tools revealed minds and technology working in a coupled system that was itself the reflection of a different kind of extended thought: namely, how imagination finds expression through language. In particular, the fantastical, magical quality that imbues Hephaestus' automata highlighted the role of language in engendering the creation of completely new and never-before-seen objects like these automata, ones that nonetheless have the clarity of lived experience that makes them "wondrous to see" (θαῦμα ἰδέσθαι) thanks to their presentation in the poet's language.

Hephaestus' robot maidens and Hesiod's two portraits of Pandora offer distinctive ways of conceptualizing cognition itself. While Hephaestus' helpers typify how forms of human-like artificial intelligence raise questions about distinguishing between minds from objects and persons from things, the two descriptions of Pandora presented in Hesiod offer a case study in the importance of language and verbal imagery for generating different ways of parsing out human cognitive processes. By drawing attention to the particular facets of human intelligence (or lack thereof) manipulated by Pandora, this section illuminated the figure of the automaton as an especially fertile locus for thinking about thinking. More generally, the emphasis placed on the vivification and lifelikeness of Pandora underscores how representations of human figures, whether in material or verbal medium, compel this kind of higher-order thinking by prompting reflection on what it means to be human.

Finally, that wonder, *thauma*, forms a recurrent theme throughout the passages discussed here establishes a sense of awe and mystery at the intersection of mind and world. The wonder elicited by automata is, I suggest, a response to the wondrously complex and mysterious nature of cognition itself.[33] Seeing inanimate objects move by themselves, seemingly by magic, confronts the viewer with the fact that the same capability is at work in equally mysterious ways in themselves. Likewise, the bundle of tools (strength, voice, mind) that animates artificial life is the same collection of faculties that defines human intelligence in particular. And contemporary neuroscience and philosophy of mind are no closer to coming to a consensus on how precisely these all work together than artificial intelligence is to producing automata with the capabilities of Hephaestus' maidens and Pandora. In short, then, figures of automata and artificial intelligence in archaic thought shed as much light on the modern mind as the ancient by making it clear how even now the precise mechanisms of human thought remain a source of wonder and mystery. At the same time, however, the examples discussed here have also illuminated how human minds both ancient and modern may capitalize on language and other external props for generating newer and newer ways of thinking about thinking.

Notes

1 Scholars have approached portrayals of ancient automata in a variety of ways. For instance, Liveley 2006 fruitfully applies Donna Haraway's reading of the cyborg figure to ancient myths, Devecka 2013 focuses on the concept of the robot in the Athenian cultural imaginary, while Berryman 2003 considers how such early accounts of automata fit into a history of Greek technology and argues that the presence of automata in Homer does not imply the conception of a mechanical (i.e. non-magical) way to create such objects. See also Cuomo 2007 for a broader perspective on the role of technology in both Greek and Roman cultures. On the mythology surrounding Hephaestus and his craftsmanship see especially Delcourt 1982.

2 A notion that dates back to Aristotle; cf. Berryman 2010: 38–9. Because my focus is automata in the literal sense, I therefore exclude from this chapter the animate but immobile dog statues crafted by Hephaestus to guard Alcinous' palace (*Od.* 7.91–4). As Faraone 1987 argues, the description of the dogs suggests that they function as magical talismans, and the idea that animated artifacts seem to acquire their lifelikeness by means of magic rather than mechanics is one that is also relevant to automata proper like Hephaestus' bellows, tripods, and robotic maidens.

3 Unless otherwise noted, all translations are the author's own.

4 Reprinted in Clark 2008: 220–32.

5 Cf. Renfrew and Zubrow 1994, which provides an illustrative model of how cognitive theories may be fruitfully applied to archaeology.

6 This work is important for the study of artificial intelligence in particular, since it examines how developments in artificial intelligence and robotics reveal how much of human cognitive processing develops not from inside the mind, but from real-world, environmental interactions.

7 On the significance of Hephaestus' disability in relation to his automata see especially Liveley 2006.

8 On Daedalus' moving statues see especially Plat. *Men.* 97d and Arist. *Pol.* 1253b. See also Morris 1995 and Frontisi-Ducroux 1975 for studies of the myths surrounding Daedalus.

9 Cf. his comments on the "material reality of language: its existence as an additional, actively created, and effortfully maintained structure in our internal and external environment . . . As a result, our cognitive relation to our own words and language . . . defies any simple logic of inner versus outer" (Clark 2008: 59).

10 Cf. Clark 2008: 58, "Rather amazingly, we are animals who can think about any aspect of our own thinking and can thus devise cognitive strategies (which may be more or less indirect and baroque) aimed to modify, alter, or control aspects of our own psychology"; and Malafouris 2013: 239 on how early mark-making affected the human cognitive architecture, "the principal role of early mark making, and later of imagery, was to provide a scaffolding device that enabled human perception gradually to become aware of itself. That is, it enabled humans to *think about thinking*" (emphasis in original).

11 See Edwards 1991: 468–73 on the problematic short alpha in ἄνοιτο here.

12 Berryman 2009: 25, "Nothing is said about internal mechanisms in the handmaidens or the bellows. Rather, they seem to be conscious and to respond directly to their master's will."

13 The bibliography on the shield of Achilles is vast and I summarize here a few of the approaches to the metapoetics of this scene: Becker 1995 offers a schema for interpreting the relation between verbal and material media in ekphrastic works and focuses especially on the shield of Achilles, Heffernan 1993: 10–22 draws attention to the "representational friction" generated by the combination of narrative's fluidity with the static nature of the shield's images, Webb 2009 provides a convincing re-interpretation of the concept of ekphrasis as a description of a work of art, de Jong 2011 sees in this ekphrasis a celebration both of Hephaestus' and the poet's craft, while Francis 2009 contextualizes the shield's description within the other accounts of lifelike or living images portrayed in archaic sources (Hesiod's Pandora and Hephaestus' maidens), and Squire 2013 considers the impact of the shield on later ekphrastic theories and strategies.

14 Cf. Snodgrass 1998: 42, "Most probably he [Homer] had seen actual examples which inspired specific details of his description: he may have compiled a picture of unattainable complexity but, as I have implied, it is too much to believe that he did so out of elements that were themselves entirely imaginary."
15 On ancient interpretations of these aspects of the shield's description see especially Culhed 2014.
16 A similar formulation appears in the description of the self-piloting Phaeacian ships at *Od.* 8.556, which "make plans with/in their minds" (τιτυσκόμεναι φρεσὶ νῆες). Cf. Clarke 1999: 72 in his study of the Homeric terms for cognition and cognitive faculties, "Knowledge and cognitive activity take place in the same way, in or by means of the apparatus." See also Padel's (1992: 19–20) take on πραπίδες, who cites it as an example of one of those Greek words that have "an uncertain physiological meaning, but nevertheless connect intuitive, mental, and emotional experience to the body."
17 On the strategies of cognitive engagement involved in written literature specifically see especially Cave 2016.
18 Cf. de Jong 2011: 10, "Homer in the ekphrasis of the Shield with its conspicuous blurring of boundaries between narration and description is not only celebrating Hephaestus' 'marvellous' visual art but at the same time his own 'marvellous' narrative art, which presents people and events in an 'enargetic'/'energetic' way."
19 Cf. Francis 2009: 16, "As Hephaestus' metallic maids demonstrate, there is no clear line between an image of life and life itself. What keeps an image in human form, endowed with power, ability, and speech, from being alive?"
20 Cf. Pelliccia 1995: 104–5, who has also observed that the presence of νόος, "mind," is also treated as a prerequisite for speech in Homeric poetry.
21 See above, n. 10. On the relationship between ancient depictions of artifacts and philosophical inquiries into the workings of nature, see especially Berryman 2010 and Summers 1999.
22 Cf. the remarks of Clark 2003: 48 on the increased integration into the human body of invisible and automatic technology, "The danger is one of loss of control. Opaque technologies were, of course, hard to use and control; that's what made them opaque. But truly invisible, seamless, constantly running technologies resist control in a subtler, perhaps even more dangerous, manner. How then can we alter and control that of which we are barely aware?"
23 My approach here is thus similar to that of Steiner 2001: xi, who also considers why statues are treated in ancient sources as "good to think with." See also Gross 1992, who delineates the history of the trope of the animated statue. For a survey of the examples of "bionic statues" in Greek literature and art see Spivey 1995.
24 Cf. Malafouris and Renfrew 2010: 4, "things can be said to have a cognitive life insofar as things are constantly implicated in networks, or better 'meshworks' of material engagement. More simply, things have a cognitive life because minds have a material life."
25 See Steiner 2001: 25–6 for a different interpretation of this concept of Pandora as both likeness and archetype that argues for Pandora's introduction of a world of resemblances rather than reality.
26 Cf. Summers 1999: 56.
27 Cf. Hunzinger 2015: 430, "Here the wonder's beauty is a snare: sign of defeat, mark of ignorance and lack of mental acuity."
28 On the ways that objects acquire agency through social relations see especially Gell 1998.
29 As Hunzinger 2015: 427 has observed, the presentation of Pandora as an artifact is also stressed in her characterization with forms of the neuter gender up until she is presented to the gaze of gods and men, where she suddenly becomes feminine (ἀγαλλομένην, 587). See also her remark on the transition from Pandora's description to that of her headband, whose description "suddenly and meaningfully substitutes for that of the creature herself" (ibid. 429).
30 Cf. Francis 2009: 16, "The description of Pandora presents not so much a visible object but instead the effects of seeing that object."
31 On the significance of this point see especially Kenaan 2008: 36, "As already noticed by numerous readers, Pandora is the first work of art, the first product of manufacture, and the first manifestation of *technē* as opposed to *phusis*. Even more importantly, *Theogony* introduces through the making of Pandora the very experience of objectification. The presentation of Pandora as an object of art results in an ekphrasis, which, by virtue of its rhetorical quality, creates two portraits: that of the object (the creation of Pandora), and that of the act of gazing at the object (the responses to Pandora)."
32 See, most famously, Prometheus' theft of fire at *Op.* 50–2 and *Theog.* 565–7 and his trick with the sacrifice at *Theog.* 538–42. See also Achilles' condemnation of duplicity at *Il.* 9.312–13 and the contrast between Odysseus' appearance and speech articulated at *Il.* 3.205–24.

33 Plato, for instance, at *Laws* 644d uses the term θαῦμα to characterize humans as the puppets of the gods precisely because the inner life of humans is characterized by a bewildering variety of conflicting desires and impulses that suggest an affinity to the strings that tug a puppet this way and that.

References

Becker, Andrew Sprague. 1995. *The Shield of Achilles and the Poetics of Ekphrasis*. Rowman & Littlefield.

Berryman, Sylvia. 2003. "Ancient Automata and Mechanical Explanation." *Phronesis* 48 (4): 344–69.

Berryman, Sylvia. 2009. *The Mechanical Hypothesis in Ancient Greek Natural Philosophy*. Cambridge University Press.

Berryman, Sylvia. 2010. "The Imitation of Life in Ancient Greek Philosophy." In *Genesis Redux: Essays in the History and Philosophy of Artificial Life*, edited by Jessica Riskin, 35–45. University of Chicago Press.

Cave, Terence. 2016. *Thinking with Literature: Towards a Cognitive Criticism*. Oxford University Press.

Clark, Andy. 1997. *Being There: Putting Brain, Body, and World Together Again*. MIT Press.

Clark, Andy. 2003. *Natural-Born Cyborgs: Minds, Technologies, and the Future of Human Intelligence*. Oxford University Press.

Clark, Andy. 2008. *Supersizing the Mind: Embodiment, Action, and Cognitive Extension*. Oxford University Press.

Clarke, Michael J. 1999. *Flesh and Spirit in the Songs of Homer: A Study of Words and Myths*. Clarendon Press.

Clay, Jenny. 1974. "Demas and Aude: The Nature of Divine Transformation in Homer." *Hermes*, 102. Bd., H. 2: 129–36.

Culhed, Eric. 2014. "Movement and Sound on the Shield of Achilles in Ancient Exegesis." *Greek, Roman, and Byzantine Studies* 54: 192–219.

Cuomo, S. 2007. *Technology and Culture in Greek and Roman Antiquity*. Cambridge University Press.

de Jong, Irene J. F. 2011. "The Shield of Achilles: From Metalepsis to Mise En Abyme." *Ramus* 40 (1) (January): 1–14. doi:10.1017/S0048671X00000175.

Delcourt, Marie. 1982. *Héphaistos, ou, La légende du magicien*. Les Belles Lettres.

Devecka, Martin. 2013. "Did the Greeks Believe in Their Robots?" *The Cambridge Classical Journal* 59: 52–69.

Faraone, Christopher. 1987. "Hephaestus the Magician and Near Eastern Parallels for Alcinous' Watchdogs." *Greek, Roman, and Byzantine Studies* 28: 257–80.

Francis, James A. 2009. "Metal Maidens, Achilles' Shield, and Pandora: The Beginnings of 'Ekphrasis.'" *American Journal of Philology* 130 (1) (April 2): 1–23. doi:10.1353/ajp.0.0038.

Frontisi-Ducroux, Françoise. 1975. *Dédale: mythologie de l'artisan en Grèce ancienne*. François Maspero.

Frontisi-Ducroux, Françoise. 2002. "Avec Son Diaphragme Visionnaire: Ἰδυίῃσι Πραπίδεσσι, Iliade XVIII, 481. À Propos Du Bouclier d'Achille." *Rev. Et. Grec.* 115: 463–84.

Gell, Alfred. 1998. *Art and Agency: An Anthropological Theory*. Clarendon Press.

Gross, Kenneth. 1992. *Dream of the Moving Statue*. Penn State Press.

Heffernan, James A. W. 1993. *Museum of Words: The Poetics of Ekphrasis from Homer to Ashbery*. University of Chicago Press.

Hunzinger, Christine. 2015. "Wonder." In *A Companion to Ancient Aesthetics*, edited by Pierre Destrée and Penelope Murray, 422–37. John Wiley & Sons, Inc.

Kenaan, Vered Lev. 2008. *Pandora's Senses: The Feminine Character of the Ancient Text*. University of Wisconsin Press.

Liveley, Genevieve. 2006. "Science Fictions and Cyber Myths: Or, Do Cyborgs Dream of Dolly the Sheep?" In *Laughing with Medusa: Classical Myth and Feminist Thought*, edited by Vanda Zajko and Miriam Leonard, 275–94. Oxford University Press.

Malafouris, Lambros. 2013. *How Things Shape the Mind*. MIT Press.

Malafouris, Lambros, and Colin Renfrew. 2010. *The Cognitive Life of Things: Recasting the Boundaries of the Mind*. McDonald Institute for Archaeological Research.

Morris, Sarah P. 1995. *Daidalos and the Origins of Greek Art*. Princeton University Press.

Padel, Ruth. 1992. *In and Out of the Mind: Greek Images of the Tragic Self*. Princeton University Press.

Pelliccia, Hayden. 1995. *Mind, Body, and Speech in Homer and Pindar*. Vandenhoeck & Ruprecht.

Renfrew, Andrew Colin, and Ezra B. W. Zubrow. 1994. *The Ancient Mind: Elements of Cognitive Archaeology*. Cambridge University Press.

Snodgrass, Anthony. 1998. *Homer and the Artists: Text and Picture in Early Greek Art.* Cambridge University Press.

Spivey, Nigel. 1995. "Bionic Statues." In *The Greek World*, edited by Anton Powell, 442–59. Routledge.

Squire, Michael. 2013. "Ekphrasis at the Forge and the Forging of Ekphrasis: The 'Shield of Achilles' in Graeco-Roman Word and Image." *Word & Image* 29 (2) (April 1): 157–91. doi:10.1080/02666286.2012.663612.

Steiner, Deborah. 2001. *Images in Mind: Statues in Archaic and Classical Greek Literature and Thought.* Princeton University Press.

Summers, David. 1999. "Pandora's Crown: On Wonder, Imitation, and Mechanism in Western Art." In *Wonders, Marvels, and Monsters in Early Modern Culture*, edited by Peter G. Platt, 45–75. University of Delaware Press.

Webb, Ruth. 2009. *Ekphrasis, Imagination and Persuasion in Ancient Rhetorical Theory and Practice.* Ashgate Publishing, Ltd.

West, Martin Litchfield. 1978. *Works & Days.* Clarendon Press.

21

Staging artificial intelligence

The case of Greek drama

Maria Gerolemou

Introduction

In general, artificial intelligence (AI) and cognitive science are partly driven by the objective to understand and "engineer" human intelligence. Specifically, AI refers to inanimate artifacts that exhibit traits of natural intelligence.[1] Natural intelligence is composed of mental functions like learning, feeling, remembering, understanding, as well as of mental habits and patterns of behavior, like obeying the law, being brave, acting according to gender norms etc. The replication of mental functions is often accompanied by an artificial reconstruction of the body and its automatisms, such as breathing, walking etc., i.e. actions that elude conscious control. The reproduction of the body could be absolute or limited; the latter is related to the mere augmentation of bodily parts, which, for instance, appears in the form of prosthetics,[2] tools or weapons, through which their users convey particular emotions or project their intentions. In this case, artificial reproduction simulates functions rather than assuming appearance. The former, i.e. absolute reproduction, includes a full reconstruction of the human body, which enables it to perform human activities.[3]

The reproduction of the body and its automatisms could be compared to the product of an engineer or a sculptor. Likewise, the replication of the mind entails too a procedure of conscious shaping that capacitates its engineering and allows for its manipulation.[4] Copying the mind and the body, though reducing human beings to their bodily reflexes and instincts, aims to overcome the vulnerability of the physical body and mind and produce a bionic body and a mind which e.g. is not concerned about the moral consequences of its deeds: i.e. it stands without consciousness and feels no affection. This issue seems to be reflected in Greek drama, where, however, the view that artificial bodies and minds can support, but not completely replace, human expertise prevails.

The integration of mind into matter and the replication of the human body and bodily automatisms were first exemplified with the intelligent artificial artifacts of the *Iliad*, products of Hephaestus' divine craftsmanship; these were built for simulating motion, human adaptive behavior and various other skills.[5] At 18.372–7, for example, in the scene where Thetis visits Hephaestus in his workshop to request a shield for Achilles, Hephaestus is described to be sweating while he labors to create twenty triple-legged tables; these have the ability to move as they automatically roll on wheels to the assembly of the gods and back home.[6] Shortly afterwards, as it is noted in the text, the god produces golden handmaidens, identical in sight to living girls, to serve and help him. They are, in addition, gifted with intellect, a speaking voice, strength and virtues, which were provided to them by all gods (18.417–20);[7] due to this, they are able to act as autonomous agents which could perceive their environment and adjust their responses according to incoming stimuli; for instance, when lame Hephaestus stands up and tries to walk toward Achilles' mother, they hurry to support him (421, αἳ μὲν ὕπαιθα ἄνακτος ἐποίπνυον).

Along with the mythical epic automata, in the works of the medical writers of the Hippocratic Corpus the idea of the body as a *natural automaton*[8] motored by various forces was first attested. In their endeavors to elucidate the mechanisms of the body, the Hippocratics argue that the inner body, which comprises various organs and fluids, operates in a spontaneous manner, *automatos* (αὐτομάτως) (cf. e.g. *Epid.* 5.1.19 L., *De articulis* 46.24–8 L), randomly, *ek tychês*, or, when the physician forcibly intervenes, according to his guidance;[9] however, although these forces in the body are hidden, a trained physician may reconstruct them, based on empirical evidence and methods of inference, and thus manipulate them with the purpose of favoring a certain treatment.[10] The fact that medical treatises supported the possibility of a conscious shaping of natural automatisms, aiming at a restoration of the body to its initial state, i.e. in accordance with its nature, seemed to ignite a discussion in literature and art on whether automatic bodily functions and form can indeed be artificially represented. In other words, since the body is composed of a series of detected activities, a craftsman or physician ought, in principle, to be able to develop a replica capable of equating, or perhaps surpassing it.[11] Such an artificial body is described in the first scene of the *Thesmophoriazusae*; here, under the influence of new philosophical trends, a cosmogony is imagined. Ether assumes the role of a craftsman, fabricating parts for living creatures whose eyes are manufactured as an *antimimon*, a counter-image, to the wheel of the sun and ears are perforated as a funnel for hearing (17f.[12]).[13]

On the other hand, the register and reproduction of cognitive functions or of certain social habits seem to be the result of a systemized education, especially of the growing interest in writing,[14] which stores knowledge and downgrades physical memory (see A. *PV.* 460f., γραμμάτων τε συνθέσεις, / μνήμην ἁπάντων, μουσομήτορ' ἐργάνην, 788f.),[15] of the sophistic focus on specialized knowledge,[16] as well as of the formal schooling system (for non-artisans)[17] that emerged during the fifth and especially fourth centuries BC in Athens. Generally, to educate someone means to transform his individual traits into a type of automatic habit for the creation of social integration and cohesion; notable factors in this procedure are mimesis and memory: through imitation, models of thoughts are replicated; through memory those models acquire duration and continuation. For example, in Aristophanes' *Th.* 155f., Agathon, proud of his mimetic faculty which works as a sine qua non for his art, argues that what we do not own by nature can be acquired by mimesis. Mimesis is defined, according to him, as a procedure that supplements the physical body and mind through the imitation of opposite manners (149–52; cf. *Ra.* 590–606).[18] No purpose seems to be served by imitating something of your own nature, since this would be unnecessary and ineffective (167; cf. *Ra.* 109f., *Pl.* 290–2, *Ec.* 278, 545).[19]

However, the codification and reproduction of thought and knowledge through education which can be used either by the individual or by the community sparked further discussions: for instance, what can be defined as useful information and merit reproduction and which is the best way to acquire it or transfer it? This is exactly what is being debated in Aristophanes' *Clouds*; on the one side, valuable information is described in the Unjust Speech as the subject of an ongoing process of construction and interpretation (1043, 1361–76, 1421f.), while in the

Just Speech, knowledge, which is actually implicit,[20] persists through time and is transmitted by the gods to humans as inflexible and internal (984–7).

Reproducing bodily automatisms[21]

Reproducing artificially the human body and its bodily functions means creating new self-moving agents. In Greek drama artificial bodies that generate the illusion of physical human presence appears to correlate with the ethereal, i.e. the incorporeal bodies of images (e.g. eidola) and with the "bodies" of animated statues; the examination of these does not intend to underline the mimetic skills of the images which are determined by the concept of the double and substitution, as Steiner's work on images and statues proposes (2001);[22] rather, animate statues and eidola are seen here as bodies that reproduce bodily automatisms, and thus challenge and reflect physical vitality by mostly mimicking humanlike motion.[23]

In *Iphigenia in Tauris*, for instance, the statue of the goddess Artemis purportedly moves and closes its eyes (1165–7, {Ιφ.} βρέτας τὸ τῆς θεοῦ πάλιν ἕδρας ἀπεστράφη. {Θο.} αὐτόματον, ἤ νιν σεισμὸς ἔστρεψε χθονός; / {Ιφ} αὐτόματον· ὄψιν δ' ὀμμάτων ξυνήρμοσεν;).[24] The audience is aware that this is an invention (*exeurêma*), a trick on behalf of Iphigenia, for convincing the barbarous Thoas of her magnitude (1180, *sophia*). Thoas is not able to recognize the "mechanisms" which determine or constitute a spectacle of artificial life; at first, he is skeptical and asks Iphigenia if perhaps an earthquake has moved the statue, but Iphigenia convinces him that it was indeed a miracle. Thus, motion and the blinking of the eyes are replicated for lending credence to a divine miracle[25] and, ultimately, for providing Iphigenia and her brother with an opportunity to escape.

A moving sculptured body motored by forces, divine or human, evokes the following question: What happens when this artificial body is mendaciously manipulated? In Euripides' *Alcestis*, after Alcestis dies, Heracles brings her body back from Hades. However, the identity of the body is initially being contested in the play; it has the same face and body as the real Alcestis, but it cannot be identified by Admetus as his wife (1037–69). A perfect replica of Alcestis could point either to a statue of her (he had ordered such a life-imitating statue in the past, 348–54) or to a *phasma*, an image sent by the gods to trick him (1126f.; cf. Euripides *Helen* 119). Even after Heracles reveals to Admetus the identity of the woman, and although she has the ability to move, Admetus is still troubled by the "unexpected duration of her silence"[26] (1143–6); after all, according to Heracles, she is, like a statue, the product of a great *mochthos* (1025). Notably, at the beginning of the play, Alcestis, who is dying in the place of her spouse, is described as if she is turning into a statue; she is barely breathing, σμικρόν εμπνέουσαν (205); death is gradually eliminating any movement from her hands and eyes (398f., βλέφαρον καὶ/ παρατόνους χέρας) and nor can she hear or see (404).[27]

Likewise, the supposedly speaking, intelligent automaton of Helen, as it interacts with Paris in Troy, is a divinely designed automaton, a product of the *mêchanê*, the skillful plans of Hera (610) who wishes to take revenge on Paris for giving the prize of the fairest goddess to Aphrodite.[28] In the play, however, the emphasis is not placed on the speech of the automaton or its ability to learn and actively adapt to its environment, i.e. not on the replication of any mental capacities. With the exception of the scene where the eidolon departs on its own accord (612f.) and frees the true Helen from all her charges (57–9), the play actually focuses on the result of the simulated physical presence, i.e. on the artificial reproduction of Helen's body and the disastrous effects of its manipulation.

Wright, with regards to Andromeda and her statue which confuses Perseus as to whether it is alive, asks (E. *Andromeda* fr. 125):[29] "Does this mean that the real Andromeda somehow lacks the lifelike quality which a real human being would possess?" (p. 322). The same, according to Wright, also applies to Helen and her eidolon. Like Perseus, Teucer and Menelaus are equally perplexed after seeing true Helen; Teucer, at the same time, is impressed by the reproduction of the body, which is identical to Helen's (71–4). Menelaus too seems perplexed with the emergence of this second Helen and doubts her true, physical presence (as the 'true' Helen was in Troy, with him); hence, his question carries great weight:[30] "What kind of craftsman can fashion a living body?" he asks (583, καὶ τίς βλέποντα σώματ' ἐξεργάζεται; cf. 585). The messenger, despite previously witnessing the departure of the eidolon into the sky, merely says in a sarcastic tone, as if he does not realize the existence of this second Helen, "ah here you are, I did not know that your body had wings" (618f., 1515f.[31]). Although this manufactured body is a living breathing body (34, *empnous*),[32] it does not possess the ability to adapt itself to the new environment and, consequently, to change its perception, while the true Helen, in contrast to the rigid behavior of the eidolon, could adjust her mind accordingly (160f., Ἑλένηι δ' ὅμοιον σῶμ' ἔχουσ' οὐ τὰς φρένας / ἔχεις ὁμοίας ἀλλὰ διαφόρους πολύ). In short, Euripides goes one step further; by discussing whether a dumb eidolon could replicate the way real human agents operate in the world, he engages with the long-standing question of what defines an original and what constitutes a replica.[33]

The artificial reproduction of the human body could also be referring to the reproduction of bodily parts, which, for instance, appear in the form of prosthetic limbs. Consider, for instance, Heracles' heroic hands in the Euripidean *Heracles*, traditionally representing the famous βίη Ἡρακλῆος (*Il.* 18.117; cf. S. *Tr.* 190–1102);[34] in this play, they are, in a sense, augmented by his bow, which, acting as prosthetics, acquires an existence outside the physical body; Malafouris (2008, 9), who draws on the perception of the Mycenaean sword as a prosthetic limb, argues that this could be explained on the basis of neuroscientific findings, according to which the systematic association between the body and inanimate objects (like tools or weapons) can result in the weapons becoming incorporated into the body.[35] As such, Heracles' bow-hand constitutes the basis of a discussion on cowardice, as opposed to courage (157–203).[36] The bow, Heracles' weapon, is, according to Lycus, a coward's weapon; for Amphitryon, who defends his son, the bow is a πάνσοφον εὕρημα (omniscient device) that gives the archer a chance to confront his enemy from a safe distance (170–94). As Lycus puts it, if Heracles is deprived of the bow, which, as a *prostheke*, is attached to his body as an extra limb, he will no longer be renowned for being the bravest man. Nevertheless, what follows in the play proves that, although the bow is a prosthetic, it is entirely controlled by Heracles' mind. As soon as Heracles becomes mad and loses control over the bow, this, as a mere technological body part, loses touch with reality and cannot distinguish between enemies or beasts and loving persons; i.e. it cannot perceive what the obstacle in front of it actually is; in order to change its behavior, it has to be "re-programmed." Therefore, the bow keeps performing tasks with the sole purpose of completing its assigned mission, which is to eliminate Heracles' every enemy, even though it actually kills Heracles' family. After Heracles kills his wife and children and recovers from insanity, the bow again becomes a part of Heracles' body, directed by his mind. Specifically, clashing against his ribs, it acquires a voice (1380–1) which warns him in a threatening manner that if he keeps it, it would be like saving his children's murderer (1379–81; cf. 1098–100).

In a consideration of the animated bow-scene of Heracles in Sophocles' *Philoctetes* 1128–39,[37] where the bow can see the shameful deceptions of his enemies, and of the fact that the bow in the play has an obvious advantage over the hands (96–105), one might prima facie argue that Sophocles was influenced by the Euripidean *Heracles*, which had already been staged.[38] Here, however, the bow's artificial liveness is adapted to Sophocles' anthropocentrism regarding the character of the heroic, i.e. in this case, the bow's heroic quality is determined by Philoctetes' *euergesia*, good deeds (662–70, esp. 670, εὐεργετῶν γὰρ καὐτὸς αὔτ' ἐκτησάμην), and it never acts by itself.

Sculpting the mind

The possibility of replicating the human mind brings to light a different and more complicated set of issues. In contrast to the artificial reproduction of bodies and body parts, the artificial reproduction of mental functions produces automata that know how to adapt their behavior to the changing conditions of an external environment.[39] However, in order to replicate thought, one would first have to map the mechanisms of the mind and intelligence, assuming that the mind can be reduced to a series of instructions, perceptual abilities and cognitive functions; the means of such reproduction, as I have already mentioned, is mimesis and memory.

Experimenting with the limits of the mind's artificial reproduction, which he presents as a threatening force to human intellect, Aristophanes in the *Clouds* portrays Strepsiades as offering both his body and mind to Socrates, the grand teacher, *didaskalos*, in the *phrontistêrion*, and allows him to manipulate it as it pleases him, hoping to learn how to deceive his enemies (439–41); the only problem is that he is old, forgetful and slow (129, πῶς οὖν γέρων ὢν κἀπιλήσμων καὶ βραδύς, cf. 414). Socrates is indeed willing to teach him how to reason and behave in response to complex tasks. More precisely, in vv. 478–80, Socrates asks Strepsiades to share the "ways of his mind" with him – for instance, if he has a good memory, if he learns easily or if he has a natural gift for speaking (482–3) – and design new components for him to replace the old ones (*kainas mêchanas*). If Strepsiades accepts these new components, he will become a new man, enhanced beyond his inborn characteristics (502f.): "You and Chaerephon are going to be very much alike," Socrates notes (503).[40] Strepsiades is upset as he does not want to look like Chaerephon. The joke here lies in the fact that Strepsiades is afraid that this process of enhancement will make him resemble a half-dead person (ἡμιθνής), like Chaerephon. Socrates, however, is referring to character improvement through imitation of a specific behavior, not to his external appearance. What is important here is that the unconditionally surrendered mind of Strepsiades is treated similarly to a machine that could be improved by replacing its defective parts. By changing the mind according to the teachers' recommendation, a bodily transformation could also occur. The latter becomes clearer later in the debate between the Just and the Unjust speech, where the Just speech instructs his candidate Pheidippides:

> If you follow my recommendations, and keep them ever in mind, you will always have a rippling chest, radiant skin, broad shoulders, a wee tongue, a grand rump and a petite dick. But if you adopt current practices, you'll start by having a puny chest, pasty skin, narrow shoulders, a grand tongue, a wee rump . . .
>
> *(1010–19, tr. by Henderson 1998, Loeb)*

However, in the play, schooling brings the opposite results, as the "improved" artificial actually fails to win over the natural. This is evident both in Strepsiades' case, who as an old man is not able to remember any newly acquired knowledge (789f.), and also later in Pheidippides' case, who is transformed into an unrestrained intellectual acting violently against his family.

The vision of mind enhancement is directly related, as we have seen, with the faculty of memory; however, memory can not only be artificially worked out; it can also be extended through mental *prostheseis*, for instance through the medium of the *deltos*, writing-tablet or letter. Athenian drama and especially Euripides discuss the possibility of extending physical memory, mostly in plays where the playwright uses a letter as a prop, as writing in general seems to be used as an analogy for describing how the mind works.[41] For example, in Euripides' *IT*, the uncertainty of Iphigenia's bodily presence related to the debate of her sacrifice in Aulis is balanced through the trustworthy, *pistos* medium of the letter written by Iphigenia and which corrects any wrong assumptions about the past, reporting that, although she was thought dead, she is alive (641f.). However, being afraid that the letter could be put aside and forgotten or destroyed, Iphigenia recites the letter to Pylades and Orestes and explains to her brother Orestes that she is alive in Tauris and needs to be rescued by him; the letter silently represents, σιγῶσα τἀγγεγραμμένα, Iphigenia's recorded words (763–5).[42] Importantly, as Iphigenia says to Pylades, if the letter vanishes, "by saving your body you are saving my words" (764f.), i.e. even though memory is expected to be mediated better through the written word, this medium is not recognized in the play as a reliable source. This is because the memory inscribed in tablets constitutes a problem, mainly because the written, material word can be destroyed.[43] Therefore, recorded and live speeches are interwoven further in the play as Iphigenia often interrupts her own recitation to answer Orestes' questions (769–87).[44]

Expanding and engineering memory through the medium of the letter reflects how the mind could generally implement cognitive processes. However, while the texts, both the *Clouds* and the *IT*, suggest the replacement and repair of natural memory in case of failure caused by sickness, old age, death etc., at the same time they underline the weakness of artificial memory related to its exposure due to its material essence.

Epilogue

To sum up, on the Athenian stage, the artificial replication of the human body and mind is represented in the form of an experiment; by portraying the overcoming of the physical body and mind, and its artificial reproduction, Aristophanes and Euripides, while they invite us to abandon the unrewarding physical condition of the human body and mind and favor its artificial replication and enhancement, actually demonstrate the problematical interrelation between artificiality and the real world, precisely by condemning or laughing at the results of the artificial simulation of the human.

Notes

1 On AI see among others Copeland 1993; Simon 1996, esp. ch. 1 on the notion of artificial vs. natural; Boden 2006, ch. 2; Frankish & Ramsey 2014.

2 On prosthetics in antiquity see Bliquez 1983; Lee 2015, 83; see further Garland 1995, 26f.; Wiesing 2008; Draycott 2018.

3 On "automatisms" first introduced by Spencer 1885, see Bublitz, Marek, Steinmann, Winkler 2010, esp. Bublitz 2010 and Siegler 2016; Gerolemou 2017.

4 Cf. Foucault's *Technologies of the Self*. In this late work, the French philosopher deals with four categories of production which seem relevant to the two categories of reproduction that I discuss here. Foucault speaks of technologies of production related to the production and manipulation of sign production technologies, denoting the way that signs are used for producing meaning, technologies of discipline which draw on relationships of domination and technologies of the self, where he deals with technologies of self-transformation.

5 See Berryman 2003 and 2009, esp. ch. 1 and 2; see further Lonie 1981; but see recently Groot 2016. Generally, on automata in antiquity see Rehm 1937; Price 1964; Sörbom 1966, 41–77; Hesberg 1987; Cambiano 1994; Pugliara 2003; Berryman 2003; Wilgaux & Marcinkowski 2004; Kalligeropoulos & Vasileiadou 2008; Francis 2009; Devecka 2013. See further von Staden 1996 and 2007.

6 τρίποδας γὰρ ἐείκοσι πάντας ἔτευχεν / ἑστάμεναι περὶ τοῖχον ἐϋσταθέος μεγάροιο, / χρύσεα δέ σφ' ὑπὸ κύκλα ἑκάστῳ πυθμένι θῆκεν, / ὄφρά οἱ αὐτόματοι θεῖον δυσαίατ' ἀγῶνα / ἠδ' αὖτις πρὸς δῶμα νεοίατο θαῦμα ἰδέσθαι.

7 ... ὑπὸ δ' ἀμφίπολοι ῥώοντο ἄνακτι/ χρύσειαι ζωῇσι νεήνισιν εἰοικυῖαι. / τῇς ἐν μὲν νόος ἐστὶ μετὰ φρεσίν, ἐν δὲ καὶ αὐδὴ / καὶ σθένος, ἀθανάτων δὲ θεῶν ἄπο ἔργα ἴσασιν. On homeric automata see Bielfeldt 2014. See further the gold and silver dogs of Hephaestus in *Od.* 7.91–4 and Faraone 1987 and Steiner 2001, 117. Likewise, in Hesiod's work, Hephaestus uses clay to create an artificial woman, Pandora, who is endowed with human traits, like speech, strength, beauty, charm, the knowledge of female activities, crafty words and a deceitful nature (*Th.* 571–84, *Op.* 61–71f.). Cf. Faraone 1992, 101; Vernant 1996; Francis 2009, 13f.

8 As termed by von Staden 1996, 93 concerning Erasistratos' conception of the body.

9 See the *Ancient Medicine*, where *technê* is portrayed as opposed to *tychê* (1.2). Likewise, the author of the *On the Places in Man* argues further that *tychê* has no place in medicine (46); but see *On Diseases*; here the author stresses the fact that there are cases where the *tychaion* accompanies the good knowledge of the physician (see e.g. 7.48, 7.108). See also Holmes 2013.

10 Holmes 2010 and 2013.

11 The issue of artificial reproduction of bodily functions in terms of objets d' art was partly raised and developed by Métraux (1995), who examined the results of the Hippocratic study of the human body, for example the representation of respiration, in relation to the naturalistic stance and movement of the fifth-century human sculptures which had replaced the unmovable image of kouros. Cf. further on that Steiner 2001, 28f. See further on the vividness of statuary in the heyday of drama, Hallett 1986, esp. 75–84; Neer 2010.

12 See on the philosophical influences implied in the vv. 16–18, Austin & Olson 2008, ad loc; Tsitsirides 2001; Clements 2014, 24–7.

13 Likening body parts to mechanical contraptions is a familiar philosophical and rhetorical modality: for the analogy between body parts and technological devices as an interpretation of various skills and processes, see e.g. Empedocles, who famously described the analogy between respiration and *klepsydra* (B100 D-K); as cited in Solmsen 1963, 477–9; cf. further Kranz 1938; Tsitsirides 2001, 62 and Webster 2014.

14 However, in the fifth century this had not replaced the traditional oral methods, see Thomas 1989: ch. 1, esp. 22–3; 1992, 130, 144; See further Robb 1994, ch. 5, 189, 191–2; Denniston 1927, 117–19 on the use of books in Aristophanes.

15 See Plato's *Phaedrus* 274b–278d on memory augmentation via writing. See Small 1997, 4 who argues that writing changes the act of remembering; Davis 1998, 12–42 (cf. "Writing is a machine," p. 30); Ong 2012, 41, 78, 82; Havelock 1988, esp. ch. 7–11; Ceccarelli 2013, 188–90; Steiner 1994, 101.

16 Marrou 1956, ch. 5 discusses the pedagogical revolution of the sophistic; see further Pfeiffer 1968, 25, 30–56.

17 Ford 2002, 85–109; Robb 1994, 184–9.

18 In vv. 52–7, Agathon is rounding, polishing, welding, hammering and polishing his plays (cf. Euripides in *Ra.* 799–802). On Agathon as craftsman see Denninston 1927, 114; Muecke 1982, 43–6; Tsitsirides 2001, 63 n. 72.

19 Schwinge 2014, 69f. Cf. further Sörbom 1966, 76; Muecke 1982, 55f.; Zeitlin 1996, 382–6; Stohn 1993 arguing that mimesis is achieved by imitating physical traits and manners (*tropos*); see, however, Lada-Richards 1999, ch. 4, esp. 169–72, Lada-Richards 2002, 402f., "mimesis cannot leave the imitator's own identity intact" (p. 403). Duncan 2006, 27–46 discusses the two kinds of mimesis: Agathon proposes the constructionist one to be based on clothing and the essentialist one to be based on the *physis* of the body. Cf. further Wyles 2011, ch. 5. Cf. further on mimesis-scenes in Greek drama S. *Ph.* 128f., E. *Hel.* 1087–9, 1186–90; Ar. *Ach.* 440f., *Ran.* 109, 463, *Ec.* 93–114, 268–79. On mimesis and automata see Reilly 2011, 5–10; on the mimetic faculty of machines see Taussig 1993.

20 See on explicit and implicit knowledge in antiquity Asper 2015; Konstan 2015.

21 This part of the chapter is based on my book-project on dramatic automata to be published by Bloomsbury Academic (2020).

22 Cf. Vernant 1983, ch. 13 and 1991, ch. 10. On substitution of the human body through images, see Steiner 2001, esp. ch. 1. See further: in Greek drama women are often associated by analogy with statues, a fact that underlines, according to scholars, their objectification by male-dominant society. Cf. e.g. Euripides *Phoenissae* 220–1, Aeshylus *Suppliants* 282f., Κύπριος χαρακτὴρ τ' ἐν γυναικείοις τύποις / εἰκὼς πέπληκται τεκτόνων πρὸς ἀρσένων. On this, see among others Philipp 1968, 26–8, 31–7; Zeitlin 1994; Steiner 2001; Hall 2006, ch. 4.

23 On animated statues through various means, see Pugliara 2003, 161–240; Johnston 2008; Hersey 2006, 1–110; Chaniotis 2014 and 2017.

24 "(Iph.) The statue of the goddess turned around (Th.) Did it move on its own or an earthquake caused it to shift? (Iph.) On its own, and it even closed its eyes."

25 On this particular false miracle see Gerolemou 2018.
26 Montiglio 2000, 179. In practice, the silence of Alcestis could also be explained by the fact that Greek playwrights never used more than three speaking actors. That is, Alcestis' silent re-entrance in the play, if the play was meant to be performed by two actors as per the claim of certain scholars (Pickard-Cambridge 1988, 145; Arnott 1989, 46f.; on the silent Alcestis cf. among others, Trammell 1941; Rabinowitz 1993, 72f.; O'Higgins 1993, 78; Wohl 1998, ch. 8; Foley 2001, esp. 317), could be explained as necessary since the same actor who was playing Alcestis is now playing Heracles. Nevertheless, Alcestis' silence, whether or not the product of technical necessity, is, as we have seen, well integrated into the dramatic plot.
27 Segal 1993, 37–50; Steiner 2001, 192f.; Hall 2006, 128; Neer 2010, 60–1; cf. further Stieber 2011, 415f.
28 Wright 2005, 278–80, 322. See further on the automaton Helen, Whitmarsh 2018 on Sappho's depiction of Helen.
29 Cf. Eur. *Prot.* where supposedly a slave mistakes the statue of Protesilaos in bed with Laodamia with a real lover; cf. Steiber 2011, 168; cf. further on "artistic apate" Steiner 2001, 44–50, 191.
30 See Kannicht 1969 ad loc.
31 She is like her father who is *ptanos* too, 1145.
32 Kannicht 1969, ad loc; cf. further Steiner 2001, 191–3; Stieber 2011, 170 on the possibility of Helen being a moving statue.
33 See Steiner 2001 passim, esp. 55f., 290f.; Hall 2006, 121. Cf. the discussion on the original and copy as well of the legitimate and illegitimate copy (*eikôn* and *phantasma*) in Plato's *Sophist* 235d–236c. See on that Deleuze 1983.
34 Cf. Nagy 1979, 318.
35 See also Clark 2003, 59–62 and 2008, ch. 2.3 on artificial bodily augmentation.
36 Cf. Holmes 2010, 266f. See further Mueller 2016, 135–40 on Ajax's shield responding to the battle with Hector in *Iliad*'s seventh book by becoming a second skin.
37 In 656f. Neoptolemus wishes to take a look at the bow, touch it and salute it as a god. Cf. further imagined animated props in A. *Ag.* 37f., E. *Hipp.* 418, *Andr.* 924, *Ph.* 1342–4, E. *Alc.* 78, 568f.
38 See Taplin 1978, 90; Fletcher 2013, 207.
39 On mind's extension see Clark and Chalmers 1998; Clark 2008, ch. 2.6.
40 Unless noted, the translations are my own.
41 Cf. the *deltographos phren*, the recording mind of Hades in Aeschylus' *Eumenides* 273–5. See on that Solmsen 1944, 39; Steiner 1994, 100–9; Mueller 2016, 233.
42 Cf. Eur. *Hipp.* 842f., 856f., 877, where the letter takes the place of Phaedra. See further letters written and read by Agamemnon in *IA* 34–40, 97–105, 107–14, 115–23, esp. 117–18, the letter in *IT* discussed above and the *deltos* in *Tr.* 46f., 157, 492–6; on *staged* letters in Euripides see Rosenmeyer 2001, 61–97; Mueller 2016, 155–89. Cf. further Torrance 2010.
43 See Gerolemou-Zira 2017, 66–8 regarding the reliability of the written word in *Iphigenia in Aulis*.
44 See Rosenmeyer 2001, 66, n. 15.

References

Aristophanes. 1998. *Clouds. Wasps. Peace.* Edited and translated by Jeffrey Henderson. Loeb Classical Library 488. Cambridge, MA: Harvard University Press.

Arnott, P. D. 1989. *Public and Performance in the Greek Theatre.* London and New York: Routledge.

Asper, M. 2015. "Explicit Knowledge." In: M. Hose, D. Schenker (eds.), *A Companion to Greek Literature.* London: Wiley. 401–14.

Austin, C., Olson, S. D. 2008. *Aristophanes: Thesmophoriazusae. Ed. with Introduction and Commentary.* Oxford: Oxford University Press.

Berryman, S. 2003. "Ancient Automata and Mechanical Explanation." *Phronesis* 48/4: 344–69.

Berryman, S. 2009. *The Mechanical Hypothesis in Ancient Greek Natural Philosophy.* Cambridge: Cambridge University Press.

Bliquez, L. J. 1983. "Classical Prosthetics." *Archaeology* 36/5: 25–9.

Boden, M. A. 2006. *Mind as Machine: A History of Cognitive Science*, 2 vols. Oxford: Oxford University Press.

Bowie, A. M. 1997. "Thinking with Drinking: Wine and the Symposium in Aristophanes." *JHS* 117: 1–21.

Bublitz, H. 2010. "Täuschend natürlich. Zur Dynamik gesellschaftlicher Automatismen, ihrer Ereignishaftigkeit und strukturbildenden Kraft." In: H. Bublitz, R. Marek, C. L. Steinmann, H. Winkler (eds.), *Automatismen.* Munich/Paderborn: Fink Verlag, 153–72.

Bublitz, H., Marek, R., Steinmann, C. L., Winkler, H. (eds.) 2010. *Automatismen*, Munich/Padeborn: Wilhelm Fink Verlag.
Cambiano, G. 1994. "Automaton." *StudStor* 35: 613–33.
Ceccarelli, P. 2013. *Ancient Greek Letter Writing: A Cultural History (600–150 BC)*. Oxford: Oxford University Press.
Chaniotis, A. 2014. "The Life of Statues [Ἡ ζωὴ τῶν ἀγαλμάτων]." *Proceedings of the Academy of Athens* 89: 246–97.
Chaniotis, A. 2017. "The Life of Statues of Gods in the Greek World." *Kernos* 30: 91–112.
Clark, A. 2003. *Natural-Born Cyborgs: Minds, Technologies, and the Future of Human Intelligence*. Oxford: Oxford University Press.
Clark, A. 2008. *Supersizing the Mind: Embodiment, Action, and Cognitive Extension*. Oxford: Oxford University Press.
Clark, A. and Chalmers, D. J. 1998. "The Extended Mind." *Analysis* 58/1: 7–19.
Clements, A. 2014. *Aristophanes' Thesmophoriazusae: Philosophizing Theatre and the Politics of Perception in Late Fifth-Century Athens*. Cambridge: Cambridge University Press.
Copeland, B. J. 1993. *Artificial Intelligence*. Oxford: Blackwell.
Davis, E. 1998. *Techgnosis: Myth, Magic + Mysticism in the Age of Information*. New York: Harmony Books.
Denniston, J. D. 1927. "Technical Terms in Aristophanes." *CQ* 21/3–4: 113–21.
Draycott, Jane. 2018. *Prostheses in Antiquity (Medicine and the Body in Antiquity)*. London: Routledge.
Duncan, A. 2006. *Performance and Identity in the Classical World*. New York.
Faraone, C. A. 1987. "Hephaestus the Magician and Near Eastern Parallels for Alcinous' Watchdogs." *GRBS* 28: 257–80.
Faraone, C. A. 1991. "Binding and Burying the Forces of Evil: The Defensive Use of 'Voodoo Dolls' in Ancient Greece." *Classical Antiquity* 10/2: 165–220.
Faraone, C. A. 1992. *Talismans and Trojan Horses: Guardian Statues in Ancient Greek Myth and Ritual*. New York: Oxford University Press.
Fletcher, J. 2013. "Weapons of Friendship: Props in Sophocles' Philoctetes and Ajax." In: G. W. M. Harrison, V. Liapis (eds.), *Performance in Greek and Roman Theatre* (Mnemosyne Supplements 353). Leiden/Boston, MA: Brill, 199–216.
Foley, H. P. 2001. *Female Acts in Greek Tragedy*. Princeton, NJ: Princeton University Press.
Ford, A. L. 2002. *The Origins of Criticism: Literary Culture and Poetic Theory in Classical Greece*. Princeton, NJ: Princeton University Press.
Ford, A. L. 2010. "Sophists Without Rhetoric: The Arts of Speech in Fifth-Century Athens." In: Y. L. Too (ed.), *Education in Greek and Roman Antiquity*. Leiden/Boston, MA/Köln: Brill, 85–110.
Foucault, M., Martin H. L., Gutman H., Hutton, P. H. (eds.) 1988. *Technologies of the Self: A Seminar with Michel Foucault*. Cambridge, MA: MIT Press.
Francis, J. A. 2009. "Metal Maidens, Achilles' Shield, and Pandora: The Beginnings of Ekphrasis." *AJPh* 130/1: 1–23.
Frankish, K., Ramsey, W. (eds.) 2014. *The Cambridge Handbook of Artificial Intelligence*. Cambridge: Cambridge University Press.
Garland, R. 1995. *The Eye of the Beholder: Deformity and Disability in the Graeco-Roman World*. Ithaca, NY: Cornell University Press.
Gerolemou, M. 2017. "Thinking of Autonomy as Automatism: The Case of Autonomy in Thucydides' History." *Araucaria* (Spring Issue, Special issue, eds. A. Tsakmakis, C. Marcaccini): 199–211.
Gerolemou, M. 2018. "Zur Auffassung des Wunders in der griechischen Tragödie." In *Mnemosyne* 71: 750–76.
Gerolemou, M., Zira, M. 2017. "The Architecture of Memory: The Case of Euripides' Iphigenia in Aulis." *Skene. Journal of Theater and Drama Studies* 3/1: 59–81.
Groot, J. de. 2016. "Motion and Energy." In: Georgia L. Irby-Massie (ed.), *A Companion to Science, Technology, and Medicine in Ancient Greece and Rome*. Chichester: John Wiley, 43–59.
Hall, E. 2006. *The Theatrical Cast of Athens: Interactions between Ancient Greek Drama and Society*. Oxford: Oxford University Press.
Hallett, C. H. 1986. "The Origins of the Classical Style in Sculpture." *JHS* 106: 71–84.
Havelock, E. 1988. *The Muse Learns to Write: Reflections on Orality and Literacy from Antiquity to the Present*. New Haven, CT: Yale University Press.
Hersey, G. L. 2006. *Falling in Love with Statues: Artificial Humans from Pygmalion to the Present*. Chicago, IL: University of Chicago Press.

Hesberg Henner v. 1987. "Mechanische Kunstwerke und ihre Bedeutung für die höfische Kunst des frühen Hellenismus." *Marburger Winckelmann-Programm*: 47–72.

Holmes, B. 2010. *The Symptom and the Subject: The Emergence of the Physical Body in Ancient Greece*. Princeton, NJ: Princeton University Press.

Holmes, B. 2013. "Causality, Agency, and the Limits of Medicine." *Apeiron* 46.3.

Johnston, S. I. 2008. "Animating Statues: A Case Study in Ritual." *Arethusa* 41.3: 445–77.

Kalligeropoulos, Dimitrios and Vasileiadou, Soultana. 2008. "The Homeric Automata and Their Implementation." In *Science and Technology in Homeric Epics* (History of Mechanism and Machine Science, vol. 6), ed. Stephanos A. Paipetis, 77–84. Dordrecht: Springer.

Kannicht, R. 1969. *Euripides Helena*. Heidelberg: Winter.

Konstan, D. 2015. "Implicit Knowledge." In: M. Hose, D. Schenker (eds.), *A Companion to Greek Literature*. London: Routledge, 415–26.

Kranz, W. 1938. "Gleichnis und Vergleich in der Frühgriechischen Philosophie." *Hermes* 73.1: 99–122.

Lada-Richards, I. 1999. *Initiating Dionysus: Ritual and Theatre in Aristophanes' Frogs*. Oxford: Oxford University Press.

Lada-Richards, I. 2002. "The Subjectivity of Greek Performance." In: P. Easterling, E. Hall (eds.), *Greek and Roman Actors: Aspects of an Ancient Profession*. Cambridge: Cambridge University Press, 395–418.

Lloyd, A. B. 1975–1976. *Herodotus: Book II. Introduction* (I) and *Commentary* (II), Leiden: Brill.

Lonie, I. M. 1981. "Hippocrates the Iatromechanist." *Medical History* 25: 113–50.

Malafouris, Lambros. 2008. "Is It 'Me' or Is It 'Mine'? The Mycenaean Sword as a Body-Part." In: Dusan Boric and John Robb (eds.), *Past Bodies: Body-Centered Research in Archaeology*. Oxford: Oxford University Press, 115–24.

Marrou, H. I. 1956. *A History of Education in Antiquity*. Translated by George Lamb. London: Sheed & Ward.

Mastronarde, D. J. 1990. "Actors on High: The Skene Roof, the Crane, and the Gods in Attic Drama." *Classical Antiquity* 9.2: 247–94.

Métraux, G. P. R. 1995. *Sculptors and Physicians in Fifth-Century Greece: A Preliminary Study*. Montreal: McGill-Queen's University Press.

Montiglio, S. 2000. *Silence in the Land of Logos*. Princeton, NJ: Princeton University Press.

Morris, S. P. 1992. *Daidalos and the Origins of Greek Art*. Princeton, NJ: Princeton University Press.

Muecke, F. 1982. "A Portrait of the Artist as a Young Woman." *CQ* 32: 41–55.

Mueller, M. 2016. *Objects as Actors: Props and the Poetics of performance in Greek tragedy*. Chicago, IL: University of Chicago Press.

Nagy, G. 1979. *The Best of the Achaeans: Concepts of the Hero in Archaic Greek Poetry*. Baltimore, MD: The Johns Hopkins University Press.

Neer, R. T. 2010. *The Emergence of the Classical Style in Greek Sculpture*. Chicago, IL: The University of Chicago Press.

O'Higgins, D. 1993. "Above Rubies: Admetus' Perfect Wife." *Arethusa* 26/1: 78–97.

Ong, W. J. 2012. *Orality and Literacy: The Technologizing of the Word* (with add. chapters by J. Hartley). London and New York: Routledge.

Pfeiffer, R. 1968. *History of Classical Scholarship from the Beginnings to the End of the Hellenistic Age*. Oxford: Clarendon Press.

Philipp, H. 1968. *Tektonon Daidala. Der bildende Kiinstler und sein Werk im vorplatonischen Schrifttum*. Berlin: B. Hessling.

Pickard-Cambridge, W. A. 1988. *The Dramatic Festivals of Athens* (revised with a new supplement by John Gould and D. M. Lewis). Oxford: Clarendon Press.

Price, D. J. S. 1964. "Automata and the Origins of Mechanism and Mechanistic Philosophy." *Technology and Culture* 5: 9–23.

Pugliara, M. 2003. *Il mirabile e l'artificio: Creature animate e semoventi nel mito e nella tecnica degli antichi*. Vol. 5. Roma: L'Erma di Bretschneider.

Rabinowitz, N. S. 1993. *Anxiety Veiled: Euripides and the Traffic in Women*. Ithaca, NY/London: Cornell University Press.

Rehm, A. 1937. "Antike Automobile." *Philologus* 92: 317–30.

Reilly, K. 2011. *Automata and Mimesis on the Stage of Theatre History*. Basingstoke, UK: Palgrave Macmillan.

Robb, K. 1994. *Literacy and Paideia in Ancient Greece*. New York: Oxford University Press.

Rosenmeyer, P. A. 2001. *Ancient Epistolary Fictions: The Letter in Greek Literature*. Cambridge: Cambridge University Press.

Schwinge, E.-R. 2014. "Aristophanes' Thesmophoriazusen - eine Homage an Euripides." *WJA* 38: 65–100.

Segal, C. 1993. *Euripides and the Poetics of Sorrow: Art, Gender, and Commemoration in Alcestis, Hippolytus, and Hecuba*. Durham, NC and London: Duke University Press.
Segal, C. 1986. *Interpreting Greek Tragedy, Myth, Poetry, Text*. Ithaca, NY: Cornell University Press.
Simon, H. A. 1996. *The Sciences of the Artificial* (3rd ed.). Cambridge, MA: The MIT Press.
Small, J.-P. 1997. *Wax Tablets of the Mind: Cognitive Studies of Memory and. Literacy in Classical Antiquity*. London, New York: Routledge.
Solmsen, F. 1944. "The Tablets of Zeus." *Classical Quarterly* 38: 27–30.
Solmsen, F. 1963. "Nature as Craftsman in Greek Thought." *Journal of the History of Ideas* 24: 473–96.
Sörbom, G. 1966. *Mimesis and Art: Studies in the Origin and Early Development of an Aesthetic Vocabulary*. Stockholm: Svenska Bokforlaget.
Staden, v. H. 1996. "Body and Machine: Interactions between Medicine, Mechanics and Philosophy in Early Alexandria." In: K. Hamma (ed.), *Alexandria and Alexandrianism*. Malibu, CA: J Paul Getty Museum, 85–106.
Staden, v. H. 2007. "Physis and Techne in Greek Medicine." In: B.-V. Bernadette, W. R. Newman (eds.), *The Artificial and the Natural: An Evolving Polarity*. Cambridge, MA: MIT Press, 21–49.
Steiner, D. 1994. *The Tyrant's Writ: Myths and Images of Writing in Ancient Greece*. Princeton, NJ: Princeton University Press.
Steiner, D. 2001. *Images in Mind: Statues in Archaic and Classical Greek Literature and Thought*. Princeton, NJ: Princeton University Press.
Stieber, M. 2011. *Euripides and the Language of Craft* (Mnemosyne Supplements, 327). Leiden: Brill.
Stiegler, B. 2016. *Automatic Society: The Future of Work*. Cambridge: Polity Press.
Stohn, G. 1993. "Zur Agathonszene in den 'Thesmophoriazusen' des Aristophanes." *Hermes* 121/2: 196–205.
Taplin, O. 1978. *Greek Tragedy in Action*. Berkeley, CA: University of California Press.
Taussig, M. 1993. *Mimesis and Alterity*. New York: Routledge.
Thomas, R. 1989. *Oral Tradition and Written Records in Classical Athens*. Cambridge: Cambridge University Press.
Torrance, I. 2010. "Writing and SelfConscious Mythopoesis in Euripides." *Cambridge Classical Journal* 56: 213–58.
Trammell, E. P. 1941. "The Mute Alcestis." *CJ* 37: 144–50.
Tsitsirides S. 2001. "Euripideische Kosmogonie bei Aristophanes (Thesm. 14–18)." Ἑλληνικά 51: 43–67.
Vernant, J.-P. 1983. *Myth and Thought among the Greeks* (tr. by Janet Lloyd with Jeff Fort). London and Boston, MA: Routledge & Kegan Paul.
Vernant, J-P. 1996 [1990]. *Myth and Society in Ancient Greece* (tr. by Janet Lloyd). New York: Zone Books.
Vernant, J.-P, Zeitlin, F. I. 1991. *Mortals and Immortals: Collected Essays*. Princeton, NJ: Princeton University Press.
Webster, C. 2014. Technology and/as Theory: Material Thinking in Ancient Science and Medicine (diss.). University of Columbia.
Whitmarsh, T. "Sappho and Cyborg Helen." In: Felix Budelmann and Tom Phillips (eds.), *Textual Events: Performance and the Lyric in Early Greece*. Oxford: Oxford University Press.
Wiesing, U. 2008. "The History of Medical Enhancement: From Restitutio ad Integrum to Transformatio ad Optimum?" In: Bert Gordijn, Ruth Chadwick (eds.), *Medical Enhancement and Posthumanity*. Dordrecht: Springer, 9–24.
Wilgaux, J., Marcinkowski, A. 2004. "Automates et créatures artificielles d'Héphaïstos: entre science et fiction." *Techniques and Culture* 43–4.
Wohl, V. 1998. *The Intimate Commerce: Exchange, Gender, and Subjectivity in Greek Tragedy*. Austin, TX: University of Texas Press.
Wright, M. 2005. *Euripides' Escape-Tragedies: A Study of Helen, Andromeda, and Iphigenia among the Taurians*. Oxford: Oxford University Press.
Wyles, R. 2011. *Costume in Greek Tragedy*. London: Bristol Classical Press.
Zeitlin, F. I. 1985. "The Power of Aphrodite: Eros and the Boundaries of the Self in the 'Hippolytus'." In: P. Burian (ed.), *Directions in Euripidean Criticism: A Collection of Essays*. Durham, NC: Duke University Press, 52–111.
Zeitlin, F. I. 1996. *Playing the Other: Gender and Society in Classical Greek Literature*. Chicago, IL: University of Chicago Press.

Part VI
Cognitive archaeology

22

Thinking with statues

The Roman public portrait and the cognition of commemoration

Diana Y. Ng

Introduction[1]

This chapter seeks to investigate public honorific portraits—a highly prevalent part of Roman urban visual culture—from a cognitive perspective. These statues and their associated rituals will be considered as scaffolds for learning and memory, a perspective derived from behavioral psychology and the philosophy of mind. Though some care must be taken in such an endeavor, it is worth the effort to bring fresh approaches to the study of a class of objects that has long been a cornerstone in Roman art history. In doing so, we may move toward a more complete understanding of how these objects functioned as commemorative monuments.[2] Turning to cognitive approaches allows us to ask how successful commemoration might have been achieved. I argue that the Roman public honorific statue is not just external storage for information about elite individuals, though it is undeniably that. Rather, I propose that the statue and the designed interactions that sometimes were associated with it actively created meaning in partnership with the viewer, so that new knowledge both resulted and was maintained.

Some caveats

It is difficult to define the related phenomena of learning and memory as a neat conceptual package, because the modalities and processes associated with them are complex and have been studied in various disciplines with widely ranging methodologies. Basic questions about the neurology of memory and the mind–body problem have been under debate. Do cognition and, therefore, learning and memory exist and occur in an internal, "brainbound" way (Clark 2011: xxvii), or do they extend beyond the brain into the different sensorimotor receptors in the human body? Are these two concepts mutually exclusive? Do learning and memory reach even beyond the boundaries of the flesh, to external objects? This chapter does not come from the fields—neuroscience, psychology, philosophy—that have grappled with these issues the longest, and therefore draws from all of them, with the view that mental or internal representations are important to memory and learning, but that cognition is enabled in a significant way by input from the rest of the body and by the body's engagement with external features. A large body of neuroscientific research does locate the neural activity of memory in the hippocampus of the brain, and has shown that neural and synaptic connections are formed, broken, and formed anew as information is encoded and retrieved (Moscovitch 2007: 19). The presence of the engram or memory trace itself—the physical artefact of information encoding—within the brain can, not surprisingly, also be found in the same brain regions (Josselyn, Köhler, and Frankland 2015: 523–524, 531). The behavior of the human subject embedded in its environment may, one might imagine, be determinative of synaptic and neural activity. If this is the case, then the stimuli to which one has been exposed, the tools with which one has interacted, the conditions within which the contact takes place, and ongoing encounters become central concerns. In addition, how the brain processes the sensorial inputs resulting from these external encounters, and not just where the neurons and neural activity reside, is key to our understanding of how objects are co-agents of sense-making with the brain.

There are certain caveats that should be noted regarding the ancient evidence under consideration in this chapter. First is my choice to focus on public portrait monuments, which are formulaic, contrived, and rhetorical (Tanner 2000: 45; Smith 1998; Stewart 2003: 46; Fejfer 2008: 279–285).[3] While formulaic or rhetorical and literary text and speech have been significant in studies of memory, distributed cognition, and extended mind, they are not usually combined with equally important, if not dominant, artistic visual elements (e.g., Johnson et al. 2014; Malafouris 2012; Small 1997).[4] Therefore, on the one hand Roman statue monuments are somewhat similar to familiar textual cognitive tools, but on the other hand they bring to bear different perceptual and cognitive opportunities. Second, it is impossible to test any hypotheses about the actual cognitive impact of these ancient objects or of the carefully designed engagements with them, as no records were kept on whether the intended audience was able to recall the honorands at some later time. However, there is a growing body of scholarship on cognitive history and cognitive archaeology focused on periods as remote as prehistory and the Aegean Bronze Age (e.g., Malafouris 2012, 2013; Malafouris and Renfrew 2010) and also as evidence-rich yet still lacking in scientifically collected data as Renaissance England (Sutton 2008: 44–47; Sutton 2010: 193; Tribble and Keene 2011; Anderson 2015: 68–80). Elsewhere I have brought behavioral studies of learning and memory to the examination of a ritual that was closely tied with statuettes and text and to the consideration of the Roman period theater as a cognitive micro-ecology (Ng 2018 and forthcoming). What the body of cognitive historical and archaeological research has shown is that ancient to early modern societies worked through the processes of learning and memory, so as to instill information of value or articles of faith in individuals.

Thinking with public portraits

The scholarship on distributed cognition, extended mind, embodied cognition, enactivism, and the science of memory and learning has blossomed in the last three decades, such that many variations and opinions coexist, sometimes uneasily, within each of these theoretical frameworks (Wheeler 2017). To answer the questions posed above of public portraits based on the available ancient evidence, there are specific lines of thinking that I believe are most relevant and useful. First is the acceptance of the use of external tools—in this case, the public portrait—as a cognitive partner. Kirsh has found that "thinking with external representations" provides cognitive cost-savings—in execution time and ease—and allows a person to perform computational tasks that are not only more onerous without external distribution but may be impossible internally (Kirsh 2013: 185). Though one would not necessarily categorize the ancient viewing of public portraits as a problem-solving task, some of the advantages of "interactions" with external representations for cognition in some ways do transfer over, specifically: 1) that it can be shared; 2) that it has a physical persistence that allows for revisiting and reinterpretation; and 3) that it permits reformulation of problems that would be difficult to do internally (Kirsh 2013: 181–184). In Kirsh's view, carrying out computation through external representations supplements and enables the work that is performed mentally; it is still brain-based if not "brainbound."

Indeed, the Roman public statue monument appears not to meet two important characteristics of the extended mind—the close coupling of the cognitive extension with the person, and the transparency of that tool. The statues, being stationary objects installed in public spaces, necessarily lack the portability and thus the consistency of access of something such as Clark and Chalmers' example of Otto's notebook as an extension of the mind (Clark and Chalmers 1998: 12–17; Clark 2011: 78). Furthermore, while the Roman urban environment was clogged with these statues and they were likely at some point after their erection taken for granted as part of the visual and physical backdrop to the daily lives of ancient city-dwellers, these public portraits were nevertheless the subject of discourse. Indeed, they could not have come into being without public deliberation or, at the very least, acknowledgment of communal values and social prestige that were held in common. As a class of objects, the statues were commented upon by various authors, as will be seen below. Unlike a pen and paper, the use of the public portrait as a cognitive tool was both remarkable and remarked upon (Clark and Chalmers 1998: 16–17; Clark 2011: 10, 80). Yet, the failure of these statues to be closely coupled and transparent does not preclude them from being considered as external representations and objects of thought. In addition, the visual perception through which these monuments furnished information to the viewer—consequently supporting epistemic activity such as recall, sorting, and comparison—invites us to examine the relationship between these sculptures and embodied cognition.

Embodied cognition prioritizes the sensorimotor perceptions of the entire body and the interaction between the body and the world that allow a being to perform sense-making (see Hutto 2013: 282; Lindblom 2015: 84–95). In a radical enactivist approach, consciousness and cognition stem from the body's reception and reaction to external stimuli; no internal representation or executive mind holds sway over these bodily perceptions (Menary 2006: 2–3; Hutto 2013: 291–292). However, enactivism does not have to be radical, and the basic notion that "cognition unfolds (is enacted) [sic] in looping interactions between an active organism and its environment" (Wheeler 2017: 460) leaves room for the kind of cognitive partnerships under consideration here.

The sticking point, when operating within an enactivist framework, is the role or indeed the existence of a content-laden internal representation (Wheeler 2017: 460–465). In order to make sense of the Roman public portrait statue, which is a social as well as a material phenomenon, there must be, I believe, some role given to both external perceptions and internal models guiding their processing. In earlier work, Clark (2011: 181–190) advocated for some caution in bypassing internal representation altogether. At the time, he cited Milner and Goodale's research (2006) on the dorsal and ventral visual processing streams as a possible bridge between a purely bodily cognition and one integrated with brainbound functioning. The visual percepts processed in the dorsal processing stream result in directed motor output—such as to aim and reach for, then grasp an object, while those processed in the ventral stream support tasks of categorization, memory, and recognition of percepts based on already stored information (Milner and Goodale 2006: 63–65, 229). As Milner and Goodale note, this latter function of visual perception is crucial in enabling "behavior . . . to be guided intelligently by events that have occurred at an earlier time" (2006: 20). The ventral stream in this model also acknowledges and accommodates internal representations—mental models including images—and temporal distance that also feature in the ancient evidence.

That an internal representation can guide sensory percepts is one of the elements at the core of the concept of predictive processing (Hohwy 2013; Friston 2014; Clark 2015, 2016).[5] In predictive processing, the brain maximizes the efficiency of sensory input processing through the guidance of preexisting internal representations and models, or prior belief. When a sensory percept conforms to prior belief, it is not appropriated attentional resources (Hohwy 2013: 196–197; Clark 2015: 57–77). When, however, the brain encounters new or unexpected information that deviates from internal models—termed in predictive processing scholarship as a "prediction error"—then attention is directed accordingly so as to resolve the difference between prediction and data. What must result when there is a prediction error is one of two things: 1) revision of the internal model, or 2) a change in sampling (Hohwy 2013: 42–46; Friston 2014: 119). Novel external inputs, as predictive errors, thus require a human subject to revise prior belief, which can extend to many hierarchical levels. What is learned guides subsequent perception in a cognitive "spiral" (Hohwy 2013: 34) that loops sequentially. The predictive brain, then, is constantly engaging with the world based on what it knows and directing attention to prediction errors, so that it can continually revise its way of understanding the world in order to minimize future errors (Clark 2016: 271). Predictive processing, of which Clark himself has been a strong proponent in his most recent work (2015, 2016), incorporates both the cognitive cost calculations that were a key aspect of Kirsch's thinking with external objects, and Milner and Goodale's ventral stream of visual processing. These collective approaches are well suited to the aims of this chapter, for our evidence of the range of Roman interactions with public portraits rests on individuals' visual percepts and other engagements with these monuments within the context of Roman social relationships and cultural values, what might be considered the highest level of internal models. At this level, these internal models can be commonly distributed among members of a group (Clark 2016: 278–280).

Though Roman portraits figure prominently in modern art-historical considerations of sculptural form, style, and technique, whether they should be approached from an aesthetic or neuroaesthetic standpoint in a study of their cognitive roles is debatable. An aesthetic, versus perceptual, approach to these statues specifically necessitates a non-goal-driven interaction (Brincker 2015: 123).[6] As will soon become clear, I argue that not only were the public portraits themselves rewards for accomplishment, but their viewing could be combined with purposeful activity. Furthermore, in ancient commentary on public portrait statues the aesthetics of a single given sculpture were not a primary concern. The appearance and material of the statuary were sometimes noted only so as to indicate the value and therefore the social prestige of the object (Stewart 2003: 81, 91; Fejfer 2008: 151–162). Nevertheless, it is true that the statues engaged not only epistemic activity, but also physical sensations and emotional arousal, and that measures were sometimes taken to maximize the stimuli presented by the public portraits. The elicited responses, as I shall argue, link back to how learning occurs and how memory is consolidated in a cognitive sense.

The fundamental importance of internal representations to processing bodily perceptions in the approaches outlined above leads one to connect theories of cognition with behavioral psychological research on memory (as well as on the impact of emotion on memory and learning) to understand how statues functioned as objects of memory and aids for recall. The especially salient aspect of ancient encounters with public portraits to behavioral cognition is the fact that they were often paired with reiterative rituals designed to engage multiple senses in specific, structured encounters. A broadly accepted idea in behavioral psychological research on learning is that the reasoning and recall that are supported by the internal processing of sensory percepts are also supported by emotionally heightened (Levine and Pizarro 2006: 38) and spaced interactions (summarized in Brown et al. 2014: 28–43). Sensory stimulation and habit therefore figure prominently across methodologies. The two approaches, behavioral and theoretical, fit together and, furthermore, are in line with the ancient evidence considered in this chapter, including statues, inscriptions, and epigraphically recorded actions.

Memory and imagery in Roman culture

Already twenty years ago, Small applied the science of memory cognition to the study of classical literary, and literacy, culture in *Wax Tablets of the Mind* (1997). Literacy allowed the creation of longer, non-formulaic, and fixed works, and training for the memorization of these texts became more important and complex as a result (1997: 83–84). Roman authors had various recommendations for ancient mnemotechnics enabling effective memorization. The most famous and oft-cited recommendation is probably the memory house and *loci*, in which different components of a set speech are keyed to specific images that are arranged in a logical spatial flow through a mental model of a home (Quint. *Inst.*11.2.18–20). The success of such a mnemotechnic relies on the ability of the rhetorician to remember correctly the paired device and invented image as progress is made mentally through either a standardized or at least familiar house plan. Even Quintilian expressed some skepticism toward this technique for all the mental work that is required, at least for remembering a set speech: one would need to invent an image for each part of the speech and place it in a specific area of the mental house, remember what topic or figure of speech each image represented, then recall the exact words of the speech, and do it all in the proper sequence. Instead, he and Cicero advised students to practice a fixed text constantly for delivery despite the tedium of such a task, and the former also noted that a speech seemed to ripen and mature overnight, to become even more firmly entrenched in memory (Quint. *Inst.* 11.2.32–36, 40–44; Cic. *De Or.* 2.87.357–358).

The *loci* are nevertheless fascinating in that they are clearly internal representations that depend on the rhetoric student's awareness of bodily movement as though enacted in external space. Furthermore, Quintilian's recommendation for constant practice of a set speech necessitated a frequent and repetitive physical interaction—seeing the image of the words (as opposed to reading for their content), following the lines, finding signs remaining from alteration or erasure—with the tablets upon which the text was written. The tablets' physical properties are thus understood to aid in memory work. Quintilian also comments on the physicality of memorization beyond the visual capacity, suggesting that practicing a recitation in a soft murmur is conducive to learning as it engages both the faculties for speech and listening. The *loci* and the precepts for rote memorization touch on the debate on mind extension fundamental to theories of cognition. Quintilian, Cicero, and the author of *ad Herrenium* would, it seems to me, acknowledge the importance of the bodily perceptions in supporting the mind as the home of memory.

However, the short-term memorization of a set text supposedly facilitated by the *loci* and constant repetition is not the kind of memory that public portraits are thought to serve. The discourse on public portraits has instead been centered on the related notions of public prestige and perpetuation of the memory of the portrait's subject (Fejfer 2002: 250–254; 2008: 63–70; Stewart 2003: 79, 120, 267). Because of the fundamental difference in the sort of memory under question and the unambiguously external nature of the statue, the role it plays in memory cognition has not been given much attention. So, we return to the simple yet crucial question: how did a Roman public portrait statue work to make its subject well and enduringly known not just to family and acquaintances, but to larger audiences?

In a letter to Macrinus, Pliny the Younger provides us with an elucidative encounter with a public portrait. In it, he relates how the Senate, having been prompted by the emperor, had voted to grant two public portraits in honor of Vestricius Spurinna—one, a portrait of the general for his victory over the Bructeri, the other, a portrait of his son Cottius, who had died while the father was on his campaign. The son was a dear friend of Pliny's and he remarks that he was glad this unusual honor was granted to the exceptionally virtuous and outstanding young man. Pliny's view is that the posthumous statue was granted not only to preserve the memory of Cottius, but also as an example for other youths to emulate and for fathers to beget and raise worthy sons. For himself, Pliny takes some consolation in the portrait, for,

> it will be a great satisfaction to me to be able to look at this figure from time to time as I pass by, contemplate it, stand underneath, and walk to and fro before it. For if having the pictures of the departed placed in our homes lightens sorrow, how much more those public representations of them which are not only memorials of their air and countenance, but of their glory and honour besides?
>
> *(Plin.* Letters, *2.7, trans. W. Melmoth)*

This account offers much to consider in terms of the different approaches to cognition outlined above. First, the portrait of Cottius is obviously an example of Kirsh's external representations, by which Pliny can think about his deceased friend. It is a persistent object, whose qualities—one should not assume an exact physical likeness of "countenance," though age, costume, and "air" probably would have been well suited to the young man—do not change even as Pliny's internally stored information about Cottius might fade or change as his grief lessened over the passage of time. One can foresee that at some point, Pliny would have trouble recalling details about his friend, and by viewing the statue rediscover them, or remember other episodes in their friendship, or even to reevaluate Cottius in comparison to other people that he would meet later. The portrait of Cottius and the one of his father (on which Pliny does not comment) also stand as a reformulation of a rather complex, and as Pliny remarked, unusual situation: Spurinna achieved a great victory and as a result, the ruling elites wished to acknowledge him and express gratitude; not only was his service to the state recognized through his own portrait, but the excellence of his public persona meant that his private virtue as a father was also deemed worthy of praise; moreover, Spurinna was so highly regarded by the emperor and Senate that they offered him emotional comfort in his personal tragedy. The circumstances around the erection of these two public portraits bind together many aspects of Roman social relationships, expectations, and cultural values; in turn, the two portraits restate this complex social and moral nexus in a concise and elegant way. This reformulation provides opportunity for reinterpretation, for Cottius' portrait granted as a consolation is immediately seen by Pliny as an exhortation to other Romans to emulate the parental example of Spurinna and the filial example of Cottius. Pliny is only able to cast the statue in such a light because it is an external representation that is by definition shared, so that anyone can use it to think about morality and duty.

But Pliny's letter tells us more than how he thought with these public portraits. Though the statue had yet to be erected at the time of his writing, Pliny already anticipates his physical encounter with the statue and his reaction to it. He does not mention that future experience with the public portrait of Cottius would entail a reading of the inscribed base that certainly would have accompanied it, but he does state that it would receive his purposeful gaze. Pliny's description of his future interaction with the statue also goes beyond sight, and emphasizes—"pass by," "stand underneath," "walk to and fro before it"—bodily movement. On an immediate level, per Milner and Goodale's model, the visual percepts processed by the dorsal stream control's Pliny's movement around the statue, so as to grasp the totality of the statue's presence. And my usage of "grasp" is only partly metaphorical; it also takes into account the real logistical problems with laying hands on a large, immobile object standing on a tall base if one desired to do so, as opposed to the ease of handling the writing tablet in Quintilian's example. Aside from special acts of dressing or cleaning statues, little is said about public portraits being touched as a part of viewing (Stewart 2003: 263–264). The movement of Pliny's eyes over the statue as he "look[s] at" and "contemplate[s]" it would provide visual sensory input, which the ventral stream of processing would cause to be sorted and compared to his preexisting internal representations of Cottius. These epistemic actions connect the moment of viewing with the past, and therefore support the higher-order cognitive tasks of remembering not just the face, but also the personality of his deceased friend.

Another cognitive dynamic is also present in Pliny's letter, as he contrasts his conjectural experience at Cottius' statue with what he believes would be more typical reactions from others. For those not so close to the young man, the statue would serve a didactic purpose. Pliny believes Cottius' portrait would inspire others to live blamelessly or to rear good children, rather than offer consolation. Any viewer who had not known Cottius or Spurinna personally would have had a different interaction with their portraits. The information furnished by these two statues, granted as extraordinary honors, is a prediction error to such a viewer, contradicting very high-level internal models of the kinds of public virtues that would merit such recognition. Cottius is granted a statue not for an outstanding act or service, which is predicted, but because he was a testament to his father's excellence. These challenges to a prior belief regarding Roman public life would cause a model revision, so as now to encompass the knowledge both of this person, Cottius, and that the Senate could decide to erect a public portrait not for one's achievements but out of respect for one's father. Having learned these things, one could view other public portraits with greater discernment.

The difference between Pliny's own and others' cognitive experience of the portrait of Cottius could be distinguished from one another as personal memory, the former, and knowledge, the latter. The distinction between memory and knowledge is not simply a matter of wordplay and nuance, as they are two cognitively different categories. Though there are now numerous subcategories and terminologies, it would suffice here to focus on two long-established basic distinctions—that of episodic (or autobiographical) memory and semantic memory (also interchangeably called knowledge). Whereas episodic memory—described by a sentence such as, "when I was ten years old, I was scared badly by a neighbor's dog that chased me down the street"—is specific to an individual's self-awareness and is important in forming one's personal and social identities, semantic memory is divorced from the place and time at which it was formed, and constitutes the basis of a person's general knowledge (Tulving 1972, 1985; Gardiner 2001: 1351). An example of semantic memory would be the order of the English alphabet; the ABCs were learned at an early point in our lives, and tested at various points, especially soon after they were learned. One probably cannot remember when one learned the alphabet, but can usually recite it on demand. So, to reframe, the statue of Cottius is for Pliny a prompt for his episodic memory ("I loved this most favored, gifted, youth"), but for most others it would be an element of the semantic memory of the Senate's decision to honor Spurinna by also recognizing his outstanding son. The public portrait in this context, then, does not work as a carrier of memories in the popular sense of a memento, nor was it understood as having that capacity in all cases. If we consider that most viewers of a public honorific portrait likely had no personal relationships with the subject, then the importance of these objects as tools for the dissemination and reinforcement of knowledge rises to the fore.

How does behavioral cognitive research on learning and memory, in addition to the distributed and embodied cognition discussed above, help us to understand public portraits in this way? In contrast to the suggestions of Cicero and Quintilian to students of oratory, the ample body of memory and learning research has, for decades now, demonstrated that cramming and repetition of texts enable only an extremely short-term retention, and that while intervals between practice do in fact reinforce encoded information, brief intervals such as one day have a minimal effect on strengthening the durability of knowledge (Logan and Balota 2008; Brown et al. 2014: 28–43). In addition, retrieval—cognitively speaking—is not merely rereading. Rather, retrieval calls on one to access with effort the encoded information, whether on a voluntary basis or through prompts or cues such as with testing. In order for encoded information to endure and become a fixed element of general knowledge, it has to be retrieved over spaced temporal intervals (Landauer and Bjork 1978; Gardiner and Richardson-Klavehn 2000: 231). It is widely acknowledged that the initial encoding of information—referring to "a neural change that accompanies a mental experience at one time" (Tulving 2007: 66)—is strengthened each time that piece of information is retrieved, as the act of retrieval appears to cause the formation of new neural changes and connections and also renders the item of information open to elaboration.

In considering the public portrait monument in terms of encoding and retrieval, we must first address the information that is being encoded. As previously stated, a public honorific portrait stands as a reformulation of social expectations that were initially expressed in the act of awarding the statue. When a public body or a private citizen erects a public portrait, the purported merits of the portrait subject and the decision to honor him or her are the encodable information. For the freshly encoded knowledge of the public honor to have any persistence in the mind of the public, it must be repeatedly retrieved. In a partially externalized cognitive process of learning, the public portraits act as retrieval prompts. These portrait monuments were set up in places such as the market square, the theaters, along streets, or near or within public buildings such as baths, gymnasia, and fountains (Stewart 2003: 136–140; Fejfer 2008: 51–63). These are obviously high-traffic areas, places that a city dweller would pass by frequently during the course of daily life. However, location by itself—though tied to prestige—does not guarantee viewing or any significant interaction with a portrait monument, especially as these same spaces were crowded with statues (Tanner 2000: 25–26; Stewart 2003: 141). There was some concern regarding the quick obsolescence of a public honorific portrait's ability to promote the knowledge of elite identity and public praise, some attestations of which we find in the speeches of Dio Chrysostom. Granted, the Rhodian oration (*Or.* 31), in which Dio inveighs against the Rhodians' recycling of honorific statues by reinscribing their bases to honor powerful Romans, is rhetoric, and may be subject to exaggeration for effect. Nevertheless, it does describe aspects of viewing that could not have been alien to the listeners. Dio's screed against the reinscription of statue bases highlights the breach of trust between community and its elite benefactors and the result of monuments bereft of their originally intended social meaning.[7] Statues would become unrecognizable and therefore pointless because not only the name but also the reason for a person's honor would have been lost (Ng 2016: 241–243; Stewart 2003: 79–80). In other situations, the sheer congestion of statues in public areas—necessitating their clearing—would suggest that there was an understanding that more than mere marble, bronze, or gilding was needed in order for a public portrait to successfully promote and reinforce the knowledge of the statue's recipient beyond intimate associates (Livy 40.51.3; Dio 60.5.4–5, 60.25.2–3; Stewart 2003: 128–136; Fejfer 2008: 63–65).

Therefore, a portrait statue must command visual attention if it is to be a viable cognitive extension that supports information retrieval. If, for example, a new physical marker is placed next to a standing statue, then perhaps it would be more likely for a passerby to pick out that specific statue from the crowd because the sight of the new marker would present a deviation from prior experience with the statue and its surroundings. For instance, a lengthy dossier of inscriptions from Oenoanda related to the founding of a musical festival called the Demosthenaia during the reign of Hadrian records that the *boule* called for the documents concerning and ratifying the terms of this gift to be inscribed on a stele to be set up next to the already standing statue of C. Iulius Demosthenes, the benefactor, in the agora (published in Wörrle 1988, English translation in Mitchell 1990: ll. 95–98). Oenoandans' attention then was redirected to Demosthenes' portrait both by the issuance of this decree, and by the epigraphic monument that was to be erected next to the statue. As the statue clearly was set up to honor Demosthenes for another service, likely the construction of a large food market that he had already built for the town (ll. 8–13), the renewed attention would prompt Oenoandans to reinterpret and reevaluate their understanding of what it represented. The information encoded when that portrait was first dedicated thus would be changed when retrieved after the installation of the stele, to incorporate the patron's latest act of generosity.

There were yet other ways of drawing notice to a public portrait. It was not uncommon in Italian towns for a *decurion* or a member of the less prestigious *augustales* to put up a portrait monument to himself and perhaps to a spouse at his own expense in the meeting place of the association or in some public spot. The decisions to erect these portrait monuments were recorded in inscriptions that sometimes additionally document ceremonies to take place at the statues. At times, there were celebratory banquets at the dedication. In other instances, statues were to be cleaned, anointed with unguents, dressed, and lit up with candles concurrently with a distribution of money, sometimes to various subsets of a town's population or to the membership of the same group of which the portrait subject had been a part (e.g., CIL 8.9052; CIL 14.367; CIL 14.353; CIL 10.5853; Laird 2015: 185–191, Appendix I inscriptions 14, 14b; Fejfer 2008: 67–69; Stewart 2003: 263–264). While the dedication banquets could be a one-time occurrence, the custodial rituals and handouts were mandated annually, such as on the birthday of the portraits' subjects, or on some religiously or socially significant date.

Cognitively speaking, what occurs in such situations is the encoding of information—a private citizen's generosity and worthiness in the context of the social system of obligation and recognition—with the initial pledge or dedication of the portrait monument, and then annual retrievals of that information when the intended recipients of the monetary distribution would have to recall the date and the reason for gathering at the statue. While at first it may seem odd to leave such a long gap between each retrieval opportunity as opposed to the very frequent rehearsals that were recommended for memorizing a speech, longer rather than shorter spacing is much more effective at promoting knowledge persistence and accessibility over the long term. Though a longer interval allows for more forgetting—of, for example, details of the portrait subject's biography—the greater effort necessitated to bring what remains to consciousness allows neural connections to be formed anew. In essence, the information is re-encoded (Logan and Balota 2008: 272–275; Stock et al. 2014: 385–388). Furthermore, the feedback provided on the occasion of retrievals—the reading of the inscribed statue base, the gifts of money—updates and reinforces the learned material. Significantly, some cognitive research suggests that, if information is retrieved at regularly spaced intervals for a relatively limited duration—one long-term study found three to five years—that information was protected from the effects of forgetting and could be recalled by a person many years, even two decades, after it was first encoded (Bahrick 2000: 351–356). That means that, even if the ritual cleaning and monetary distributions took place for only a few years as funds and piety allowed, the cognitive impact on participants' knowledge of the statue recipient was profound and perhaps indelible.

Emotion and memory cognition

The cleaning and dressing, distributions of money, and banquets that took place at the statues on anniversaries of dedication or birth also highlight the impact of emotion on information retention. Though Pliny the Younger's letter regarding Cottius' statue conveys the author's sorrow, love, and hope for consolation, such feelings stemming from autobiographical memory are not so relevant to learning and knowledge consolidation in people who were not also deeply connected to the portrait subject. Rather, the capacity of the public portraits themselves to elicit emotion from their viewers through the presentation of both arousing stimuli and a specific goal to be attained is much more pertinent. Studies have shown that subjects are able to recall more details of emotionally arousing, as opposed to emotionally neutral, information. In addition, researchers have also identified an emotion-related phenomenon, memory narrowing, in which details thematically central to the emotional arousal *and* spatially central visual details are better remembered than peripheral data. Interestingly for the present discussion of public portraits, scholars of emotion and memory cognition have found that memory narrowing does not occur in all cases when there is emotional arousal, but can be linked to a positive emotional state (Reisberg 2006: 17–20, 31). In fact, it appears that memory consolidation and the retention of certain details are not only connected to a subject's emotional arousal, but are also affected by motivation. Memory narrowing results because attention is focused on central information related to the accomplishment of a currently active goal—the achievement of which would produce a positive emotional state—facilitating its encoding (Levine and Edelstein 2010; Bower 1992). The arousal-biased competition theory sheds further light on memory narrowing, with research demonstrating that emotional arousal enhances the encoding and consolidation of "high-priority" or goal-related information, such as "attention-grabbing" things that dominate initial perception or things that one is directed to recall at a later time (Mather and Sutherland 2011: 122–128). Biased competition—whether goal-driven in its classic sense or with the added factor of emotional arousal as just noted—has been noted as being convergent with aspects of predictive processing in that "the impact of specific prediction error signals can be systematically varied" such that the impact of "task-relevant" information is heightened (Clark 2015: 11–12; 2016: 61, 83). The ephemeral changes made to the statues on specified anniversaries are, of course, prediction error signals in that they disrupt the predicted perception of an unadorned monument. Indeed, emotional arousal itself also "is considered prediction error" in bringing a subject outside normal physiological states (Hohwy 2013: 243). Emotional effects on memory cognition and learning via predictive processing are thus intertwined.

It seems logical, then, that the physical maintenance of public portraits and the handouts taking place at their feet afforded physical and emotional engagements that made the information conveyed by the monuments themselves both more easily encoded and better consolidated. The freshly decorated sculpture offered heightened sensory stimuli in the form of flickering candlelight, smells, and decoration, which very likely aroused feelings such as pleasure or awe. Furthermore, the handouts of money or gifts of food in the form of banquets can be seen as defined goals to be accomplished by the people who were invited to participate. Not only is goal-pursual central to memory narrowing, but achieving these particular objectives of free food and money undoubtedly further resulted in positive emotions—satisfaction, for instance—linked to better recollection.

The designed experience of these public portraits was ideally suited to attentional focus and memory narrowing. The actions and interactions centered on the statue imparted a stronger impression of the specific identity and qualities of its subject in the people who gathered around the monument each year. At the same time, the effects of mandated decoration and distributions of food or money would cause a filtering out of peripheral information such as, perhaps, other buildings that were close by or events happening around the same time, or aspects of the portrait subject's persona that were not specifically referenced by the statue and its base. The resulting knowledge of the statue's honorand and/or donor would therefore be limited but predictably favorable. The annual retrieval of this same information—that is, recalling the generosity and gift of the patron as a reason to come to the portrait—then consolidates that knowledge for the long term. I would also suggest that the peripheral information that is cognitively edited out from recall should not include other public portraits. The social competition fueling the granting of these statues continues after their installation. When different but conceptually similar items of information are interleaved with each other throughout the learning process, there is a greater facility on the part of the learner to make meaningful comparisons and discernments among similar types of things (Kornell and Bjork 2008: 590–591; Kang and Pashler 2012: 101–102). That is, perhaps it could be said that a learner-participant at periodic statue-associated rituals who also had exposure to other public portraits in the meantime—whether through looking or through getting a handout or banqueting—would not only have to exert some cognitive labor in order to differentiate one distinguished portrait subject from another and to compare their qualities according to the criteria of honor and munificence, but also become better at this kind of evaluation. By learning about individuals through their public portraits, an ancient observer was in fact becoming ever more proficient in the social system of benefaction, service, and prestige that was a mainstay of Roman urban life.

Conclusion

The entirety of a public portrait monument—the granting of one, the physical object, ritual, and distributions—is connected to cognitive processes that promote effective long-term learning about the portrait's subject, all of which is predicated on the notion that the portrait is an external representation with which one can think. I have argued that the Roman public honorific statue presented an array of external sensory inputs, the processing of which was guided by internal models, which in turn were revised through learning so as to shape future perception. The emotional arousal caused by the statue's physical maintenance and the reiterative, goal-driven encounters with the monument consolidated the acquired knowledge in memory. Cognition was thus enacted in the interaction between statue and viewer in a mode that is different from that supporting the rote memorization of text.

This chapter is a first attempt to understand the relationship between public portraits and the cognitive processes of learning and memory. Obviously many more questions remain. The cognitive modes and behaviors that define other informal kinds of interactions with statues, such as when they were used as public notice boards or landmarks, bear investigating. In addition, the model I have laid out in this chapter may be well suited for investigating the rather common phenomena of reworking and reuse in Roman sculpture and architecture. Physical change to a publicly displayed statue or structure, or the incorporation of elements that one expects in another location and context, is an obvious example of a prediction error that compels attention and drives the processing of new information. Does the great variance in the visual evidence of reuse and reworking, from barely perceptible to intentionally obvious, induce different kinds or degrees of internal model revision? How does the revision of high-level internal models—of social relationships and censure, political personas and ideologies, how a statue or monument is meant to look—guide subsequent perceptions of objects both reworked and left whole? By engaging with theories of cognition and behavioral cognition, we may find new ways to explore and understand the interdependence of static monuments and dynamic cognitive and social actions.

Notes

1 I am very thankful to the editors and readers for their feedback and am especially grateful to Peter Meineck for his patience. Any errors are my own.

2 One must keep in mind that commemoration was not always the purpose of a public portrait; as Tanner (2000: 26) has noted, portraits were granted by governing bodies in order to hold a citizen to pledged action.
3 Tanner's discussion of verism in Republican portraiture connects the expression of Roman cultural and social norms with the influence of Hellenistic artistic and political culture on elite Romans. Smith argues that the selection of certain poses and hairstyles for public portraits in the Greek East was a conscious choice to articulate one's Roman or Greek cultural identity. Fejfer notes the importance of imperial portraiture in the evolution of portrait faces, even as there was relatively little change in the body types used.
4 Most of the behavioral psychology experiments in the studies cited in this chapter are based on word recall experiments.
5 I thank Peter Meineck for pointing out that the dorsal and ventral processing streams are conceptual precursors to the guided perception of the predictive processing model.
6 Tanner (2000: 48) also believes that most Romans would have had "casual, but not inconsequential, glance and inference" viewing of public portraits, and that their interest would have been "detached, affectively uninvolved."
7 The phenomenon of reinscribed bases is especially fascinating when considered from the predictive processing perspective, though that is beyond the scope of this chapter. The physical traces and the content of reinscription are prediction errors that would force a revision of high-level internal models of social obligation and recognition. The resulting knowledge—that Romans who have invested and participated little in Rhodes are usurping the honor of local benefactors—is, on a civic level, awkward to acknowledge. Yet this learning process and its social implications are exactly what Dio discusses at length.

References

Anderson, Miranda, *The Renaissance Extended Mind* (New York: Palgrave Macmillan, 2015).

Bahrick, Henry, "Long-Term Maintenance of Knowledge," in *Oxford Handbook of Memory*, eds. Endel Tulving and Fergus I. M. Craik (New York: Oxford University Press, 2000), 347–362.

Bower, Gordon, "How Might Emotions Affect Learning," in *The Handbook of Emotion and Memory, Research and Theory*, ed. Sven-Ake Christianson (Hillsdale, NJ: Lawrence Erlbaum Associates, Inc., 1992), 3–31.

Brincker, Maria, "The Aesthetic Stance: On Conditions and Consequences of Becoming a Beholder," in *Aesthetics and the Embodied Mind: Beyond Art Theory and the Cartesian Mind-Body Dichotomy*, ed. Alfonsina Scarinzi (New York: Springer, 2015), 117–138.

Brown, Peter, Henry Roediger, and Mark McDaniel, *Make It Stick: The Science of Successful Learning* (Cambridge, MA: Belknap Press, 2014).

Clark, Andy, *Supersizing the Mind: Embodiment, Action, and Cognitive Extension* (New York: Oxford University Press, 2011).

—, "Embodied Prediction," in *Open MIND*, eds. Thomas Metzinger and Jennifer Windt (Frankfurt am Main: MIND Group, 2015), 1–21.

—, *Surfing Uncertainty: Prediction, Action, and the Embodied Mind* (New York: Oxford University Press, 2016).

Clark, Andy, and David Chalmers, "The Extended Mind," *Analysis* 58, no. 1 (January 1998): 7–19.

Fejfer, Jane, "Ancestral Aspects of the Roman Honorary Statue," in *Images of Ancestors*, ed. Jakob Munk Højte (Aarhus: Aarhus University Press, 2002), 247–257.

—, *Roman Portraits in Context* (Berlin: Walter de Gruyter, 2008).

Friston, Karl, "Active Inference and Agency," *Cognitive Neuroscience* 5, no. 2 (2014): 119–121.

Gardiner, John, "Episodic Memory and Autonoetic Consciousness: A First-Person Approach," *Philosophical Transaction: Biological Sciences* 356, no. 1413 (2001): 1351–1361.

Gardiner, John, and Alan Richardson-Klavehn, "Remembering and Knowing," *in Oxford Handbook of Memory*, eds. Endel Tulving and Fergus I. M. Craik (New York: Oxford University Press, 2000), 229–244.

Hohwy, Jakob, *The Predictive Mind* (New York: Oxford University Press, 2013).

Hutto, Daniel, "Enactivism, from a Wittgensteinian Point of View," *American Philosophical Quarterly* 50, no. 3 (July 2013): 281–302.

Johnson, Laurie, John Sutton, and Evelyn Tribble, eds., *Embodied Cognition and Shakespeare's Theater: The Early Modern Body-Mind* (New York: Routledge, 2014).

Josselyn, Sheena A., Stefan Köhler, and Paul W. Frankland, "Finding the Engram," *Nature Reviews Neuroscience* 16 (September 2015): 521–534.
Kang, Sean, and Harold Pashler, "Learning Painting Styles: Spacing Is Advantageous When It Promotes Discriminative Contrast," *Applied Cognitive Psychology* 26 (2012): 97–103.
Kirsh, David, "Thinking with External Representations," in *Cognition Beyond the Brain: Computation, Interactivity and Human Artifice*, eds. Stephen Cowley and Frédéric Vallée-Tourangeau (London: Springer-Verlag, 2013), 171–194.
Kornell, Nate, and Robert Bjork, "Learning Concepts and Categories: Is Spacing the 'Enemy of Induction'?" *Psychological Science* 19, no. 6 (2008): 585–592.
Laird, Margaret, *Civic Monuments and the Augustales in Roman Italy* (New York: Cambridge University Press, 2015).
Landauer, Thomas, and Robert Bjork, "Optimum Rehearsal Patterns and Name Learning," in *Practical Aspects of Memory*, eds. Michael Gruneberg, Peter Morris, and Robert Sykes (New York: Academic Press, 1978), 625–632.
Levine, Linda, and Robin Edelstein, "Emotion and Memory Narrowing: A Review and Goal-Relevance Approach," in *Cognition and Emotion: Reviews of Current Research and Theories*, eds. Jan de Houwer and Dirk Hermans (New York: Psychology Press, 2010), 168–210.
Levine, Linda, and David Pizarro, "Emotional Valence, Discrete Emotions, and Memory," in *Memory and Emotion, Interdisciplinary Perspectives*, eds. Bob Uttl, Nobuo Ohta, and Amy Siegenthaler (Malden, MA: Blackwell Publishing, 2006), 37–58.
Lindblom, Jessica, *Embodied Social Cognition* (New York: Springer, 2015).
Logan, Jessica, and David Balota, "Expanded vs. Equal Interval Spaced Retrieval Practice: Exploring Different Schedules of Spacing and Retention Interval in Younger and Older Adults," *Aging, Neuropsychology, and Cognition* 15 (2008): 257–280.
Malafouris, Lambros, "Linear B as Distributed Cognition: Excavating a Mind Not Limited by the Skin," in *Excavating the Mind: Cross-Sections through Culture, Cognition, and Materiality*, eds. Niels Johannsen, Mads Jessen, and Helle Jensen (Aarhus: Aarhus University Press, 2012), 69–84.
—, *How Things Shape the Mind: A Theory of Material Engagement* (Cambridge, MA: MIT Press, 2013).
Malafouris, Lambros, and Colin Renfrew, eds., *The Cognitive Life of Things: Recasting the Boundaries of the Mind* (Cambridge: McDonald Institute for Archaeological Research, 2010).
Mather, Mara, and Matthew Sutherland, "Arousal-Biased Competition in Perception and Memory," *Perspectives on Psychological Science* 6, no. 2 (March 2011): 114–133.
Menary, Richard, "What Is Radical Enactivism?" Introduction to *Radical Enactivism: Intentionality, Phenomenology, and Narrative. Focus on the Philosophy of Daniel D. Hutto*, ed. Richard Menary (Philadelphia, PA: John Benjamins Publishing Company, 2006), 1–12.
Milner, A. David, and Melvyn Goodale, *The Visual Brain in Action*, 2nd ed. (Oxford: Oxford University Press, 2006).
Mitchell, Stephen, "Festivals, Games, and Civic Life in Roman Asia Minor," *Journal of Roman Studies* 80 (1990): 183–193.
Moscovitch, Morris, "Why the Engram Is Elusive," in *Science of Memory: Concepts*, eds. Henry Roediger III, Yadin Dudai, and Susan Fitzpatrick (New York: Oxford University Press, 2007), 17–22.
Ng, Diana Y., "Monuments, Memory, and Status Recognition in Roman Asia Minor," in *Memory in Ancient Rome and Early Christianity*, ed. Karl Galinsky (Oxford: Oxford University Press, 2016), 235–260.
—, "The Salutaris Foundation: Monumentality through Periodic Rehearsal," in *Roman Artists, Patrons and Public Consumption: Familiar Works Reconsidered*, eds. Brenda Longfellow and Ellen Perry (Ann Arbor, MI: Michigan University Press, 2018), 63–87.
—, "The Roman-Period Theatre as Cognitive Micro-Ecology," in *Distributed Cognition in Classical Antiquity*, eds. Douglas Cairns, Mark Sprevak, and Miranda Anderson (Edinburgh: Edinburgh University Press, forthcoming), 117–131.
Pliny the Younger, *Letters of Pliny*, trans. William Melmoth (Teddington, UK: Echo Library, 2006).
Reisberg, Daniel, "Memory for Emotional Episodes: The Strengths and Limits of Arousal-Based Accounts," in *Memory and Emotion, Interdisciplinary Perspectives*, eds. Bob Uttl, Nobuo Ohta and Amy Siegenthaler (Malden, MA: Blackwell Publishing, 2006), 15–36.
Roediger, Henry (III), Franklin Zaromb, and Andrew Butler, "The Role of Repeated Retrieval in Shaping Collective Memory," in *Memory in Mind and Culture*, eds. Pascal Boyer and John Wertsch (New York: Cambridge University Press, 2009), 138–170.

Small, Jocelyn, *Wax Tablets of the Mind: Cognitive Studies of Memory and Literacy in Classical Antiquity* (New York: Routledge, 1997).

Smith, R. R. R., "Cultural Choice and Political Identity in Honorific Portrait Statues in the Greek East in the Second Century A.D.," *Journal of Roman Studies* 88 (1998): 56–93.

Stewart, Peter, *Statues in Roman Society: Representation and Response* (New York: Oxford University Press, 2003).

Stock, Ann-Kathrin, Hannah Gajsar, and Onur Güntürkün, "The Neuroscience of Memory," in *Memory in Ancient Rome and Early Christianity*, ed. Karl Galinsky (Oxford: Oxford University Press, 2016), 369–391.

Sutton, John, "Distributed Cognition: Domains and Dimensions," *Pragmatics & Cognition* 14, no. 2 (2006): 235–247.

—, "Material Agency, Skills and History: Distributed Cognition and the Archaeology of Memory," in *Material Agency: Towards a Non-Anthropocentric Approach*, eds. Carl Knappett and Lambros Malafouris (Berlin: Springer, 2008), 37–55.

—, "Exograms and Interdisciplinarity: History, the Extended Mind, and the Civilizing Process," in *The Extended Mind*, ed. Richard Menary (Cambridge, MA: MIT Press, 2010), 189–225.

Tanner, Jeremy, "Portraits, Power, and Patronage in the Late Roman Republic," *Journal of Roman Studies* 90 (2000): 18–50.

Tribble, Evelyn, and Nicholas Keene, *Cognitive Ecologies and the History of Remembering: Religion, Education and Memory in Early Modern England* (New York: Palgrave Macmillan, 2011).

Tulving, Endel, "Episodic and Semantic Memory," in *Organization of Memory*, eds. Endel Tulving and Wayne Donaldson (New York: Academic Press, 1972), 381–403.

—, "How Many Memory Systems Are There?" *American Psychologist* 40, no. 4 (1985): 385–398.

—, "Coding and Representation: Searching for a Home in the Brain," in *Science of Memory: Concepts*, eds. Henry Roediger III, Yadin Dudai, and Susan Fitzpatrick (New York: Oxford University Press, 2007), 65–69.

Wheeler, Michael, "The Revolution Will Not Be Optimised: Radical Enactivism, Extended Functionalism and the Extensive Mind," *Topoi* 36 (2017): 457–472.

Wörrle, Michael, *Stadt und Fest im kaiserzeitlichen Kleinasien: Studien zu einer agonistischen Stiftung aus Oinoanda* (München: C.H. Beck, 1988).

23

Animal sacrifice in Roman Asia minor and its depictions

A cognitive approach

Günther Schörner

Animal sacrifice was the most important ritual in the Roman Empire and therefore representations of animal sacrifice figure prominently in Roman visual culture (Italy: Ryberg 1955; Huet and Scheid 2004; Huet 2005; Asia minor and Northern Africa: Schörner 2006; Schörner 2009), but research on both the ritual of sacrifice and particularly on the iconography of that ritual runs the risk of facing a dead end. In recent scholarship on Roman religion the grand theories on sacrifice, especially animal sacrifice, established by scholars like W. Burkert (1972), R. Girard (1972), M. Detienne and J.-P. Vernant (1979) and J. Z. Smith (1987), have been abandoned or massively questioned, and no alternatives or explanatory models have been created to replace them (recent studies on sacrifice: McClymond 2008; Petropoulou 2008; Knust and Varhelyi 2011; Faraone and Naiden 2012; Naiden 2013; Ullucci 2015). That has led to the present situation in which the ritual of animal sacrifice is understood as self-evident and/or is lacking a religious explanation.

In Classical Archaeology – as a subject without a large body of its own theories – this move away from religious concepts had profound consequences for the study of the iconography of ritual: visual depictions of animal sacrifice, especially in the Roman Empire, are currently analysed either in extremely positivistic ways by iconographically scrutinizing ever-smaller figurative elements, or are only seen as a means for representation in a purely political and social discourse (e.g. Meyer 2011; Aldrete 2014; Lennon 2015). Both approaches neglect religious content completely.

Omitting religious content is a deeply unsatisfactory situation because ignoring religion in religious iconography surely cannot be reasonable. New theories developed in the Cognitive Science of Religion (CSR) have the ability to break the current deadlock by helping us better comprehend both the conception and implementation of ancient rituals, to understand the motivations for their visual representation and to explain the particular relevance of visual depictions (fundamental: Lawson 2000; Pyysiäinen 2001; McCauley and Lawson 2002; Geertz 2004; Whitehouse 2004; Tremlin 2006; Barrett 2007; CSR and archaeology: McCauley and Lawson 2007). Therefore, the following study adopts CSR's theories in the research of ancient rituals and visual culture of the Roman Empire. The areas (and problems) of applying this completely new approach (at least for Classical Archaeology) shall be discussed using two examples found on friezes depicting animal sacrifice in the theatres of two cities in Asia minor – Perge and Hierapolis – but they are applicable also to literary descriptions of sacrifice, as will be demonstrated by one example, a passage from the Ephesian tale of Anthia and Habrocomes by Xenophon of Ephesus.

The frieze at Perge (Fig. 23.1) decorated the central doorway of the stage building and can be dated to the late Antonine or Severan time (end of second/beginning of third century CE) (Öztürk 2009). The entire frieze is composed in strict axial symmetry with the seated female personification of the polis, Perge, in the centre holding the aniconical xoanon of Artemis Pergaia, the main goddess of the city, in her right hand, and in her left a cornucopia (Lindner 1994, 105; Öztürk 2009, 51, 85, 156, Nr. 339, 340, 343; Schörner 2014, 355–357). From both sides six men clad in chitons and himations approach, with each man guiding a zebu, thus forming a sacrificial procession which continues on the two projecting reliefs. None of the men's images is individualized as a portrait but it is evident that different persons are depicted. The frieze concludes on both ends with the depiction of the ritual killing: two bulls tied up by rope killed by sacrificial attendants using knives.

The next depiction of animal sacrifice (Fig. 23.2a–b) is found at the theatre of Hierapolis built during the reign of the emperor Hadrian in the first half of the second century CE and renovated at the beginning of the third century CE (De Bernardi Ferrero 2007; D'andria 2008, 124–126). At this time, the front of the stage was rebuilt as an impressive three-story façade with elaborate sculptural decoration. Besides mythological friezes, a procession and sacrifices to the goddess Artemis Ephesia are depicted (description of the frieze: D'Andria and Ritti 1985, 143–160; Schörner 2013). This section of the frieze is situated on both sides of a doorway and the left front of one of the gates. It starts with three standing young women in a gesture of salutation and adoration. The following slab is decorated with a relief depicting two women and the statue of Artemis. The goddess shows the typical features of the Ephesian Artemis, a special form of skirt (*ependytes*) and the *polymasteia*, an ornament on her chest consisting of ovoid objects. At the left a woman with covered head disperses incense in a fire burning on an altar. At the right a second woman sprinkles water on Artemis' head using a twig, followed by men guiding bulls or oxen to the right. The servants are accompanied by a woman in long garments presenting a garland, a young servant in a chiton holding a basket with fruit in his raised hands and two female flute players. On the next slab, the sacrifice is executed: a second attendant in exomis and broad girdle lifts his arms ready to strike a blow. The following sequence is very similar to that at the beginning: a person at a flaming altar offers incense, followed by a woman adoring the statue of Ephesian Artemis which stands in an architectural setting.

Figure 23.1 The Perge frieze: sacrifice in honour of Artemis Pergaia

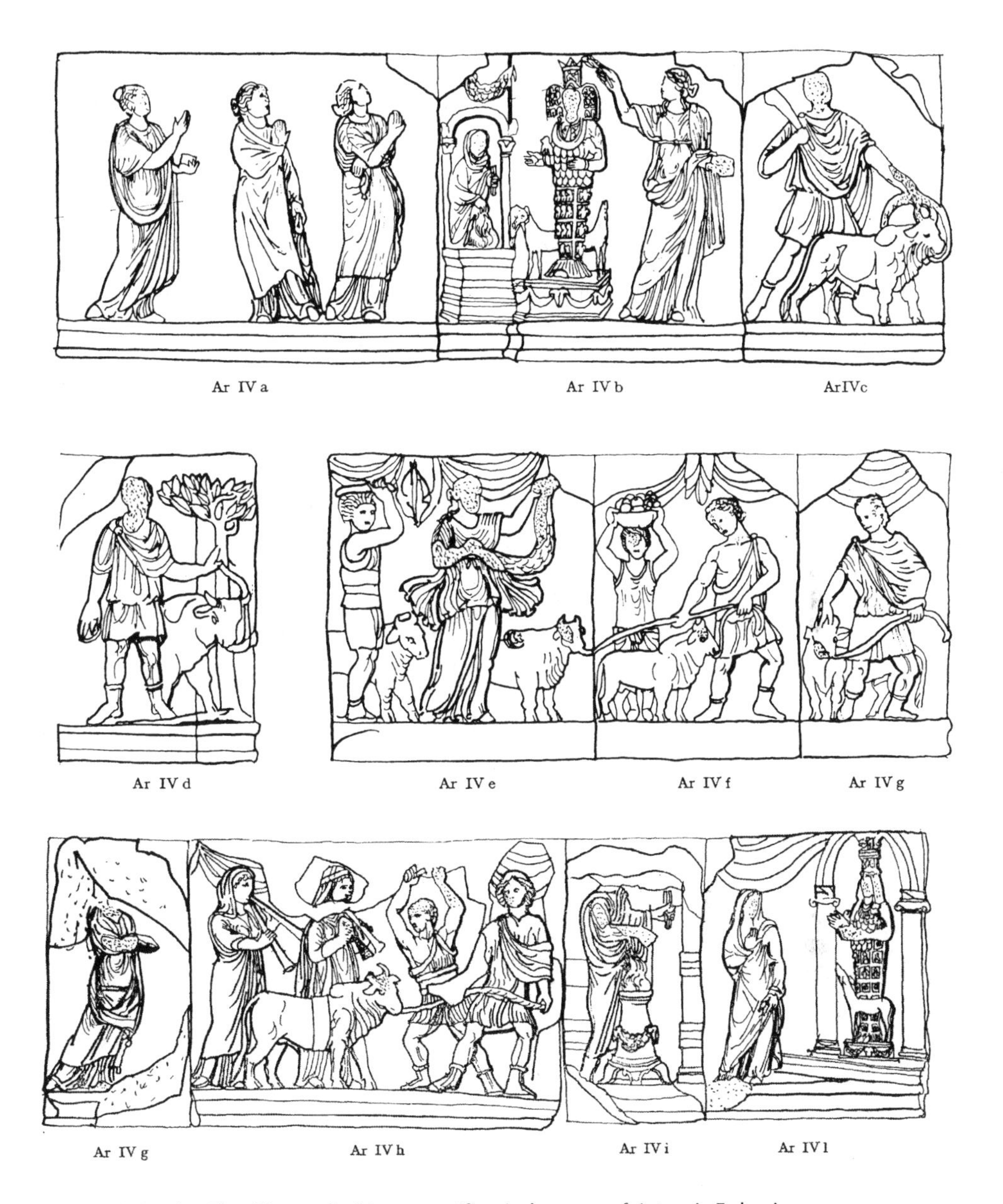

Figure 23.2a–b The Hierapolis frieze: sacrifice in honour of Artemis Ephesia

These examples of depictions of animal sacrifices – only two out of several hundred – have some common features, but more astonishing are the differences given the fact that the same ritual sequences are depicted.

In conventional archaeological practice focus is laid on details like garments, tools and other details without regarding the main fact, namely that religious rituals are depicted. Of interest as we move forward is that although in both friezes sumptuous animal sacrifices are depicted, they differ considerably in iconography and composition.

CSR enables a deeper understanding of animal sacrifice. The seminal works by J. Barrett, P. Boyer, R. McCauley and E. T. Lawson (Lawson and McCauley 1993; Boyer 2001; McCauley and Lawson 2002; Barrett 2004; see also the References below) gave fundamental insights into the structure and purpose of rituals, but the more specific Cognitive Resource Depletion (CRD) model developed by U. Schjoedt, J. Sørensen and J. Bulbulia is especially pertinent to a better understanding of ritual depictions (Schjoedt et al. 2013a; discussion: Alcorta 2013; Brown 2013; Eilam and Mort 2013; Legare and Herrmann 2013; Lienard, Martinez and Moncrieff 2013; McCauley 2013; McClenon 2013; Sousa and White 2013; Whitehouse 2013; response: Schjoedt et al. 2013b).

There is a long tradition of scholars studying religion in cognitive terms to attribute to rituals their own set of distinct features, separating them from other types of behaviour. Such features include: formality, repetition, redundancy, stereotypy, invariance, causal opaqueness and goal demotion (Boyer and Liénard 2006; Kapitány and Nielsen 2017; Legare and Souza 2012; Rappaport 1999; Rossano 2012). Collective rituals – such as animal sacrifice – are seen as an ideal means for the transmission of symbolic ideas. The starting point for establishing the CRD model was based on the observation that little is known about the cognitive mechanisms that mediate this symbolic-transmission hypothesis at the level of cognitive processing. That hypothesis states that collective rituals constitute an ideal setting for the transmission of contents, which symbolize religious knowledge and myths, as Durkheim already postulated (1912/1995). This is in harmony with the observation that collective rituals are both pervasive and ancient but – as Schjoedt and colleagues emphasize – "few researchers have proposed plausible cognitive models that explain how, at the level of cognitive processing, specific aspects of ritual affect participants' susceptibility to collective ideas" (Schjoedt et al. 2013a, 39). The core of the problem leads to the question: why do participants not perform a meaningful play or use oral communication to communicate meaning but instead (and very often) rituals which are characterized as goal-demoted and causally opaque actions (Schjoedt et al. 2013a, 40), denying clear answers to the questions "by what mechanism is an effect being caused?" (ritual opacity) and "why does the actor act?" (goal demotion) (Kapitány and Nielsen 2017, 28)? One answer could be that when we perceive an action as opaque and goal-demoted, we recognize it as deliberate (not incidental or accidental) and adopt the so-called ritual stance (for the term: Du Bois 2007), i.e. seeking out a rationale for actions based on social convention in addition to an instrumental stance, i.e. seeking out a rationale for actions based on physical causation. Further, recent studies suggest that perceiving the goal-demoted and causally opaque actions that characterize ritualized behaviours demands precious cognitive resources, which would otherwise be available for processing symbolic content (Boyer and Liénard 2006). With these hypotheses the cognitive processes of perception come to the fore: the brain processes information through an "executive system" referring to integrative processes by the so-called frontal executive network, which consists of the dorsolateral, prefrontal cortex, medial prefrontal cortex and anterior cingulate cortex (Miyake et al. 2000; Alvarez and Emory 2006). That frontal executive network is recruited for different tasks depending on context and situation, for example volitional control of thought and movements (Hallett 2007), conflict monitoring (Carter, Botvinick and Cohen 1999; Carter et al. 1998), action perception (Zalla, Pradat-Diehl and Sirigu 2003), autobiographical memory (Svoboda, McKinnon and Levine 2006) and decision-making (Bechara, Damasio and Damasio 2000). Because executive functions and attention recruit the same frontal region of the brain, these regions have been found to compete for cognitive resources in demanding situations (Alvarez and Emory 2006). When a strong increase of cognitive load from attention is necessary, it inevitably detracts from executive processes because it depletes a common pool of resources. This phenomenon is called cognitive resource depletion (Schjoedt et al. 2013a, 41) and is the basis for a recent explanatory model for why rituals affect the susceptibility of the participants to collective ideas by creating an attributional and inferential gap which we are motivated to fill with meaning, interpretation or behaviour (Kapitány 2017, 56), whereby this model can be used for a better understanding of ancient rituals and their representations, as shall be demonstrated.

The effects of resource depletion in more general terms of brain function conform with the axiom of predictive coding as the principal model to understand perception and cognition (Bar 2009; Schjoedt et al. 2013a, 41; Clark 2016). That principle concludes that the way the brain makes sense of the world is by constantly generating predictive models which are continuously compared and updated according to input from the sensory apparatus. A. Clark, for example, sees the predictive brain as "an action-oriented engagement machine, adept at finding efficient embodied solutions that make the most of body and world" (Clark 2016, 300; see also Colombo 2017, 381s.). If the brain detects a mismatch between its own predictions and the sensory input, the predictions will be changed. As Carter, Frith and others stated, both prediction error-monitoring and model updating is associated with the same brain regions and attentional resources that are involved in processing other executive functions (Carter et al. 1999; Frith 2007; wording: Schjoedt et al. 2013a, 41). Against this background, cognitive resource depletion may also have a negative impact on the brain's capacity for error-monitoring and updating predictive models. This model is especially valid in the context of two features of ritual behaviour: causal opaqueness and goal demotion.

Why is this the case? In general, humans process all actions in a hierarchical system by parsing them into low-level units which are then integrated into coherent action representations based on their intentional and causal specification.

Kapitány and Nielsen have suggested that there are three hierarchical levels of this kind of action-parsing (Kapitány and Nielsen 2017, 29): scripts, behaviour and gesture. According to that theory, behaviour is the default level of action-parsing, contingent upon our understanding of the causal processes and goals that bring about the end-state. Gestures are the sub-units of actions which make up behaviour, and appear to be guided by observers recognizing the hierarchical necessity of sub-goals.

Gestures are a necessary component within the entire behaviour. This, of course, is somewhat dependent on our understanding of schemas or scripts. How we understand a series of actions and gestures depends upon our familiarity with them (Kapitány 2017, 37s.). For example: a script would be *cleaning the kitchen*, consisting of various behaviours, for example *cleaning a glass, cleaning a dish* and so on, whereby the behaviour *cleaning a glass* includes the gestures *raising a glass, grasping a cloth, rubbing the glass with a cloth* (example: Kapitány and Nielsen 2017, 29). Research on the perception of ordinary actions suggests that we automatically link gestures into a causally coherent sequence without a large allocation of attention to the low-level units of gesture and so forth. Our perception, however, is altered when cognitive predictions are disrupted by error-checking processes: When actions are goal-demoted and causally opaque we are forced to parse at a gestural level (Schjoedt et al. 2013a, 44–46). The causal opaqueness and goal demotion of rituals transform ordinary actions with relatively low attentional loads on basic action perception into less comprehensible behaviour with a much higher attentional load, because they obstruct the automatic linking of low-level units and goal prediction. Thus the updating of the top-down models of a ritual action is disrupted and the construction of coherent action structures is hampered. This means – as the core of the CRD model – that participating in or merely observing such rituals may cause depletion and limit the capacity of the participants to construct their own narrative accounts of the event, and to form causally and intentionally meaningful representations of these interactions (Schjoedt et al. 2013a, 45s.).

These hypotheses and assumptions were tested in a series of behavioural studies. Schjoedt and colleagues found that exposure to non-functional behaviour, i.e. behaviour that was not causally integrated and lacked a transparent goal structure, resulted in an increasing segmentation rate compared to functional behaviour (Nielbo and Sørensen 2011; Schjoedt et al. 2013a, 46). The second test was still more important: in that experiment, the effect of familiarization was probed, because participants often practise or watch the same ritual several times or even on a regular basis and may learn to parse rituals more easily (experiment description and analysis: Kapitány and Nielsen 2017, 30–38). It was found that familiarization had no effect on action segmentation. Thus, rituals prevent individuals from forming ordinary interpretations of these actions by depleting the cognitive resources required for such constructions. This depletion leaves participants with an inferential gap, which amplifies a search for meaningful interpretations of the ritual after the event and which facilitate the communication of collective propositions delivered before and after the event (Schjoedt et al. 2013a, 46). Thus rituals could be seen as blanks which can be filled with meaning by authorities. These collective propositions can be communicated by means of verbal narratives but also – most importantly – by means of visual depictions (Schjoedt et al. 2013a, 51s.). So although the Roman depictions of sacrifices are of course not photographs of ritual action but artistic representations, that model applies here as well because the artisan has to have made decisions about how he wants to depict the sacrifice, which stage of the ritual, which participants and so on. These decisions had to be taken either by the sculptors or – by far more probable – by the principals and initiators of the work. Because of the inferential gap caused by depletion, a generally and collectively perceptible meaning is lacking and could be substituted by a more targeted interpretation.

Animal sacrifice should be seen as such a causally opaque and goal-demoted action because it does not (only) consist of killing and butchering an animal in a straightforward way. According to the typology proposed by McCauley and Lawson, sacrifices are "special patient rituals" (in contrast to "special agent rituals" or "special instrument rituals") because the most direct connection with the gods is through the animal victim which is acted upon (typology of rituals: McCauley and Lawson 2002, 1–37). So it is the animal that is at the centre of the ritual. This means that the general purpose of the script of "animal sacrifice" is to provide food to the gods and please them, in a plethora of single gestures which are centred on the animal as the patient and which are highly opaque and not focused on the goal, from choosing the right kind of animal, taking into account its type, age, sex and colour, to cutting locks from the forehead, to the burning of fat and bones on the altar instead of the better meat cuts. Thus, animal sacrifice meets all the requirements for applying the CRD model by the causal opaqueness and goal demotion of its gestures (see above).

The images from Perge and Hierapolis used as examples depict animal sacrifice as a common script consisting of common behaviours like *driving a bull* or *killing an animal* and also several common gestures. Differences exist at the level of gestures like *killing the bull by using a knife* or *killing the bull by using an axe* but they differ also at the higher level of "symbolic" interpretation that is the meaning ascribed: thus the frieze from the theatre at Perge depicts in a very clear-cut manner the decisive role of the leading families from which the magistrates of the polis originate. Members of this highest social group of the city appear as the leading persons and main actors of the sacrifice directly beside the goddess of Perge, while the killing, although the main action of the ritual, is relegated to the less visible subsidiary side. Therefore, the frieze from Perge commemorates the ritual as a reflection of the social order and by these means perpetuates the social hierarchy of the polis. The frieze from Hierapolis significantly contrasts in its basic message. Here the animal sacrifice is part of a larger narrative sequence picturing the ritual veneration of Artemis Ephesia at two different locations. Much more weight is laid on the various ritual acts and the narrative character of the frieze is much more pronounced by giving more space to the driving of the bulls, the carrying of garlands and baskets with plant offerings, the killing of the victim and the accompanying rites like playing music. By the depiction of the easily recognizable Artemis Ephesia with her polymastia, the ritual is also interpreted as part of the city's history and seen as a means to connect Hierapolis with other Poleis in Asia minor, for example Ephesus. Be that as it may, the message of the representation is different to that of the Perge frieze and this ambiguity is enabled by and a result of the uncertain meaning of the animal sacrifices themselves. Depictions of animal sacrifices are to a certain extent "empty spaces" to which an interpretation has to be added.

This lack of inherent or intrinsic meaning could be exemplified by many more pictorial representations like the so-called Parthian Altar from Ephesus or on individually offered votive altars, which although depicting the same subject – animal sacrifice – differ evidently in their message (Schörner 2006; Schörner 2013), but it is a literary source which most clearly shows a completely different reading of that ritual.

Xenophon of Ephesus, a local writer living in the second century AD, gives in his novel, the Ephesian tale of Anthia and Habrocomes, a sketch of a festival to venerate the city goddess (O'Sullivan 1995; Elsner 2007, 233 s.). This episode is part of the first book, when the two protagonists meet for the first time:

> The local festival of Artemis was in progress, with its procession from the city to the temple nearly a mile away. All the local girls had to march in the procession, richly dressed, as well as all the young men . . . There was a great crowd of Ephesians and visitors alike to see the festival, for it was the custom at this festival to find husbands for the young girls and wives for the young men. So the procession filed past – first the sacred objects, the torches, the baskets and the incense; then horses, dogs and hunting equipment . . . And when the procession was over, the whole crowd went into the temple for the sacrifice, and the files broke up; and men and women, girls and boys, came together.
>
> *(Xen.* Eph. *I 2.2–I 3.3; text and translation: Henderson 2009, 213–221)*

Here a different meaning or goal has been attributed to the ritual by a single observer, highlighting the function of the festival for making contacts and finding partners, and nearly completely neglecting the ritual core, the sacrifice, to which there is only a passing reference.

To sum up: a cognitive approach to the study of animal sacrifice in Asia minor and pertinent depictions produces important new results. Animal sacrifice as a collective ritual had evidently no fixed detailed meaning for all participants and was not interpreted by them in the same way, as we know through the plethora of representations which convey not a firm statement but differing messages and which can be demonstrated by the examples studied so far. This lack of fixed personal interpretation could be explained according to the Cognitive Resource Depletion model and allows for (or even makes necessary) various collective interpretations. These interpretations could be provided – *inter alia* – by depictions which filled an interpretative gap. This line of argument explains one essential characteristic of the depictions of sacrifice in Roman Asia minor: their lack of compositional, iconographical and interpretational homogeneity or similarity, because no explicit, steadfast meaning was given by the ritual itself but the interpretation of the ritual was open and could be moved in various directions. This, however, was not the case because sacrifice has been considered a-religious and its depictions regarded as a means for representation in a purely political and social discourse – two current ways of seeing sacrifices, as I stated in the introduction.

The Cognitive Resource Depletion model, however, offers the possibility to explain this situation without the necessity of renouncing the religious content of the ritual. The cognitive processes during the perception of ritual gestures which are goal-demoted and causally opaque cause a kind of depletion which affects the susceptibility of the participants to collective ideas by creating an attributional and inferential gap, which we are moved to fill with meaning, interpretation or behaviour. Thus the pictorial representations of animal sacrifice in Asia minor are not "dereligionized" variables filled afterwards with meaning, but purposeful interpretations of the complex ritual of animal sacrifice, the polysemy of which is rooted in the processes during the perception of that ritual itself.

References

Alcorta, C. S. (2013) "Religious Ritual and Modes of Knowing: Commentary on the Cognitive Resource Depletion Model of Ritual," *Religion, Brain & Behavior* 3(1): 55–58.

Aldrete, G. S. (2014) "Hammers, Axes, Bulls, and Blood: Some Practical Aspects of Roman Animal Sacrifice," *Journal of Roman Studies* 104: 28–50.

Alvarez, J. A. and E. Emory (2006) "Executive Function and the Frontal Lobes: A Meta-Analytic Review," *Neuropsychology Review* 16: 17–42.

Bar, M. (2009) "Predictions: A Universal Principle in the Operation of the Human Brain," *Philosophical Transactions of the Royal Society B: Biological Sciences* 364(1521): 1181–1182.

Barrett, J. L. (2004) *Why Would Anyone Believe in God?*, Lanham, MD: AltaMira Press.

Barrett, J. L. (2007) "Cognitive Science of Religion: What Is It and Why Is It?," *Religion Compass* 1(6): 768–786.

Bechara, A., H. Damasio and A. R. Damasio (2000) "Emotion, Decision Making and the Orbitofrontalcortex," *Cerebral Cortex* 10: 295–307.

Boyer, P. (2001) *Religion Explained: The Human Instincts That Fashion Gods, Spirits and Ancestors*, New York: Basic Books.

Boyer, P. and P. Liénard (2006) "Why Ritualized Behavior? Precaution Systems and Action Parsing in Developmental, Pathological and Cultural Rituals," *Behavioral and Brain Sciences* 29(6): 595–650.

Brown, S. (2013) "Religious Ritual and the Loss of Self," *Religion, Brain & Behavior* 3(1): 58–60.

Burkert, W. (1972) Homo Necans. Interpretationen altgriechischer Opferriten und Mythen. *Religionsgeschichtliche Versuche und Vorarbeiten* 32, Berlin: Walter de Gruyter.

Carter, C. S., M. M. Botvinick and J. D. Cohen (1999) "The Contribution of the Anterior Cingulate Cortex to Executive Processes in Cognition," *Reviews in Neuroscience* 10: 49–57.

Carter, C. S., T. S. Braver, D. M. Barch, M. M. Botvinick, D. Noll and J. D. Cohen (1998) "Anterior Cingulate Cortex, Error Detection, and the Online Monitoring of Performance," *Science* 280: 747–749.

Clark, A. (2016) *Surfing Uncertainty: Prediction, Action, and the Embodied Mind*, Oxford and New York: Oxford University Press.

Colombo, M. (2017) "Review of A. Clark, Surfing Uncertainty: Prediction, Action, and the Embodied Mind," *Minds and Machines* 27(2): 381–385.

D'Andria, F. (ed.) (2008) *Atlante di Hierapolis di Frigia*, Hierapolis II, Istanbul: Ege Yayinlari.

D'Andria, F. and T. Ritti (1985) *Hierapolis Scavi e ricerche II: Le sculture del teatro. I rilievi con i cicli di Apollo e Artemide*, Rome: Bretschneider.

De Bernardi Ferrero, D. (ed.) (2007) *Il teatro di Hierapolis di Frigia. Restauro, architettura ed epigrafia*, Genova: de Ferrari.

Detienne, M. and J.-P. Vernant (eds.) (1979) *La cuisine du sacrifice en pays grec*, Paris: Gallimard.

Du Bois, J. W. (2007) "The Stance Triangle," in R. Englebretson (ed.), *Stancetaking in Discourse: Subjectivity, Evaluation, Interaction*, Amsterdam: Benjamins, pp. 139–182.

Durkheim, E. (1995/1912) *The Elementary Forms of Religious Life*, New York: Free Press (French original 1912: *Les formes élémentaires de la vie religieuse*, Paris: Alcan).

Eilam, D and J. Mort (2013) "Adding Mist to the Fog Surrounding Collective Rituals: What Are They, Why, When and How Often Do They Occur?," *Religion, Brain & Behavior* 3(1): 60–63.

Elsner, J. (2007) *Roman Eyes: Visuality & Subjectivity in Art & Text*, Princeton, NJ: Princeton University Press.

Faraone, C. A. and F. S. Naiden (eds.) (2012) *Greek and Roman Animal Sacrifice: Ancient Victims, Modern Observers*, Cambridge: Cambridge University Press.
Frith, C. (2007) *Making Up the Mind: How the Brain Creates Our Mental World*, Oxford: Blackwell.
Geertz, A. W. (2004) "Cognitive Approaches to the Study of Religion," in P. Antes, A. W. Geertz and R. R. Warne (eds.), *New Approaches to the Study of Religion Volume 2: Textual, Comparative, Sociological, and Cognitive Approaches*, Berlin: Walter de Gruyter, pp. 347–399.
Girard, R. (1972) *La Violence et le sacré*, Paris: Grasset.
Hallett, M. (2007) "Volitional Control of Movement: The Physiology of Free Will," *Clinical Neurophysiology* 118: 1179–1192.
Henderson, J. (2009) *Longus, Daphnis and Chloe: Xenophon of Ephesus, Anthia and Habrocomes*, Loeb Classical Library 69, Cambridge, MA and London: Harvard University Press.
Huet, V. (2005) "La mise à mort sacrificielle sur les reliefs romains. Une image banalisée et ritualisée de la violence?," in J. M. Bertrand (ed.), *La violence dans les mondes grec et romain. Actes du colloque international*, Paris 2–4 mai 2002, Paris: Publications de la Sorbonne, pp. 91–119.
Huet, V. and J. Scheid (2004) "Sacrifice, Romain," in J. C. Balty et al. (eds.), *Thesaurus Cultus et Rituum Antiquorum I*, Los Angeles, CA: The Paul Getty Museum, pp. 183–235.
Kapitány, R. (2017) *Ritual Cognition: Ritualized Action and Artefact*. PhD Thesis, University of Queensland, Brisbane: University of Queensland (https://espace.library.uq.edu.au/view/UQ:640510) (accessed 10–10-2017).
Kapitány, R. and M. Nielsen (2017) "The Ritual Stance and the Precaution System: The Role of Goal-Demotion and Opacity in Ritual and Everyday Actions," *Religion, Brain & Behavior* 7(1): 27–42.
Knust, J. W. and Z. Varhelyi (eds.) (2011) *Ancient Mediterranean Sacrifice*, Oxford: Oxford University Press.
Lawson, E. T. (2000) "Toward a Cognitive Science of Religion," *Numen* 47(3): 338–349.
Lawson, E. T. and R. N. McCauley (1993) *Rethinking Religion: Connecting Cognition and Culture*, Cambridge: Cambridge University Press.
Legare, C. H. and A. L. Souza (2012) "Evaluating Ritual Efficacy: Evidence from the Supernatural," *Cognition* 124(1): 1–15.
Legare, C. H. and P. A. Herrmann (2013) "Cognitive Consequences and Constraints on Reasoning About Ritual," *Religion, Brain & Behavior* 3(1): 63–65.
Lennon, J. J. (2015) "Victimarii in Roman Religion and Society," *Papers of the British School at Rome* 83: 65–89.
Lienard, P., M. Martinez and M. Moncrieff (2013) "What Are We Measuring?," *Religion, Brain & Behavior* 3(1): 65–68.
Lindner, R. (1994) *Mythos und Identität. Studien zur Selbstdarstellung kleinasiatischer Städte in der römischen Kaiserzeit*, Stuttgart: Steiner.
McCauley, R. N. (2013) "Functions, Mechanisms, and Contexts: Comments on 'Cognitive Resource Depletion in Religious Interactions'," *Religion, Brain & Behavior* 3(1): 68–71.
McCauley, R. N. and E. T. Lawson (2002) *Bringing Ritual to Mind: Psychological Foundations of Cultural Form*, Cambridge: Cambridge University Press.
McCauley, R. N. and E. T. Lawson (2007) "Cognition, Religious Ritual, and Archaeology," in E. Kyriakidis (ed.), *The Archaeology of Ritual*, Los Angeles, CA: Cotsen Institute of Archaeology Publications, pp. 209–254.
McClenon, J. (2013) "Cognitive Resource Depletion and the Ritual Healing Theory," *Religion, Brain & Behavior* 3(1): 71–73.
McClymond, K. (2008) *Beyond Sacred Violence: A Comparative Study of Sacrifice*, Baltimore, MD: Johns Hopkins University Press.
Meyer, H. (2011) Kunst und Politik. Religion und Gedächtniskultur. Von der späten Republik bis zu den Flaviern. *Römische Kunstgeschichte in Fallstudien*, 2, 1, München: Biering und Brinkmann.
Miyake, A, N. P. Friedman, M. J. Emerson, A. H. Witzki, A. Howerter and T. D. Wager (2000) "The Unity and Diversity of Executive Functions and Their Contributions to Complex 'Frontal Lobe' Tasks: A Latent Variable Analysis," *Cognitive Psychology* 41(1): 49–100.
Naiden, F. S. (2013) *Smoke Signals for the Gods: Ancient Greek Sacrifice from the Archaic through Roman Periods*, Oxford: Oxford University Press.
Nielbo, K. L. and J. Sørensen (2011) "Spontaneous Processing of Functional and Non-Functional Action Sequences," *Religion, Brain & Behavior* 1(1): 18–30.
O'Sullivan, J. (1995) *Xenophon of Ephesus: His Compositional Technique and the Birth of the Novel*, Berlin: Walter de Gruyter.

Petropoulou, M.-Z. (2008) *Animal Sacrifice in Ancient Greek Religion, Judaism, and Christianity, 100 B.C.–A.D. 200*, Oxford: Oxford University Press.

Öztürk, A. (2009) *Die Architektur der Scaenae frons des Theaters in Perge*, Berlin: Walter de Gruyter.

Pyysiäinen, I. (2001) *How Religion Works: Towards a New Cognitive Science of Religion*, Leiden and New York: Brill.

Rappaport, R. A. (1999) *Ritual and Religion in the Making of Humanity*. Cambridge: Cambridge University Press.

Rossano, M. J. (2012) "The Essential Role of Ritual in the Transmission and Reinforcement of Social Norms," *Psychological Bulletin* 138(3): 529–549.

Ryberg, I. S. (1955) *Rites of the State Religion in Roman Art: Memoirs of the American Academy in Rome 22*, Rome: American Academy.

Schjoedt, U., J. Sørensen, K. L. Nielbo, D. Xygalatas, P. Mitkidis and J. Bulbulia (2013a) "Cognitive Resource Depletion in Religion Interactions," *Religion, Brain & Behavior* 3(1): 39–55.

Schjoedt, U., J. Sørensen, K. L. Nielbo, D. Xygalatas, P. Mitkidis and J. Bulbulia (2013b) "The Resource Model and the Principle of Predictive Coding: A Framework for Analyzing Proximate Effects of Ritual," *Religion, Brain & Behavior* 3(1): 76–86.

Schörner, G. (2006) "Sacrifices and Their Representation in Roman Asia Minor: Reconsidering the Core-Periphery Concept," in C. Mattusch, A. Donohue and A. Brauer (eds.), *Common Ground: Archaeology, Art, Science, and Humanities*, Proceedings of the 16th International Congress of Classical Archaeology, Boston, MA, 23–26 August 2003, Oxford: Oxbow, pp. 71–74.

Schörner, G. (2009) "Bild und Vorbild: Nordafrika – Rom – Kleinasien," in H. Cancik and J. Rüpke (eds.), *Die Religion des Imperium Romanum. Koine und Konfrontationen*, Tübingen: Mohr Siebeck, pp. 249–271.

Schörner, G. (2013) "Depicting Sacrifice in Roman Asia Minor: Narratives of Ritual in Classical in Classical Archaeology," in V. Nünning, J. Rupp and G. Ahn (eds.), *Ritual and Narrative: Theoretical Explorations and Historical Case Studies*, Bielefeld: Transcript-Verlag, pp. 103–130.

Schörner, G. (2014) "Bilder vom Opfern: Zur Erzählweise von Ritualdarstellungen im römischen Kleinasien," in M. Meyer and D. Klimburg-Salter (eds.), *Visualisierungen von Kult*, Wien, Köln and Weimar: Böhlau, pp. 356–389.

Smith, J. Z. (1987) "The Domestication of Sacrifice," in R. G. Hamerton-Kelly (ed.), *Violent Origins: Walter Burkert, René Girard, and Jonathan Z. Smith on Ritual Killing and Cultural Formation*, Stanford, CA: Stanford University Press, pp. 191–205.

Sousa, P. and C. White (2013) "Problems for the Cognitive-Depletion Model of Religious Interactions," *Religion, Brain & Behavior* 3(1): 73–76.

Svoboda, E., M. C. McKinnon and B. Levine (2006) "The Functional Neuroanatomy of Autobiographical Memory: A Meta-Analysis," *Neuropsychologia* 44: 2189–2208.

Tremlin, T. (2006) *Minds and Gods: The Cognitive Foundations of Religion*, Oxford: Oxford University Press.

Ullucci, D. (2015) "Sacrifice in the Ancient Mediterranean: Recent and Current Research," *Currents in Biblical Research* 13(3): 388–439.

Whitehouse, H. (2004) *Modes of Religiosity: A Cognitive Theory of Religious Transmission*, Lanham, MD: AltaMira Press.

Whitehouse, H. (2013) "Ritual and Acquiescence to Authoritative Discourse," *Religion, Brain & Behavior* 3(1): 76–79.

Zalla, T., P. Pradat-Diehl and A. Sirigu (2003) "Perception of Action Boundaries in Patients with Frontal Lobe Damage," *Neuropsychologia* 41: 1619–1627.

24

Art, architecture, and false memory in the Roman Empire

A cognitive perspective

Maggie L. Popkin

Memory mattered to the ancient Romans, and monuments mattered to memory. The Latin language itself makes this connection clear: *memoria* (memory), *monumenta* (monuments), and *monere* (to remind) all derive from the same root (Varro *Ling.* 6.49). Romans also recognized and exploited a close link between visualization and memory in the *ars memoriae* (the so-called art of memory), in which an orator memorized his speech by visualizing a house and putting different elements of his speech in each room (Cic. *De or.* 2.350–60; Quint. *Inst.* 11.2; *Rhetorica ad Herennium* 3.16–24; Yates 1966; Bergmann 1994; Small 1997). According to Cicero, a person training his memory "must select localities and form mental images of the facts they wish to remember and store those images in the localities" (Cic. *De or.* 2.354).[1]

Following in Cicero's footsteps, unconsciously or not, scholars have often viewed monuments and works of art as repositories for human memory, places that can "capture" memories or where memories can "dwell" (e.g., Edwards 1996: 29; Favro 1999; Alcock 2002: 28; Bastéa 2004: 1). But monuments and objects were much more than just storage facilities for already-formed memories. Scholars have begun to complicate the relationship between visual culture and memory in antiquity, ascribing more and more agency to the former in its impact on the latter (e.g., Galinsky 2014; Popkin 2016). This chapter asks how art and architecture actively shaped how ancient Romans remembered and knew their world by generating and manipulating memories and knowledge of places, events, and institutions. Through three case studies—a sports souvenir from Roman Britain, travel souvenirs from the Bay of Naples, and the Arch of Septimius Severus in Rome—it will demonstrate how the cognitive science of memory enriches our understanding of the complex relationships among visual culture, memory, and knowledge construction in the Roman Empire. I will begin with a brief overview of cognitive theories about the impact of visual imagery on how humans remember, before presenting each case study and its relationship to memory and knowledge. I will conclude by analyzing how all the objects and monuments under consideration, despite their different scales and geographic origins, combine image and narrative in ways that made them particularly efficacious, cognitively speaking, at shaping memories and knowledge of their depicted subjects in the Roman world.

Cognitive theories of memory and visual imagery

Memory has provided fertile ground for cognitive scientists for decades. It is widely accepted today that human remembering is a highly (re)constructive process, one that is subject to illusions, distortions, and even falsifications (for overviews see Schacter 2001; Brainerd and Reyna 2005; Loftus 2005; Nash and Ost 2017). From the initial perception and encoding of an event to subsequent storage, retrieval, and report, our memories are subject to transformation. In fact, neuroscience is increasingly suggesting that our memories are ceaselessly dynamic rather than ever permanently consolidated (Hupbach et al. 2007; Dudai 2012). Talking about "memory" rather than "remembering" can obscure the fundamentally processual quality of the action, eliding "the fact that there are agents here, people who actively listen, select, modify, or rigidify particular versions of their past" (Boyer and Wertsch 2009: 113).

Remembering is such a malleable process that it can result in the distortion and even falsification of memories (see Loftus 2005). Cognitive research has proven that people can come to hold false memories; that is, they remember experiencing an event or seeing something that they did not experience or see. In a seminal study, Roediger and McDermott showed subjects a list of words related to the idea of sleep (Roediger and McDermott 1995; Roediger and McDermott 2000). Subjects then falsely remembered seeing the word "sleep" on the list, because they perceived the gist of the list to be "sleep." Subsequent studies have shown that people can remember childhood experiences (such as getting lost in a shopping mall) that are entirely fabricated. These false memories can be held with as much conviction as "true" memories—and people can continue to remember things falsely even after they have been corrected (Lewandowsky et al. 2005). Functional magnetic resonance imaging (fMRI) of people's brains has shown that many of the same regions of the brain are active during the retrieval of true and false memories, which may help explain why false memories can feel so real to subjects (Schacter and Loftus 2013: 121).

Memories can be distorted and false memories introduced by suggestive and/or misleading questions or statements. Even just thinking about an experience can "contaminate memory with falsehoods and distortions," while imagining a fictitious event increases the likelihood that one will falsely remember it. False memories can develop as a result of imagination because the act of imagining an event often involves the creation of an image rich in sensory, perceptual, and contextual details (Zaragoza et al. 2011: 18–19; Morris et al. 2006). Similar brain activity is involved in remembering and imagining, which may additionally explain how imagination can lead to false memories (Schacter and Loftus 2013: 121). Of particular relevance to ancient art historians is the extent to and efficacy with which visual images can generate false memories: a representation of an event can come to serve as a more powerful source of memory than the actual event itself (see, e.g., Davis and Loftus 2009). In one study, subjects were shown a doctored photograph of themselves as a child in a hot air balloon. Many subjects then remembered having participated in this activity as a child, even though family members corroborated that it was not true (Wade et al. 2002). Showing subjects a true photograph associated with a suggested (false) event doubles the rate of false memory reports (Lindsay et al. 2004), while photographs, true or doctored, influence the way that people remember current events (Garry and Gerrie 2005: 323; Frenda et al. 2013).

Since Wade et al.'s initial study (2002), they have revisited their theory that photographs help construct false memories more effectively than narratives. A subsequent study (Garry and Wade 2005) demonstrated that textual narratives resulted in a higher rate of false memory report than doctored photographs, although the photographs still caused significant false memory report. They theorized that "false narratives gave more free rein to the imagination and permitted more infusing of personal knowledge" than photographs, which "impose more constraints on imagination because they depict specific details, people, and settings" (Garry and Wade 2005: 360, 363). Research has continued to show, however, that photographs (both true and doctored) effectively generate false memories, because they make untrue or distorted events feel more familiar, they encourage us to create mental imagery that intertwines with imagery from real memories, and people tend to perceive visual images as credible sources of information (Wade, Nightingale, and Colloff 2017).

False memories and the "misinformation effect"—that is, the distortion of memories through suggestive (mis)information—are widely documented (Loftus 2005). The misinformation effect has even been observed in animals as well as people, so its basis seems at least partly biological, not just cultural (Loftus 2005: 362–3). Thus, while culture has an impact on how cognitive processes play out in our minds, certain cognitive processes appear across cultures, even if in somewhat different instantiations (see Park and Huang 2010 and Ross and Wang 2010 on the neurological impact of culture on cognition). It is therefore justified to use the cognitive theory behind the misinformation effect to interrogate the memory practices of a historically distinct culture such as that of the Roman Empire, as long as scholars keep an open mind to what is distinct about Roman mnemonic practices. I will elaborate on the cognitive research over the course of this chapter, but the underlying point is that cognitive research has proven that images affect how we remember—not just metaphorically, but literally. It is difficult to overestimate the impact that art and architecture had on memory and knowledge construction in the classical world.

Sports memorabilia from Roman Britain

Portable objects could have profound effects on the way knowledge was generated during the Roman imperial period, when there were relatively few ways to "see" distant sites and monuments—and events that unfolded within—unless one visited in person. The sports memorabilia from Roman Britain illustrate this point well.

In the 1860s, a remarkable glass beaker was discovered in the western cemetery of Colchester (ancient Camulodunum) in England (Figure 24.1). The body of the beaker is divided into three registers, two of which contain scenes of chariot racing, and the third of which is inscribed with the charioteers' names. The Colchester beaker is representative of a broader class of fascinating if understudied objects from the Roman Empire: sports memorabilia. Objects such as the Colchester beaker did not commemorate a specific chariot race that the owner had witnessed himself but instead commemorated the sport of chariot racing generally, as well as the owner's fandom. Chariot racing was widely popular in ancient Rome, and fan loyalty to circus factions, or racing teams, could reach epic proportions; Pliny the Elder reports that one fan immolated himself on the funeral pyre of his favorite charioteer, so distraught was he at his hero's passing (Plin. *HN* 7.186).

The Colchester beaker represents a chariot race in the Circus Maximus in Rome, the largest and best-known circus of the Roman world. In the lower register, four *quadrigae* (chariots drawn by a team of four horses) sprint from left to right. The middle register depicts the monuments that decorated the *spina*, or central barrier, of the Circus Maximus. Finally, in the top register, an inscription gives the names of the four charioteers racing below, each name accompanied by an acclamation: *va(de)* or *va(le)* for the losers, and *av(e)* for the victor (see Futrell 2006: 75).

The Colchester beaker was manufactured in the middle of the first century in northern Italy or Gaul and imported to Britain, probably for Roman soldiers stationed there as well as legionary veterans (Humphrey 1986: 92; Landes 1998: 17). Most of the soldiers stationed in Roman Britain both before and after the Boudican uprising of 60 CE would not have been from Rome; many would not even have been from Italy. It is unlikely most of them would have seen a

Figure 24.1a–b The Colchester beaker. Glass beaker with scenes of a chariot race in the Circus Maximus, from Colchester. First century CE

London, British Museum, inventory number 1870,0224.3. Photograph © Trustees of the British Museum chariot race in the Circus Maximus in person. A cup such as the Colchester beaker was therefore an important generator of semantic memory of the Circus Maximus, a monument truly worthy of commemoration—the original, historic home of the wildly popular circus factions.

In his seminal 1972 essay, Endel Tulving distinguished between episodic and semantic memory. Episodic memory refers to memory for events that a person has actually experienced, located in a particular time and place, whereas semantic memory refers to the brain's system for encoding and retrieving facts and concepts—general and historical knowledge, in other words (Tulving 1972; Schacter 2001: 27). Most people living in Roman Britain—whether they were Roman soldiers or native Britons—would not have witnessed a chariot race in the Circus Maximus and would not, therefore, have stored episodic memories of these events located in the Circus. Memorabilia such as the Colchester beaker, however, would have enabled people in Britain to encode and retrieve semantic memories of the Circus Maximus and its races. They enabled knowledge of this building and these events in the absence of personal experience and episodic memory.

At the same time that the Colchester beaker could generate and disseminate semantic knowledge of the Circus Maximus and its chariot races in Roman Britain, it could also emphasize certain aspects of the represented subject over others. Somebody living in Roman Colchester or London in the early and high imperial periods could not watch a chariot race in Rome remotely, as somebody in New York today can watch a soccer match in England on television or the Internet. But neither were Romans or native Britons likely to encounter a large-scale relief or mosaic of a chariot race in the Circus Maximus, as one might have done in Italy or elsewhere in the empire. Only one circus mosaic is known from Roman Britain, from Horkstow, and it dates to the fourth century CE (Dunbabin 1999: 99). At the time the Colchester beaker was manufactured and imported into Britain in the first century CE, glass circus vessels may have been the *only* images of the Circus Maximus some people living in Roman Britain ever saw. The aspects of chariot racing that they emphasize would, therefore, have had a major impact on people's semantic memories of the Circus Maximus and its chariot races.

The Colchester beaker does not depict the façade or *cavea* of the circus itself, which would not have been fully monumentalized in stone by the time the beaker was produced. Instead, it emphasizes the *metae*, or conical turning posts grouped in triads at either end of the *spina*, which are shown larger in scale and which extend through to the lower register of racing teams. The lap counters contribute to the cup's allusions to the temporal unfolding of the race, evoking the seven laps chariots would have raced around the Circus Maximus's track. The assortment of columnar monuments, shrines or altars, and statues on the *spina* contribute to the impression given by the cup of the Circus Maximus as one of Rome's most impressive structures, filled with artistic wonders as well as exciting races (see Dion. Hal. 3.68.1; Plin. *Pan.* 51.4). The obelisk of Augustus also features prominently. Augustus had brought this obelisk back to Rome from Heliopolis as a spoil of war to stand on the Circus Maximus' *spina*. It evoked Rome's conquest of Egypt and subsequent annexation of Egypt as a province. The obelisk was a victory monument (Davies 2011: 364-6; Swetnam-Burland 2015: 65-104), and its inclusion on the Colchester beaker would have been appropriate for the cup's audience of Roman soldiers, who participated in the campaigns that brought new lands, such as Egypt before and now Britannia, into the Roman fold.

Travel souvenirs from Ancient Campania

Other Roman souvenirs could shape episodic as well as semantic memories. In the late third to early fourth centuries CE, glassmakers in the city of Puteoli (modern Pozzuoli) on the Bay of Naples produced a series of glass *ampullae* (globular flasks for holding liquids). At least thirteen of these flasks survive, discovered across the western Roman Empire from Italy and North Africa to the Iberian Peninsula, Great Britain, and Germany (Popkin 2017, 2018b, both with bibliography). These flasks are remarkable for the detailed engravings of cityscapes of Puteoli and the neighboring resort town of Baiae (modern Baia) that decorate their bodies. Some flasks show Puteoli alone (Figures 24.2 and 24.3), while others combine representations of the two cities (Figures 24.4 and 24.5). Puteoli and Baiae were both major tourist destinations in Roman Italy (D'Arms 1970), and these flasks were purchased as souvenirs by travelers to the region, who brought them home as reminders of their visit to Campania, much as modern travelers might do.

The flasks that depict Puteoli alone represent many of the city's most striking monuments, including the harbor mole, surmounted by monumental decorations; spectacle buildings that include, depending on the individual flask, a theater, amphitheater, and stadium; and a large distyle pedimented temple with a cult statue between the columns. The flasks that combine views of Puteoli and Baiae feature Puteoli's harbor mole along with domed or pedimented buildings and checkered squares representing Baiae's baths and *ostriaria* (artificial oyster beds), respectively. One flask discovered in Rome may represent Baiae alone, but I believe the prow-like structure

Figure 24.2 The Pilkington flask. Glass flask with scene of Puteoli, found in North Africa. Third to fourth century CE

St. Helens, United Kingdom, the World of Glass, inventory number SAHGM.1974.002. Photograph courtesy of the World of Glass

Figure 24.3 Drawing of the engraved scene on the Pilkington flask

Drawing by author

upon which the reclining personification rests should be identified as Puteoli's harbor mole. On all the flasks, the engraved images are accompanied by inscriptions that identify the depicted structures and locations, as well as some monuments that are not visually represented.

Scholars have generally approached these flasks as documents providing topographical information about Puteoli and Baiae (Painter 1975; Ostrow 1979), and the flasks can helpfully attest to the existence of certain monuments that have left no archaeological traces. But they do not strive for cartographic accuracy in a modern sense. This is perhaps nowhere more apparent than in the flasks that combine views of Puteoli and Baiae, even though in reality the two cities were several kilometers apart. Instead, the flasks present a carefully curated view of the monuments that their producers deemed the most significant in the late imperial period, namely those that would have attracted tourists and redounded most gloriously on the cities' reputations as longstanding and prestigious tourist destinations favored by Rome's emperors themselves. Consequently, the engraved scenes of Puteoli emphasize monuments, such as the temple celebrating Augustus and the amphitheater emulating the Flavians' Colosseum, that linked Puteoli to Roman emperors who had shown favor to the city, as well as singular landmarks, such as the harbor mole, that would have featured prominently in tourists' views of the city (cf. Lynch 1960: 78–83). The scenes of Baiae emphasize the leisure activities that lay at the root of the city's attraction to tourists and emperors alike, namely its baths, oyster beds, and *stagna*, or pools, patronized by emperors such as Nero and Alexander Severus (Popkin 2018b).

The souvenir flasks thus shaped how people came to visualize, remember, and hence know Puteoli and Baiae around the western Roman Empire. I am not suggesting that the aim of the Campanian souvenir flasks was to create false memories. Their goal was to present the two cities in a particular, and particularly favorable, light. Whether intentionally or not, however, the flasks raised the genuine possibility, even probability, that people who had visited Puteoli and Baiae in person would remember their visits as much based on the souvenir flasks as on their actual experiences in Campania. This process can be explained first by the fact that memories become susceptible to suggestion and transformation each time they are retrieved, and second by the source monitoring framework (SMF), in which "people mistake internally generated information for genuine experience" (Foster and Garry 2012: 226; Wade, Nightingale, and Colloff 2017: 50). According to the SMF, a person recalling a memory of Puteoli evoked by looking at the flask could misattribute the source of that memory to his in-person experience of the city. Moreover, "[t]he more the thoughts and images that come to mind have characteristics of an actually witnessed event, the more likely they are to be experienced as a memory of an actually-experienced event" (Zaragoza et al. 2011: 19). The rich detail of the flasks' images and their focus on landmarks that would have been most memorable to a tourist unfamiliar with Puteoli or Baiae could have caused the resulting thoughts and memories to be misattributed to the actual experience of visiting these cities, leading somebody to remember the memories evoked by the flasks as genuine, personal experience. One can imagine a person beginning to remember he had spent a lot of time looking at Puteoli's harbor mole and Augustan temple, and feeling very well acquainted with those monuments, even if he had spent most of his time conducting business in the city's banking establishments. This person would have, in essence, constructed a memory perceived as episodic to ease any disjuncture or gaps between his personal experience and the image of his experience presented to him on his souvenir flask.

Additionally, as seen already with the Colchester beaker, the souvenir flasks of Puteoli and Baiae could generate semantic memories and knowledge of these cities for people who had never traveled there. There was no mass media in the Roman Empire, no mechanically reproduced and easily distributable prints, let alone photographs or digital images. For many people living in Rome's western provinces, these souvenir flasks would likely have been the only images of

Figure 24.4a–b The Populonia flask. Glass flask with combined scene of Puteoli and Baiae, found near Piombino, Italy

Corning, NY, Corning Museum of Glass, inventory number 62.1.31. Photograph courtesy of the Corning Museum of Glass

Figure 24.5 Drawing of the engraved scene on the Populonia flask
Drawing by author

Puteoli and Baiae they ever saw. (Two wall-paintings from Italy may represent Puteoli, but these would not have been portable, or accessible even to people beyond those admitted to their private display contexts.) Literary descriptions of Puteoli and Baiae would have been accessible only to literate Romans, and, in any event, they tend to eschew ekphrastic descriptions of how these cities looked (Popkin 2018b). The images on the souvenir flasks, therefore, would have been the primary means for generating semantic knowledge of Puteoli and Baiae.

The Arch of Septimius Severus in Rome

Portable objects such as the Colchester circus beaker and the Campanian souvenir flasks could project images of cities and monuments around the Roman Empire that served as the basis for generating memories of the represented cities and monuments, sometimes even creating false memories. Public monuments in the Roman Empire could also manipulate memory. Though obviously not portable, temples, arches, and other monument types reached wide audiences, situated, as they often were, in high-traffic areas. These structures and their decoration—for example, relief sculptures and inscriptions—projected selective images and messages about historical events and persons; they attempted to present their narrative of history as factual (see Popkin 2016: 176). Public monuments could therefore have a profound impact on how Romans remembered historical events and how scholars today continue to remember Roman history, sometimes even falsifying memories of events that never happened. They thus illustrate a memory function of Roman art and architecture that is related to but distinct from the souvenirs' generation of semantic memory about places and sports: namely, the capacity to manipulate historical memory of persons and events. Perhaps nowhere is the impact of a public monument on ancient *and* modern historical memory clearer than in the case the Arch of Septimius Severus in Rome.

In 203 CE, the Senate dedicated a monumental triumphal arch to the emperor Septimius Severus (r. 193–211 CE) in the northwest corner of the Roman Forum, in honor of his two Parthian campaigns (Brilliant 1967). This massive, triple-bay arch still towers above the Forum between the Curia and the Rostra (Figure 24.6). Although the arch was dedicated by the SPQR, it likely was designed in close consultation with Septimius Severus or his court. There is no contemporary literary evidence that Septimius Severus or his eldest son, Caracalla, celebrated a triumph in 202, the year preceding the arch's dedication. Cassius Dio was probably in Rome

Figure 24.6 The Arch of Septimius Severus in the Roman Forum, view of the west façade
Photograph by author

in 202 but does not mention a triumphal procession in the epitomized version of his Roman history that survives (76.1.1–5). Herodian, who may not have been in Rome in 202 but likely visited shortly thereafter, also does not mention a triumphal procession, although he attributes several triumphal celebrations to Septimius that year (3.8.9, 3.10.2). However, beginning with the late-fourth-century *Historia Augusta* and continuing into modern scholarship, one sees references to a Severan triumph of 202. The author of the *Historia Augusta* writes that Caracalla celebrated a triumph that year, in lieu of his father (*SHA Sev.* 16.6–7; *SHA M. Ant.* 9.6). By the time one reaches many modern biographies of Septimius and histories of the Severan era, the triumph has morphed into one celebrated by Septimius himself (Southern 2001: 43; Hekster 2008: 39; Hinterhöller 2008: 15; Lusnia 2014: 54).

I cannot say with certainty whether Septimius (or Caracalla) celebrated a triumph in Rome in 202 or not. Cassius Dio (76.1.1–5) describes in great detail the events of that year, including the emperor's *decennalia*, the marriage of Caracalla, and various spectacles that Septimius staged in Rome. He does *not* include a triumph in this list. Why, then, do a number of judicious scholars state that Septimius triumphed in 202, offering as evidence this passage in Cassius Dio (e.g., Chastagnol 1987: 97–102; Southern 2001: 43; Hekster 2008: 39; Hinterhöller 2008: 15; Lusnia 2014: 54, 60, 125, 148)? The cognitive science of memory can help us answer this otherwise perplexing question. The gist of an arch on the triumphal route in Rome (the path generally followed by triumphal processions) was that it celebrated a military victory of the sort that could have earned its honorand a triumph. Whether or not a triumphal procession had actually been celebrated, people going forward would see these arches and assume that the general or emperor honored had, in fact, triumphed. The gist-memory (that is, memory encapsulating the "essential semantic meaning" of something: Davis and Loftus 2009: 196–7) associated with the monumental form of the arch and the topographical location of the triumphal route could cause future generations to look at the Arch of Septimius Severus and see "triumph."

The arch also presented images implicitly and explicitly connected to the Roman triumph. It was originally surmounted by a gilded bronze statue of Septimius Severus in a triumphal chariot, flanked by subjugated Parthians (Figure 24.7). The triumphal procession was the sole occasion where an emperor would ride in this chariot, so the depiction of Septimius in this statue format would have contributed to the arch's triumphal gist. The massive panel reliefs show Septimius and his army besieging and capturing Parthian cities (Figure 24.8). Originally brightly painted, these reliefs illustrated the military accomplishments on which any Severan triumph would have been based and also recalled the paintings carried in triumphal processions and then displayed on the Curia. The relief panels might have sought to evoke false memories of Septimius's campaigns themselves, for, although they show Septimius vanquishing Parthian cities, in reality one of Septimius's notable challenges during his Parthian wars was his failure to capture the enemy city of Hatra (Birley 1988: 130–2). The small frieze running below the panel reliefs depicts a procession of Roman soldiers, Parthian prisoners, and wagons loaded with goods (perhaps spoils of war?). Finally, the pedestal reliefs depict Roman soldiers leading captive Parthians—a reenactment in stone of the segment of triumphal processions in which prisoners were paraded before the eyes of Roman spectators.

Figure 24.7 Denarius of Septimius Severus, 201–210 CE, reverse showing the Arch of Septimius Severus in the Roman Forum, including the arch's attic statuary

London, British Museum, inventory number R.15321. Photograph © Trustees of the British Museum

Figure 24.8 Northwest panel relief from the Arch of Septimius Severus in the Roman Forum, showing the capture of a Parthian city

Photograph by author

There are precedents for arches built along the triumphal route that commemorate victories that were never followed by a triumphal procession. The first Roman freestanding arches documented in the annalistic tradition were erected by L. Stertinius in 196 BCE to commemorate his victories in Spain (Livy 33.27.4). Stertinius did not receive a triumph, but Livy makes clear that the general intended his arches as an ersatz triumph. Nearly 200 years later, the Senate honored Augustus with an arch in the Roman Forum commemorating the return of the Parthian standards: a largely diplomatic coup that the arch presented as worthy of a triumph. The Arch of Nero on the Capitoline Hill was, according to Tacitus (*Ann.* 15.18), erected not to commemorate an actual military victory of the emperor but rather to compensate for the troubles that Rome's army was experiencing in Armenia. Following Septimius, the Arch of Constantine similarly depicts a triumph even though literary sources suggest that Constantine did not celebrate one (*Pan. Lat.* 4[10].30.5, 31.1–2; *Chronography of 354* Oct. 29). Rome's emperors and senators seem to have understood well that monuments could effectively memorialize non-events, transforming diplomatic coups or even military challenges into celebration-worthy achievements before the public eye, and the Severan arch stands in this tradition of implying triumphs through architecture (for further discussion of this phenomenon and the examples mentioned here, see Popkin 2018a).

Cognitive science demonstrates that even people who were in Rome in 202 and had not experienced Septimius celebrating a triumph could still have been susceptible to the development of false memories. When we process false information or suggested information, we can misattribute its source to real memories, as discussed above with the SMF. Psychologists have found that misinformation about modern military endeavors can create false memories even after the misinformation has been corrected, such is its initial power (Lewandowsky et al. 2005). Whether the Arch of Septimius exaggerated or falsified a triumph of Septimius Severus, it certainly presented this misinformation or misleading information as "fact" in Rome's public record, and some Romans may have accepted this message in their memories, particularly those who might have been favorably inclined toward Septimius Severus (Popkin 2016: 177–9). Counterfactual narrative devices might also have entered into people's conception of the arch. One might imagine, for example, a Roman imagining how Septimius's triumph would have looked, were he to have had one, given how great his victory must have been as presented by the arch.

The Arch of Septimius Severus's suggestive imagery could have generated distorted memories, leading to shared semantic memory of a Severan triumph that never occurred. Whether the arch distorted the episodic memories of any Romans is impossible to know. It seems unlikely that an adult who lived in Rome in 202 would come to remember falsely that a triumph was celebrated that year, although cognitive research reminds us it is not out of the question (Nash and Ost 2017). Triumphal processions during the imperial period were relatively rare events (no triumphs were celebrated during the long reigns of Hadrian and Antoninus Pius, for example), but most Romans would have been able to witness at least one triumph during their lifetime. The decreased number of triumphs in the imperial period arguably made them more outstanding than frequent republican triumphs, and some people might have desired to place themselves, whether consciously or not and correctly or not, at these potentially "once-in-a-lifetime" events. Research has shown that children and the elderly are more susceptible to false memories (Loftus 2005: 362); one can imagine, albeit speculatively, that a young child who was alive in Rome in 202 came later in life to recall having seen a triumphal procession that year.

That false memories could enter into Romans' conceptions of deeds and events related to historical individuals is suggested by Livy. He writes that mendacious funeral orations and statue inscriptions caused Romans incorrectly to attribute great achievements and political offices to men whose *res gestae* must have been less illustrious in reality (8.40.3–5; 22.23.11). Would a deceptive triumphal arch have been so different in effect? According to Cicero, what people see and hear can "lead astray" their beliefs, manipulating and possibly falsifying popular opinion (*Leg*. 1.47; see Wiseman 2014: 50). Juvenal (7.12–16) alludes to people who give false testimony in court; misinformation that could lead to false memories apparently could come from legal testimony as well as monuments, statues, and inscriptions.[2] Unsurprisingly, Roman authors do not use cognitive theory's language of false memories, and we cannot discern from literary references whether Romans in fact held false memories, but there seems to have been a general awareness among Roman authors that memories could be falsified. Regardless, the Arch of Septimius Severus does seem to have had an impact on semantic memories of Septimius's putative triumph. That the author of the *Historia Augusta* and the modern scholars cited above ascribe a Severan triumph to the year 202 in the absence of literary testimony of such an event suggests that the arch's manipulation of semantic memory has left an imprint not only in ancient imperial biographies but also in modern scholarship (Popkin 2018a).

Image, narrative, and memory

The objects and monuments considered here differ greatly in many ways. The Colchester beaker and the flasks from Puteoli are small, portable objects made of glass, while the Arch of Septimius Severus is a monumental, freestanding stone arch. The glass vessels traveled to the far reaches of the western Roman Empire, while the Arch of Septimius was firmly rooted in Rome. The Campanian souvenir flasks depict the Italian cities of Puteoli and Baiae, while the Colchester beaker represents a major monument in Rome itself. The Arch of Septimius, though standing in Rome, depicts far-off Parthian cities. Despite differences in scale, material, subject matter, and contexts of display and use, all three case studies share two important similarities in terms of their impact on remembering and constructing knowledge: they all would have encouraged frequent viewing, and they all combined image and narrative.

The manner in which the vessels and arch invited repeated viewing differed. The Colchester beaker was a drinking cup, and it would have been used as such, probably in social settings (see below). Consuming wine was a common activity in the Roman world, and the cup, while fragile, is hardly the sort of extremely luxurious and expensive object that might have been reserved for special occasions. The glass flasks of Puteoli and Baiae were not drinking vessels, but they would likely have been displayed in domestic contexts. That one flask has been excavated in a thermal complex in Astorga, Spain (Tafalla, Palomar, and Peralta 2003) and another in the area of the *capitolium* in Brescia, Italy (Roffia 2002) suggests that they were sometimes also taken about town and used, perhaps for carrying oil. The Arch of Septimius Severus was a major monument in the Roman Forum, a buzzing center of social, political, and religious activity. People living in Rome would have encountered the arch as often as they visited or passed through the Forum.

By inviting frequent viewing, the Colchester beaker, the Campanian flasks, and the Arch of Septimius Severus all would have encouraged repeated retrieval of episodic memories, semantic memories, or both—of chariot races in the Circus Maximus, of travels to the Bay of Naples, of Septimius's Parthian wars and putative triumph. Repeated viewing, suggestions, and imagining of events results in a greater incidence of false memory creation, likely because our memories become susceptible to transformation each time we recall them. The more we retrieve a memory, the more it can change. At the same time, repeated retrieval of a memory strengthens that memory regardless of how (in)accurate it is (Roediger, Zaromb, and Butler 2009: 163). Repetition, in other words, increases "the illusion of truth" (Zaragoza and Mitchell 1996: 294; Zaragoza et al. 2011: 19). Thus, for example, the Arch of Septimius encouraged viewers to recall a Severan military victory and triumph, thereby making memories of those events subject to distortion; by encouraging *repeated* viewing and memory retrieval, it would have presented the resulting semantic memory of a Severan triumph as increasingly true.

The arch and glass vessels effectively manipulated memory not only through repeated viewing but also through their combinations of image and narrative. Accompanied by the acclamatory inscription in the upper register, which mimics the cheers shouted by the crowd and thus recalls the auditory experience of a chariot race in the Circus Maximus, the scenes on the Colchester beaker recall the monumental decoration of the Circus Maximus, the chariot races that took place on its track, and the crowd experience of cheering on the racing teams. The beaker thus elicited a narrative of watching an exciting chariot race in the Roman Empire's most hallowed venue for the sport. The souvenir flasks of Puteoli and Baiae likewise invited a narrative of a traveler's visit to these cities: sites seen, leisure activities experienced, and so on. According to Susan Stewart, souvenirs actually become complete only when their owners invent a narrative about the objects' origins (Stewart 1993: 135); souvenirs exist for us to tell stories about and with them. As for the Severan arch, many of the images that decorate it are inherently narrative, most obviously the great panel reliefs with their episodic narratives of the capture of cities. The other sculptures could also have generated narrative recollection in viewers. The pedestal reliefs, for example, invite a narrative explanation to questions such as: Who are those prisoners? Where are they from? Where are the Roman soldiers leading them?

Cognitive experiments in laboratories have tended to separate narrative (generally defined there as words and text) from visual images to discern the relative impact of each on creating false memories (Garry and Wade 2005). But this is not how the real world often works. All the case studies presented in this chapter combine word and image. They have representational scenes that include architecture and, sometimes, people, but they also include inscriptions. The Colchester beaker gives the names of the charioteers with accompanying acclamations that evoke the excitement of a race in progress. The flasks from Puteoli include labels of monuments and also, in some instances, personalized inscriptions running above the engraved scenes. The Arch of Septimius features a prominent dedicatory inscription that narratively suggests the emperor's accomplishments through its references to his foreign titles and his restoration of the republic and expansion of the empire. These objects and monuments suggest that word and image *together* might manipulate and distort memories even more effectively than each on its own. Cognitive researchers have long recognized "picture memory"; that is, the ability of pictorial representations to facilitate "recall, recognition, and associative learning" (Madigan 2014: 75). Some neuroimaging research on how people integrate new words into their lexicon suggests that people ultimately remember words better when they encode a word with the phonological form *and* an associated picture of the word than when they encode a word only via the phonological form (Takashima et al. 2014). However, most laboratory experiments of the impact of word and image on memory necessarily use relatively simple word and object stimuli (e.g., a single word, or a picture of a single object). Richer narratives elicit more elaborate brain processes, and the same should be true for more complex images (Jouen et al. 2015: 73).

I would suggest that the rich narratives evoked by the combination of images and words on the souvenir vessels and triumphal arch would have encouraged conceptual elaboration; that is, elaboration of the meaning and implications of the suggested event. Conceptual elaboration increases false memory, in part because it serves to link "true and suggested events *to each other* as well as to the overall representation of the story" (Zaragoza et al. 2011: 29, emphasis in original). In other words, the suggested events become plausibly embedded in a person's recollection of real events. In the case of the Arch of Septimius Severus, the suggestion of a triumphal procession could easily have been embedded in recollections of Septimius's Parthian wars, which were historical events, creating a plausible, coherent narrative in which Septimius's battlefield victories in Parthia led to a triumph in Rome. In fact, the Arch of Septimius could have led to semantic memories of a Severan triumph without exposing the emperor to the potential for public embarrassment that Beard has observed (Beard 2007: 34; Popkin 2016: 173).

Part of the conceptual elaboration that would have taken place around these objects and monuments would have occurred through conversation. The beaker, flasks, and arch would all have been talked about by people, with other people. Conversation is a critical means for distorting memories, generating false memories, and sharing memories among individuals (Hirst and Echterhoff 2012). People remember selectively in conversation; when recalling an event one has experienced, one does not tell listeners *everything* about that event, but only what one thinks listeners will find most interesting or entertaining, for example. When people selectively remember in conversation, two things happen: they reinforce the memories that they do recall, and they forget the memories that they do not recall. This is called retrieval-induced forgetting (RIF) (Hirst and Echterhoff 2012: 66–7). RIF is highly relevant to the impact Roman souvenirs and memorabilia had on remembering. In the case of the Campanian souvenir flasks, their focus on particular landmarks and leisure activities would have prioritized those experiences in people's memories and facilitated the forgetting of other, unrepresented experiences. A person holding his souvenir flask and passing it around to friends as he talked about his trip to Puteoli might have talked about watching the ships come in from the city's great harbor mole or watching a show in the city's imperial amphitheater. In conversing in this manner, he would have reinforced his memories of these activities, but he would also have increasingly forgotten experiences whose narratives were not evoked by the flask, such as dining out at a local tavern.

Although the souvenir flasks of Puteoli and Baiae would have been viewed and discussed, perhaps no object considered in this chapter would have been more used in such a conversational context than the Colchester beaker, which was a drinking cup. It would have been used on occasions of social drinking, where people would have gathered in revelry and conversation. Beakers decorated with images of chariot racing would have encouraged conversation about this popular Roman sport, just as certain Greek vases would have provided topics of conversation at a Greek symposium. Here, the function of the Colchester beaker as a drinking cup enabled it to be an instigator of conversations that could literally shape how its owner and companions conceived of and knew of the Circus Maximus in Rome and the chariot races that unfolded within it. This knowledge of the races would even have included the identities of star charioteers; the inscribed names on the beaker—Cresce(n)s, the victor, and Hierax, Olympus, and Antiocus—correspond to at least some known charioteers, as is common on Roman circus cups (see Allason-Jones 2011: 226). When people remember collaboratively (e.g., through conversation), they tend to recall shared information and tune their story to their audience. When a listening member of a conversation hears misleading, or simply new, information, he or she can adopt that information as his or her own memory, through a process called social contagion. A conversational speaker "can impose a new memory onto the listener, that is, a memory of something that the listener did not experience" (Hirst and Echterhoff 2012: 63). As discussed earlier, many people living in Roman Britain, even Roman soldiers, would never have visited Rome and its Circus Maximus. Hearing tales of the Circus and its chariot races would have provided new information that listeners would have internalized, creating new semantic, if not episodic, memories.

Conclusions

The Arch of Septimius is monumental, public, and metropolitan Roman. The Colchester beaker and Campanian flasks are small, largely private, and were manufactured and/or used in provincial contexts. Yet all these objects and monuments—through their carefully curated images, accompanying inscriptions, and the narratives they evoked—projected false memories as a form of cultural and social connectivity, a means of creating a shared cultural or historical narrative. In the case of portable souvenir vessels that could generate false memories, possessing the representation of a Campanian city or a race in the Circus Maximus could in a sense be almost as good as having been to those sites in person; one gained the cultural capital of possessing knowledge about chariot racing, or about the prestigious destinations of Puteoli and Baiae. The souvenirs thus had the potential to link the center of Rome's empire with its periphery and collapse the distance—both geographic and cultural—between them. In the case of the Arch of Septimius, the monument presented a particular, and probably distorted, narrative of the emperor's accomplishments that became the shared cultural history of Romans, both inhabitants of the city of Rome and visitors who would have seen the arch in the Forum and carried its stories back to their homes.

Rome's emperors, from Augustus onward, were keenly aware of the power of monuments and images to shape historical memory (Popkin 2016: 173–4; 2018a), even though their conception of this phenomenon would not have been framed in terms of false memory and other cognitive theories. The manufacturers of the Roman souvenirs were likely unaware of their ability to generate false memories, but they were surely aware of the souvenirs' value as markers of Roman culture; this was part of the objects' commercial appeal, after all. And this value enabled the objects to be used and perceived in ways that heightened their ability to construct false or distorted memories. The souvenirs and Arch of Septimius were highly enactive, as it were: active agents of memory and knowledge construction in the Roman Empire and beyond. The cognitive science of memory enables us to recognize the agency of these monuments and objects, to theorize how they served not as repositories but instead as generators and shapers of memory.

The Severan arch and the souvenirs evoked memories that guided people's memories, semantic and sometimes episodic, in certain directions. Through repeated viewing and use, and through collaborative remembering through conversation, these objects and monuments could create distorted and even false memories, as well as shared semantic memories. The Colchester beaker could shape how people living in Roman Britain knew of and perceived Roman chariot racing. The Campanian souvenirs could manipulate travelers' episodic memories of visits to Puteoli and Baiae and also shape general knowledge of these cities across the western Mediterranean. And the Arch of Septimius Severus in the Roman Forum could create a misleading impression of a Severan triumph in 202 that persists in some scholarship to this day.

Thinking through ancient art and architecture with cognitive science enriches our understanding of the impact of visual culture on individual and collective memory and knowledge in antiquity. It is possible, however, that thinking through cognitive science with ancient art and architecture can enrich our conception of theories of memory as well. As scholars have noted, memory is best understood by examining cognitive *and* social processes (Hirst and Manier 2008: 189; Blank, Walther, and Isemann 2017). Scholars of the classical world are well positioned to take up this challenge, to test the cognitive theories developed in laboratory settings against our complex picture of ancient societies. Classicists can integrate cognitive and socio-cultural approaches to understand more fully the agency of art and architecture in shaping individual and collective memory and knowledge in the "real" Roman world, where literacy and means of mass communication were limited. In studying the interrelationship of visual culture and human remembering in complex, historical situations, we can theorize about this interrelationship today. Researchers have investigated how imagery (both mental and tangible) and words affect recall memory; that is, how words and images help people "correctly" remember things they have encountered (e.g., Sadoski and Paivio 2001: 67–78; Madigan 2014; Takashima et al. 2014; Caplan and Madan 2016). The case studies presented here, however, suggest that cognitive researchers might further investigate the impact of rich narratives composed of word *and* image on the creation specifically of false or distorted memories. When looking at the Arch of Septimius Severus, the Colchester beaker, or one of the flasks of Puteoli and Baiae, a viewer would have been confronted by images and inscriptions that intentionally invited narrative completion. Cognitive approaches to memory illuminate how this combination of image and text united in a narrative context made these examples of Roman art so effective at manipulating memories in antiquity, and even in the present.

Notes

1 All translations are adapted from the Loeb Classical Library unless stated otherwise.
2 I thank J. Matthew Harrington for this reference.

References

Alcock, S. E. 2002. *Archaeologies of the Greek Past: Landscape, Monuments, and Memories*. Cambridge: Cambridge University Press.

Allason-Jones, L. 2011. "Recreation." In *Artefacts in Roman Britain: Their Purpose and Use*, ed. L. Allason-Jones, 219–42. Cambridge: Cambridge University Press.

Bastéa, E. 2004. "Introduction." In *Memory and Architecture*, ed. E. Bastéa, 1–21. Albuquerque, NM: University of New Mexico Press.

Beard, M. 2007. *The Roman Triumph*. Cambridge, MA: Belknap Press of Harvard University Press.

Bergmann, B. 1994. "The Roman House as Memory Theater: The House of the Tragic Poet in Pompeii." *Art Bulletin* 76: 225–56.

Birley, A. R. 1988. *The African Emperor: Septimius Severus*. Updated, rewritten, expanded and reillustrated ed. London: Batsford.

Blank, H., E. Walther, and S. D. Isemann. 2017. "The Past Is a Social Construction: Susceptibility to Social Influence in (Mis)remembering." In *False and Distorted Memories*, ed. R. A. Nash and J. Ost, 55–71. London: Routledge.

Boyer, P., and J. V. Wertsch, eds. 2009. *Memory in Mind and Culture*. New York: Cambridge University Press.

Brainerd, C. J., and V. F. Reyna. 2005. *The Science of False Memory*. Oxford: Oxford University Press.

Brilliant, R. 1967. *The Arch of Septimius Severus in the Roman Forum*. Rome: American Academy in Rome.

Caplan, J. B., and C. R. Madan. 2016. "Word Imageability Enhances Association-Memory by Increasing Hippocampal Engagement." *Journal of Cognitive Neuroscience* 28: 1522–38.

Chastagnol, A. 1987. "Aspects concrets et cadre topographique des fêtes décennales des empereurs à Rome." In L'Urbs: Espace urbain et histoire (Ier siècle av. J.-C. – IIIe siècle ap. J.-C.): Actes du colloque international organisé par le Centre national de la recherché scientifique et l'École française de Rome (Rome, 8–12 mai 1985), 491–507. Rome: École française de Rome.

D'Arms, J. H. 1970. *Romans on the Bay of Naples: A Social and Cultural Study of the Villas and Their Owners from 150 B.C. to A.D. 400*. Cambridge, MA: Harvard University Press.

Davies, P. J. E. 2011. "Aegyptiaca in Rome: Adventus and Romanitas." In *Cultural Identity in the Ancient Mediterranean*, ed. E. S. Gruen, 354–370. Los Angeles, CA: Getty Research Institute.

Davis, D., and E. F. Loftus. 2009. "Expectancies, Emotion, and Memory Reports for Visual Events." In *The Visual World in Memory*, ed. J. R. Brockmole, 178–214. Hove: Psychology Press.

Dudai, Y. 2012. "The Restless Engram: Consolidations Never End." *Annual Review of Neuroscience* 35: 227–47.

Dunbabin, K. M. D. 1999. *Mosaics of the Greek and Roman World*. Cambridge: Cambridge University Press.

Edwards, C. 1996. *Writing Rome: Textual Approaches to the City*. New York: Cambridge University Press.

Favro, D. 1999. "The City Is a Living Thing: The Performative Role of an Urban Site in Ancient Rome, the Vallis Murcia." In *The Art of Ancient Spectacle*, ed. B. Bergmann and C. Kondoleon, 205–19. Washington, DC: National Gallery of Art.

Foster, J. L., and M. Garry. 2012. "Building False Memories Without Suggestions." *American Journal of Psychology* 125: 225–32.

Frenda, S. J., E. D. Knowles, W. Saletan, and E. F. Loftus. 2013. "False Memories of Fabricated Political Events." *Journal of Experimental Social Psychology* 39: 280–6.

Futrell, A. 2006. *The Roman Games: Historical Sources in Translation*. Malden, MA: Blackwell.

Galinsky, K., ed. 2014. *Memoria Romana: Memory in Rome and Rome in Memory*. Ann Arbor, MI: University of Michigan Press.

Garry, M., and M. P. Gerrie. 2005. "When Photographs Create False Memories." *Current Directions in Psychological Science* 14: 321–5.

Garry, M., and K. A. Wade. 2005. "Actually, a Picture Is Worth Less than 45 Words: Narratives Produce More False Memories Than Photographs Do." *Psychonomic Bulleting & Review* 12: 359–66.

Hekster, O. 2008. *Rome and Its Empire, AD 193-284*. Edinburgh: Edinburgh University Press.

Hinterhöller, M. 2008. *Der Triumphbogen des Septimius Severus und die historischen Reliefs der Partherkriege: Ein Triumphalmonument am Beginn der späten Kaiserzeit*. Munich: Grin.

Hirst, W., and G. Echterhoff. 2012. "Remembering in Conversations: The Social Sharing and Reshaping of Memories." *Annual Review of Psychology* 63: 55–79.

Hirst, W., and D. Manier. 2008. "Towards a Psychology of Collective Memory." *Memory* 16: 183–200.

Humphrey, J. H. 1986. *Roman Circuses: Arenas for Chariot Racing*. London: B.T. Batsford.

Hupbach, A., R. Gomez, O. Hardt, and L. Nadel. 2007. "Reconsolidation of Episodic Memories: A Subtle Reminder Triggers Integration of New Information." *Learning & Memory* 14: 47–53.

Jouen, A. L., T. M. Ellmore, C. J. Madden, C. Pallier, P. F. Dominey, and J. Ventre-Dominey. 2015. "Beyond the Word and Image: Characteristics of a Common Meaning System for Language and Vision Revealed by Functional and Structural Imaging." *NeuroImage* 106: 72–85.

Landes, C. 1998. "Verreries et spectacles romains du 1er siècle." In *Les verres romains à scenes de spectacales trouvés en France*, ed. G. Sennequier, A. Hochuli-Gysel, B. Rütti, S. Fünfschilling, L. Berger, J. Nelis-Clément, and C. Landes, 9–18. Rouen: Association française pour l'archéologie du verre.

Lewandowsky, S., W. G. K. Stritzke, K. Oberauer, and M. Morales. 2005. "Memory for Fact, Fiction, and Misinformation: The Iraq War 2003." *Psychological Science* 16: 190–5.

Lindsay, D. S., L. Hagen, J. D. Read, K. A. Wade, and M. Garry. 2004. "True Photographs and False Memories." *Psychological Science* 15: 149–54.

Loftus, E. F. 2005. "Planting Misinformation in the Human Mind: A 30-Year Investigation of the Malleability of Memory." *Learning & Memory* 12: 361–6.

Lusnia, S. S. 2014. *Creating Severan Rome: The Architecture and Self-Image of L. Septimius Severus (A.D. 193-211)*. Brussels: Latomus.

Lynch, K. 1960. *The Image of the City*. Cambridge, MA: MIT Press.

Madigan, S. 2014. "Picture Memory." In *Imagery, Memory and Cognition: Essays in Honor of Allan Paivio*, ed. J. C. Yuille, 65–89. London: Psychology Press.

Morris, E. K., C. Laney, D. M. Bernstein, and E. F. Loftus. 2006. "Susceptibility to Memory Distortion: How Do We Decide It Has Occurred?" *American Journal of Psychology* 119: 255–74.

Nash, R. A., and J. Ost, eds. 2017. *False and Distorted Memories*. London: Routledge.

Ostrow, S. E. 1979. "The Topography of Puteoli and Baiae on the Eight Glass Flasks." *Puteoli, studi di storia antica* 3: 77–140.

Painter, K. S. 1975. "Roman Flasks with Scenes of Baiae and Puteoli." *Journal of Glass Studies* 17: 54–67.

Park, D. C., and C.-M. Huang. 2010. "Culture Wires the Brain: A Cognitive Neuroscience Approach." *Perspectives on Psychological Science* 5: 391–400.

Popkin, M. L. 2016. *The Architecture of the Roman Triumph: Monuments, Memory, and Identity*. New York: Cambridge University Press.

Popkin, M. L. 2017. "Souvenirs and Memory Manipulation in the Roman Empire: The Glass Flasks of Ancient Pozzuoli." In *Materializing Memories in Art and Popular Culture*, ed. L. Munteán, L. Plate, and A. Smelik, 45–61. London: Routledge.

Popkin, M. L. 2018a. "The Parthian Arch of Augustus and Its Legacy: Memory Manipulation in Imperial Rome and Modern Scholarship." In *Afterlives of Augustus, AD 14-2014*, ed. P. J. Goodman, 271–93. Cambridge: Cambridge University Press.

Popkin, M. L. 2018b. "Urban Images in Glass from the Late Roman Empire: The Souvenir Flasks of Puteoli and Baiae." *American Journal of Archaeology* 122: 427–62.

Roediger, H. L. III, and K. B. McDermott. 1995. "Creating False Memories: Remembering Words Not Presented in Lists." *Journal of Experimental Psychology: Learning, Memory, and Cognition* 21: 803–14.

Roediger, H. L. III, and K. B. McDermott. 2000. "Tricks of Memory." *Current Directions in Psychological Science* 9: 123–7.

Roediger, H. L. III, F. M. Zaromb, and A. C. Butler. 2009. "The Role of Repeated Retrieval in Shaping Collective Memory." In *Memory in Mind and Culture*, ed. P. Boyer and J. V. Wertsch, 138–70. New York: Cambridge University Press.

Roffia, E. 2002. "Alcuni vetri incise." In *Nuove ricerche sul Capitolium di Brescia: scavi, studi e restauri*, ed. F. Rossi, 413–34. Milan: Et.

Ross, M., and Q. Wang. 2010. "Why We Remember and What We Remember: Culture and Autobiographical Memory." *Perspectives on Psychological Science* 5: 401–9.

Sadoski, M., and A. Paivio. 2001. *Imagery and Text: A Dual Coding Theory of Reading and Writing*. Mahwah, NJ: Lawrence Erlbaum.

Schacter, D. L. 2001. *The Seven Sins of Memory: How the Mind Forgets and Remembers*. Boston, MA: Houghton Mifflin.

Schacter, D. L., and E. F. Loftus. 2013. "Memory and Law: What Can Cognitive Neuroscience Contribute?" *Nature Neuroscience* 16: 119–23.

Small, J. P. 1997. *Wax Tablets of the Mind: Cognitive Studies of Memory and Literacy in Classical Antiquity*. London: Routledge.

Southern, P. 2001. *The Roman Empire from Severus to Constantine*. London: Routledge.

Stewart, S. 1993. *On Longing: Narratives of the Miniature, the Gigantic, the Souvenir, the Collection*. Durham, NC: Duke University Press.

Swetnam-Burland, M. 2015. *Egypt in Italy: Visions of Egypt in Roman Imperial Culture*. New York: Cambridge University Press.

Tafalla, M. T. A., M. E. O. Palomar, and J. A. P. Peralta. 2003. "Un 'Souvenir' de *Baiae* en *Asturica Augusta* (Provincia Tarraconense, *Hispania*)." *Journal of Glass Studies* 45: 105–13.

Takashima, A., I. Bakker, J. G. van Hell, G. Janzen, and J. M. McQueen. 2014. "Richness of Information about Novel Words Influences How Episodic and Semantic Memory Networks Interact during Lexicalization." *NeuroImage* 84: 265–78.

Tulving, E. 1972. "Episodic and Semantic Memory." In *Organization of Memory*, ed. E. Tulving and W. Donaldson, 381–403. New York: Academic Press.

Wade, K. A., S. J. Nightingale, and M. F. Colloff. 2017. "Photos and Memory." In *False and Distorted Memories*, ed. R. A. Nash and J. Ost, 39–54. London: Routledge.

Wade, K. A., M. Garry, J. D. Read, and D. S. Lindsay. 2002. "A Picture Is Worth a Thousand Lies: Using False Photographs to Create False Childhood Memories." *Psychonomic Bulletin & Review* 9: 597–603.

Wiseman, T. P. 2014. "Popular Memory." In *Memoria Romana: Memory in Rome and Rome in Memory*, ed. K. Galinsky, 43–62. Ann Arbor, MI: University of Michigan Press.

Yates, F. A. 1966. *The Art of Memory*. Chicago, IL: University of Chicago Press.

Zaragoza, M. S., and K. J. Mitchell. 1996. "Repeated Exposure to Suggestion and the Creation of False Memories." *Psychological Science* 7: 294–300.

Zaragoza, M. S., K. J. Mitchell, K. Payment, and S. Drivdahl. 2011. "False Memories for Suggestions: The Impact of Conceptual Elaboration." *Journal of Memory and Language* 64: 18–31.

Index